Compendium of Plant Genomes

Series editor

Chittaranjan Kole, Raja Ramanna Fellow, Department of Atomic Energy, Government of India, Kalyani, India

Whole-genome sequencing is at the cutting edge of life sciences in the new millennium. Since the first genome sequencing of the model plant Arabidopsis thaliana in 2000, whole genomes of about 70 plant species have been sequenced and genome sequences of several other plants are in the pipeline. Research publications on these genome initiatives are scattered on dedicated web sites and in journals with all too brief descriptions. The individual volumes elucidate the background history of the national and international genome initiatives; public and private partners involved; strategies and genomic resources and tools utilized; enumeration on the sequences and their assembly; repetitive sequences; gene annotation and genome duplication. In addition, synteny with other sequences, comparison of gene families and most importantly potential of the genome sequence information for gene pool characterization and genetic improvement of crop plants are described.

Interested in editing a volume on a crop or model plant? Please contact Dr. Kole, Series Editor, at ckole2012@gmail.com

More information about this series at http://www.springer.com/series/11805

Tapan K. Mondal · Robert J. Henry
Editors

The Wild Oryza Genomes

Springer

Editors
Tapan K. Mondal
IARI Campus
ICAR-National Research Center on Plant Biotechnology
New Delhi, Delhi
India

Robert J. Henry
Queensland Alliance for Agriculture and Food Innovation
University of Queensland
St Lucia, QLD
Australia

ISSN 2199-4781 ISSN 2199-479X (electronic)
Compendium of Plant Genomes
ISBN 978-3-319-71996-2 ISBN 978-3-319-71997-9 (eBook)
https://doi.org/10.1007/978-3-319-71997-9

Library of Congress Control Number: 2017962553

Printed on acid-free paper

This Springer imprint is published by Springer Nature
The registered company is Springer International Publishing AG
The registered company address is: Gewerbestrasse 11, 6330 Cham, Switzerland

This book series is dedicated to
my wife Phullara, and our children Sourav,
and Devleena

Chittaranjan Kole

Foreword

Rice is a staple crop for mankind, and all concerted efforts are to improve rice with yields of more than 10 tons/ha possible under favorable environments. The increasing world population coupled with climate change and reduced availability of cultivable land demand continually improved rice cultivars. This calls for extensive research efforts to decode rice genome helping to exploit the maximum genetic potential of the domesticated rice gene pool to improve rice. However to move past the current genetic limits of rice, we must look for some alternative genomic resources.

Wild species are a potential source of many useful genes that may not be present in the gene pool of the domesticated species. They are great sources of many alternative useful alleles. Although there are 24 species available in the genus *Oryza*, only two are currently domesticated species. The other species are largely unexplored, though every species has agronomically useful traits that could be introduced in rice. Several of the species have high drought, salinity or lodging tolerance, and disease and pest resistance, the degree of which is much higher than that in the most tolerant or resistant genotype of rice. Some of the traits such as acid soil tolerance, shade or low light intensity tolerance, and high micronutrient content are unique to wild species. However to exploit these traits, the prerequisite is to generate large-scale genetic and genomic resources. Some international initiatives have been taken up such as OMAP, IOMAP to decode the genomes of these species, to accelerate breeding efforts. To date, 11 species have been sequenced and released into the public domain. Further, several breeding techniques to transfer genes from wild species have been developed and some improved cultivars of rice with the DNA of these wild species have been commercialized. Several genes from wild species have been cloned for various traits and are in process to make transgenic rice which will come to the market in the future. These efforts are collectively generating huge amounts of information on breeding, genetics, genomics, and in other OMICS areas.

I am also pleased to note the effort of the series editor, Prof. Chittaranjan Kole, an internationally acclaimed scientist, in identifying Dr. Tapan K. Mondal, Principal Scientist, ICAR-National Research Centre on Plant Biotechnology, India, and Prof. Robert J. Henry, Director of Queensland Alliance for Agriculture and Food Innovation, The University of Queensland, Australia, who have contributed significantly to wild rice genome sequencing, for this synthesis volume. I am sure that this book will be very useful to researchers not only working on wild species of rice and rice itself, but also in the field of wild crop genomics, in addition to having utility among science managers and policy makers. I feel humble to write the foreword for this important book.

New Delhi, India
August 2017

Trilochan Mohapatra, Ph.D., FNA, FNASc. FNAAS
Secretary and Director General
Department of Agricultural Research and Education
and Indian Council of Agricultural Research
Ministry of Agriculture and Farmers Welfare
Government of India

Preface to the Series

Genome sequencing has emerged as the leading discipline in the plant sciences coinciding with the start of the new century. For much of the twentieth century, plant geneticists were only successful in delineating putative chromosomal location, function, and changes in genes indirectly through the use of a number of "markers" physically linked to them. These included visible or morphological, cytological, protein, and molecular or DNA markers. Among them, the first DNA marker, the RFLPs, introduced a revolutionary change in plant genetics and breeding in the mid-1980s, mainly because of their infinite number and thus potential to cover maximum chromosomal regions, phenotypic neutrality, absence of epistasis, and codominant nature. An array of other hybridization-based markers PCR-based markers, and markers based on both facilitated construction of genetic linkage maps, mapping of genes controlling simply inherited traits and even gene clusters (QTLs) controlling polygenic traits in a large number of model and crop plants. During this period a number of new mapping populations beyond F_2 were utilized and a number of computer programs were developed for map construction, mapping of genes, and for mapping of polygenic clusters or QTLs. Molecular markers were also used in studies of evolution and phylogenetic relationship, genetic diversity, DNA-fingerprinting, and map-based cloning. Markers tightly linked to the genes were used in crop improvement employing the so-called marker-assisted selection. These strategies of molecular genetic mapping and molecular breeding made a spectacular impact during the last one and a half decades of the twentieth century. But still they remained "indirect" approaches for elucidation and utilization of plant genomes since much of the chromosomes remained unknown and the complete chemical depiction of them was yet to be unraveled.

Physical mapping of genomes was the obvious consequence that facilitated development of the "genomic resources" including BAC and YAC libraries to develop physical maps in some plant genomes. Subsequently, integrated genetic-physical maps were also developed in many plants. This led to the concept of structural genomics. Later on, emphasis was laid on EST and transcriptome analysis to decipher the function of the active gene sequences leading to another concept defined as functional genomics. The advent of techniques of bacteriophage gene and DNA sequencing in the

1970s was extended to facilitate sequencing of these genomic resources in the last decade of the twentieth century.

As expected, sequencing of chromosomal regions would have led to too much data to store, characterize, and utilize with the-then available computer software could handle. But development of information technology made the life of biologists easier by leading to a swift and sweet marriage of biology and informatics and a new subject was born—bioinformatics.

Thus, evolution of the concepts, strategies, and tools of sequencing and bioinformatics reinforced the subject of genomics—structural and functional. Today, genome sequencing has traveled much beyond biology and involves biophysics, biochemistry, and bioinformatics!

Thanks to the efforts of both public and private agencies, genome sequencing strategies are evolving very fast, leading to cheaper, quicker and automated techniques right from clone-by-clone and whole-genome shotgun approaches to a succession of second-generation sequencing methods. Development of software of different generations facilitated this genome sequencing. At the same time, newer concepts and strategies were emerging to handle sequencing of the complex genomes, particularly the polyploids.

It became a reality to chemically—and so directly—define plant genomes, popularly called whole-genome sequencing or simply genome sequencing.

The history of plant genome sequencing will always cite the sequencing of the genome of the model plant *Arabidopsis thaliana* in 2000 that was followed by sequencing the genome of the crop and model plant rice in 2002. Since then, the number of sequenced genomes of higher plants has been increasing exponentially, mainly due to the development of cheaper and quicker genomic techniques and, most importantly, development of collaborative platforms such as national and international consortia involving partners from public and/or private agencies.

As I write this preface for the first volume of the new series "Compendium of Plant Genomes", a net search tells me that complete or nearly complete whole-genome sequencing of 45 crop plants, eight crop and model plants, eight model plants, 15 crop progenitors and relatives, and three basal plants are accomplished, the majority of which are in the public domain. This means that we nowadays know many of our model and crop plants chemically, i.e., directly, and we may depict them and utilize them precisely better than ever. Genome sequencing has covered all groups of crop plants. Hence, information on the precise depiction of plant genomes and the scope of their utilization is growing rapidly every day. However, the information is scattered in research articles and review papers in journals and dedicated web pages of the consortia and databases. There is no compilation of plant genomes and the opportunity of using the information in sequence-assisted breeding or further genomic studies. This is the underlying rationale for starting this book series, with each volume dedicated to a particular plant.

Plant genome science has emerged as an important subject in academia, and the present compendium of plant genomes will be highly useful both to students and teaching faculties. Most importantly, research scientists involved in genomics research will have access to systematic deliberations on the plant genomes of their interest. Elucidation of plant genomes is not only of interest for the geneticists and breeders, but also for practitioners of an array of plant science disciplines, such as taxonomy, evolution, cytology, physiology, pathology, entomology, nematology, crop production, biochemistry, and obviously bioinformatics. It must be mentioned that information regarding each plant genome is ever-growing. The contents of the volumes of this compendium are therefore focusing on the basic aspects of the genomes and their utility. They include information on the academic and/ or economic importance of the plants, description of their genomes from a molecular genetic and cytogenetic point of view, and the genomic resources developed. Detailed deliberations focus on the background history of the national and international genome initiatives, public and private partners involved, strategies and genomic resources and tools utilized, enumeration on the sequences and their assembly, repetitive sequences, gene annotation, and genome duplication. In addition, synteny with other sequences, comparison of gene families, and, most importantly, potential of the genome sequence information for gene pool characterization through genotyping by sequencing (GBS) and genetic improvement of crop plants have been described. As expected, there is a lot of variation of these topics in the volumes based on the information available on the crop, model, or reference plants.

I must confess that as the series editor it has been a daunting task for me to work on such a huge and broad knowledge base that spans so many diverse plant species. However, pioneering scientists with life-time experience and expertise on the particular crops did excellent jobs editing the respective volumes. I myself have been a small science worker on plant genomes since the mid-1980s and that provided me the opportunity to personally know several stalwarts of plant genomics from all over the globe. Most, if not all, of the volume editors are my longtime friends and colleagues. It has been highly comfortable and enriching for me to work with them on this book series. To be honest, while working on this series I have been and will remain a student first, a science worker second, and a series editor last. And I must express my gratitude to the volume editors and the chapter authors for providing me the opportunity to work with them on this compendium.

I also wish to mention here my thanks and gratitude to the Springer staff, Dr. Christina Eckey and Dr. Jutta Lindenborn in particular, for all their constant and cordial support right from the inception of the idea.

I always had to set aside additional hours to edit books besides my professional and personal commitments—hours I could and should have given to my wife, Phullara, and our kids, Sourav, and Devleena. I must mention that they not only allowed me the freedom to take away those hours from them but also offered their support in the editing job itself. I am really not sure whether my dedication of this compendium to them will suffice to do justice to their sacrifices for the interest of science and the science community.

Kalyani, India Chittaranjan Kole

Preface

Economically, rice is the most important staple cereal feeding the world from the ancient times. Academically, it is a model crop plant, which is used to address several basic questions in plant biology, yet its wild relatives are an untapped reservoir of agronomically important alleles that are absent in the rice gene pool. Although the genus *Oryza* is well known due to the importance of rice, the genus includes 25 species, among which only the Asiatic rice (*Oryza sativa*) and the African rice (*Oryza glaberrima*) are used to feed the world. The genus is diverse as indicated by a wide range of chromosome numbers ($2n = 24$ to 48), different ploidy levels, and genome sizes. Although research efforts from conventional breeding to functional genomics are advanced in rice, comparatively little is known about the wild species. However, many of the wild species of rice are already known for their tolerance to biotic and abiotic stress. Additionally, some of them harbor specific growth and developmental attributes such as profuse tillering, low shattering of mature seeds, presence of a salt gland to pump excessive salt, which upon transfer to cultivated rice could lead to increased productivity and profitability of rice cultivation. However, a prerequisite to understand the genetic factors, which are responsible for such desirable characters present in wild species, is sequencing of the genomes to identify the genes involved. Following the first sequencing of rice in 2002, the genome of 11 species has been sequenced so far and made available for public use recently. The genomes of the remaining species are expected to be sequenced soon. These vast genomic resources will be extremely useful for addressing some of the basic questions about the origin of the genus, evolutionary relationship among the species, domestication, and environmental adaptation and also will be useful to substantiate molecular breeding as well as pre-breeding work to introgress useful characters horizontally from those wild species to cultivated rice.

This book collates the latest state-of-the-art information from a wide range of research domains such as cytology, breeding, physiology, genomics, and proteomics. The volume places emphasis on the latest genomic related works of the 23 species of the genus *Oryza*, public as well as private genomic resources and their impact on genetic improvement research which will be useful to the international research community at large helping to feed 7 billion people in a sustainable manner. The current volume entitled, "The Wild Oryza Genomes" covers genomics and its application for the improvement of rice breeding. It includes 25 chapters comprising three general topics with the rest on specific species. Chapter 1 deals with an overview of the different species of *Oryza*, their genomic and genetic resources and conservation, and is written by two eminent rice breeders. Due to availability of genome sequence of several wild species, the requirements to manage the vast sequence-related data have also increased and to make them more user-friendly, development of databases has become essential. Chapter 2 describes the various Web resources that are available for wild species of *Oryza*. Speciation remains always attractive to biologist as it is a source of the variation that is the basis for genetic improvement. Chapter 3 dealt with various aspects of the evolutionary relationships among the different species of *Oryza* based on the conserved nuclear-encoded genes and organelle genomes.

Chapters 4–20 have been written on individual species including a brief account of academic interest, the trait they have, and discussion on their chromosomes, breeding potential, availability of genomic and genetic resources, progress on the genome sequence, organelle genome sequences. Each of the chapters also provides a photograph of the adult plant of a particular species and its geographical distribution. Finally, each chapter ends by discussing the research gaps that need to be filled in future. These chapters are presented in the order of the alphabetic name of the species as follows: *O. alta* (Chap. 4), *O. australiensis* (Chap. 5), *O. barthii* (Chap. 6), *O. brachyantha* (Chap. 7), *O. coarctata* (Chap. 8), *O. glaberrima* (Chap. 9), *O. glumaepatula* (Chap. 10), *O. grandiglumis* (Chap. 11), *O. granulate* (Chap. 12), *O. latifolia* (Chap. 13), *O. longiglumis* (Chap. 14), *O. longistaminata* (Chap. 15), *O. meridionalis* (Chap. 16), *O. meyeriana* (Chap. 17, *O. minuta* (Chap. 18), *O. neocaledonica* (Chap. 19), *O. nivara* (Chap. 20).

The officinalis complex consists of ten different species but Chap. 21 describes seven perennial species (*O. officinalis, O. rhizomatis, O. eichengeri, O. minuta, O. malampuzhaensis, O. punctate*, and *O. schweinfurthiana*) in details as the other three species have been described in individual chapters primarily due to the availability of more information about these three species. Considerable success has been achieved in terms of developing genetic stocks through wide hybridization and mapping agronomically important genes from the species of *O. officinalis* complex which have been depicted in the chapter.

Further, four chapters have been deliberated species-wise. They are *O. perenni* (Chap. 22), *O. rhizomatis* (Chap. 23), *O. ridleyi* (Chap. 24) and *O. rufipogon* (Chap. 25). Among these, *O. rufipogon*—the progenitor of present-day cultivated rice, *O. sativa* is one of the most studied wild species of rice. These four chapters provide more details with a special emphasis on genomics and breeding. Every effort has been taken to include all the published reports on these species and to discuss their future potential.

The last chapter, i.e., Chap. 26 with the name, "An account of unclassified species (*Oryza schlechteri*), subspecies (*Oryza indandamanica* Ellis, *Oryza sativa* f. *spontanea* Baker) and ortho-group species (*Leersia perrieri*) of *Oryza*," describes one unclassified species, i.e., *O. schlechteri*, ortho-group species, i.e., *Leersia perrieri*, and two subspecies, i.e., *O. indandamanica* and

O. sativa f. spontanea. The purpose of this chapter is to sensitize the readers with the fact that the *Oryza* genome is very dynamic, still evolving, and ultimately will give birth to new species with useful trait in future.

We also thank our family members for bearing with us throughout the process of editing and finalization of this book. We are also thankful to all 67 authors from 14 different countries who despite their busy schedules have shared their research experience on their respective species in the form of a chapter. We also express our gratitude to Springer-Verlag and its entire staffs particularly Ms. Abirami Purushothaman and Mr. Naresh Kumar for their kind help and understanding in publication and promotion of this book. We made every effort to include all the published research papers in the area of genomics but apologies for those works, if any, which did not appear in this volume despite a detailed search worldwide. Finally, we also thank Prof. C. R. Kole to give an opportunity to edit this important volume.

We are confident that this book will be useful to researchers, both in academia and industry, and policy makers working on not only wild species of rice but also domesticated rice and cereals as a whole.

New Delhi, India — Tapan K. Mondal
St Lucia, Australia — Robert J. Henry

Contents

Abbreviations

AAC	Apparent amylose content
ABC	Advanced backcross
ADB	Africa Development Bank
Adh	Alcohol dehydrogenase
AFLP	Amplified fragment length polymorphism
AILs	Alien introgression lines
AMPs	Advanced mapping populations
ASH	Asymmetric somatic hybridization
BAC	Bacterial artificial chromosome
BADH	Betaine aldehyde dehydrogenase
BB	Bacterial blight
BES	BAC end sequences
BGI	Beijing Genomics Institute
BILs	Backcross inbred lines
BLB	Bacterial leaf blight
BPH	Brown planthopper
BSA	Bulked segregant analysis
CGH	Comparative genomic hybridization
CGIAR	Consultative Group for International Agricultural Research
C_i	Intercellular CO_2 concentration
CIAT	International Center for Tropical Agriculture
CMS	Cytoplasm male sterility
CSSLs	Chromosome segment substitution lines
csSSR	Candidate gene-based SSR
CTAB	Cetyltrimethyl ammonium bromide
CWR	Common wild rice
DArT	Diversity Arrays Technology
DDC	Degeneration divergence complementation
DGE	Digital gene expression technology
DH	Double haploid
EBI	The European Bioinformatics Institute
eIF	Eukaryotic translation initiation factor
ERF	Ethylene responsive factor
ESTs	Expressed sequence tags
FAO	Food and Agriculture Organization
FISH	Fluorescent in situ hybridization
FLcDNA	Full-length cDNA
FPC	Fingerprint contig

GISH	Genome in situ hybridization
g_m	Mesophyll conductance
GMS	Genealogy Management System
GRH/GLH	Green rice/leafhopper
g_s	Stomatal conductance
GSV	Grassy stunt virus
GWAS	Genome-wide association studies
H_e	Genetic diversity
HL-CMS	Honglian cytoplasmic male sterility
H_o	Heterozygosity
IAfRIC	International African Rice Improvement Consortium
ICIS	International Crop Information System
IGS	Intergenic spacer
ILs	Introgression lines
IOMAP	International *Oryza* Map Alignment Project
IRs	Inverted repeats
IRGSP	International Rice Genome Sequencing Project
ISSR	Inter-simple sequence repeat
ITS	Internal transcribed spacer
KASPar	Competitive allele-specific PCR
LD	Linkage disequilibrium
LINEs	Long interspersed nuclear elements
LRR	Leucine-rich repeat
LSC	Large single copy
LTRs	Long terminal repeats
MAALs	Monosomic alien addition lines
MAB	Marker-assisted breeding
MALDI-TOF	Matrix-assisted laser desorption ionization time of flight
MAPK	Mitogen-activated protein kinase
MARs	Matrix attachment regions
MAS	Marker-assisted selection
MENERGEP	Methodologies and new resources for genotyping and phenotyping
miRNA	microRNA
MITEs	Miniature inverted-repeat transposable elements
MLS	Multilateral system
NBS-LRR	Nucleotide binding sites and leucine-rich repeats
NCBI	National Center for Biotechnology Information
NCGR	National Center for Genome Resource
NCGRP	National Center for Genetic Resources Preservation
NERICA	New Rice for Africa
NGS	Next-generation sequencing
NIG	National Institute of Genetics
NILs	Near isogenic lines
OEC	Oxygen-evolving complex
OGEP	*Oryza* genome evolution
OMAP	*Oryza* Map Alignment Project
OSCA	Hyperosmolality-gated Ca-permeable channels
PAGE	Polyacrylamide gel electrophoresis

PAL	Phenylalanine ammonia lyase
PGRC	Plant Genetic Resources Centre
Pn	Net photosynthetic rate
P_n/c_i	Carboxylation efficiency
P_n/g_s*(WUEi)*	Intrinsic water use efficiency
P_n/T	Transpiration efficiency
PReDA	Plant Repeat Database
PS	Photosystem
PV	Phenotypic variance
QTL	Quantitative trait loci
RAM	Rapid Alleles Mobilization
RAP	Rice Annotation Project
RAPD	Random amplified polymorphic DNA
RAP-DB	Rice Annotation Project Database
RcbL	Rubisco large subunit
Rf	Fertility restoration
RGKbase	Rice Genome Knowledgebase
RILs	Recombinant inbred lines
RiTE-db	Rice TE database
RT-PCR	Real-Time polymerase chain reaction
RYMV	Rice yellow mottle virus
SGSV	Svalbard Global Seed Vault
SINEs	Short interspersed nuclear elements
SNP	Single nucleotide polymorphism
SOAP	Short Oligonucleotide Analysis Package
SSC	Small single copy
STMS	Sequence-tagged microsatellite
TDM	Total dry matter
TEs	Transposable elements
TFs	Transcription factors
TRF	Tandem Repeat Finder
TSS	Transcription start site
VCF	Variant Call Format
WA-CMS	Wild abortive cytoplasmic male sterility
WBPH	White-backed Planthopper
WGS	Whole-genome shotgun
WPM	Woody Plant Medium

Wild Relatives of Rice: A Valuable Genetic Resource for Genomics and Breeding Research

1

Darshan Singh Brar and Gurdev S. Khush

Abstract

Worldwide, more than 3.5 billion people depend on rice for more than 20% of their daily calories. Global rice demand is estimated to rise from 723 million tons in 2015 to 763 million tons in 2020 and to further increase to 852 million tons in 2035, an overall increase of 18% or 129 million tons in the next 20 years. World rice production has more than doubled from 257 million tones in 1966 to 680 million tons in 2010. This was mainly achieved through the application of principles of classical Mendelian genetics and conventional plant breeding. Further, rice productivity is continually threatened by several diseases (bacterial blight, blast, tungro virus, rice yellow mottle virus, sheath blight, etc.) and insects (plant hoppers, stemborer, and gall midge) including many abiotic stresses (drought, salinity, submergence, cold, heat, soil toxicities, etc.). To overcome these constraints particularly in the context of global climatic changes, there is urgent need to broaden the gene pool of rice; one of the options is to exploit wild species of *Oryza* which are reservoirs of useful genes/QTLs for rice improvement. Interspecific hybrids, alien introgression lines (AILs), chromosomal segmental substitution lines (CSSLs) have been produced. Several genes/QTLs governing agronomic traits have been transferred from wild species into rice, and a few of these tagged with molecular markers and used in marker-assisted selection (MAS). Some breeding lines of rice derived from wide crosses have been released as varieties.

1.1 Wild Relatives of Rice

The genus *Oryza* has two cultivated and 24 wild species ($2n = 24$, 48 chromosomes) representing 11 genomes: AA, BB, CC, BBCC, CCDD, EE, FF, GG, HHJJ, KKLL (Vaughan 1989, 1994; Aggarwal et al. 1997; Ge et al. 1999; Table 1.1). Of the two cultivated species, *O. sativa* ($2n = 24$, AA) commonly referred as 'Asian rice' is high yielding and cultivated worldwide, whereas *O. glaberrima* ($2n = 24$, AA) known as 'African rice' is low yielding and grown in a limited area in West Africa. The wild species are grass-like plants that are weedy and inferior in morphological traits, having poor plant type, poor grain characteristics, low grain yield and are shattering in nature. These wild species differ markedly in morphological characteristics such as growth

D. S. Brar (✉)
School of Agricultural Biotechnology, Punjab Agricultural University, Ludhiana, Punjab, India
e-mail: darshanbrar@pau.edu

G. S. Khush
University of California, Davis, CA, USA

T. K. Mondal and R. J. Henry (eds.), *The Wild Oryza Genomes*,
Compendium of Plant Genomes, https://doi.org/10.1007/978-3-319-71997-9_1

habit, height, flowering, leaf size, morphology, and panicle size, panicle branching, awning, and seed size and adaptation to different habitats (Fig. 1.1; Table 1.1). In spite of their weedy nature, these wild species are important genetic resource for breeding and genomics research and are reservoir of useful genes/QTLs for tolerance to major biotic (diseases, insects) and abiotic (drought, salinity, heat) stresses, yield-related traits including weed-competitive ability, new source of cytoplasmic male sterility (CMS), and other traits related to rice improvement.

Oryza probably originated at least 130 million years ago and spread as a wild grass in Gondwanaland, the super continent that eventually broke up and drifted apart to become Asia, Africa, the Americas, Australia, and Antarctica. The genus *Oryza* has been classified into four species complexes: (i) *sativa* complex, (ii) *officinalis* complex, (iii) *meyeriana* complex, and (iv) *ridleyi* complex (Table 1.1). *O. schlechteri* and *O. coarctata* are placed in an unclassified group.

1.1.1 *O. sativa* Complex

This complex consists of 8 diploid species ($2n = 24$) belonging to AA genome, of which two are cultivated (*O. sativa* and *O. glaberrima*) and six are wild species (*O. nivara, O. rufipogon, O. breviligulata, O. longistaminata, O. meridionalis,* and *O. glumaepatula*) (Table 1.1). These species form the primary gene pool and are easily crossable with rice commonly used in transfer of genes into rice cultivars. *O. rufipogon* is distributed in South and Southeast Asia including Northern Territory and Queensland of Australia, commonly found in swamps, marshes, swampy grasslands, and in deepwater rice fields. *O. nivara* is an annual form distributed in India, Nepal, Cambodia, Laos, Thailand, whereas *O. rufipogon* exists as perennial populations. The three taxa–*O. sativa, O. nivara,* and *O. rufipogon*–together with the weedy race (*O. sativa* f. *spontanea*) form a large species complex.

O. longistaminata, a native of Africa, is closely related to its annual relative *O. barthii* which grows in swampy areas, river, or stream sides. These species, closely related to *O. glaberrima,* are somewhat easier to distinguish from each other. *O. glaberrima* is cultivated in a limited area in West Africa in upland and rain-fed ecology and also in deepwater fields. Perennial and annual relatives of *O. glaberrima* are *O. longistaminata* and *O. breviligulata,* respectively. However, now many scientists consider that *O. barthii* was domesticated to produce *O. glaberrima. O. glaberrima* is distinguished from *O. sativa* by its short, rounded ligule, panicle-lacking secondary branches, and almost glabrous lemma and palea. *O. glaberrima* is not as variable as *O. sativa.*

1.1.2 *O. officinalis* Complex

The *O. officinalis* complex consists of 12 species: 6 diploid and 6 allotetraploid species (*O. schweinfurthiana, O. minuta, O. malampuzhaensis, O. latifolia, O. alta,* and *O. grandiglumis*) (Table 1.1). Some of the species grow in partial shade or moist soil; others are adapted to swamps and seasonal pools of water and open habitat. This complex has related species groups in Asia, Africa, and Latin America. In Asia, the most common species is *O. officinalis,* widely distributed in South and Southeast Asia and South and Southwest China. *O. officinalis* thrives in partial shade or full sun. In the Philippines, it is called bird rice.

O. minuta is distributed in the Philippines and is sympatric with *O. officinalis.* It grows in shade or partial shade along stream edges. Only a few populations of tetraploid *O. officinalis, now classified as O. malampuzhaensis,* have been found localized in neighboring parts near the town of Malampuzha of Kerala and Tamil Nadu in South India. A new species from Sri Lanka, *O. eichingeri,* was distributed in Sri Lanka (Vaughan 1989).

In Africa, the two species of the *O. officinalis* complex are *O. punctata* and *O. eichingeri.* The American species of this complex (*O. latifolia, O. alta,* and *O. grandiglumis)* are allotetraploid ($2n = 48$) with a CCDD genome. *O. latifolia* is widely distributed, growing in Central and South America as well as in the Caribbean islands;

Table 1.1 Chromosome number, genomic composition and distribution of *Oryza* species, and their useful traits

Species	2*n*	Genome	No. of accessions[a]	Distribution	Useful traits
O. sativa complex					
O. sativa L.	24	AA	116,751	Worldwide	Cultigen, highly productive, grown worldwide
O. glaberrima Steud.	24	A^gA^g	1655	West Africa	Cultigen; low yielding but has tolerance to drought, acidity, iron toxicity; resistance to blast, RYMV, African gall midge, nematodes; weed competitiveness
O. nivara Sharma et Shastry	24	AA	1503	Tropical and subtropical Asia	Resistance to grassy stunt virus, BB, blast, BPH, yield-enhancing loci(QTLs)
O. rufipogon Griff.	24	AA	1034	Tropical and subtropical Asia, tropical Australia	Resistance to BB, BPH, tungro virus; moderately tolerant to Shb, tolerance to aluminum and soil acidity; source of CMS, yield-enhancing loci (QTLs)
O. breviligulata A. Chev.et Roehr (O. barthii)	24	A^gA^g	207	Africa	Resistance to GLH, BB; drought avoidance
O. longistaminata A. Chev. et Roehr .	24	A^lA^l	216	Africa	Resistance to BB, nematodes, stemborer, drought avoidance
O. meridionalis Ng	24	A^mA^m	53	Tropical Australia	Elongation ability; drought avoidance
O. glumaepatula Steud.	24	$A^{gp}A^{gp}$	54	South and Central America	Resistance to blast, elongation ability; source of CMS
O. officinalis complex					
O. punctata Kotschy ex Steud	24	BB	46	Africa	Resistance to BPH, zigzag leafhopper
O. schweinfurthiana Prodoehl	48	BBCC	35	Africa	Resistance to BPH, zigzag leafhopper
O. minuta J.S. Presl. ex C.B. Presl.	48	BBCC	62	the Philippines and Papua New Guinea	Resistance to BB, blast, BPH, GLH
O. malampuzhaensis Krishnasw. & Chandrasekh	48	BBCC	13	India	Resistance to thrips, BPH, GLH, WBPH, BB, stem rot
O. officinalis Wall ex Watt	24	CC	275	Tropical and subtropical Asia, tropical Australia	Resistance to thrips, BPH, GLH, WBPH, BB, stem rot
O. rhizomatis Vaughan	24	CC	20	Sri Lanka	Drought avoidance
O. eichingeri A. Peter	24	CC	22	South Asia and East Africa	Resistance to BPH, WBPH, GLH
O. latifolia Desv.	48	CCDD	58	South and Central America	Resistance to BPH, high biomass production
O. alta Swallen	48	CCDD	12	South and Central America	Resistance to striped stemborer; high biomass production

(continued)

Table 1.1 (continued)

Species	2*n*	Genome	No. of accessions[a]	Distribution	Useful traits
O. grandiglumis (Doell) Prod.	48	CCDD	10	South and Central America	High biomass production
O. australiensis Domin.	24	EE	36	Tropical Australia	Resistance to BPH, BB, blast; drought avoidance
O. brachyantha A. Chev. et Roehr.	24	FF	17	Africa	Resistance to BB, yellow stemborer, leaf folder, whorl maggot; tolerance to laterite soil
O. meyeriana complex					
O. granulata Nees et Arn. ex Watt	24	GG	23	South and Southeast Asia	Shade tolerance; adaptation to aerobic soil
O. meyeriana (Zoll. et (Mor. ex Steud.) Baill.	24	GG	9	Southeast Asia	Shade tolerance; adaptation to aerobic soil
O. neocaledonica Morat	24	GG	1	New Caledonia	–
O. ridleyi complex					
O. longiglumis Jansen	48	HHJJ	6	Irian Jaya, Indonesia, and Papua New Guinea	Resistance to blast, BB
O. ridleyi Hook. F.	48	HHJJ	14	South Asia	Resistance to blast, BB, blast, stemborer, whorl maggot
Unclassified					
O. schlechteri Pilger	48	HHKK	1	Papua New Guinea	Stoloniferous
O. coarctata Roxb.	48	KKLL	1	South Asia	Salt tolerance

BPH brown planthopper; *GLH* green leafhopper; *WBPH* white-backed planthopper; *BB* bacterial blight; *Shb* sheath blight; *CMS* cytoplasmic male sterility; *RYMV* rice yellow mottle virus

[a]Accessions maintained in rice genebank at IRRI, the Philippines

Modified from Khush (1997), Brar and Singh (2011). Taxonomy changes harmonized with recent literature and GRIN (Courtesy of Dr. S. Hamilton, IRRI, the Philippines)

- *O. breviligulata* is now *O. barthii*
- Tetraploid *O. punctata* is now *O. schweinfurthiana*
- Tetraploid *O. officinalis* is now *O. malampuzhaensis*
- *Porteresia coarctata* is now *O. coarctata*

O. alta and *O. grandiglumis* grow only in South America, primarily in the Amazon basin. A diploid species of this complex, *O. australiensis*, (EE) occurs in northern Australia in isolated populations and grows under wet places, seasonally dry pools, river levees, and also in open habitats.

O. brachyantha (2*n* = 24, FF) is distributed in the African continent. It grows in the Sahel zone and in East Africa in ponds, in shallow water, and in open habitats in granite/laterite soils. It is often sympatric with *O. longistaminata*. Of all the species, it is most closely related to the genus *Leersia*. This species has a small, narrow spikelet with long awns (6–17 cm).

1.1.3 *O. ridleyi* Complex

This complex comprises two tetraploid species, (*O. ridleyi* and *O. longiglumis* 2*n* = 48, HHJJ),

Fig. 1.1 Plants of some wild species of *Oryza*

which usually grow in shaded habitats beside rivers, streams, or pools. The *O. ridleyi* complex is primarily found in Southeast Asia (Cambodia, Malaysia, Myanmar) and New Guinea, whereas *O. longiglumis* is distributed in Indonesia (Iran Jaya) and Papua New Guinea. *O. ridleyi* and *O. longiglumis* are very similar in morphology and ecology.

1.1.4 *O. meyeriana* Complex

Species of *O. meyeriana* complex differ from other three complexes (i.e., *O. sativa*, *O. officinalis, O. ridleyi*) in that they are small-sized plants and have unbranched panicles with small spikelets. This complex has two species (*O. meyeriana* and *O. granulata,* $2n = 24$, GG). *O. granulata* grows in South Asia, Southeast Asia, and Southwest China, whereas *O. meyeriana* is found in Southeast Asia (Ellis 1985). Unlike other complexes, *O. meyerina* and *O. granulata* species do not grow in permanently or seasonally flooded water. This species complex grows in the shade or partial shade in forests. *O. granulata* is called forest rice by tribal people of Kerala, South India; peacock rice in parts of Vietnam; and bamboo rice in the Philippines. Recently, a new diploid species, *O. neocaledonica,* has been included in this complex (Table 1.1)

1.1.5 Unclassified Complex

O. schlechteri: This is a tetraploid species ($2n = 48$, HHKK) distributed in Indonesia (Irian Jaya) and Papua New Guinea. Richard Schlechter first collected it in 1907 from Northeast New Guinea. Vaughan and Sitch (1991) recollected it as living material from the same location. Naredo et al. (1993) reported that the presence of a sterile lemma and a striated spikelet epidermal (abaxial) surface lacking siliceous triads in *O. schlechteri* allies this species with other *Oryza* species. It is a tufted perennial, 30–40 cm tall, with an erect, 4- to 5-cm panicle and small, unawned spikelets. It is a stoloniferous species and can grow either under full or partial shade.

Porteresia coarctata is now classified as *O. coarctata*; it is a tetraploid species ($2n = 48$ KKLL) commonly found in coastal areas of South Asia. It has unusual anatomy, including glands to secrete salts. It has rough, erect leaves and occurs in brackish water. The species is characterized by large caryopses with a somewhat bent apex, a large embryo relative to the endosperm, and a short petiole attachment at the base. The leaf blade is coriaceous with prickly tuberculate margins and has a peculiar arrangement of vascular bundles; each rib contains one smaller vascular bundle and below it a larger one.

1.1.6 Related Genera

Besides *Oryza,* the tribe *Oryzeae* contains 10 other genera. Vaughan (1989, 1994) has given a brief description of these different genera. These genera include *Chikusichloa, Hygroryza, Leersia, Luziola, Prosphytochloa, Rhynchoryza, Zizania, Zizaniopsis, Potamophila,* and *Porteresia.*

1.2 Exploration and Conservation of Wild Species of Oryza

Rice genetic resources comprise different landraces, modern and obsolete varieties, genetic stocks, and the wild *Oryza* species. The International Rice Gene Bank at the International Rice Research Institute (IRRI), the Philippines, conserves the largest and most diverse collection of rice germplasm comprising 116,751 accessions of rice (*O. sativa*), 3728 accessions of wild species, and 1655 accessions of *O. glaberrima* (Table 1.1). Seeds of wild species including other rice accessions can be obtained from IRRI. The facilities of the gene bank ensure the long-term conservation of this valuable gene pool. The seeds are stored at −20 °C for long-term storage and 2–4 °C as an active collection for distribution and use in research. The IRRI gene bank has 3728 accessions of 24 wild species of *Oryza*, 1655 accessions of *O. glaberrima* (cultivated African rice), and 116,751 accessions of *O. sativa* (cultivated Asian rice). The largest number of wild species (3067) accessions is of the *O. sativa* complex followed by 606 (*O. officinalis* complex), 33 (*O. meyerina* complex), 20 accessions of the *O. ridleyi* complex and one accession each of *O. schlechteri* and *O. coarctata* (Table 1.1).

Since the establishment of IRRI, over the last 50 years, these genetic resources have been maintained and shared with the global scientific community. These wild species have also been characterized based on morphological, cytological, biochemical, and molecular markers.

Besides IRRI, other centers also maintain and conserve seeds of some selected wild *Oryza* species in their respective gene banks including Africa Rice Center, Cotonou, Benin (West Africa); International Institute of Tropical Agriculture (IITA), Nigeria; National Bureau of Plant Genetic Resources (NBPGR), New Delhi, India; National Institute of Agrobiological Resources (NIAR), Japan; China National Rice Research Institute, Huangzou, China; Cuu Long Delta Rice Research Institute (CLRRI) Omon, Cantho, Vietnam; Kaesart University, Bangkok, Thailand; Malaysian Agricultural Research and Development Institute (MARDI), Malaysia; University of Queensland, Australia. Some of these institutes maintain wild species germplasm for active use or only for short-term storage.

1.2.1 Exploration and Collection of Germplasm

Records of early occasional exploration and collection for wild species of rice in different countries can be found in various publications (Oka 1988). Specimens of wild *Oryza* species can be found in many herbaria, particularly in India, China, Indonesia, Thailand, and Malaysia. Efforts for collection and conservation of wild *Oryza* species were initiated in the late 1950s by many National Agricultural Research Systems (NARS) on small scales, along with conservation programs for cultivated rice varieties and landraces. India and China are the world's two largest rice-producing and rice-consuming countries, and wild *Oryza* species are found abundantly in both countries. Exploration and collection of wild species in China can be traced back to as early as 1917 when Dr E. D. Merrill and colleagues first found *O. rufipogon* at Lofu Mountain and the Shilong Plain in Guangdong

Province (Wu 1990). Professor Ding Ying initiated the systematic exploration and collection of wild *Oryza* species in China. In 1926, he found *O. rufipogon* and collected its samples in many more sites, such as those in Guangzhou, Heiyang, Zhengcheng, Qingyuan, and Sashui in Guangdong. In the 1960s, many Chinese scientists from various agricultural research stations and universities organized several expedition trips, which covered larger areas of the wild *Oryza* species 'natural range.' Several exploration and collection missions at different scales were undertaken in the late 1980s and early 1990. Lu and Sharma (2003) have reviewed the exploration and collection of wild *Oryza* species in different countries/regions.

IRRI has been actively involved in collaboration with NARS and International Board for Plant Genetic Resources in organizing workshops on collection, conservation, and utilization of rice genetic resources. Two workshops were organized in 1977 and 1983. The third international workshop established new initiatives for international cooperation and conservation activities of wild *Oryza* species. Since then collection of wild *Oryza* species has gradually received more attention by the NARS, particularly in Asian countries, and more intensive and systematic collecting activities have been conducted in different countries. Based on reports by several NARS programs, a nominal number of seed samples of wild *Oryza* specks were collected in the early 1980s. Lu (1998, 1999) emphasized conservation of wild species of rice in Nepal and other countries.

Much germplasm exploration for rice was completed by the early 1990s. By the end of 1962, the IRRI varietal collection contained 6867 accessions from 73 countries. By 1972, the collection had grown to 14,600 accessions (Chang 1972, 1976). By the early 1980s, the number of accessions in the IRRI gene bank reached 49,027. More than 200 accessions of rice were collected during the second half of 1995 from Lao People's Democratic Republic (PDR). The IRRI gene bank now contains 3728 accessions of 24 wild species of *Oryza,* besides 1655 accessions of African cultivated rice (*O. glaberrima*).

1.2.2 Genetic Erosion

Wild species of *Oryza* are important reservoirs of useful genes for rice improvement. Through millions of years of evolution and genetic adaptation to variable environments, wild species have accumulated abundant biodiversity. Genetic erosion or loss of biodiversity of rice varieties has been recognized as a problem since the 1960s. Factors such as the adoption of high-yielding rice varieties, farmer's increased integration into the markets, change of farming systems, industrialization, human population increases, and cultural change have significantly accelerated the continual erosion of the rice gene pool (Bellon et al. 1998). A similar situation has also been observed for wild *Oryza* species. In many places of Asia, populations of wild *Oryza* species are becoming extinct or are threatened because their natural habitats are seriously damaged by extension of cultivation areas, expansion of communication systems such as road construction, and urban pressures. According to unpublished data collected by the Chinese Academy of Sciences in 1994, nearly 80% of the common wild rice (*O. rufipogon*) sites recorded during the 1970s have already disappeared (cited by Lu and Sharma 2003). The size of some surviving *O. rufipogon* populations was also found to be significantly reduced. A similar situation has been observed in other countries such as the Philippines, Vietnam, Thailand, Nepal, Indonesia, Malaysia, India, and Bangladesh.

The problems of genetic erosion are severe, but international efforts to conserve rice genetic resources, in which IRRI has taken a leading role, have led to the establishment of several gene banks in Asia. These joint efforts between national, regional, and international organizations ensure the long-term conservation of the biodiversity of the rice gene pool.

1.2.3 Conservation Strategy

For many plant species, ex situ conservation of seeds is safe and cost-efficient, provided proper attention is paid to seed drying and storage conditions. Fortunately, rice seeds exhibit orthodox storage behavior and can be dried to a low moisture content of ca. 6% and stored at −20 °C, retaining their viability for decades, if not longer. Vaughan (1994) has elaborated on Herbarium specimens of various wild *Oryza* species preserved in many herbaria of different countries. Jackson (1997) has suggested strategies for conservation of genetic resources. There are two basic approaches to germplasm conservation (ex situ and in situ conservation).

1.2.3.1 Ex Situ Conservation

In this approach, genetic resources are actually removed from their original habitat or natural environment. Ex situ conservation provides efficient means for germplasm preservation, utilization, exchange, and information generation through effective management and value-added research of the conserved wild rice species. However, seed samples placed under ex situ conservation in a gene bank become isolated from the *Oryza* ecosystem where they originated and grew. The expected microevolution of these *Oryza* species in their original environment is stopped, particularly the adaptive variations that could occur during change in environmental conditions. Therefore, in evolutionary terms, ex situ conservation is static (Bellon et al. 1998). Concerns have been raised following the observation that static conservation may reduce the adaptive potential of wild *Oryza* species and their populations in the future. Thus, ex situ conservation cannot be considered the only approach for conserving biodiversity of wild *Oryza* species. Complementary dynamic approaches such as in situ conservation are also necessary.

The long-term conservation of rice genetic resources is the principal aim of the International Rice Gene Bank (IRG). The gene bank has operated since 1977, although genetic conservation activities started in the early 1960s. For several countries, including Sri Lanka, Cambodia, Lao PDR, and the Philippines, the germplasm conserved in the IRG represents a more or less complete duplicate of their national collections. For other countries, such as India and the People's Republic of China, only part of their national collections are duplicated at IRRI. Nevertheless, the IRG has provided an important safety net for national conservation efforts. On several occasions, it has been possible to restore rice germplasm that had been lost in national gene banks with accessions already conserved at IRRI. IRRI maintains an active collection for medium-term storage and distribution of rice germplasm, at +2 °C in sealed, laminated aluminum foil packets, and long-term (50–100 years) conservation, at −20 °C, each with two vacuum-sealed aluminum cans.

The germplasm collection is held in trust by IRRI under the auspices of FAO in an International Network of ex situ collections. Duplicate storage of the IRG collection is maintained at the National Seed Storage Laboratory (NSSL), Fort Collins, USA, and about 75% of the collection is currently stored under black-box conditions. Duplicate storage of African rices is shared between IRRI, the International Institute of Tropical Agriculture (IITA) in Nigeria, and the Africa Rice Center (earlier named as WARDA) in Benin, Africa.

1.2.3.2 In Situ Conservation

This method attempts to preserve the integrity of genetic resources by conserving them within the evolutionary dynamic ecosystems of their original habitat or natural environment. Under in situ conservation, local control of traditional rice varieties will ensure that benefits accrue to farmers and communities that have developed them. For long-term and dynamic conservation, the in situ approach has great value. However, for some reasons, in situ conservation has, in general, received the least attention and even been rejected. Limited scientific and financial inputs are constraints in in situ conservation and its design and management for wild *Oryza* species.

1.3 Gene Transfer from Wild Species into Rice

Recent advances in tissue culture, genetic engineering, molecular cytogenetics, comparative genetics, and genomics, particularly in rice genome sequencing, have opened new opportunities to develop improved rice germplasm with novel genetic properties, and in understanding the function of rice genes. Breeders have successfully used conventional breeding methods and exploited the rice (*O. sativa*) gene pool to develop high-yielding improved rice varieties resistant to pests and abiotic stresses with improved quality characteristics. The major emphasis has been on utilizing indica, japonica, and javanica germplasm through intraspecific hybridization (indica × indica, japonica × japonica, indica × japonica). In several cases, genetic variability for target agronomic traits is limited in the cultivated rice gene pool. Under such situations, interspecific hybridization is an important plant-breeding approach to introduce novel genes for different agronomic traits from wild species into rice.

The genus *Oryza* has 24 wild species ($2n = 24, 48$). The number has been 22 for many years; later 2 were reclassified (tetraploid *O. punctata* and teterploid *O. officinalis*), added as *O. schweinfurthiana* and *O. malampuzhaensis,* respectively, and became 24. These wild species are reservoirs of many useful genes, particularly for resistance to major biotic and abiotic stresses (Table 1.1). However, these wild species are associated with several weedy traits, such as grain shattering, poor plant type, poor grain characteristics, and low seed yield. Besides, several incompatibility barriers such as pre- and post-fertilization barriers, hybrid sterility, limited recombination between the chromosomes of cultivated and wild species, hybrid breakdown and linkage drag limit the transfer of useful genes from wild species into cultivated species (Brar and Khush 1986, 1997). The major consideration in alien gene transfer is to selectively transfer agronomically important genes from wild species, avoiding linkage drag. To achieve precise transfer of genes from wild to cultivated species, strategies involving a combination of conventional plant-breeding methods with tissue culture and molecular approaches have become important (Brar and Khush 2002, 2006). Advances in tissue culture, molecular marker technology, genomics, and genomic in situ hybridization (GISH) have opened new opportunities to tap and characterize alien genetic variability even from distant genomes of *Oryza* through interspecific hybridization. Useful genes for resistance to BPH, BB, blast, and tungro and tolerance to acid sulfate conditions and cytoplasmic male sterility have been transferred from wild species of rice. Some of the breeding lines derived from wide crosses have been released as varieties.

1.3.1 Strategies for Alien Gene Transfer

Strategy to transfer genes from wild species into rice depends on the nature of the target trait(s), relatedness of the wild species and incompatibility barriers. Several protocols are available to overcome such barriers (Brar and Khush 2002). Some of the steps involved are described below:

Growing of wild species: The main source of wild species seed is the International Rice Gene Bank located at IRRI, Manila, the Philippines. Seeds can also be obtained from other institutes involved in rice research who have their own gene banks in rice-growing countries. Wild species are usually maintained and multiplied by vegetative (tiller) propagation. However, growing wild species from seeds needs careful handling during germination and raising of seedlings to maturity. Also, special care is needed to bag the panicle and to collect seeds as spikelet shattering is the most common problem. Vaughan (1994) has given details on: (i) how to obtain seeds, (ii) how to store seeds, (iii) where, when, and how to grow wild species (iv) breaking seed dormancy, (v) care during germination and raising of seedlings to maturity and selfing and harvesting of panicles/seeds.

Search for useful genetic variability for target traits: Wild species are known to be natural reservoir of useful genes for rice improvement; however, it is not uncommon that some accessions of a particular species may have limited genetic variability for the target agronomic trait of interest to breeders. Thus, it is essential to thoroughly screen and phenotype several accessions of different species to ensure sufficient genetic variability. As an example, out of 6000 accessions of cultivated rice and wild species screened, none was found to be resistant to grassy stunt virus except one accession (IRGC 101508) of *O. nivara* collected from Uttar Pradesh (UP), India; in fact, only few plants of these accessions were resistant. Priority should be placed on identifying variability in the closely related AA genome species (*O. sativa* complex $2n = 24$, AA), followed by *O. officinalis* complex CC genome species, and later on distantly related species such as *O. brachyantha* (FF), *O. granulata* (GG), and *O. ridleyi* (HHJJ), which show limited homoeologous pairing with the AA genome of rice.

Production of hybrids and alien introgression lines (AILs): Interspecific hybrids are produced between elite breeding lines with the wild species accessions carrying useful genes for target traits of immediate interest to the breeder. Such hybrids are produced through direct crosses between rice and AA genome wild species. However, embryo rescue is required to produce interspecific hybrids and AILs between rice and all the other 17 wild species of *Oryza* except the AA genome wild species. Backcrossing with the recurrent rice parent is used to develop introgression lines. MAS is practiced to accelerate breeding through transfer of wild species genes/QTLs for traits linked with molecular markers.

Evaluation of introgression lines for transfer of target traits: Advanced introgression lines generated through backcrossing are evaluated for the transfer of target traits. This involves extensive laboratory, screen house, and field testing using various screening and inoculation protocols and testing in hotspot nurseries for major biotic and abiotic stresses. Evaluation for target trait(s) is carried out at multilocations and also across regions/countries in collaboration with NARS partners.

Molecular characterization of alien introgression: Molecular markers are used to characterize introgression from wild species during backcross breeding. The availability of dense molecular maps of rice comprising simple sequence repeat (SSR) and more recently SNP markers has facilitated large-scale analysis to determine the extent and process of alien introgression. SNPs are proving to be great value in construction of CSSL.

Mapping of introgressed alien genes/QTL: Monosomic alien addition lines (MAALs $2n = 25$) can be used to locate the wild species gene on a specific chromosome. Different types of mapping populations are generated through wide crosses such as BILs, recombinant inbred lines (RIL), doubled haploid (DH), and near-isogenic alien introgression lines, including other segregating population F_2, F_3, etc. Introgressed alien genes/QTL are mapped and tagged with molecular markers for use in MAS. Genome in situ hybridization (GISH) has become popular in characterization of parental genomes in interspecific progenies and for locating introgressed segments on rice chromosomes.

1.3.2 Alien Introgression Lines (AILs) and Chromosomal Segmental Substitution Lines (CSSLs)

Interspecific hybrids have been produced between rice (AA genome) and wild species of *Oryza,* representing 10 of the 11 genomes (AA, BB, CC, BBCC, CCDD, EE, FF, GG, HHJJ, KKLL) either through direct crosses or through embryo rescue (Brar et al. 1991; Brar and Khush 2002, 2006). A large number of AILs have been produced from crosses of rice with different wild species of *Oryza* by IRRI and several other NARS institutes (Brar and Singh 2011). As an example, at the Punjab Agricultural University (PAU), India, large sets of AILs have been produced from crosses of rice varieties (PR 114, Pusa 44) with six (AA genome) wild species: *O. glaberrima, O. nivara, O. rufipogon, O. longistaminata, O. glumaepatula,* and *O. barthii.* A large number of wide cross derivatives have been used by different institutes to transfer useful genes into rice (Table 1.2). Some examples include transfer of resistance to BPH, BB, blast, and tungro and tolerance to acid sulfate conditions, including introgression of genes for cytoplasmic male sterility from different wild species into rice. Besides these traits, AILs have also been important genetic resources for the identification of QTL's for yield-related traits. One of the QTL, qSP2.2 from *O. longistaminata,* governing increased number of spikelets per panicle is being introgressed into Basmati rice which otherwise has panicle with a few spikelets.

Besides AILs, monosomic alien addition lines (MAALs $2n = 25$) representing 24 chromosomes of rice and a single extra chromosome of wild species have been produced. MAALs have been established from 7 wild species (CC, BBCC, CCDD, EE, FF, GG, HHJJ genome) (Jena and Khush 1990; Brar and Khush 2002). Similarly, CSSL having 24 chromosomes of rice but with one or two substituted segments of wild species chromosomes has also been produced. These substitution lines are being characterized using molecular markers, and a number of a genes/QTLs introgressed from wild species have been mapped. Doi et al. (2003) developed a set of CSSL with segments of *O. glaberrima, O. glumaepatula,* and *O. meridionalis* in the background of japonica cultivar (Taichung 65) and identified genes governing several traits. Surapaneni et al. (2017) developed CSSLs having segments of *O. nivara* substituted in the rice variety Swarna to identify major effect QTLs for yield-related traits.

Yoshimura et al. (2010) reported construction of CSSL from different AA genome species in the background of Taichung 65. Similarly, many institutes in China, Japan, India, USA, and France have developed CSSL and mapped and cloned genes/QTLs Ramos et al. (2016), produced CSSL of *O. longistaminata* in the background of Taichung 65 and reviewed development of CSSL of wild species by various other institutes.

1.3.3 Introgression from AA Genome Species of *O. sativa* complex

Crosses between rice and six other wild species of the *O. sativa* complex having the AA genome can be easily made and the genes have been transferred through conventional crossing and backcrossing procedures. Among the classical examples are the introgression of a gene for grassy stunt virus resistance from *O. nivara* to cultivated rice varieties (Khush 1977) and the transfer of a cytoplasmic male sterile (CMS) source from wild rice, *O. sativa* f. *spontanea*, (Lin and Yuan 1980). Other useful genes, such as *Xa21* for bacterial blight resistance, were transferred into rice from *O. longistaminata*, and new CMS sources from *O. perennis* and *O. glumaepatula*. More recently, genes for BPH, green leafhopper (GLH), green rice leafhopper (GRH) blast, tungro tolerance and tolerance to salinity, and acid sulfate soil conditions have been transferred from *O. sativa* complex species into indica rice cultivars (Table 1.2). Some breeding lines with genes transferred from wild species have been released for commercial cultivation in rice-growing countries (Table 1.3). Some examples on the specific traits transferred from wild species into rice are discussed below.

Introgression of gene(s) for resistance to grassy stunt virus: The grassy stunt virus is a serious disease transmitted by the vector brown planthopper (*Nilaparvata lugens*) BPH. Severe yield losses or even total loss may occur under epidemic conditions. Of the 6000 accessions of cultivated rice and several wild species screened, only one accession of *O. nivara* (accession 101508) was found to be resistant (Ling et al. 1970). A single dominant gene *Gs* confers resistance in *O. nivara*. Following four backcrosses with improved rice varieties, the gene for grassy stunt resistance was transferred into cultivated germplasm (Khush 1977). The first set of grassy stunt resistant varieties, IR28, IR29, and IR30, was released for cultivation in 1974. Subsequently, many such varieties, e.g., IR34, IR36, IR38, IR40, IR48, IR50, IR56, and IR58, have been released. These varieties have been widely grown and grassy stunt-infected plants are rarely seen in farmer's field now.

Introgression of gene(s) for resistance to tungro disease: Rice tungro disease (RTD) is another most serious viral disease in South and Southeast Asia. It is transmitted by the vector

Table 1.2 Introgression of genes from wild *Oryza* species into rice

Trait (s) transferred into *O. sativa* (AA)	Wild species (donor)	Gene/QTL's	Genome
Grassy stunt resistance	*O. nivara*	*Gs*	AA
Bacterial blight resistance	*O. logistaminata*	*Xa21*	AA
	O. rufipogon	*Xa23*	AA
	O. nivara	*Xa38*	AA
	O. officinalis	*Xa 29(t)*	CC
	O. minuta	*Xa27*	BBCC
	O. latifolia	Unknown	CCDD
	O. australiensis	Unknown	EE
	O. brachyantha	Unknown	FF
Blast resistance	*O. glaberrima*	Unknown	AA
	O. rufipogon	Unknown	AA
	O. nivara	Unknown	AA
	O. glumaepatula	Unknown	AA
	O. barthii	Unknown	AA
	O. minuta	*Pi9*	BBCC
	O. australiensis	*Pi40*	EE
Brown plant hopper (BPH) resistance	*O. nivara*	*Bph33(t)*	AA
	O. officinalis	*bph11,bph12,Bph14,Bph15,qBph1,qBph2*	CC
	O. eichingeri	*Bph13*	CC
	O. minuta	*Bph20,Bph21*	BBCC
	O. latifolia	*Unknown*	CCDD
	O. australiensis	*Bph10,Bph18*	EE
White-backed plant hopper (WBPH) resistance	*O. officinalis*	*Wbph7(t),Wbph8(t)*	CC
	O. latifolia	Unknown	CCDD
Green leaf hopper (GLH) resistance	*O. glaberrima*	*Glh11*	AA
Green rice leaf hopper(GRH) Resistance	*O. rufipogon*	*Grh5*	AA
	O. glaberrima	*QGrh9*	AA
Cytoplasmic male sterility (CMS)	*O. sativa f spontanea*	Wild abortive (WA)	AA
	O. perennis	Unknown	AA
	O. glumaepatula	Unknown	AA
	O. rufipogon	Unknown	AA
Tungro tolerance	*O. rufipogon*	Unknown	AA
Tolerance to iron toxicity	*O. rufipogon*	Unknown	AA
	O. glaberrima	Unknown	AA
Drought-related traits	*O. glaberrima*	QTLs	AA
Tolerance to aluminum toxicity	*O. rufipogon*	QTLs	AA
Tolerance to acidic conditions	*O. glaberrima*	Unknown	AA
	O. rufipogon	Unknown	AA
Tolerance to phosphorus deficiency	*O. glaberrima*	Unknown	AA
	O. rufipogon	Unknown	AA
Yield-enhancing loci (wild species alleles)	*O. rufipogon*	QTLs	AA
	O. nivara	QTLs	AA
	O. grandiglumis	QTLs	CCDD
Earliness, abiotic stress tolerance, weed competitiveness	*O. glaberrima*	Unknown	AA
Increased elongation ability	*O. rufipogon*	Unknown	AA

Modified from Brar and Khush (2006) and Brar and Singh (2011)

Table 1.3 Rice varieties developed through interspecific hybridization

Key trait (s) introgressed	Wild species (donor)	Varieties released	Country
Grassy stunt resistance	*O. nivara*	Many rice varieties	Several rice-growing countries in Asia
BPH resistance	*O. officinalis*	MTL98, MT103, MTL105, MTL110	Vietnam
	O. australiensis (bph18)	Suweon 523	South Korea
BPH and BB resistance	*O. rufipogon*	Dhanarasi	India
Tungro tolerance	*O. rufipogon*	Matatag 9	the Philippines
Acid sulfate tolerance	*O. rufipogon*	AS 996	Vietnam
Salinity tolerance	*O. rufipogon* *O. rufipogon*	BRRI Dhan 55(AS996) Jaraya	Bangladesh India
Bacterial blight (BB) resistance	*O. logistaminata (Xa 21)* *Xa 21* pyramided along with other genes for BB resistance	NSIc Rc 112 More than 12 BB resistant varieties released through MAS using *Xa* 21 and other pyramided genes	the Philippines India, China, the Philippines
Blast resistance	*O. glaberrima*	Yun Dao	China
High yield, earliness, weed-competitive ability and tolerance to abiotic stresses	*O. glaberrima*	Many NERICA lines/varieties	African countries
Yield-enhancing loci (QTL's): salinity tolerance	*O. nivara*	DRR Dhan 40	India (Maharashtra, Tamil Nadu, West Bengal states)
Salinity tolerance	*O. rufipogon*	Chinsurah Nona 2	India (Coastal saline area of West Bengal)

Modified from: Brar and Singh (2011)

green leafhopper (*Nephotettix virescens*) GLH. The disease is caused either by a single infection or by a double infection with two viral particles, the rice tungro bacilliform virus (RTBV), a double-stranded DNA virus, and the rice tungro spherical virus (RTSV), a single-stranded RNA virus. There is limited variability in cultivated rice germplasm for resistance to RTBV, the main cause of tungro symptoms. *O. longistaminata* and *O. rufipogon* have shown tolerance to RTBV. From the crosses of IR64 with *O. rufipogon*, many tungro-tolerant lines have been developed. One of the elite breeding lines (IR73885-1-4-3-2-6) resistant to tungro has been released as a variety (Matatag 9) for cultivation in tungro-prone areas of the Philippines (Table 1.3). Many tungro-tolerant backcross breeding lines from crosses with *O. longistaminata* and *O. rufipogon* have been selected.

Introgression of gene(s) for resistance to bacterial blight: The bacterial blight (BB) caused by *Xanthomonas oryzae* pv. *oryzae* is one of the most destructive diseases of rice in Asia. As many as 40 genes for BB resistance have been identified from the cultivated and wild species. Of these, six genes, *Xa21, Xa23, Xa27, Xa29, Xa33, and Xa38,* have been derived from wild species. A dominant gene (*Xa21*) with a wide spectrum of resistance to BB from *O. longistaminata* was transferred through backcrossing into IR 24 (Khush et al. 1990). *Xa21* was mapped on chromosome 11 close to the RG103 marker (Ronald et al. 1992). Jiang et al. (1995) used BAC clones and FISH and physically mapped the *Xa21* locus to chromosome 11 of rice. Zhang et al. (1998) transferred BB resistance gene *Xa 23*(t) from *O. rufipogon* into rice. This gene also conferred a very wide spectrum of resistance and showed resistance to all races of BB of the Philippines. Two new genes one each from *O. glaberrima* and *O. barthii* have been identified and transferred to *O. sativa*. Cheema et al. (2008a) have transferred *Xa38* (t) from *O. nivara* into cultivated rice. The *Xa21* has been incorporated into elite breeding lines of rice through MAS (Sanchez et al. 2000; Singh et al. 2001). This gene has also been pyramided with other bacterial blight resistance genes such as *Xa4*, X*a5* and *Xa13*, and more than 12 varieties have been developed through MAS (Brar and Singh (2011). These include PR106, improved Sambha Mahsuri and improved Pusa Basmati 1 in India, NSICRc142 and NSICRc154 in the Philippines, and 8 commercial hybrids in China (Xieyou 218, Zhongyou 218, Guodao 1, Guodao 3, Neizyou, Ilyou 8006, Ilyou 218, and ZhongbaiYou 1). Several parental lines of rice hybrids in India and China also carry the *Xa21* gene.

Two genes, *Xa21* from *O. longistaminata* and *Xa27* from *O. minuta,* have been cloned (Song et al. 1995; Gu et al. 2003). *Xa 21* is developmentally regulated while *Xa 27* involves induction by the pathogen. *Xa21* has also been transferred into IR72 through transformation (Tu et al. 2000). Transgenic rice lines carrying *Xa21* were field-tested in the Philippines and had shown a wide spectrum of resistance. A new resistance gene, *Xa38*, introgressed from *O. nivara* has been mapped to a region of 38.4 kb of chromosome 4 (Cheema et al. 2008a). Bhasin et al. (2012) developed PCR-based site sequence-tagged marker (OSo-4g53050-1) for *Xa38* to facilitate its use in MAS. Ellur et al. (2016) compared NILs of Basmati rice variety PB1121 carrying *Xa38* with PB1121 NILs carrying *xa13* + *Xa21*. Both NILs showed resistance against Xoo races 1, 2, 3, and 6. Additionally, *Xa38* also showed resistance to Xoo race 5 to which *xa13* + *Xa21* was susceptible. *Xa38* is now used in rice-breeding programs in India.

Introgression of genes for resistance to blast: Blast disease caused by *Magnaporthe oryzae* is another major rice disease. Genes for blast resistance have been introgressed from four AA genome wild species (*O. rufipogon, O. nivara, O. glumaepatula, and O. barthii)* including *O. glaberrima* into rice (Rama Devi et al. 2015). Of the 326 AILs tested, 4 were found to be resistant to blast. Multilocation testing confirmed field resistance to both leaf and neck blast. Ram et al. (2007) evaluated AILs derived from *O. sativa* × *O. rufipogon* and *O. sativa* × *O. glaberrima* which showed introgression for blast resistance. The derived breeding lines with introgression for blast resistance have been released as varieties (Dhanarasi in India and Yun Dao in China) (Table 1.3).

Introgression for resistance to Brown plant hopper (BPH): Genes for BPH resistance have been introgressed from *O. rufipogon* and *O. nivara* into rice. A mapping population was generated from a cross between indica rice cultivar PR 122 and *O. nivara* (IRGC104646). Two SSR markers, RM16994 and RM 17007, co-segregated with the BPH resistance gene. *Bph33(t)* was located on long arm of chromosome 4. High-resolution mapping of BPH resistance locus in *O. nivara* spanning 91 Kb of genomic region revealed 11 candidate genes (Kumar et al. 2017, under review).

Introgression for tolerance to abiotic stresses: Several introgression lines derived from crosses *O. sativa* × *O. rufipogon* and *O. sativa* × *O. glaberrima* have been tested for tolerance to abiotic stresses under field conditions in the Philippines. Several breeding lines with good agronomic traits and moderate tolerance to iron toxicity, aluminum toxicity, and acid sulfate conditions have been identified. Three promising breeding lines were selected from a cross of IR 64 × *O. rufipogon* and tested in the field under acid sulfate conditions in Vietnam in cooperation with Cuu Long Delta Rice Research Institute (CLRRI). One breeding line, IR73678-6-9-B, was released as a variety (AS996) for cultivation under acid sulfate conditions in Vietnam. Nguyen et al. (2003) analyzed RIL population from a cross of IR 64 × *O. rufipogon* and mapped QTLs for tolerance to aluminum toxicity. A major effect QTL with 24.9% phenotypic variation for relative root length under stress was found on chromosome 3 which is conserved across cereal species. Introgression for salinity tolerance from crosses of rice with two wild species (*O. nivara* and *O. rufipogon*) has been obtained. Elite breeding lines derived from cross of rice variety Swarna with *O. nivara* showed introgression for tolerance to salinity, and a variety DRR Dhan 40 has been released in India. Similarly, another variety Chinsurah Nona 2 from the cross of KMR 3 with *O. rufipogon* has been released for cultivation in coastal saline area of West Bengal, India (Table 1.3).

Incorporation of cytoplasmic male sterility from wild species: The AA genome of wild species has been an important genetic resource for cytoplasmic male sterility (CMS). The most commonly used CMS source in hybrid rice breeding is derived from the wild species *O. sativa* f. *spontanea* (Lin and Yuan 1980). This cytoplasmic source has been designated as wild abortive (WA). More than 90% of the male sterile lines used in commercial rice hybrids grown in China and other counties have the WA type of cytoplasm. From crosses of several accessions of *O. rufipogon* and *O. perennis* with IR 64, followed by backcrossing with IR64, a set of CMS lines were selected. Of these, a new CMS source from *O. perennis* (accession 104823) was transferred into rice (Dalmacio et al. 1995). Molecular analysis using mt-specific DNA probes and progeny test with several restorers of WA cytoplasm revealed that the male sterility source of this line (IR 66707A) is different from WA. Another CMS line (IR69700A) having the cytoplasm of *O. glumaepatula* and the nuclear genome of IR64 has been developed (Dalmacio et al. 1996). Extensive efforts were also made at the Directorate of Rice Research (now named as Indian Institute of Rice Research-IIRR), Hyderabad, India, to develop CMS lines with cytoplasm source from different AA genome wild species. However, none of these CMS lines could be used for developing commercial varieties due to lack of good restorers.

Identification and introgression of yield-enhancing loci/QTLs: Wild species are phenotypically inferior to the cultivated species having poor plant type and poor seed characteristics but are proving valuable genetic resource not only for stress tolerance but also for yield-related traits. Transgressive segregation for yield in crosses of cultivated and wild species suggests that, despite inferior phenotypes, wild species contain genes that can improve quantitative traits, such as yield. Tanksley and Nelson (1996) proposed advanced-backcross (AB) QTL analysis to discover and transfer valuable QTL alleles from unadapted germplasm, such as wild species into elite breeding lines.

Since the mapping of yield QTLs, *yld1.1* and *yld 2.1* from *O. rufipogon* that could increase yield by 17 and 18%, respectively (Xiao et al. 1996), in a number of studies, QTLs/yield-enhancing loci 'wild species alleles' have been identified (Shakiba and Eizenga 2014). Xiao et al. (1998) identified 68 QTLs, of which 35 had trait-improving alleles from wild species. Cheema et al. (2008b) identified genomic regions of *O. rufipogon* that were associated with *qyldp2* for yield increase. Swamy et al. (2014) identified yield-enhancing QTLs such as *qyldp2.1, qyldp3.1 qyldp8.1 qyldp9.1, and qyldp11.1.*

Neelamraju (2014) from a cross of *O. sativa* × *O. nivara* population identified two QTL's *yld9.1* for yield and *nfg 9.1* for number of filled grains. Surapaneni et al. (2017) analyzed CSSLs derived from a cross of rice variety Swarna with *O. nivara* to identify major effect QTLs for yield-related traits. Ten QTLs from wild species were identified which contributed to yield enhancement. Ramos et al. (2016) reported CSSLs of *O. longistaminata* in the background of Taichung 65. A set of 40 CSSLs was evaluated for different yield-related traits and identified 10 QTLs for such traits. One of the QTL showed a pleiotropic effect.

At Punjab Agricultural University (PAU), India, *qsqSPP2.2* from *O. longistaminata* has been identified for increased spikelets per panicle. This QTL is being introgressed into Basmati rice for increasing spikelets per panicle which otherwise has few spikelets per panicle. Genomic selection is being used to transfer the desirable alleles from a core collection of *O. rufipogon* into cultivated rice. Bhatia et al. (2017a) evaluated 1781 backcross inbred lines (BILs) derived from crosses of two indica rice cultivars with 70 different accessions of five different AA genome wild species. These BILs (BC_2F_5-BC_2F_{10}) were evaluated in an augmented design over 3 years. Of the 108 BILs possessing superior agronomic traits, 15 BILs showed >10% yield increase; the most significant traits were number of spikelets per panicle and 1000-grain weight. In another study, Bhatia et al. (2017b) evaluated a set of 63 BILs derived from *O. rufipogon* and 122 BILs of *O. glumaepatula* for mapping yield-related traits using genotyping by sequencing (GBS). A custom-designed approach for GBS data analysis identified 3322 informative SNPs in 55 *O. rufipogon* BILs and 3437 informative SNPs in 79 *O. glumaepatula*-derived BILs. QTL mapping identified *qtgw5.1* for 1000-grain weight; *qgw5.1*, *qgw5.2* for grain width; and *qgl7.1* for grain length in BILs of *O. rufipogon*. Similarly, 5 QTLs were identified from BILs derived from *O. glumaepatula*. The QTLs identified in the above studies from low-yielding wild species of *Oryza* are indicative of increasing rice yield through introgression of alleles from wild species to broaden the gene pool of rice cultivars.

Introgression from O. glaberrima into O. sativa: Cultivated African species *O. glaberrima* has very low yielding but has several desirable traits such as resistance to rice yellow mottle virus (RYMV), African gall midge, and nematodes, as well as tolerance to drought, acid soils, and iron toxicity. It also has strong weed-competitive ability. Scientists at West Africa Rice Development Association (WARDA) now called Africa Rice Center have transferred earliness, weed-competitive ability into high-yielding breeding lines of *O. sativa* (Jones et al. 1997). Africa Rice Center has made extensive crosses of *O. sativa* with *O. glaberrima* and released a series of breeding lines as NERICA varieties for commercial cultivation in African and other countries. A large number of advanced introgression lines have also been produced from crosses of *O. sativa* and several accessions of *O. glaberrima* at IRRI. These progenies have been evaluated in collaborative projects with Africa Rice Center for introgression for tolerance to RYMV, African gall midge, and abiotic stresses. Promising lines tolerant to iron toxicity and resistance to blast have also been identified. Soriano et al. (1999) identified *O. glaberrima* to be highly tolerant to root-knot nematode. A few AILs showed introgression for nematode resistance from *O. glaberrima*.

Bimpong et al. (2011) produced backcross progenies from cross of IR64 and *O. glaberrima* which were evaluated for yield and other traits under lowland drought stress conditions at IRRI. As many as 33 AILs had higher yield than recurrent parent. A set of 200 AILs was selectively genotyped with SSRs and sequence-tagged site markers. Molecular analysis showed that on average 4.5% of the *O. glaberrima* genome was introgressed into the AILs derived from *O. sativa* × *O. glaberrima* where *O. sativa* was used as recurrent parent. Molecular analysis of BILs revealed 33 QTLs for different traits. One of the QTL at RM208 marker on chromosome 2 positively affected yield under drought stress, accounting for 22% of the genetic variation.

Production of doubled haploids (DH) from crosses of O. sativa × O. glaberrima: Hybrids between *O. sativa* and *O. glaberrima* can be produced easily through direct crosses; however, the F_1 in spite of complete chromosome pairing is highly sterile. Several sterility genes differentiate these two species (Oka 1988). Anther culture is an important technique to produce homozygous lines from sterile F_1, fix recombinants, and DH lines as mapping populations to locate genes governing agronomic traits. Enriquez et al. (2000) cultured 45,400 anthers from 75 F_1s; no calli were produced from 34 crosses. The other 41 F_1s showed, on average, 1.3% callus formation from 144,160 cultured anthers. The anther-derived calli from 16 F_1s showed no plant regeneration. Strong genotypic differences for anther culturability for both callus induction and plant regeneration were observed. Furthermore, callus induction and plant regeneration from anther culture were found to be independent of each other. Although 562 DH lines could be produced from different crosses, the majority showed high seed sterility (56.2–100%). Such high sterility of DH lines is indicative of the presence of several loci for sterility that differentiate the Asian and African rice species.

Genotyping of DH lines using microsatellite markers indicated frequent exchange of chromosome segments between *O. sativa* and *O. glaberrima*. Some of the markers located on chromosomes 1, 6, 9, 11, and 12 showed distorted segregation favoring alleles of either parents (Enriquez 2001).

1.3.4 Introgression from Wild Species of *O. officinalis* Complex

Among the four complexes, *O. officinalis* is the largest complex comprising of 6 diploid and 6 allotetraploid species. Interspecific hybrids have been successfully produced between rice and wild species of this complex. Also, MAALs have been developed in 7 species involving CC, BBCC, CCDD, EE, and FF genomes representing 6–12 chromosomes (Jena and Khush 1990; Brar and Singh 2011). A few genes for resistance to BPH, WBPH, BB, and blast have been transferred from wild species of this complex into rice (Table 1.2).

Introgression from the CC genome species: A number of introgression lines have been produced from the crosses between *O. sativa* and *O. officinalis* (Jena and Khush 1990). Four genes for resistance to BPH, e.g., *bph11, bph12, Bph14, and Bph15,* and two QTLs, *qBph1 and qBph2,* have been introgressed into the progenies. Four breeding lines have been released as varieties (MTL95, MTL98, MTL103, and MTL110) for commercial cultivation in the Mekong Delta of Vietnam (Table 1.3). Hirabayashi et al. (2003) analyzed recombinant inbred lines (RILs) from the cross between Hinohikari (susceptible japonica) and a BPH-resistant introgression line derived from *O. sativa* × *O. officinalis*. Two genes for BPH resistance, *bph11*(t) and *bph12*(t), were identified and mapped to chromosomes 3 and 4 of rice, respectively. Huang et al. (2001) also transferred BPH resistance from *O. officinalis* into Zhensheng 97B. At IRRI, we produced advanced-backcross progenies from crosses of an elite breeding line of NPT rice (IR65600-81-5-3-2) with two accessions of *O. officinalis* and another with tetraploid species *O. minuta*. The NPT line is highly susceptible to bacterial blight. More than 1053 BC_2F_3 progenies derived from NPT × *O. officinalis* (accession 101399) were screened after inoculation with race 1. Several progenies resistant to bacterial blight have been identified.

Introgression from the BBCC genome species: Interspecific hybrids have been produced between *O. sativa* and the tetraploid wild species *O. minuta* (BBCC). Advanced introgression lines were produced through embryo rescue of F_1 hybrid followed by backcrossing with *O. sativa* parent. Genes for resistance to bacterial blight and blast were transferred into rice (Amante-Bordeos et al. 1992). Blast resistance gene, *Pi9,* was introgressed from *O. minuta* into rice. It has a wide spectrum resistance and has been used in breeding programs in India. High-resolution mapping indicates that *Pi-z* and *Pi 9* are physically tight-linked and may even be allelic.

BPH resistance from *O. minuta* has also been transferred to rice (Brar et al. 1996), and derived lines have shown a wide spectrum of resistance to BPH in the Philippines and Korea. Two major QTLs and two significant digenic epistatic interactions between marker intervals were identified for BPH resistance. One QTL was mapped to 193.4 kb region located on the short arm of chromosome 4, and the other QTL was mapped to a 194.0 kb region on the long arm of chromosome 12 (Rahman et al. 2008). These major QTLs are new BPH resistance loci and have been designated as *Bph20 (t)* on chromosome 4 and *Bph21 (t)* on chromosome 12. Both of these genes are being used in pyramiding of genes for BPH resistance.

Shim et al. (2010) developed *O. minuta* specific clones by representational difference analysis (RDA), a subtractive cloning method. Hybridization of 105 clones could identify BB- and CC-genome-specific clones. *OmSc45,* a highly repetitive clone, upon hybridization with *O. minuta* and *O. australiensis,* was found in 6000 and 9000 copies, respectively. This clone also showed introgression from *O. minuta* in the derived line. Different hybridization patterns observed with other species showed conservation of *O. minuta* fragments across *Oryza* species.

Introgression from the CCDD genome species: A number of workers have produced hybrids between rice and *O. latifolia* (CCDD) (Sitch 1990; Brar et al. 1991). Several introgression lines derived from this cross have been evaluated for introgression of useful traits (Multani et al. 2003). Ten allozymes of *O. lafifolia*, such as *Est5, Amp1, Pgi1, Mdh3, Pgi2, Amp3, Pgd2, Est9, Amp2*, and *Sdh1*, located on 8 of the 12 chromosomes were observed in introgression lines. Alien introgression was also detected for morphological traits such as long awns, earliness, black hull, purple stigma, and apiculus. Genes for resistance to bacterial blight, BPH and white-backed plant hopper have been introgressed into elite breeding line from *O. latifolia.* Yield-enhancing loci in the population derived from crosses of japonica cultivar Hwaseongbyeo × *O. grandiglumis* (CCDD) have been identified. Of the 39 QTLs, *grandiglumis* contributed desirable alleles in 18 QTLs (Yoon et al. 2006).

Introgression from EE genome species: Hybrids between cultivated rice and EE genome species *O. australiensis* were produced by Multani et al. (1994). Disomic as well as $2n + 1$ progenies were examined. Of the 600 BC_2F_4 progenies, four were resistant to BPH. Introgression was observed for morphological traits such as long awns and earliness and *Amp3* and *Est2* allozymes. Resistance to BPH was found to be under monogenic recessive control in two progenies, and dominant genes conveyed resistance in the other two. Dominant gene in one of the progenies was designated as *Bph10* which conferred resistance to three biotypes of BPH in the Philippines. MAAL analyses showed that the gene for BPH resistance is located on chromosome 12 of *O. australiensis.* All 14 probes were polymorphic in recurrent parent and wild species; however, only RG457 detected introgression from *O. australiensis* into the introgression line. Co-segregation for BPH reaction and molecular markers showed a gene for BPH resistance linked to RG457, with a distance of 3.68 cM (Ishii et al. 1994). Jena et al. (2006) identified another gene for resistance to BPH-designated *Bph18* introgressed from the same accession of *O. australiensis.* Jeung et al. (2007) introgressed a major gene *Pi40* (t) for blast resistance from this species.

Introgression from FF genome species: A hybrid between cultivated rice and FF genome species *O. brachyantha* was produced and 149 backcross progenies were obtained. Out of these, 27 showed introgression for resistance to the Philippines bacterial blight races 1, 4, and 6 (Brar and Khush 1997). Gene transfer was not associated with any undesirable traits of *O. brachyantha.*

Introgression from species of O. meyeriana complex: Earlier, this complex has only 2 diploid species (*O. meyerina* and *O. granulata*, $2n = 24$, GG). One of the species, *O. indandamanica*, once considered as a separate species is now classified under *O. meyeriana.* The third species, *O. neocaledonica,* has been added to this complex (Table 1.1). The species of this complex grow under full shade. Hybrids have been produced from the cross between *O. sativa* and *O. granulata* (Brar et al. 1991). Some MAALs have also been isolated. Backcross progenies resembled *O. sativa* parent in morphological traits. No introgression was observed for any of the trait of *O. granulata* in the derived progenies (Brar et al. 1996).

Introgression from species of O. ridleyi complex: The complex comprises two allotetraploid species ($2n = 48$, HHJJ), i.e., *O. ridleyi* and *O. longiglumis. O. ridleyi* shows strong resistance to all 10 Philippine races of BB. Hybrids between rice cv. IR56 and O. *ridleyi* (accession 100821) have been produced. However, the cross shows strong necrosis. Thus, only a few introgression lines (BC_3F_3) from this cross have been produced, but no introgression could be detected.

Introgression from species of unclassified group: Two tetraploid species, *O. schlechteri* ($2n = 48$, HHKK) and *O. coarctata* ($2n = 48$, KKLL), have been included in the unclassified group. So far no interspecific hybrids have been reported from the cross of *O. sativa* × *O. schlechteri. O. coarctata* was earlier designated as *Porteresia coarctata.* Interspecific hybrids between *O. sativa* and *O. coarctata* have been produced both though sexual crosses following embryo rescue (Brar et al. 1997) and protoplast fusion (Jelodar et al. 1999). The hybrid ($2n$ = 36) is sterile and shows no chromosome pairing as well as no elimination of chromosomes of either parent. Due to strong incompatibility barriers, no backcross progenies could be obtained.

A tissue culture cycle of the wide hybrids of rice (F_1, MAALs) could enhance the frequency of genetic exchange of chromosome segments, particularly among distant genomes (AA × FF, AA × GG, AA × HHJJ, AA × HHKK) of *Oryza* species, which otherwise lack or show very limited homoeologous pairing, and where gene transfer through conventional methods is difficult. However, so far the technique has not been employed for enhancing exchanges among cultivated rice and distant genomes of *Oryza*.

1.4 Characterization of Parental Genomes in Interspecific Hybrids, MAALs, and Homoeologous Pairing in *Oryza* Through GISH

In situ hybridization is a powerful technique to characterize parental genomes in interspecific hybrids, identify alien chromosome(s) in MAALs, locate introgressed segments, and detect homoeologous pairing, thus leading to the precise understanding of alien introgression into the rice genome. Genomic in situ hybridization (GISH) protocols have been successfully used to characterize parental genomes and extra chromosome in MAALs (Brar and Khush 2002, 2006). GISH was used successfully to characterize homoeologous pairing in F_1 hybrids during meiosis involving rice and wild species genomes: AA × CC, AA × EE, AA × BBCC, AA × FF, AA × HHJJ, and AA × GG. Autosyndetic and allosyndetic pairing have been characterized among chromosomes involving different genomic combinations (Hue 2004).

Asghar et al. (1998) applied fluorescence in situ hybridization (FISH) for characterizing the chromosomes of *O. sativa* and *O. officinalis* and located rDNA loci on somatic chromosomes of both *O. sativa* and *O. officinalis*. Yan et al. (1999) used FISH to characterize A and C genome chromosomes in F_1 and BC_1 of *O. sativa* × *O. eichingeri*. Abbasi et al. (1999) used total genomic DNA of *O. australiensis* as a probe and hybridized it with the meiotic chromosomes of the F_1 hybrid (*O. sativa* × *O. australiensis*). Both autosyndetic and allosyndetic pairing among AA and EE genomes could be detected. Meiotic analysis using GISH showed three types of pairing: (i) between A and E genome chromosomes, (ii) within A genome chromosomes, and (iii) within E genome chromosomes. Of the paired chromosomes, 78.8% involved A and E genomes, 16.8% showed pairing within A genome chromosomes, and 4.3% within E genome chromosomes. (Hue 2004) Pairing was also observed using in situ hybridization between HJ genomes in a hybrid involving (AHJ genomes) cross of *O. sativa* × *O. ridleyi*, indicating sufficient homoeology between unknown H and J genomes. The results demonstrate the usefulness of FISH for precisely characterizing homoeologous pairing and highlighting over estimation resulting from autosyndetic pairing, which is otherwise difficult to detect through conventional cytogenetic analysis.

1.5 Molecular Characterization of Alien Introgression

Both isozyme and molecular markers have been used for characterization of introgression from wild species into rice (Jena et al. 1992; Multani et al. 1994, 2003). Jena et al. (1992) analyzed 52 BC_2F_8 introgression lines derived from the cross *O. sativa* × *O. officinails*. Of the 177 RFLP markers studied, 174 were polymorphic between two parents when one or two enzymes were used. Most markers were polymorphic when studied with multiple enzymes. Of the 174 RFLP markers, 28 (16.1%) identified putative *O. officinalis* introgressed segments in one or more introgression lines. Individual introgression lines contained 1.1–6.8% introgressed *O. officinails* segments. Introgressed segments were found on 11 of the 12 rice chromosomes. In a majority of cases, *O. sativa* alleles were replaced by *O. officinails* alleles. Single RFLP markers detected most introgressed segments, and the flanking markers were negative for introgression. Brar et al. (1996) analyzed 29 derivatives of *O. sativa* × *O. brachyantha* and 40 derivatives of *O. sativa* × *O. granulata.* Extensive polymorphism between rice and the two wild species was observed. Of the six chromosomes surveyed, no introgression was detected for chromosomes 7, 9, 10, and 12 of *O. granulata* and chromosomes 10 and 12 of *O. brachyantha.* In each of the remaining chromosomes, introgression was observed for one or two RFLP markers. Although the extent of introgression is extremely low, the results show possibilities of introgressing genes from even the most distantly related species of *Oryza* to cultivated rice.

The alien introgression lines derived from crosses of rice and distantly related wild species have shown complete resemblance to the recurrent rice parent in morphological traits. BC_2 progenies obtained from *O. sativa* × *O. officinalis* cross resembled rice parents so much that they were released as varieties (Jena and Khush 1990). Similarly, BC_3 and BC_4 progenies derived from *O. sativa* × *O. australiensis* and *O. sativa* × O. *latifolia* had complete resemblance to rice (Multani et al. 1994, 2003). The rapid recovery of recurrent parent phenotype after two or three backcrosses is indicative of extremely limited recombination between the distant genome of cultivated rice and wild species genomes. Furthermore, replacement of *O. sativa* alleles with the wild species alleles supports the conclusion that alien introgression involves classical crossing over as the mechanism for recombination and reciprocal exchange of segments.

1.6 Wild Species of Oryza: Genomics

Rice breeding is showing paradigm shift with the advances in genomics research particularly with the advent of new molecular marker technologies and genome sequencing of indica and japonica rice. This has facilitated precise characterization of germplasm, identification and mapping of large number of genes/QTLs for various agronomic traits, accelerated breeding using MAS and pyramiding of genes/QTLs, genome selection, allele mining, map-based cloning of genes/QTLs, and advancing the frontiers of 'Omics' and functional genomics to understand function of agronomically important genes.

Genomics research is also becoming powerful tool to precisely characterize alien genetic variation, dissect the hidden genetic variability in the wild species germplasm, find favorable wild species alleles to broaden the gene pool of rice for tolerance to major biotic, abiotic stresses and identification of new QTLs for yield-related traits. Furthermore, genome sequencing of different wild species has provided valuable information to understand the evolutionary relationship in *Oryza* and in applying this knowledge to identify new alleles for use in rice breeding. The International *Oryza* Map Alignment Project (I-OMAP) at Arizona Genomics Institute, Tucson, USA, has focused research on three activities as summarized by Sanchez et al. (2013): (i) generation of Ref-seqs and transcriptome data on AA genome and other species belonging to distant genomes, (ii) development of molecular maps, RILs, CSSLs involving different teams for use in functional and breeding research, and (iii) applying genomics on new collections of naturally occurring *Oryza* populations for diversity, conservation, and evolutionary studies. The group has sequenced the genomes of several wild species such as *O. nivara, O. rufipogon, O. barthii, O. glumaepatula, O. meridionalis, O. punctata, O. brachyantha,* and *Leersia perrieri.* These assemblies are publicly available on NCBI genome. In collaboration with partners in China, Japan, Taiwan, France, the Philippines, assembly of *O. officinalis, O. rhizomatis, O. glaberrima* is also available. The genome sequence data of *Oryza* species has become an important genetic resource for future research in genomics and breeding of rice including understanding of molecular mechanisms for genetic diversity and evolution of *Oryza.* Extensive data and information have become available on comparative genomics of wild species and divergence at molecular level during evolution of genus *Oryza.*

Ammiraju et al. (2006) constructed BAC libraries representing different species: for *O. nivara* 55,296 BAC clones, *O. rufipogon* 64,512, *O. glaberrima* 55,296, *O. punctata* 36,864, *O. officinalis* 92,160, *O. minuta* 129,024, *O. alta* 92,160, *O. australiensis* 92,160, *O. brachyantha* 36,864, *O. granulate* 73,728, *O. ridleyi* 129,024, *O. coarctata* 147,456.

BAC analysis revealed long terminal repeat (LTR) retrotransposons to be the predominant class of repeat elements in *Oryza,* and roughly linear relationship of these elements with genome size was observed. It is evident that LTR retrotransposable element amplifications increased the size of *O. australiensis* and *O. granulata* as 400 Mb and 200 Mb, respectively. Piegu et al. (2006) reported that *O. australiensis* ($2n = 24$, EE) has undergone recent bursts of three LTR-retrotransposon families. This genome has accumulated more than 90,000 retrotransposon copies more than its size. Three LTR-retrotransposon families (*RIRE1, Kangourou,* and *Wallabi*) compose 60% of the *O. australiensis* genome. Ammiraju et al. (2007) discovered LTR-retrotransposon family named RWG in the genus *Oryza*. The basal GG genome of *Oryza* (*O. granulata*) has expanded by nearly 25% by a burst of the RWG lineage *Gran3* subsequent to speciation. The causes for massive RWG family proliferation in some lineages (i.e., *O. granulata* and *O. australiensis*) and almost sudden death or attenuation in *O. brachyantha* and *O. coarctata* are not clearly understood.

The AA genome of *O. nivara* and *O. rufipogon* has contracted by 10% of their genome size as compared with rice reference genome. *Adh1* region (100–300 Kb) across *Oryza* showed significant shift in synteny, changes in genome size, rearrangements, transposable-mediated gene movement (Ammiraju et al. (2008). These genomics studies would lead to the better understanding of evolutionary relationships in *Oryza*, domestication of traits, and identification of novel wild species alleles for different agronomic traits to further accelerate rice breeding. Available data on genomics of *Oryza* species suggests that a single reference genome of *Oryza* is insufficient to explore and understand the allelic diversity and natural variation hidden in the genomes of wild relatives of rice.

1.7 Utilization of Wild Relatives of Oryza: Future Priorities

Future research should focus on broadening the gene pool of rice by identifying and introgressing new genes/QTLs from different wild species, particularly traits for which there is limited variability in *O. sativa* such as sheath blight, false smut, neck blast, stemborer, salinity, drought, and high temperature tolerance. Emphasis should be on developing pre-breeding populations with desirable genes from wild species.

1. Molecular markers and new tools of genomics should be used to precisely characterize introgression of alien chromosome segments, allele mining of wild species germplasm, cloning of alien genes, and pyramiding of alien genes/QTLs preferably with different mechanisms of tolerance to biotic and abiotic stresses.
2. Identification and introgression of yield-enhancing loci/QTLs 'wild species alleles' into elite breeding lines to further increase the yield potential of indica and japonica rice cultivars. Develop CSSL from different wild species for mapping genes/QTLs and use in breeding and functional genomics.
3. Search for genes promoting homoeologous pairing to transfer genes from distant genomes of *Oryza* which is a major bottleneck in alien introgression from secondary and tertiary gene pool.
4. Understanding evolutionary mechanisms in genus *Oryza* using new tools of genomics.
5. Intensify exploratory research on C4 ness and apomixis in wild species.
6. Exploratory research to identify endophytes in wild species as novel source for nitrogen fixation is emphasized.
7. Explore large-scale production of haploids particularly in anther culture recalcitrant

indica rice through wide hybridization in *Oryza* similar to the existing chromosome elimination systems for production of haploids in wheat and barley.

8. Collection of new wild species germplasm from different countries particularly in hotspots is emphasized to enhance genetic diversity in rice gene pool.
9. Extinction of wild species is a threat to genetic diversity, and international efforts are needed to overcome the trend in loss of biodiversity. Integration of new tools of genomics to accelerate alien gene introgression is emphasized. With the advances in tissue culture, molecular markers, and genomics, the scope for utilization of wild species in genetic enhancement of rice seems more promising than before.

References

Abbasi FM, Brar DS, Carpena AL, Fukui K, Khush GS (1999) Detection of auto-syndetic and allosyndetic pairing among A and E genomes of *Oryza* through genomic in situ hybridization. Rice Genet Newsl 16:24–25

Aggarwal RK, Brar DS, Khush GS (1997) Two new genomes in the *Oryza* complex identified on the basis of molecular divergence analysis using total genomic DNA hybridization. Mol Gen Genet 254:1–12

Amante-Bordeos A, Sitch LA, Nelson RJ, Dalmacio RD. Oliva NP, Aswidinnoor H, Leung H (1992) Transfer of bacterial blight and blast resistance from the tetraploid wild rice, *Oryza minuta* to cultivated rice. Theor Appl Genet 84:345–354

Ammiraju JSS, Luo M, Goicoechea JL, Wang W, Kudrna D, Mueller C, Talag J, Kim H, Sisneros NB, Blackmon B, Fang E, Tomkins JB, Brar DS, MacKill DJ, McCouch S, Kurata N, Lambert G, Galbraith DW, Arumuganathan K, Rao K, Walling JG, Gill N, Yu Y, San Miguel P, Soderlund C, Jackson SA, Wing RA (2006) The *Oryza* bacterial artificial chromosome library resource: Construction and analysis of 12 deep-coverage large-insert BAC libraries that represent the 10 genome types of the genus *Oryza*. Genome Res 16:140–147

Ammiraju JSS, Zuccolo A, Yu Y, Song X, Piegu B, Chevalier F, Walling JG, Ma J, Talag J, Brar DS, San Miguel PJ, Jiang N, Jackson SA, Panaud O, Wing RA (2007) Evolutionary dynamics of an ancient retrotransposon family provides insights into evolution of genome size in the genus *Oryza*. Plant J 52:342–351

Ammiraju JSS, Lu F, Sanyal A, Yu Y, Song X, Jiang N, Pontaroli AC, Rambo T et al (2008) Dynamic evolution of *Oryza* genomes is revealed by camparative genomic analysis of a genus wide vertical data set. Plant Cell 20:3191–3209

Asghar M, Brar DS, Hernandez JE, Ohmido N, Khush GS (1998) Characterization of parental genomes in a hybrid between *Oryza sativa* L and *O. officinalis* Wall ex Watt. through fluorescence in situ hybridization. Rice Genet Newsl 15:83–84

Bellon MR, Brar DS, Lu BR, Pham JL (1998) Rice genetic resources. In: Dwoling NG, Greenfield SM, Fischer KS (eds) Sustainability of rice in the global food system. Pacific Basin Study Center and International Rice Research Institute, Manila, pp 251–283

Bhasin H, Bhatia R, Raghuvanshi S, Lore JS, Sahi GK, Kaur B, Vikal Y, Singh K (2012) New PCR based sequence-tagged site markers for bacterial blight resistance gene *Xa 38* of rice. Mol Breed 30:607–611

Bhatia D, Joshi S, Das A, Vikal Y, Sahi GK, Neelam K, Kaur K, Singh K (2017a) Introgression of yield component traits in rice (*Oryza sativa* ssp *indica*) through interspecific hybridization. Crop Sci 57:1–17. https://doi.org/10.2135/cropsci2015.11.0693

Bhatia D, Wing RA, Yu Y, Chougle K, Kudrna D, Lee S, Rang A, Singh K (2017b) Genotyping by sequencing of rice interspecific backcross inbred lines identifies QTLs for grain weight and grain length. Euphytica (In press)

Bimpong IK, Serraj R, Chin JH, Ramos J, Mendoza EMT, Hernandez J, Mendioro MS, Brar DS (2011) Identification of QTLs for drought related traits in alien introgression lines derived from crosses of rice (*O. sativa* cv. IR 64) × *O. glaberrima* under low land drought stress. J Plant Biol 54:237–250

Brar DS, Khush GS (1986) Wide hybridization and chromosome manipulation in cereals. In: Evans DH, Sharp WR, Ammirato PV (eds) Handbook of plant cell culture, vol 4. Techniques and Applications. MacMillan Publ. Co., New York, pp 221–263

Brar DS, Khush GS (1997) Alien introgression in rice. Plant Mol Biol 35:35–47

Brar DS, Khush GS (2002) Transferring genes from wild species into rice. In: Kang MS (ed) Quantitative genetics, genomics and plant breeding. CABI Publishing, Wallingford, pp 197–217

Brar DS, Khush GS (2006) Cytogenetic manipulation and germplasm enhancement of rice (*Oryza sativa* L.). In: Singh RJ, Jauhar PP (eds) Genetic resources, chromosome engineering and crop improvement. CRC Press, Boca Raton, pp 115–158

Brar DS, Singh K (2011) *Oryza*. In: Kole C (ed) Wild crop relatives: genomic and breeding resources, cereals. Springer, Berlin, pp 321–365

Brar DS, Elloran R, Khush GS (1991) Interspecific hybrids produced through embryo rescue between cultivated and eight wild species of rice. Rice Genet Newsl 8:91–93

Brar DS, Dalmacio R, Elloran R, Aggarwal R, Angeles R, Khush GS (1996) Gene transfer and molecular characterization of introgression from wild *Oryza* species into rice. In: Khush GS (ed) Rice genetics III. International Rice Research Institute, Manila, pp 477–486

Brar DS, Elloran RM, Talag JD, Abbasi F, Khush GS (1997) Cytogenetic and molecular characterization of an intergeneric hybrid between *Oryza sativa* L. and *Porteresia coarctata* (Roxb) Tateoka. Rice Genet Newsl 14:43–44

Chang TT (1972) International cooperation in conserving and evaluation of rice germplasm resources. Rice Breeding. IRRI, Manila, Philippines, pp 177–185

Chang TT (1976) The origin, evolution, cultivation, dissemination and diversification of Asian and African rices. Euphytica 25:425–441

Cheema KK, Grewal NK, Vikal Y, Das A, Sharma R, Lore JS, Bhatia D, Mahajan R, Gupta V, Singh K (2008a) A novel bacterial blight resistance gene from *Oryza nivara* mapped to 38 Kbp region on chromosome 4L and transferred to *O. sativa* L. Genet Res 90:397–407

Cheema KK, Bains NS, Mangat GS, Das A, Brar DS, Khush GS, Singh K (2008b) Introgression of quantitative trait loci for improved productivity from *Oryza rufipogon* into *O. sativa*. Euphytica 160:401–409

Dalmacio RD, Brar DS, lshii T, Sitch LA, Virmani SS, Khush GS (1995) Identification and transfer of a new cytoplasmic male sterility source from *Oryza perennis* into indica rice (*O. sativa*). Euphytica 82:221–225

Dalmacio RD, Brar DS, Virmani S, Khush GS (1996) Male sterile line in rice (*Oryza sativa*) developed with O. *glumaepatula* cytoplasm. Int Rice Res Notes 21:22–23

Doi K, Sobrizal IK, Sanchez PL, Kurakazu Y, Yoshimura A (2003) Developing and evaluating rice chromosome segment substitution lines. In: Mew TW, Brar DS, Peng S, Dawe D, Hardy H (eds) Rice science: innovations and impact livelihood. International Rice Research Institute and CASE and CAAS, Beijing, pp 289–296

Ellis JL (1985) *Oryza indandamanica*. Ellis, a new rice plant from islands of Andamans. Bot Bull Surv India 27:225–227

Ellur RK, Khanna A, Gopala Krishnan S, Bhowmick PK, Vinod KK, Nagarajan M, Mondal KK, Singh NK, Singh K, Prabhu KH, Singh AK (2016) Marker-assisted incorporation of *Xa 38*, a novel bacterial blight resistance gene in PB1121 and comparison of its resistance spectrum with *xa13* + *Xa21*. Scientific Reports 6 (Nature) http://www.nature.com/articlessrep29188

Enriquez EC (2001) Production of doubled haploids from *Oryza sativa* L. × *O. glaberrima* Steud. and their characterization using microsatellite markers. PhD thesis, UPLB, Laguna, Philippines, p 116

Enriquez EC, Brar DS, Rosario TL, Jones M, Khush GS (2000) Production and characterization of doubled haploids from anther culture of the F_1s of *Oryza sativa* L. × *O. glaberrima* Steud. Rice Genet Newsl 17:67–69

Ge S, Sang T, Lu BR, Hong DY (1999) Phylogeny of rice genomes with emphasis on origin of allotetraploid species. Proc Natl Acad Sci USA 96:14400–14405

Gu K, Tian D, Yang F, Wu L, Sreekala C, Wang GL, Yin Z (2003) High-resolution genetic mapping of *Xa27*(t), a new bacterial blight resistance gene in rice, *O. sativa* L. Theor Appl Genet 108:800–807

Hirabayashi H, Kaji R, Okamoto M, Ogawa T, Brar DS, Angeles ER, Khush GS (2003) Mapping QTLs for brown planthopper (BPH) resistance introgressed from *O. officinalis* in rice. In: Khush GS, Brar DS, Hardy B (eds) Advances in rice genetics. International Rice Research Institute, Manila, Philippines, pp 268–270

Huang Z, He G, Shu L, Li X, Zhang Q (2001) Identification and mapping of two brown planthopper resistance genes in rice. Theor Appl Genet 102:929–934

Hue NTN (2004) Homoeologous chromosome pairing and alien introgression analyses in wide-cross derivatives of *Oryza* through fluorescence in situ hybridization. PhD thesis, UPLB, Laguna, Philippines, p 160

Ishii T, Brar DS, Multani DS, Khush GS (1994) Molecular tagging of genes for brown planthopper resistance and earliness introgressed from *Oryza australiensis* into cultivated rice, *O. sativa*. Genome 37:217–221

Jackson MT (1997) Conservation of rice genetic resources: the role of the international rice gene bank at IRRI. Plant Mol Biol 35:61–67

Jelodar NB, Blackhall NW, Hartman TPV, Brar DS, Khush GS, Davey MR, Cocking EC, Power JB (1999) Intergeneric somatic hybrids of rice [*Oryza sativa* L. (+) *Porteresia coarctata* (Roxb.) Tateoka]. Theor Appl Genet 99:570–577

Jena KK, Khush GS (1990) Introgression of genes from *Oryza officinalis* Well ex Watt to cultivated rice, *O. sativa* L. Theor Appl Genet 80:737–745

Jena KK, Khush GS, Kochert G (1992) RFLP analysis of rice (*Oryza sativa* L.) introgression lines. Theor Appl Genet 84:608–616

Jena KK, Jeung JU, Lee JH, Choi HC, Brar DS (2006) High-resolution mapping of a new brown planthopper (BPH) resistance gene, *Bph18*(t), and marker-assisted selection for BPH resistance in rice (*Oryza sativa* L.). Theor Appl Genet 112:288–297

Jeung JU, Kim BR, Cho YC, Han SS, Moon HP, Lee YT, Jena KK (2007) A novel gene, *Pi40*(t), resistance in rice. Theor Appl Genet 115:1163–1177

Jiang J, Gill BX, Wang GL, Ronald PC, Ward DC (1995) Metaphase and interphase fluorescence in situ hybridization mapping of the rice genome with bacterial artificial chromosomes. Proc Natl Acad Sci USA 92:4487–4491

Jones MP, Dingkuhn M, Aluko GK, Semon M (1997) Interspecific *Oryza sativa* L. × *O. glaberrima* Steud. progenies in upland rice improvement. Euphytica 92:237–246

Khush GS (1977) Disease and insect resistance in rice. Adv Agron 29:265–341

Khush GS (1997) Origin, dispersal, cultivation and variation of rice. Plant Mol Biol 35:25–34

Khush GS, Bacalangco E, Ogawa T (1990) A new gene for resistance to bacterial blight from *O. longistaminata*. Rice Genet Newsl 7:121–122

Kumar K, Sarao PS, Bhatia D, Kumari N, Kaur A, Mangat GS, Brar DS, Singh K (2017) High resolution genetic mapping of a novel brown planthopper resistance locus, *Bph33* in *Oryza sativa* L. × *Oryza nivara* (Sharma & Shastry) derived interspecific F_2 population. Theor Appl Genet (under review)

Lin SC, Yuan LP (1980) Hybrid rice breeding in China. Innovative approaches to rice breeding. International Rice Research Institute, Manila, Philippines, pp 35–51

Ling KC, Aguiero VM, Lee SH (1970) A mass screening method for testing resistance to grassy stunt disease of rice. Plant Dis Rep 56:565–569

Lu BR (1998) Diversity of rice genetic resources and its utilization and conservation. Chin Biodivers 6:63–72 (In Chinese with English abstract.)

Lu BR (1999) Need to conserve wild rice species in Nepal. Int Rice Res Notes 24:41

Lu B, Sharma SD (2003) Exploration, collection and conservation of wild *Oryza* species. In: Nanda JS, Sharma SD (eds) Monograph on genus *Oryza*. Science Publishers, Enfield, pp 263–283

Multani DS, Jena KK, Brar DS, delos Reyes BC, Angeles ER, Khush GS (1994) Development of monosomic alien addition lines and introgression of genes from *Oryza australiensis* Domin. to cultivated rice. Theor Appl Genet 88:102–109

Multani DS, Khush GS, Delos Reyes BG, Brar DS (2003) Alien genes introgression and development of monosomic alien additional lines from *Oryza latifolia* Desv. to rice. Theor Appl Genet 107:395–405

Naredo EB, Vaughan DA, Cruz FS (1993) Comparative spikelet morphology of *Oryza schlechteri* Pilger and related species of *Leersia* and *Oryza* (Poaceae). J Plant Res 106:109–112

Neelamraju S (2014) DRR Dhan40 with yield enhancing QTL's from wild species. DRR Newsl 12:2

Nguyen BD, Brar DS, Bui BC, Nguyen TV, Pham LN, Nguyen HT (2003) Identification and mapping of the QTL for aluminum tolerance introgressed from new source, *Oryza rufipogon* Griff. into indica rice, (*Oryza sativa* L.). Theor Appl Genet 106:583–593

Oka HI (1988) Origin of cultivated rice. Developments in crop science, vol 14. Japan Scientific Society Press, Tokyo

Piegu B, Guyot R, Picault N, Roulin A, Saniyal A, Kim H, Collura K, Brar DS, Jackson SA, Wing RA, Panaud O (2006) Doubling genome size without polyploidization: dynamics of retrotransposition-driven genomic expansions in *Oryza australiensis*, a wild relative of rice. Genome Res 16:1262–1269

Rahman Md, Jiang W, Chu SH, Qiao Y, Ham T-H, Woo M-O, Lee J, Sakina Khanam M, Chin J-H, Jeung J-U, Brar DS, Jena KK, Koh HJ (2008) High-resolution mapping of two rice brown planthopper resistance genes, *Bph20(t)* and *Bph21(t)*, originating from *Oryza minuta*. Theor Appl Genet 119:1237–1246

Ram T, Majmudar ND, Mishra B, Ansari AM, Padavathi G (2007) Introgression of broad spectrum blast resistance gene (s) into cultivated rice (*O. sativa* ssp indica) from wild species, *O. rufipogon*. Curr Sci 92:225–230

Rama Devi SJS, Singh K, Umakanth B, Renuka P, Sudhakar KV, Prasad MS, Viraktamath BC, Babu V, Madhav MS (2015) Development and identification of novel rice blast resistant sources and their characterization using molecular markers. Rice Sci 22:300–308

Ramos JM, Furuta T, Uchara K, Chihiro N, Angles-Shim RB, Shim J, Brar DS, Ashikari M, Jena KK (2016) Development of chromosome segmental substitution lines (CSSLs) of *Oryza longistaminata* A. Chev. & Rohr in the background of japonica cultivar, Taichung 65 and their evaluation for field trials. Euphytica 210:151–163

Ronald PC, Albano B, Tabien R, Abenes L, Wu K, McCouch S, Tanksley SD (1992) Genetic and physical analysis of rice bacterial blight resistance locus. *Xa21*. Mol Gen Genet 236:113–120

Sanchez AC, Brar DS, Huang N, Li Z, Khush GS (2000) Sequence tagged site marker-assisted selection for three bacterial blight resistance genes in rice. Crop Sci 40:792–797

Sanchez PL, Wing RA, Brar DS (2013) The wild relative of rice: genomes and genomics. In: Zhang Q, Wing R (eds) Genetics and genomics of rice. Springer, New York, pp 9–25

Shakiba, E., Eizenga EC (2014) Unraveling the secrets of rice wild species. In: Yan W, Bao J (eds) Rice germplasm, genetics and improvement, pp 1–58. https://www.google.co.in/search?q=10.5772%2F58393&ie=utf-8&oe=utf-8&client=firefox-b-ab&gfe_rd=cr&dcr=0&ei=h6tEWtDYComW8QfW9paYBg

Shim J, Panaud O, Vitte C, Mendoro MS, Brar DS (2010) RDA derived *Oryza minuta* specific clones to probe genomic conservation across *Oryza* and introgression into rice (*O. sativa* L.). Euphytica 176:269–279

Singh S, Sidhu JS, Huang N, Vikal Y, Li Z, Brar DS, Dhaliwal HS, Khush GS (2001) Pyramiding three bacterial blight resistance genes (*xa5*, *xa13* and *Xa21*) using marker assisted selection into indica rice cultivar PR106. Theor Appl Genet 102:1011–1015

Sitch LA (1990) Incompatibility barriers operating in crosses of *Oryza sativa* with related species and genera. In: Gustafson JP (ed) Genetic manipulation in plant improvement II. Plenum Press, New York, pp 77–94

Song WY, Wang GL, Chen LL, Kim HS, Pi YL, Holsten T, Gardner J, Wang B, Zhai WX, Zhu LH, Fauquet C, Ronald PA (1995) A receptor kinase-like protein encoded by the rice disease resistance gene, *Xa21*. Science 270:1804–1806

Soriano IR, Schmit V, Brar DS, Prot JC, Reversat G (1999) Resistance to rice root-knot nematode *Meloidogyne graminicola* identified in *Oryza longistaminata* and *O. glaberrima*. Nematology 1:395–398

Surapaneni M, Balakrishnan D, Mesapogu S, Addanki KR, Yadavalli VR, Venkata TVGN, Neelamraju S (2017) Identification of major effect QTLs for agronomic traits and CSSL *in* rice from Swarna/

Oryza nivara derived backcross inbred lines. Front Plant Sci 8:1027–1036

Swamy BPM, Kaladhar K, Ramesha MS, Viraktamath BC, Neelamraju S (2014) Mapping and introgression of QTL for yield and related traits in two backcross populations derived from *Oryza sativa* cv Swarna and two accessions of *O. nivara*. J Genet 93:643–653

Tanksley SD, Nelson JC (1996) Advanced backcross QTL analysis: a method for the simultaneous discovery and transfer of valuable QTLs from unadapted germplasm into elite breeding lines. Theor Appl Genet 92:191–203

Tu J, Datta K, Khush GS, Zhang G, Datta SK (2000) Field performance of *Xa21* transgenic indica rice (*Oryza sativa* L.) IR 72. Theor Appl Genet 101:15–20

Vaughan DA (1989) The genus *Oryza* L.: current status of taxonomy, IRRl Research Paper Series 138. International Rice Research Institute, Manila, Philippines, p 21

Vaughan DA (1994) The wild relatives of rice: a genetic resources handbook. International Rice Research Institute, Manila, Philippines, pp 137

Vaughan DA, Sitch LA (1991) Gene flow from the jungle to farmers: wild-rice genetic resources and their uses. Bioscience 44:22

Wu MS (1990) Proceedings of wild rice resources & researches. China Science and Technology Press, Beijing (In Chinese)

Xiao J, Grandillo S. Ahn SN, McCouch SR, Tanksley SD, Li J, Yuan L (1996) Genes from wild rice improve yield. Nature 384:223–224

Xiao J, Li J, Grandillo S, Ahn SN, Yuan L, Tanksley SD, McCouch SR (1998) Identification of trait-improving quantitative trait loci alleles from a wild rice relative, *Oryza rufipogon*. Genetics 150:899–909

Yan H, Min S, Zhu L (1999) Visualization of *Oryza eichingeri* chromosomes in hybrid plants from *O. sativa* × O. *eichingeri* via fluorescence in situ hybridization. Genome 42:48–51

Yoon DB, Kang KH, Kim KJ, Ju HG, Kwon SJ, Suh JP et al (2006) Mapping quantitative trait loci for yield components and morphological traits in an advanced back cross population between *Oryza grandiglumis* and the *O. sativa* japonica cultivar Hwaseongbyeo. Theor Appl Genet 112:1052–1062

Yoshimura A, Nagayama H, Sobrizal et al (2010) Introgression lines of rice (*Oryza sativa* L.) carrying donor genome from the wild species, *O. glumaepatula* Steud. and *O. meridionalis* Ng. Breed Sci 60(5):507–603

Zhang Q, Lin SC, Zhao BY, Wang CL, Wang WC, Zhou YL, Li DY, Chen CB, Zhu LH (1998) ldentification and tagging of a new gene for resistance to bacterial blight (*Xanthomonas oryza*e pv. *Oryza*e) from *O. rufipogon*. Rice Genet Newsl 15:138–142

2 Informatics of Wild Relatives of Rice

Deepak Singh Bisht, Amolkumar U. Solanke and Tapan K. Mondal

Abstract

The wild species of rice are expected to harbour novel beneficial alleles which have been lost from cultivated rice during the process of domestication. This unchecked loss of alleles has made cultivated rice varieties more susceptible to the changing climatic conditions. Therefore, to develop climate-resilient rice crops, it is desirable to transfer the beneficial alleles from the wild rice species to the cultivated ones. In this direction, efforts are underway to completely unlock the genetic blueprint of the wild relatives of rice, which should facilitate the identification of novel alleles and their subsequent use in various rice-breeding programs. Although the sequence information is now generated with ease, it is equally imperative to develop user-friendly informatics resources harbouring the information that could be made easily accessible to users with diverse research interest. Despite the availability of a plethora of informatics resources for cultivated rice varieties, very few are available for the wild relatives of rice. Nevertheless, with the unprecedented surge in exploratory genomics efforts of wild relatives of rice and concurrent advancement in the frontier areas of bioinformatics, it is foreseeable that the informatics resources for wild relatives of rice will rise too. These resources will definitely spur new avenues of research in rice biology and thus should ensure a stable global food supply.

D. S. Bisht · A. U. Solanke · T. K. Mondal (✉)
ICAR-National Research Centre on Plant Biotechnology, LBS Building, IARI, New Delhi 110012, India
e-mail: mondaltk@rediffmail.com

2.1 Introduction

Rice is a unique crop particularly for its importance as a staple food for half of the world's population and simultaneously for its utility as a model system in plant biology. Rice belongs to the family *Poaceae* and the genus *Oryza* (Vaughan et al. 2003). The sequencing of rice, *Oryza sativa* ssp. *japonica* cv. Nipponbare, genome was a scientific marvel achieved by toiling efforts of the International Rice Genome Sequencing Project (IRGSP) in 2005 (IRGSP 2005). Before that a draft genome from indica rice appeared in 2002 (Yu et al. 2002). Both genomes were sequenced by whole-genome shotgun sequencing. For these draft genomes, the size of the indica genome was 466 Mb with 92% coverage while it was 420 Mb in the case of japonica with 93% genome coverage. The availability of high-quality sequence information provided a concrete genomic framework for

T. K. Mondal and R. J. Henry (eds.), *The Wild Oryza Genomes*,
Compendium of Plant Genomes, https://doi.org/10.1007/978-3-319-71997-9_2

understanding the genetic control of agronomically important traits in rice. Moreover, the high-quality reference genome sequence for *O. sativa* ssp. *japonica* is extremely useful for identification of novel sequence variant(s) associated with beneficial phenotypes from wild rice germplasm. Comprehensive information on structural and functional genomics of wild relatives will further facilitate the transfer of important traits from the wild races to cultivated ones.

The advent of next-generation sequencing (NGS) technologies has driven many large-scale genome sequencing projects in crops of economic importance resulting in large sets of data representing an extraordinary source of information (Nagamura and Antonio 2010). Although the sequence information is generated with ease, the major challenge is to develop a unifying repository for the generated data with inventorying of data on genes, gene families and functional annotation that can be made easily accessible to researchers with diverse interest. Additionally, these databases should also be able to integrate the deluge of data generated from other 'omics' studies like transcriptome, metabolome and proteome studies. Incidentally, for rice where large-scale genome sequencing projects of cultivated as well as wild accession are underway, it is compelling to develop databases that should also be equipped with tools for comparative genomics studies. Other than international consortia, many countries have their own repositories on rice, particularly focusing on the native cultivars. These databases should be integrated so that a universally accepted vocabulary can be formulated for categorising the information. This will not only avoid repetition of data but will also facilitate transfer of germplasm as well as DNA material for worldwide collaborative efforts towards improvement of rice crop (Antonio et al. 2007).

In this chapter, we surveyed a collection of biological databases that host information on various aspects of wild relatives of rice (Table 2.1). Despite the unavailability of any exclusive database on wild rice, we gathered information scattered across various other databases and presented it here.

2.2 Exploring Rice Diversity

The rice genus, having a rich genetic diversity, consists of 24 species. Amongst which *O. sativa* (Asian rice) and *Oryza glaberimma* (African rice) are the only two cultivated species (Zhang et al. 2014). The *Oryza* species have ten different genome types, including six diploid genome types (AA, BB, CC, EE, FF and GG) and four allotetraploid genome types (BBCC, CCDD, HHKK and HHJJ). Comparative sequence analysis of complete genomes as well as of the amplification of selected loci has yielded significant insight into the process of diversification and genome evolution (Ammiraju et al. 2008, 2010; Zhang et al. 2014). Evolutionary studies have revealed that the genus *Orzya* has undergone a rapid diversification within a short evolutionary time span of 15 million years (Ammiraju et al. 2010). This diversity of cultivated and wild species provides a powerful system for studying evolutionary and phylogenetic relationship, and can contribute to the genetic improvement of rice, which is pivotal in addressing the imminent problem of food security (Zhang et al. 2014).

2.3 Sequencing Wild Relatives of Rice

The NGS techniques of genome sequencing have been used to further explore the possibility of mining wild rice germplasm for identification of novel beneficial allelic variant(s) embedded in the genome of wild rice germplasm. The sequencing of wild rice relatives revealed interesting facts. For instance, the genome compactness of *Oryza brachyantha,* the first wild rice genome to be sequenced (Chen et al. 2013), was attributed to the reduced activity of long-terminal retrotransposons and massive internal deletions of ancient long-terminal repeat elements. The comparative genomic analyses of five wild relatives of rice, viz *Oryza nivara*, *Oryza glaberrima*, *Oryza barthii*, *Oryza glumaepatula*, and *Oryza meridionalis*, empirically revealed that wild relatives with AA genome are reservoirs for novel disease resistance

Table 2.1 List of databases having information on wild *Oryza* genus

S. No	Database	Important feature	URL	References
1	International Rice Information System	Whole-genome sequence data, transcriptomics and proteomics studies, genetic experiments and functional genomics studies data from Oryza specie	http://irri.org/tools-and-databases/international-rice-information-system	McLaren et al. (2005)
2	Oryzabase	Database on genetic and genomic resources of wild rice	http://www.shigen.nig.ac.jp/rice/oryzabase/	Kurata and Yamazaki (2006)
3	OryzaGenome	Wild rice genotype–phenotype information	http://viewer.shigen.info/O.genome/	
4	Rice Genome Knowledgebase (RGKbase)	Database on comparative genomics of wild and cultivated rice	http://rgkbase.big.ac.cn/RGKbase/	Wang et al. (2013)
5	Rice Annotation Project Database (RAP-DB)	*O. rufipogon* full length cDNA sequence	http://rapdb.dna.affrc.go.jp/	Sakai et al. (2013)
6	Oryza Map Alignment Project	Genomics and transcriptomic resources of wild rice accession, advance mapping population	http://www.omap.org/	Jacquemin et al. (2013)
7	Gramene	Genomics data on wild relatives of rice	http://www.gramene.org/releasenotes	Tello-Ruiz et al. (2016)

alleles (Zhang et al. 2014). Currently, 12 reference genome sequences for 11 rice accessions are available. Amongst which three are of cultivated rice and nine are of wild relatives (Table 2.2). Evolution of species is often driven by genome polyploidisation events (Wendel 2000; Jackson and Chen 2010; Ammiraju et al. 2010). To understand the evolutionary consequences of polyploidisation, large-scale efforts are underway to sequence the polyploid wild rice relatives (Jacquemin et al. 2013). Moreover, the origin and domestication of rice, monophyletic or diphyletic has been debated for a long time (Second 1982; He et al. 2011; Molina et al. 2011). Comprehensive genome wide studies could identify genomic structural variants, recording ancient diversification events, that can be used for resolving the ambiguity related to the origin and domestication of rice (Zhang et al. 2014).

Organelle genomes being structurally stable, uniparental and non-recombinant provide a vital resource for plant identification and establishing the evolutionary relatedness between species which is an important consideration for selection in pre-breeding programs. Whole-genome sequencing of organellar DNA, particularly chloroplast DNA, has circumvented the problem of DNA barcoding in plants which was otherwise debated for the lack of a universally accepted standard barcoding loci (Nock et al. 2011). The principle of DNA barcoding using the chloroplast genome has been recently demonstrated in wild relatives of rice, wherein chloroplast genomes of *O. barthii, O. ryza longistaminata, O. glaberrima, O. glumaepatula and O. ryza officinalis* were sequenced and compared for restructuring the phylogeny and evolutionary relationships (Nock et al. 2011; Wambugu et al. 2015). The sequences of assembled reference quality chloroplast genomes can be retrieved from GenBank accession numbers listed in Table 2.3. Similarly, molecular phylogenies can also be established, by comparing mitochondrial genome sequences across the species. Recently, on the basis of mitochondiral genome sequence information, evolutionary divergence of the wild rice, *O. minuta,* was analysed with reference to other sequenced rice accessions (Asaf et al. 2016). A tabular depiction

Table 2.2 List of reference quality sequenced genomes of *Oryza* species

Species	Genome type	Genome size	Genbank Assembly ID	Genbank bioproject	References
O. sativa ssp. indica	AA	~400	GCA_000004655.2	PRJNA361	Yu et al. (2002).
O. sativa ssp. japonica	AA	~400	GCA_000005425.2	PRJNA13141	IRGSP (2005)
O. rufipogon	AA	~445	GCA_000817225.1	PRJEB4137	Huang et al. (2012)
O. nivara	AA	~375	GCA_000576065.1	PRJNA48107	Jacquemin et al. (2013)
O. barthii	AA	~335	GCA_000182155.2	PRJNA30379	Jacquemin et al. (2013)
O. glaberrima	AA	~354	GCA_000147395.2	PRJNA13765	Wang et al. (2014)
O. glumaepatula	AA	~334	GCA_000576495.1	PRJNA48429	Jacquemin et al. (2013)
O. meridionalis	AA	~340	GCA_000338895.2	PRJNA48433	Jacquemin et al. (2013)
O. longistaminata	AA	~347	NA	PRJNA245492	Jacquemin et al. (2013)
O. punctata	BB	~423	GCA_000573905.1	PRJNA13770	Jacquemin et al. (2013)
O. brachyantha	FF	~261	GCA_000231095.2	PRJNA70533	Chen et al. (2013)

of mitochondrial genomes sequenced so far is presented in Table 2.4. Comparative analysis of mitochondrial genomes can provide insight for agronomically important traits like cytoplasmic male sterility (CMS) especially for hybrid breeding (Fuji et al. 2010). Hence, collectively these genomics resources provide a wonderful opportunity for conducting pre-breeding studies aiming at identification of novel alleles/variants useful in breeding programs.

2.4 Database of *Oryza* Genus

2.4.1 International Rice Information System

The global germplasm repositories serve as an invaluable resource for various crop improvement programs. To facilitate unambiguous identification, unified nomenclature and genealogies, easy accessibility and proper cataloguing of the deposited entries, it is imperative to develop databases integrating the global information of worldwide genetic resources. To address this concern the International Crop Information System (ICIS, www.icis.cgiar.org), an 'open source' database system for the management and integration of global information on genetic resources and crop improvement for any crop was developed by Consultative Group for International Agricultural Research (CGIAR) and collaborative partners. The International Rice Information System (www.iris.irri.org) is the rice implementation of the ICIS (Bruskiewich et al. 2003; McLaren et al. 2005). The major goal of IRIS is to build a well-organised repository of the latest information about rice diversity, particularly focusing on providing unambiguous information linked to specific rice germplasm which can be accessed through a single Web portal.

The Genealogy Management System (GMS), the core database of IRIS, harbours information for more than one and half million varieties, breeding lines and accessions of rice. Several new modules have now been added to IRIS for incorporation of information generated from whole-genome sequence data, transcriptomics and proteomics studies, genetic experiments, and functional genomics studies. Additionally, links to other freely available specialised software packages and interfaces for efficient retrieval of data on germplasm and species are also provided. The IRIS database serves as an invaluable tool for gaining a deeper insight into various aspects of rice biology.

Table 2.3 List of sequenced chloroplast genome of rice and its wild relatives

Species	Chloroplast genome size (bp)	Genbank accession	References
O. sativa indica	134,496	AY522329	Tang et al. (2004)
O. sativa japonica	134,551	GU592207	Nock et al. (2011)
O. meridionalis	134,551	GU592208; JN005831	Nock et al. (2011), Waters et al. (2012)
O. australiensis[a]	134,549	GU592209	Nock et al. (2011)
O. longistaminata	134,567	KM881641	Wambugu et al. (2015)
O. barthi	134,674	KM881634	Wambugu et al. (2015)
O. glumaepatula	134,583	KM881640	Wambugu et al. (2015)
O. officinalis	134,911	KM881643	Wambugu et al. (2015)
O. glaberrima	134,606	KM881638	Wambugu et al. (2015)
O. rufipogon	134,557	JN005833; N005832	Waters et al. (2012)
O. nivara	134,494	AP006728	Masood et al. (2004)

[a]Partial sequence

Table 2.4 Sequenced mitochondrial genomes of rice and its wild relatives

Species	Mitochondrial genome size (bp)	GenBank accession	References
O. sativa indica	491,515	DQ167399	Tian et al. (2006)
O. sativa japonica	490,520	BA000029	Notsu et al. (2002)
O. minuta	515,022	KU176938	Asaf et al. (2016)
O. rufipogon	559,045	AP011076	Fujii et al. (2010)

2.4.2 Oryzabase

Oryzabase is a comprehensive rice database sponsored by National Bio Resource Project-Rice of Japan and hosted by National Institute of Genetics (NIG), Japan. The database is accessible at URL:http://www.shigen.nig.ac.jp/rice/oryzabase/. It is an integrated database which combines biological information collected from all the developmental stages along with the genomic information of rice, with a major focus on wild relatives. The database contains information on rice anatomy, morphological characteristics, growth and developmental features, habitat distribution, genes and genetic resources. The information in the database can be broadly divided into three subheadings: (i) biological data, (ii) genetic resources, and (iii) genome map and sequences (Kurata and Yamazaki 2006).

2.4.3 Genetic Resources: Wild Rice Core Collection

The Oryzabase wild rice core collection has information on 1729 wild rice accessions which is a representative set of 18 species from 9 genomes (Fig. 2.1). To maintain uniformity with IRRI gene bank accessions, the Oryzabase accessions are provided with an IRIS identifier. For users, the core collection of 289 accessions is chosen, which further subdivided into three different ranks (Fig. 2.2 and https://shigen.nig.ac.jp/rice/O.base/strain/wildCore/about). Rank 1 contains highly desirable accessions, with two or three accessions from 18 wild species. Rank 2 is the next recommended collection with 64 accessions taken from all species. Rank 3 is a supplementary collection that includes 171 accessions (Kurata and Yamazaki 2006)

(Fig. 2.1). The accessions are chosen to cover as much as genetic variation as possible. Additionally, the information on four RIL populations generated by crossing indica and japonica cultivars are also present in Oryzabase. For functional genomic studies, chromosomal segment substitution lines (CSSLs) of AA genome wild species in a japonica background are also available. The information is freely accessible to global scientific community at http://www.shigen.nig.ac.jp/rice/oryzabase/strains/summary.jsp. In addition to specifying information on various aspects of rice biology, sequence information on *Oryza* species can be accessed through the OryzaGenome hyperlink on Oryabase. OzyzaGenome is discussed in following section.

2.4.4 OryzaGenome

OryzaGenome is a genomic information database of rice allied to Oryzabase. It is a united repository of genotype–phenotype information of wild relatives of rice, ensuring access to the information about geographical origins, phenotypic traits and genetic stocks. This database is primarily designed to visualise the genome wide variant identified in various wild rice accessions by massive scale sequencing by NGS platforms (Ohyanagi et al. 2016) (Fig. 2.2). In addition to the specifying database scheme, reference sequences for several other cultivated rice accessions are also available at OryzaGenome.

The current release of OryzaGenome 1.0 (19 June 2017) includes SNP information of 446 *Oryza rufipogon* accessions derived by imputation and of 17 accessions derived by deep sequencing. *Oryza sativa* ssp. *japonica* sp. Nipponbare genome IRGSP version 1.0 was used for alignment and mapping of SNPs. For visualisation of genomic variants, two different modes are provided in OryzaGenome: (a) SNP viewer and (b) variant table. SNP viewer provides a global visualisation of the variant distribution; however, the variant table is a text-based browser wherein the precise location of variants can be identified. Additionally, portable variant call format (VCF) file or tab-delimited file download is also available. The raw data generated for the variant detection can be retrieved from publically available databases like DDBJ of EBI. The tools and resources present in OryzaGenome database can be accessed through URL http://viewer.shigen.info/O.genome/. The identification of genome wide SNPs is a perquisite in population genomics; thus this kind of resources are obvious choices for conducting genome wide association studies (GWAS) for identification of genotypic associations with beneficial traits.

The future goal of OryzaGenome is to develop a pan-Oryza repository of variants identified in different rice germplasm. Towards this objective, a project ORYZA200 was initiated in 2014 with a major focus on sequencing and variant identification of 200 wild rice germplasm. Although the project is in its infancy, the magnitude of data generated will surely contribute in solving the imminent global food security issue.

2.4.5 Rice Genome Knowledgebase (RGKbase)

With the accumulation of sequence information on a large number of rice accession including both cultivated and wild relatives, it becomes imperative to develop databases facilitating the comparison of genomes across species and genera (Wang et al. 2013). Genome comparison provides computational derived evidence of phylogeny and pedigree of the species that assists the rice breeders in selecting the desirable parents in pre-breeding experiments. Rice Genome Knowledgebase (RGKBase) is one such database which provides comparative genome analysis of five different genomes of rice including four cultivated, Nipponbare (japonica), 93-11 (indica), PA64 s (indica), African rice (*O. glaberrima*) and a wild rice species (*O. brachyantha*) (http://rgkbase.big.ac.cn/RGKbase/). RGKBase is an integrated database where information on rice genomics, transcriptomics, proteomics, comparative genomics, a epigenome analysis and QTLs are chronologically arranged for easy access to

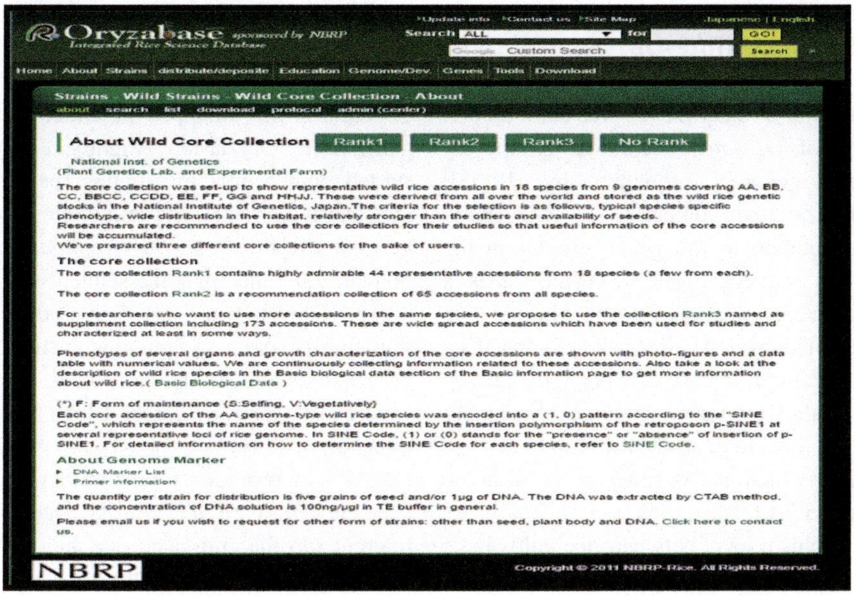

Fig. 2.1 Snapshot of Web interface of Oryzabase

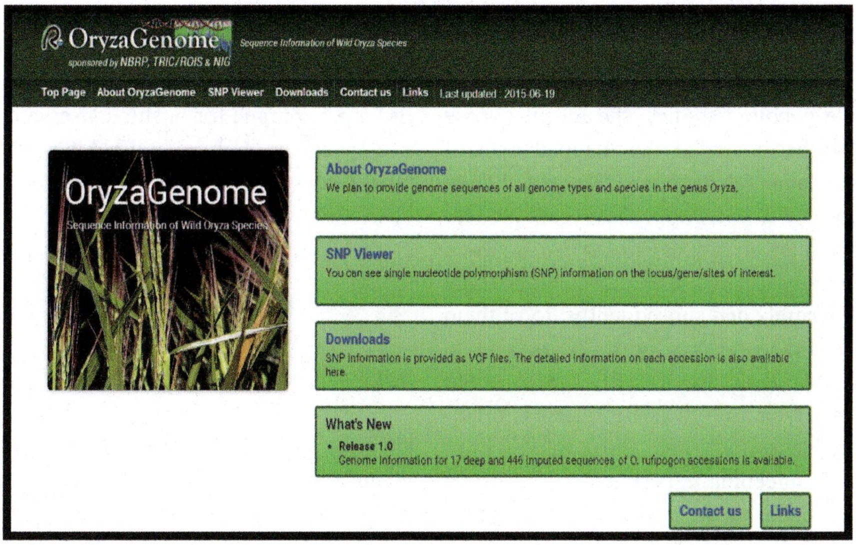

Fig. 2.2 Web interface of OryzaGenome

researchers working in diverse aspects of rice biology (Wang et al. 2013). Moreover, RGKbase has integrated many user-friendly interfaces for data browsing and bioinformatics tools for genome annotation, gene family classification, pathway analysis and study of co-expression networks. The database is constantly evolving with the addition of new datasets. The sequence data from 1000 rice varieties are likely to be added in a future release of the database (Wang et al. 2013).

2.4.6 Rice Annotation Project Database (RAP-DB)

Rice Annotation Project (RAP) was started upon completion of *O. sativa* (japonica group) cv. Nipponbare genome sequence by IRGSP (Fig. 2.3). The main objective of RAP is to provide annotation to the genes present in the rice genome sequence. The RAP-DB hosts a comprehensive annotation information for rice genes identified by RAP (http://rapdb.dna.affrc.go.jp/). The current release of RAP-DB is based on the latest *Oryza* genome assembly Os-Nipponbare-Reference-IRGSP-1.0 (IRGSP-1.0) which is 10 times more accurate than the previous release (Sakai et al. 2013). As the polymorphism level between rice cultivars is limited, it is indispensable to use a highly accurate genome assembly for annotation as well as comparison of closely related rice cultivars (Kawahara et al. 2013). The major interest of RAP-DB is to provide annotation information on evolving rice genome sequence assemblies. However, realising the importance of wild rice in comparative genomic studies, the recent release of RAP-DB has included the mapped data of publicly available full length cDNA (FLcDNA) of *O. rufipogon* to the Nipponbare genome (Lu et al. 2008). Interestingly, 87 FLcDNA sequence of *O. rufipogon* did not match the Nipponbare reference assembly and subsequently, 15 of them were confirmed to be specific to *O. rufipogon*. These *O. rufipogon* specific FLcDNAs could carry genes which are otherwise missing in cultivated rice accessions and thus could be of importance in breeding experiments. The list of unmapped FLcDNAs is available in RAP-DB. A snapshot of Web interface depicting *O. rufipogon* specific genes is provided in Fig. 2.4. Additionally, the hyperlink for other databases like Oryzabse, RiceXpro, Q-TARO, SALAD database and Gramene is also included in RAP-DB from which more information on wild rice could also be obtained.

2.4.7 *Oryza* Map Alignment Project (OMAP)

To harness the untapped potential of wild rice germplasm, OryzaMap Alignment project was started with a goal of aligning the BAC-based physical map of 18 wild and one cultivated rice accession to the *O. sativa* ssp. *japonica* cv Nipponbare reference genome sequence (Fig. 2.5).

The OMAP was started in 2007 as a joint initiative of Arizona Genomic Institute, Purdue University, Cold spring Harbor Lab and National Centre for Gene Research, China (http://www.omap.org/nsf.html). The basic purpose for OMAP is to provide a robust framework of detail characterisation of the wild rice genomes with reference to the cultivated one and to utilise the information for unlocking the genetic potential of wild rice for the identification of novel agriculturally important genes and QTLs (Wing et al. 2005). The overarching goals of OMAP are: (i) development of a reference genome and transcriptome for all 23-rice species, (ii) development of advanced mapping populations and (iii) a programme for in situ conservation of wild rice species in their natural habitat. OMAP is so far the most comprehensive genomic repository of wild rice germplasm which is expected to have a profound impact in future research on rice biology.

2.4.8 Genomic Resources

Genome sequencing for sixteen of the 23 species is underway. The draft assembly of a few is present in GenBank for early access. Annotation efforts for all the sequenced species are also in progress. The future goals of OMAP include the sequencing of polyploid rice species which will undoubtedly assist in resolving the enigmatic polyploidy in rice. The data on genus level comparative studies of rice accessions could be utilised for answering fundamental questions of

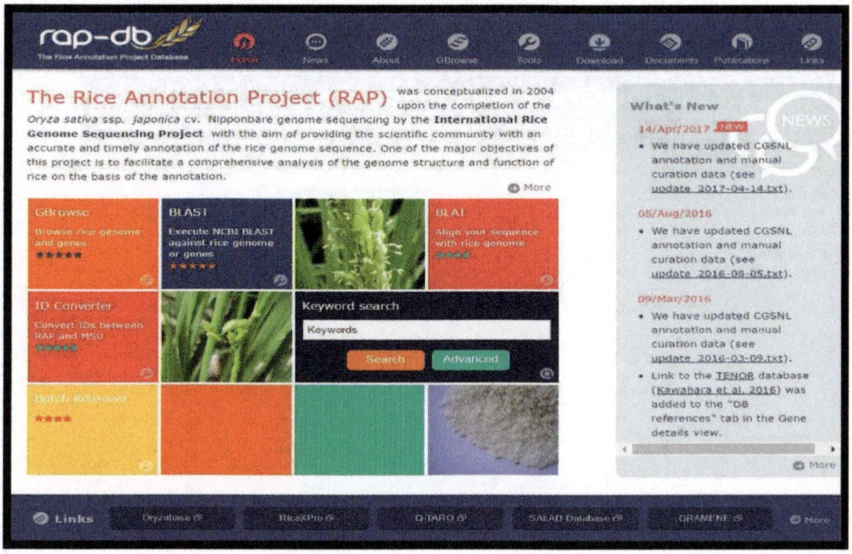

Fig. 2.3 Snapshot of Web interface of RAP-DB

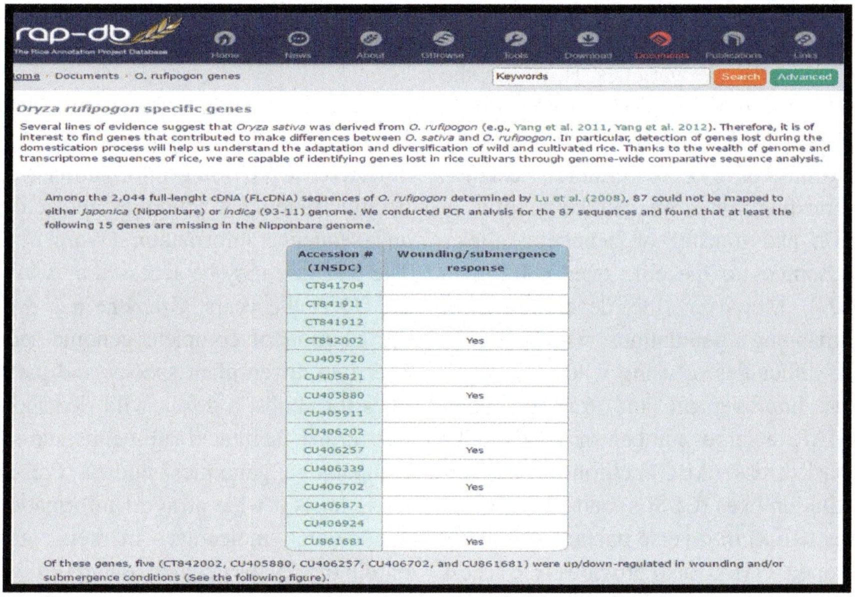

Fig. 2.4 Snapshot of Web interface of RAP-DB having information on *O. rufipogon*

evolutionary biology and also for knowledge-based breeding experiments. The BAC libraries, physical maps, genome and transcriptome assemblies for the wild rice relatives are available on the OMAP server.

2.4.9 Advance Mapping Population and In Situ Conservation

Wild rice accessions being under natural selection are valuable sources of favourable

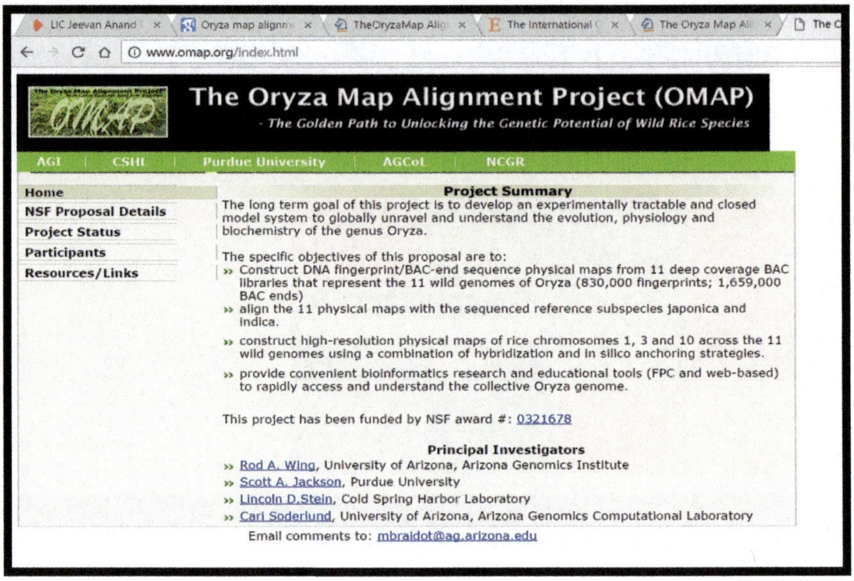

Fig. 2.5 Front Web page of OMAP

genes/variants which are presumed to have been lost in the cultivated rice varieties. However, the deployment of these genes in cultivated varieties is fraught with challenges like incompatibility barriers and linkage drag. Nevertheless, exhaustive 'pre-breeding' research can be helpful in identification and transfer of beneficial alleles from wild sources to the elite ones (McCouch et al. 2007). Moreover, the development of advance mapping populations (AMPs) can address the difficulties of using wild germplasm for cultivar improvement and trait mapping. Under OMAP, a large number of AMPs, like advance back cross (ABC), chromosome segment substitution lines (CSSLs) and recombinant inbred lines (RILs) in diverse backgrounds have been developed. OMAP also provides free access to AMPs for the AA genome species for conducting functional genomics and pre-breeding studies. Additionally, OMAP has also started in situ conservation of wild rice population threatened by the risk of climate change, habitat destruction and human activities (Jacquemin et al. 2013).

2.4.10 Gramene

Gramene is a curated 'open source' data resource launched in 2000 as a repository with a major objective of providing information on rice comparative functional genomics, molecular markers and sequence information (Ware et al. 2002). The database may be accessed at www.gramene.org. Over the years, Gramene has evolved with the addition of complete genome sequences of over two dozen plant species and partial assemblies for almost a dozen wild rice species which has further facilitated sub-genus and sub-species comparative genomics studies. Presently, Gramene hosts a wide array of information on gene annotation, molecular markers and QTLs, ontologies, plant pathway databases, genetic and physical maps from various studies. The database content is updated five times per year and the changes incorporated are provided in the Gramene release note (www.gramene.org).

Gramene has genome sequence and variant information of more than 39 plant species which serve as a quality aid for comparative and functional genomics studies. Gramene has imported reference quality genome information of wild rice accessions (Table 2.1), and SNP data for three *Oryza* species, (i) 5418373 SNPs of *O. sativa* ssp. *indica* (NCBI dbSNP), (ii) 5512746, SNPs of *O. sativa* ssp. *japonica* (McNally et al. 2009; Zhao et al. 2010, NCBI dbSNP) and (iii) 7172036 SNPs of *O. glaberrima* (Oryza Genome Evolution Project). In Gramene, for comparative genomics, genomes are aligned for selected pairs of species and on the basis of protein coding region, phylogenetic trees are made using EnsemblCompara Gene Tree pipeline. Additionally, the orthologous relationship between genes is also identified that facilitated the building of synteny maps between sufficiently related species.

The current version of Gramene has additional Web portals like Genome Browser, Plant Reactome Broswer, BLAST and Gramene Mart (Tello-Ruiz et al. 2016). Plant Reactome is a freely accessible database of plant metabolic and regulatory pathways (http://plantreactome.gramene.org/). It contains 200 curated rice reference pathways. These pathways are used as a reference for homology-based pathway projection for plant species having a sequence genome or transcriptome (Tello-Ruiz et al. 2016).

2.4.11 Genebank Project, NARO

National Agriculture and Food Research Organisation (NARO) is the key institution of Japan involved in conservation of biodiversity and DNA material. It has been functioning since 1985 and presently has more than 224,000 registered entries. NARO has developed core collection diversity panels for crops like rice, wheat, maize, soybean and azuki bean. For rice, it has two core collections: (1) NIAS world rice core collection and (2) NIAS core collection of Japanese landraces. The core collection of rice also contains some of the unadaptable rice species. In addition, NARO also has information on morphological, physiological and habitat data of wild germplasm. For instance, the information about *Oryza barthii* can be accessed from the URL: https://www.gene.affrc.go.jp/databases-plant_images_detail_en.php?plno=3110390012. Moreover, the genetic resources of the NARO Genebank are freely distributed to researchers for breeding and educational purposes.

2.4.12 RiTE-db: Database for Rice Repetitive Sequence and Transposable Elements

In plants, repetitive DNA sequences constitute the majority of the plant genome and may for account up to 90–95% of the nuclear DNA. Broadly, repetitive DNA sequence can be categorised into coding (rRNA) and non-coding (tandem and interspersed repeats) sequences (Shcherban et al. 2015). The presence of huge amount of repetitive sequences in the genome is conjectured to be important for the maintenance and stabilisation of the chromosome sequence. Unfortunately, the repeat sequences are poorly understood. The main reason being the difficulties is cloning and sequencing of the repetitive DNA sequences. Indeed, most of the gaps in the genome assemblies are due to the presence of highly repetitive sequences in the genome. However, now there are various computational tools which can identify and annotate the repeat sequences accurately and with high specificity (Treangen and Salzberg 2012). In rice, many dedicated efforts have been done to specifically identify the repeats in the rice genome, resulting into many specialised repeat specific databases (Chaparro et al. 2007). Rice TE database (RiTE-db) is once such publically available database which has been developed from the sequence information of 11 Oryza species and the orthogroup *Leersia perrieri* (Copetti et al. 2015). The main objective of RiTE-db is to provide highly curated annotation for all the repetitive sequence in the rice genome which will be of immense utility in conducting the comparative and evolutionary studies in rice. Three structured datasets form the backbone of RiTE-db: (i) the plant repeat database (PReDA), (ii) de novo repeat assemblies and (iii) full-length

(FL) transposable elements. Altogether, the database consists of 17,000 entries that included the information on publically available repeat sequence, repeat sequence identified from sequencing surveys of 12 rice species and full length TEs, identified from whole-genome assembly of 12 rice species. The information furnished in database can be freely accessed by the users.

2.5 Conclusion

Database or systematic information of the genome for end users is an important aspect of genome sequencing. Although few databases by various groups have been developed yet there are several gaps in this aspect. Next-generation sequencing has really speeded up the information generation as a consequence there is a growing interest in sequence driven research. This has resulted in generation of several types of new genomic resources such as transcriptomic, non-coding RNAs, e.g. miRNA, long non-coding RNA, circular RNA etc. Databases on this information are scanty for rice as well as its wild relatives. Similarly, an exclusive database on organelle genomes of wild rice is also needed. Therefore, genomics of wild relative of rice will be fruitful with the simultaneous support of informatics and bioinformatics, and researchers should come forward to fulfil this gap.

References

Ammiraju JS, Lu F, Sanyal A, Yu Y, Song X, Jiang N, Pontaroli AC, Rambo T, Currie J, Collura K, Talag J, Fan C, Goicoechea JL, Zuccolo A, Chen J, Bennetzen JL, Chen M, Jackson S, Wing RA (2008) Dynamic evolution of *Oryza* genomes is revealed by comparative genomic analysis of a genus-wide vertical data set. Plant Cell 20(12):3191–3209

Ammiraju JS, Fan C, Yu Y, Song X, Cranston KA, Pontaroli AC, Lu F, Sanyal A, Jiang N, Rambo T, Currie J, Collura K, Talag J, Bennetzen JL, Chen M, Jackson S, Wing RA (2010) Spatio-temporal patterns of genome evolution in allotetraploid species of the genus *Oryza*. Plant J 63(3):430–442

Antonio BA, Buell CR, Yamazaki Y, Yap I, Perin C, Bruskiewich R (2007) Informatics resources for rice functional genomics. rice functional genomics. Springer, New York, pp 355–394

Asaf S, Khan AL, Khan AR, Waqas M, Kang SM, Khan MA, Shahzad R, Seo CW, Shin JH, Lee IJ (2016) Mitochondrial genome analysis of wild rice (*Oryza minuta*) and its comparison with other related species. PLoS ONE 11(4):e0152937

Bruskiewich R, Cosico A, Eusebio W, Portugal A, Ramos LR, Reyes T, Sallan MAB, Ulat VJM, Wang X, McNally KL, Sackville Hamilton R, McLaren CR (2003) Linking genotype to phenotype: The International Rice Information System (IRIS). Bioinformatics 19(Suppl. 1):i63–i65

Chaparro C, Guyot R, Zuccolo A, Piegu B, Panaud O (2007) RetrOryza: a database of the rice LTR-retrotransposons. Nucleic Acids Res. 35:D66–D77

Chen J, Huang Q, Gao D, Wang J, Lang Y, Liu T, Li B, Bai Z, Luis Goicoechea J, Liang C, Chen C, Zhang W, Sun S, Liao Y, Zhang X, Yang L, Song C, Wang M, Shi J, Liu G, Liu J, Zhou H, Zhou W, Yu Q, An N, Chen Y, Cai Q, Wang B, Liu B, Min J, Huang Y, Wu H, Li Z, Zhang Y, Yin Y, Song W, Jiang J, Jackson SA, Wing RA, Wang J, Chen M (2013) Whole-genome sequencing of *Oryza brachyantha* reveals mechanisms underlying *Oryza* genome evolution. Nature Commun 4:1595

Copetti D, Zhang J, El Baidouri M, Gao D, Wang J, Barghini E, Cossu RM, Angelova A, Maldonado LCE, Roffler S, Ohyanagi H, Wicker T, Fan C, Zuccolo A, Chen M, Costa de Oliveira A, Han B, Henry R, Hsing YI, Kurata N, Wang W, Jackson SA, Panaud O, Wing RA (2015) RiTE database: a resource database for genus-wide rice genomics and evolutionary biology. BMC Genom 16:538

Fujii S, Kazama T, Yamada M, Toriyama K (2010) Discovery of global genomic re-organization based on comparison of two newly sequenced rice mitochondrial genomes with cytoplasmic male sterility-related genes. BMC Genom 11:209

He Z, Zhai W, Wen H, Tang T, Wang Y, Lu XM et al (2011) Two evolutionary histories in the genome of rice: the roles of domestication genes. PLoS Genet 7:e1002100. https://doi.org/10.1371/journal.pgen.1002100

Huang X, Kurata N, Wei X, Wang ZX, Wang A, Zhao Q, Guo Y (2012) A map of rice genome variation reveals the origin of cultivated rice. Nature 490(7421):497–501

Jackson S, Chen ZJ (2010) Genomic and expression plasticity of polyploidy. Curr Opin Plant Biol 13 (2):153–159

Jacquemin J, Bhatia D, Singh K, Wing RA (2013) The International Oryza Map Alignment Project: development of a genus-wide comparative genomics platform to help solve the 9 billion-people question. Curr Opin Plant Biol 16(2):147–156

Kawahara Yoshihiro et al (2013) Improvement of the *Oryza sativa* Nipponbare reference genome using next generation sequence and optical map data. Rice 6(1):4

Kurata N, Yamazaki Y (2006) Oryzabase. An integrated biological and genome information database for rice. Plant Physiol 140(1):12–17

Lu T, Yu S, Fan D, Mu J, Shangguan Y, Wang Z, Minobe Y, Lin Z, Han B (2008) Collection and comparative analysis of 1888 full-length cDNAs from wild rice *Oryza rufipogon* Griff. W1943. DNA Res 15 (5):285–295

McCouch SR, Sweeney M, Li J, Jiang H, Thomson M, Septiningsih E, Tai T (2007) Through the genetic bottleneck: *O. rufipogon* as a source of trait-enhancing alleles for *O. sativa*. Euphytica 154(3):317–339

McLaren CG, Bruskiewich RM, Portugal AM, Cosico AB (2005) The international rice information system. A platform for meta-analysis of rice crop data. Plant Physiol 139:637–642

McNally KL et al (2009) Genomewide SNP variation reveals relationships among landraces and modern varieties of rice. Proc Natl Acad Sci U S A 106 (30):12273–12278

Molina J, Sikora M, Garud N, Flowers JM, Rubinstein S, Reynolds A et al (2011) Molecular evidence for a single evolutionary origin of domesticated rice. Proc Natl Acad Sci 108:8351–8356. https://doi.org/10.1073/pnas.1104686108

Nagamura Y, Antonio BA (2010) Current status of rice informatics resources and breeding applications. Breed Sci 60:549–555

Nock CJ, Waters DL, Edwards MA, Bowen SG, Rice N, Cordeiro GM, Henry RJ (2011) Chloroplast genome sequences from total DNA for plant identification. Plant Biotech J 9:328–333

Notsu Y, Masood S, Nishikawa T, Kubo N, Akiduki G, Nakazono M, Hirai A, Kadowaki K (2002) The complete sequence of the rice (*Oryza sativa* L.) mitochondrial genome: frequent DNA sequence acquisition and loss during the evolution of flowering plants. Mole Genet Genom 268(4):434–445

Ohyanagi H, Ebata T, Huang X, Gong H, Fujita M, Mochizuki T, Toyoda A, Fujiyama A, Kaminuma E, Nakamura Y, Feng Q, Wang ZX, Han B, Kurata N (2016) OryzaGenome: genome diversity database of wild *Oryza* species. Plant Cell Physiol 57(1):e1–e1

Project, I. R. G. S. (2005) The map-based sequence of the rice genome. Nature 436:793–800

Sakai H, Lee SS, Tanaka T, Numa H, Kim J, Kawahara Y, Wakimoto H, Yang CC, Iwamoto M, Abe T, Yamada Y, Muto A, Inokuchi H, Ikemura T, Matsumoto T, Sasaki T, Itoh T (2013) Rice Annotation Project Database (RAP-DB): an integrative and interactive database for rice genomics. Plant Cell Physiol. 54(2):e6

Second G (1982) Origin of the genetic diversity of cultivated rice (*Oryza* spp.): study of the polymorphism scored at 40 isozyme loci. Jpn J Genet 57:25–57

Shahid Masood M et al (2004) The complete nucleotide sequence of wild rice (*Oryza nivara*) chloroplast genome: first genome wide comparative sequence analysis of wild and cultivated rice. Gene 340(1):133–139

Shcherban AB (2015) Repetitive DNA sequences in plant genomes. Russian J Genet Appl Res 5(3):159–167

Tang J, Xia H, Cao M, Zhang X, Zeng W, Hu S, Tong W, Wang J, Wang J, Yu J, Yang H, Zhu L (2004) A comparison of rice chloroplast genomes. Plant Physiol 135(1):412–420

Tello-Ruiz MK, Stein J, Wei S, Preece J, Olson A, Naithani S, Amarasinghe V, Dharmawardhana P, Jiao Y, Mulvaney J, Kumari S, Chougule K, Elser J, Wang B, Thomason J, Bolser DM, Kerhornou A, Walts B, Fonseca NA, Huerta L, Keays M, Tang YA, Parkinson H, Fabregat A, McKay S, Weiser J, D'Eustachio P, Stein L, Petryszak R, Kersey PJ, Jaiswal P, Ware D (2016) Gramene 2016: comparative plant genomics and pathway resources. Nucleic Acid Res 44(D1):D1133–D1140

Tian X, Zheng J, Hu S, Yu J (2006) The rice mitochondrial genomes and their variations. Plant Physiol 140 (2):401–410

Treangen TJ, Salzberg SL (2012) Repetitive DNA and next-generation sequencing: computational challenges and solutions. Nat Rev Genet 13(1):36

Vaughan DA, Morishima H, Kadowaki K (2003) Diversity in the *Oryza* genus. Current Opin Plant Biol 6 (2):139–146

Wambugu PW, Brozynska M, Furtado A, Waters DL, Henry RJ (2015). Relationships of wild and domesticated rice (*Oryza* AA genome species) based upon whole chloroplast genome sequences. Sci Rep 5

Wang D, Xia Y, Li X, Hou L, Yu J (2013) The Rice Genome Knowledgebase (RGKbase): an annotation database for rice comparative genomics and evolutionary biology. Nucl Acids Res 41(D1):D1199–D1205

Wang M, Yu Y, Haberer G, Marri PR, Fan C, Goicoechea JL, Zuccolo A, Song X, Kudrna D, Ammiraju JS, Cossu RM, Maldonado C, Chen J, Lee S, Sisneros N, de Baynast K, Golser W, Wissotski M, Kim W, Sanchez P, Ndjiondjop MN, Sanni K, Long M, Carney J, Panaud O, Wicker T, Machado CA, Chen M, Mayer KF, Rounsley S, Wing RA (2014) The genome sequence of African rice (*Oryza glaberrima*) and evidence for independent domestication. Nat Genet 46:982–988

Ware DH, Jaiswal P, Ni J, Yap IV, Pan X, Clark KY, Teytelman L, Schmidt SC, Zhao W, Chang K, Cartinhour S, Stein LD, McCouch SR (2002) Gramene, a tool for grass genomics. Plant Physiol 130:1606–1613

Waters DL, Nock CJ, Ishikawa R, Rice N, Henry RJ (2012) Chloroplast genome sequence confirms distinctness of Australian and Asian wild rice. Ecol Evol 2:211–217

Wendel JF (2000) Genome evolution in polyploids. Plant Mol Biol 42:225–249

Wing RA, Ammiraju JS, Luo M, Kim H, Yu Y, Kudrna D, Goicoechea JL, Wang W, Nelson W, Rao K, Brar D, Mackill DJ, Han B, Soderlund C, Stein L, SanMiguel P, Jackson S (2005) The Oryza map alignment project: the golden path to unlocking the genetic potential of wild rice species. Plant Mol Biol 59:53–62

Yu J, Hu S, Wang J, Wong GK, Li S, Liu B, Deng Y, Dai L, Zhou Y, Zhang X, Cao M, Liu J, Sun J, Tang J, Chen Y, Huang X, Lin W, Ye C, Tong W, Cong L, Geng J, Han Y, Li L, Li W, Hu G, Huang X, Li W, Li J, Liu Z, Li L, Liu J, Qi Q, Liu J, Li L, Li T, Wang X, Lu H, Wu T, Zhu M, Ni P, Han H, Dong W, Ren X, Feng X, Cui P, Li X, Wang H, Xu X, Zhai W, Xu Z, Zhang J, He S, Zhang J, Xu J, Zhang K, Zheng X, Dong J, Zeng W, Tao L, Ye J, Tan J, Ren X, Chen X, He J, Liu D, Tian W, Tian C, Xia H, Bao Q, Li G, Gao H, Cao T, Wang J, Zhao W, Li P, Chen W, Wang X, Zhang Y, Hu J, Wang J, Liu S, Yang J, Zhang G, Xiong Y, Li Z, Mao L, Zhou C, Zhu Z, Chen R, Hao B, Zheng W, Chen S, Guo W, Li G, Liu S, Tao M, Wang J, Zhu L, Yuan L, Yang H (2002) A draft sequence of the rice genome (*Oryza sativa* L. ssp. *indica*). Science 296(5565):79–92

Zhang QJ, Zhu T, Xia EH, Shi C, Liu YL, Zhang Y, Liu Y, Jiang WK, Zhao YJ, Mao SY, Zhang LP, Huang H, Jiao JY, Xu PZ, Yao QY, Zeng FC, Yang LL, Gao J, Tao DY, Wang YJ, Bennetzen JL, Gao LZ (2014) Rapid diversification of five Oryza AA genomes associated with rice adaptation. Proc Nat Aca Sci 111(46):E4954–E4962

Zhao K, Wright M, Kimball J, Eizenga G, McClung A, Kovach M, Tyagi W, Ali ML, Tung CW, Reynolds A, Bustamante CD, McCouch SR (2010) Genomic diversity and introgression in *O. sativa* reveal the impact of domestication and breeding on the rice genome. PloS One 5:e10780

Evolutionary Relationships Among the *Oryza* Species

3

Peterson W. Wambugu, Desterio Nyamongo, Marie-Noelle Ndjiondjop and Robert J. Henry

Abstract

Despite being highly studied, the evolutionary relationships in the *Oryza* genus have remained inconsistent and inconclusive. The origin and domestication history of Asian rice has particularly remained contentious. This chapter discusses the evolutionary relationships between various species in the *Oryza* genus, with a special focus on the application of current advances in genomics in understanding the various evolutionary dynamics. Advances in genomics are offering opportunities for resolving the origin of cultivated rice and clarifying phylogenetic and evolutionary relationships between the various *Oryza* species. Analysis of genomes of cultivated rice and their putative progenitors is providing useful information such as unusually diverged genomic regions, which provides vital insights into rice evolution, domestication and demographic history. However, with the increase in whole genome sequence data, it appears that data analysis and subsequent interpretation may now present the next challenge in efforts aimed at resolving this hot debate. The use of nuclear and whole chloroplast genome sequences is helping define the relationships between the recent newly discovered Australian taxa which are believed to be novel gene pools. In this chapter, we have also discussed the challenges faced in efforts aimed at resolving evolutionary relationships in the *Oryza* genus.

P. W. Wambugu (✉) · D. Nyamongo
Kenya Agricultural Livestock Research Organization (KALRO), Genetic Resources Research Institute, P.O. Box 30148-00100, Nairobi, Kenya
e-mail: p.wambugu@uq.edu.au

M.-N. Ndjiondjop
Africa Rice Centre (Africa Rice), 01 B.P. 2031 Cotonou, Benin

R. J. Henry
Queensland Alliance for Agriculture and Food Innovation, University of Queensland, Brisbane, QLD 4072, Australia

3.1 Introduction

Rice is the most important food crop in the world, acting as the staple food for over half of the world's population (IRGSP 2005). The evolutionary relationships between cultivated rice and wild rice species are perhaps the most studied of the crop plants due to the importance of rice in world agriculture and particularly in food security. A variety of approaches have been used in elucidating the evolution and demographic history of rice ranging across archaeological, biochemical, molecular and genetic studies. In addition to the two cultivated rice species, the

T. K. Mondal and R. J. Henry (eds.), *The Wild Oryza Genomes*,
Compendium of Plant Genomes, https://doi.org/10.1007/978-3-319-71997-9_3

Oryza genus has 22 wild relatives, all of which have undergone a wide ecological diversification (Vaughan et al. 2003) within an evolutionary time span of about 15 million years (Tang et al. 2010).

However, despite great efforts, the evolutionary relationships determined in the *Oryza* genus have been inconsistent among studies, perhaps due to different evolutionary signatures among the studied loci (Ammiraju et al. 2008; Ge et al. 1999; Sanyal et al. 2010). Previously, phylogenetic and evolutionary relationships have been studied using various molecular markers. Data generation has largely involved the laborious process of amplifying selected loci. With continued advances in genomics especially in DNA sequencing, it is increasingly becoming easier to cost effectively and cheaply generate sequence data across the whole genome.

Phylogenomics which is one of the rapidly expanding fields of genomics is finding wide application in reconstructing the phylogenetic and evolutionary history of species by analysing a large number of loci across the genome (Eisen and Fraser 2003; Frédéric et al. 2005). Analysis of multiple loci is based on the assumption that the predominant phylogenetic and evolutionary information inherent in the data set will weed out any inconsistencies and incongruence, thus resulting in a robust phylogeny (Rakshit et al. 2007; Molina et al. 2011). The availability of whole genome sequences presents particular advantages in evolutionary genomic studies as it reduces the need for extensive within-species sampling. Due to the independent evolutionary histories of unlinked genes, analysis of genomes of single representative of each species allows for species-wide population genomic inference (Gronau et al. 2011). Inadequate sampling has been identified as one of the challenges that faces studies aimed at elucidating evolutionary relationships.

The increased availability of whole genome sequences is providing opportunities for undertaking comparative genomic analysis. Compared to single genome analysis, comparative analysis of multiple genomes is providing greater understanding of rice genome evolution and other related inferences (Tieyan and Mingsheng, 2014). The *Oryza* genus is an excellent system in plants for these comparative genomic studies (Ammiraju et al. 2008). In this chapter, we discuss the evolutionary relationships between the various species in the *Oryza* genus. We have put special focus on the application of the current advances in genomics in analysing the origin, domestication and demographic history of rice as well as gene and genome evolution of this economically important genus. Challenges in studying rice evolution are also highlighted.

3.2 Origin and Domestication of Cultivated Rice Species

Domestication is key in shaping the genetic diversity present in a crop and provides a good model to study the evolutionary and demographic history of cultivated species. Knowledge of the origin and domestication of crops is therefore key in not only understanding patterns and level of genetic diversity that may be available for crop improvement but also how to manipulate it. Such knowledge may, for example, be important in efforts aimed at reintroducing useful genetic variation that may have been lost during the process of domestication. In the next section of this chapter, we will discuss the various theories on the domestication and demographic history of the two independently domesticated rice species. The domestication of the two rice-cultivated species is a story of parallel evolutionism, where different cultures selected different wild species targeting the same set of genes/traits (Purugganan 2014; Wang et al. 2014). Previously, these domestication genes have mainly been identified through candidate gene analysis and quantitative trait loci (QTL) mapping (Li et al. 2006; Tan et al. 2008; Zhang et al. 2009). As earlier stated, advances in genomics are offering opportunities for in-depth understanding of genetic architecture of domestication process through genome-wide association studies and whole genome resequencing studies (Meyer and Purugganan 2013). As will be highlighted later, the origin and domestication history of rice has primarily been studied through phylogenetic analysis, which can be misleading. However, analysis of genomes of cultivated and wild rice is providing useful information such as unusually diverged genomic regions between closely related species, which provides vital insights into rice evolution and domestication.

3.2.1 Domestication of Asian Rice

Despite being widely studied, the origin and domestication history of *Oryza sativa* has remained highly controversial. For several decades, the origin and demographic history of the two subspecies of *O. sativa, japonica and indica*, has been the subject of numerous studies conducted using both archaeological and genetic approaches. Among the issues that have generated a lot of controversy in this debate is the geographic origin of this cultivated species and the number of times it has been domesticated (Huang et al. 2012). The number of times that a crop has been domesticated is important to evolutionary biologists as it determines not only the amount of genetic diversity present but also the genetic architecture of important traits (Morrell et al. 2012). While majority of studies agree that there is deep genetic differentiation between the two subspecies of rice, the evolutionary and demographic factors explaining this differentiation remain contentious. Historically, debate on the origin of the two subspecies has revolved around two theories: monophyletic and polyphyletic origin.

Support for the multiple origin theory is primarily obtained from molecular phylogenetic analysis which gives a clustering pattern that shows both *japonica* and *indica* to be more closely related to distinct lineages of *Oryza rufipogon* or *Oryza nivara* than to each other (Cheng et al. 2003; Zhu and Ge 2005; Jason et al. 2006). A population genetic analysis of genome-wide data is conducted by He et al. (2011) who suggest that though the genomes of *indica* and *japonica* generally appear to be of independent origin, regions of low diversity present in these genomes appear to have originated once. These low diversity regions which are associated with selection during domestication may have been spread through introgression. A multi-locus analysis of various gene regions obtained from nuclear, chloroplast and mitochondrial genomes suggests that the two subspecies were independently domesticated from different populations of *O. rufipogon* (Wei et al. 2012b). Recent genome-wide (Civan et al. 2015; Choi et al. 2017) and whole chloroplast genome-based analysis (Wambugu et al. 2015) gave topologies which suggest that the various rice subpopulations originated from pre-differentiated ancestral gene pools, thus pointing to a polyphyletic origin. As will be highlighted later in this section, Civan et al. (2015) hypothesized a multiple origin theory where *indica*, *japonica* and *aus* subpopulations had three separate origins with independent domestication. Choi et al. (2017) proposed yet another model where the different subspecies had separate origins but with only one domestication event. According to this model, *japonica* was the first to be domesticated followed by introgressive hybridization from *japonica* to proto-*indica* and proto-*aus* which led to domesticated *indica* and *aus* rice. Other studies have similarly reported introgression from *japonica* into *indica* to have occurred during the domestication process (Wei et al. 2012a; Yang et al. 2012). This gene flow has, however, not been detected in some studies especially those analysing organelle DNA (Wei et al. 2012b). Rather than resolving the debate on the origin of cultivated rice, it appears that with the increase in availability of whole genome data, more divergent theories and models are being proposed.

On the other hand, various genetic studies have supported the single origin theory, with various pathways being suggested to explain the differentiation between *indica* and *japonica*. The sharing of domestication genes and alleles, among them *sh4* and *prog1* loci which are responsible for reduction in grain shattering and for erect habit, respectively, remains one of the strongest evidences in support of single origin theory (Li et al. 2006; Tan et al. 2008; Zhang et al. 2009; Zhu et al. 2011). However, divergent views have been raised on this, with proponents of the multiple origin theory arguing that the sharing of genes could be as a result of hybridization between the subspecies after their independent domestication (Fuller et al. 2010; Sang and Ge 2007a, b). Taking gene flow into account in their demographic modelling study, Molina et al. (2011), however, ruled out post-domestication introgression and concluded that the sharing of domestication loci reflects the single origin. Based on molecular clock analysis, the divergence between the two subspecies of rice has been estimated to have occurred approximately 86–440 ky ago which by far predates any known archaeological evidence of domestication (Ma and Bennetzen 2004; Vitte et al. 2004; Zhu and Ge 2005). This divergence has been used to argue in support of the independent origin of *japonica* and *indica*.

Among the suggested pathways that have been used to explain the differentiation between *indica* and *japonica* is that *japonica* was derived from *indica* (Lu et al. 2002) and an incipient domesticate differentiating into *indica* and *japonica* under different climatic conditions (Oka and Morishima 1982). In another single origin model, developed through a population scale whole genome-based study, it is postulated that *japonica* was domesticated from a specific *O. rufipogon* population found in Southern China, while *indica* is the product of crosses between *japonica* and local wild rice in South East and South Asia (Huang et al. 2012). Despite this study analysing the largest data set so far, it seems to have suffered from an analytical artefact as its conclusions are not entirely supported by the results of the analysis (Civan et al. 2015). Civan et al. (2015) observed that based on the conclusions of the study, it would be expected that genomic regions in the identified domestication sweeps would be passed from *japonica* to *indica* due to crossing. Consequently, contrary to the findings of the study where *japonica* is nested within *indica*, it would be expected that *indica* would be nested within *japonica* in the neighbour joining tree. These authors further noted that if the single origin theory as suggested by the original study was correct, the various rice groups, namely *japonica*, *indica* and *aus,* would be expected to have identical selective sweeps. After reanalysing the data set by Huang et al. (2012), Civan et al. (2015) concluded that the three groups of rice, namely *japonica*, *indica* and *aus,* were domesticated in three geographically separate regions. However, defending their earlier findings and conclusions, Xuehui and Bin (2015) dismissed the criticism levelled by Civan et al. (2015), terming the reanalysis of their data and review of their study as technically flawed.

These studies clearly show that despite the continued advances in genomics, coupled with equally revolutionary advances in bioinformatics, the debate on the origin and domestication history of rice is far from over. With the increase in whole genome sequence data, it appears that data analysis and subsequent interpretation may now present the next challenge in efforts aimed at resolving this hot debate. Other factors that may cause bias which may lead to contradictory conclusions include differences in sequencing technologies and sequencing coverage.

3.2.2 Domestication of African Rice

With a total of eight species of both cultivated and wild rice species, representing six out of the ten known genome types, the African region arguably has the greatest diversity of the rice gene pool, in terms of number of species and genome types. Apart from the two cultivated species, the other species include *Oryza barthii*, *Oryza longistaminata*, *Oryza brachyantha*, *Oryza punctata*, *Oryza eichingeri* and *Oryza schweinfurthiana.* African rice was domesticated about 3500 years ago, with several theories on the origin of this species being proposed. An Asian origin of African rice has previously been advanced but has been rejected (Li et al. 2011). For over four decades, one author (Nayar 1973, 2010, 2012, 2014) who can be likened to a lone ranger in this debate has consistently argued that African rice was domesticated from Asian rice in West Africa. His proposal has received very little support as the arguments seem less convincing. The major theory around which many studies and opinions appear to converge postulates that

African rice was domesticated from *O. barthii* in West Africa.

Using a population genetics approach, Wang et al. (2014) recently provided arguably the strongest evidence so far on the origin and domestication history of African rice. By resequencing 94 *O. barthii* and 20 *Oryza glaberrima* accessions as well as comparative genetic analysis of selected domestication genes, these authors concluded that African rice was domesticated from *O. barthii* in a convergent but independent selection process to that of *O. sativa*. This population genetics approach was able to map the actual domestication centre along the Niger River, consistent with original proposals from Porteres (1962) and later supported by Li et al. (2011). A recent phylogenetic analysis (Fig. 3.1) based on whole chloroplast sequences of cultivated and wild *Oryza* species in the AA genome group representing the *Oryza* primary gene pool provided support for an *O. barthii* origin of African rice. This study resulted in a monophyly of all *O. barthii* and *O. glaberrima* accessions (Wambugu et al. 2015). The sharing of some traits between *O. glaberrima* and *O. barthii* such as resistance to some biotic stresses (Thottappilly and Rossel 1993) as well as unique starch structure (Wang et al. 2015) may be a sign of inheritance which lends further support to this theory.

Like other crops, domestication process in African rice has been associated with a reduction in genetic diversity compared to the wild progenitors. A multi-loci study involving 14 genes showed an unusually low nucleotide diversity in both African rice and its progenitor, with African rice having 70% less diversity than its progenitor (Li et al. 2011). This extremely low nucleotide diversity which is suggestive of a severe domestication bottleneck was recently confirmed by transcriptome population genomic analysis (Nabholz et al. 2014), a deep coverage population genetic resequencing study (Wang et al. 2014), a large-scale genotyping study (Orjuela et al. 2014) and a SNP map analysis (Meyer et al. 2016). Vaughan et al. (2008) observed that *O. glaberrima* may have gone through a double evolutionary bottleneck, thereby leading to the low diversity. The first bottleneck may have been associated with the divergence from the ancestors of Asian *Oryza*, while the second may have been due to domestication.

As already highlighted, African rice was domesticated in an independent but convergent process to that of Asian rice where the same traits were targeted. While these domestication traits are well studied in Asian rice, relatively little is known about these traits in African rice. Specifically, the molecular basis of domestication traits remains poorly studied in African rice. Grain size is one of the traits that was under selection during the domestication process, with the selection process leaning towards bigger seeds. African rice exhibits an unusual phenomenon where the progenitor, *O. barthii*, has a larger seed size compared to the domesticated crop. Recently, the genetic basis of seed size in African rice was reported (Wu et al. 2017). This study identified a SNP in the GL4 gene which led to a premature stop codon resulting in small seeds and also reduced seed shattering. These findings provide useful highlights that can be used in breeding for high yield as grain size is one of the determinants of yield.

3.3 Evolutionary Relationships Within *Oryza* Species Complexes

Most of the *Oryza* species can be grouped into four species complexes, namely *O. sativa*, *O. officinalis*, *Oryza ridleyi* and *O. meyeriana* species complexes (Vaughan 1989) As already highlighted, evolutionary relationships within the *Oryza* genus have previously been studied using molecular markers. However, with the continued release of complete chloroplast sequences (Table 3.1), *Oryza* chloroplast phylogenomics is finding great application in elucidating phylogenetic and evolutionary relationships within this genus.

Though evolutionary relationships between the various genome groups and species complexes are generally well established, there exist some gaps in knowledge. One of the most

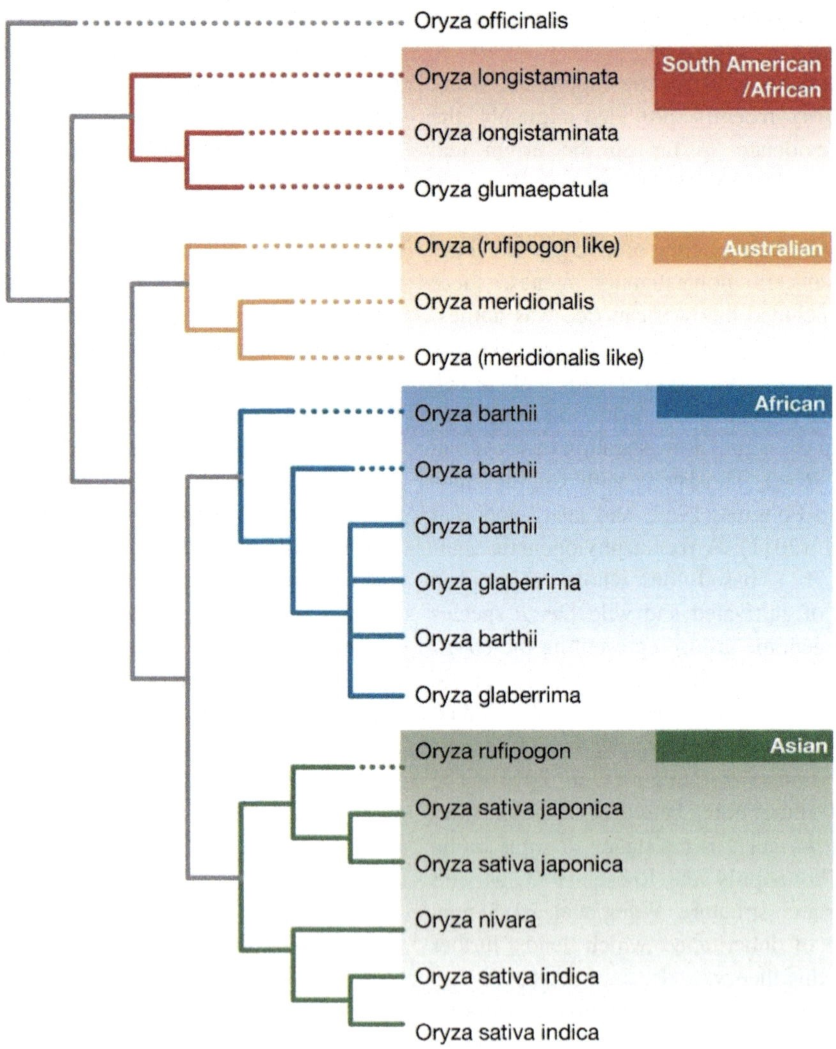

Fig. 3.1 Neighbour joining tree showing phylogenetic and evolutionary relationships among the AA genome species based on whole chloroplast sequences. *Source* Wambugu et al. (2015)

comprehensive studies on the evolutionary relationships between *Oryza* genome groups was conducted by Ge et al. (1999). This study sampled all the extant *Oryza* species representing the various genome groups and species complexes and obtained a monophyly of each of the genome group. Nevertheless, due to incongruence between gene trees, relationships between some genome groups were not resolved. Specifically, the relationship among the A (*O. sativa* complex), B (*O. officinalis* complex) and C (*O. officinalis* complex) genome types and also among the F genome, G (*O. officinalis* complex) genome and the rest of the genus remained unresolved. Using a phylogenomics approach that studied a total of 142 genes, Zou et al. (2008) resolved the relationships between all the diploid genomes. This study shows that the A and B genomes are sisters, with the C genome being a sister to that clade. This phylogenomic study also identified two rapid diversification events that are responsible for almost all the genome diversity that is found in the genus.

Table 3.1 Released *Oryza* whole chloroplast genomes

Species name	Genome size (bp)	Remarks	References
O. sativa ssp. *japonica*	134551	This genome was assembled through reference-guided assembly of massively parallel sequencing data. An error in an earlier published reference sequence was detected, thus demonstrating the robustness of this approach	Nock et al. (2011)
O. sativa ssp. *indica*	134496	Using whole chloroplast genomes, this study estimated the divergence between the two subspecies of *Oryza* to have occurred approximately 86,000–200,000 years ago	Tang et al. (2004)
O. longistaminata	134657	*O. longistaminata* formed the basal clade with *O. glumaepatula* suggesting an African or South American origin of *Oryza* genus	Wambugu et al. (2015)
O. meridionalis	134558	This study defined the genetic relationships between Asian and Australian wild rice. It suggested that the taxa currently identified as *O. rufipogon* from Australia may be a perennial form of *O. meridionalis*	Waters et al. (2012)
O. nivara	134494	The *O. nivara* chloroplast was assembled and compared with that of *O. sativa* revealing their high similarity	Shahid Masood et al. (2004)
O. rufipogon	134544	Confirmed the distinctness of Australian *O. rufipogon* populations compared to those from Asia	Waters et al. (2012)
O. glaberrima	132629	A cost-effective enrichment hybridization protocol for assembling chloroplast genomes was developed	Mariac et al. (2014)
O. barthii	134674	The analysis supported the hypothesis that *O. barthii* is the progenitor of *O. glaberrima*	Wambugu et al. (2015)
O. glumaepatula	134583	*O. glumaepatula* shared the basal clade with *O. longistaminata* suggesting an African or South American origin of *Oryza* genus	Wambugu et al. (2015)
O. minuta	135094	*O. minuta* formed the same clade with *O. punctata* which is its maternal parent	Asaf et al. (2017)
O. punctata	134604	–	Unpublished
O. officinalis	134911	*O. officinalis* was used as the out-group. This study obtained a well-resolved and strongly supported phylogeny of the AA genome group	Wambugu et al. (2015)
O. australiensis	134549	This study assembled whole chloroplast genomes through massive parallel sequencing of total DNA. Data from this reference-guided assembly accurately determined phylogenetic relationships between species	Nock et al. (2011)
O. brachyantha	134604	This study confirmed *O. brachyantha* as an early diverging taxa	Liu et al. (2016)

3.3.1 *Oryza sativa* Species Complex

Species within the *O. sativa* complex constitute the *Oryza* primary gene pool. They have a pan tropical distribution, with their occurrence being reported in various parts of Africa, South America, Asia and Australia. The *O. sativa* complex was the most recently diverged. Based on whole chloroplast genome sequences, Wambugu et al. (2015) recently found that the AA genome group diverged about 0.69–1.11 million years ago (Mya). This divergence time between the species in the AA genome group was too close to be explained by pan-Gondwana distribution. Long dispersal remains the most plausible explanation for their pan tropical distribution. Evolution of the *Oryza* genus has probably involved repeated processes of dispersal (sometimes long distance), diversification and introgression between closely related populations.

The phylogenetic and evolutionary relationships within the *O. sativa* complex remain the most studied of all the *Oryza* species complexes. This great research interest in these species could be due to their importance as a readily accessible source of valuable diversity for rice improvement. The evolutionary relationships between *O. barthii* and *Oryza longistaminata* have been contentious. While some studies have suggested that *O. longistaminata* is the immediate ancestor of *O. barthii* (Khush 1997), results of recent phylogenetic analysis based on whole chloroplast genome analysis do not support this hypothesis (Asaf et al. 2017; Wambugu et al. 2015). In these whole chloroplast genome-based studies, *O. longistaminata* appeared in the basal clade of the AA genome group where it clustered with *O. glumaepatula* suggesting that the AA genome group originated either in Africa or South America. An African origin of the AA genome group had earlier been suggested by Cheng et al. (2002) and Iwamoto et al. (1999) where *O. longistaminata* was found to be the most ancestral species. The morphological features of *O. longistaminata* seem to further lend credence to this theory. Its perennial habit and other unique morphological features such as self-incompatibility, presence of rhizomes and unique ligule characteristics point to this species having undergone significant differentiation during evolution. This is consistent with the conventional wisdom in evolutionary biology that perenniality is ancestral to annual growth habit. The annual habit acts as a useful strategy for adapting to various seasonal changes, and the change in growth habit is important in the evolution of cereal crops from their wild progenitors (Hu et al. 2003).

Oryza meridionalis which has been considered to be endemic to Australia is the other species in the *O. sativa* complex. However, *O. meridionalis* has been reported from New Guinea and may extend further north, thus putting to question the theory of its endemism to Australia. Recently, molecular and morphological analysis of rice wild relatives in Australia has led to the discovery of distinct perennial taxa which could be novel *Oryza* gene pools (Waters et al. 2012; Sotowa et al. 2013; Brozynska et al. 2014). *Oryza rufipogon* has been considered to extend from Asia into northern Australia. Populations originally identified as *O. rufipogon* in Australia were shown to be distinct from Asian *O. rufipogon* (Brozynska et al. 2014). The phylogenetic and evolutionary relationships of these newly discovered taxa have been studied using both chloroplast and nuclear genome sequences (Wambugu et al. 2015; Brozynska et al. 2017). The *O. rufipogon* like taxa (taxa A) has a chloroplast closer to *O. meridionalis* and a nuclear genome closer to *O. rufipogon* from Asia (Brozynska et al. 2017). Widespread perennial taxa (taxa B) that resemble *O. meridionalis* have also been studied. *Oryza meridionalis* has been described as an annual, however, recent evidence suggests that taxa B may not be distinct from *O. meridionalis* and that *O. meridionalis* should be considered to include annual and perennial populations.

The evolutionary history of these taxa currently remains unclear but taxa A may have diverged from *O. meridionalis* around 3 Mya followed by divergence of taxa A from the Asian *O. rufipogon* (around 1.6 Mya) but with ongoing introgression between the wild taxa (Sotowa et al. 2013). An admixture test revealed possible introgression with Asian wild populations (Brozynska et al. 2017). Though it remains contentious, analysis of genome scale data shows *O. meridionalis* to be the most ancestral species among the AA genome species, thus suggesting an Australia origin of this genome group (Zhu et al. 2014). These newly discovered Australian taxa provide a valuable resource for studying rice evolution as their signatures of gene and genome evolution have not been eroded through gene flow.

3.3.2 *Oryza officinalis* Complex

With ten species representing five genome types, the *O. officinalis* complex is the largest species complex in the *Oryza* genus. It is closely related to the *O. sativa* complex (Kumagai et al. 2010). Relationships between species within the CC genome have remained unclear and inconsistent between studies. Using an extended data set of 68 single copy genes, Zang et al. (2011) resolved these relationships by showing a close sister relationship between *O. officinalis* and *Oryza rhizomatis*, with *O. eichingeri* being much more diverged. With *O. eichingeri* being the only wild *Oryza* species found in Asia and Africa (Vaughan 1989), this distribution has drawn considerable research interest. Due to the significant differentiation between the *O. eichingeri* races from Africa and those from Sri Lanka, it has previously been suggested that they should be treated as distinct species (Federici et al. 2002). However, based on a population genetic analysis, Zhang and Ge (2007) suggested that these geographically isolated populations should be treated as geographic races rather than distinct species. These authors attributed this geographic isolation to long-distance dispersal between the two regions. A species delimitation analysis conducted by Zang et al. (2011) using genome scale data supported the distinct species status of the two geographical races. This study, however, recommended further detailed morphological and molecular analysis with extensive sampling in the two regions. The chloroplast sequence for *O. minuta* was recently released where it was used to study phylogenetic relationships between various *Oryza* species (Asaf et al. 2017). In this study, *O. minuta* formed a clade with *O. punctata,* thereby confirming *O. punctata* as its maternal parent. Based on analysis of the chloroplast genome, *Oryza australiensis* which is endemic in Australia was the first to diverge suggesting that it is the most ancestral in this complex (Kumagai et al. 2010; Asaf et al. 2017). The chloroplast genomes of *Oryza latifolia, Oryza alta, Oryza. grandiglumis, O. rhizomatis and O. eichingeri* have not been released.

3.3.3 *Oryza meyeriana* Complex

This species complex has two diploid species, namely *O. meyeriana* and *O. granulata.* The complex has received less attention perhaps due to its genetic distance from the *Oryza* primary gene pool. The two species are closely related, with some authors even suggesting the need for a revision of the taxonomic classification of the two species so that they are treated as one species (Gong et al. 2000). This recommendation seems to gain support from analysis conducted by Kumagai et al. (2010) using selected variable regions of the chloroplast genome where the two species had a monophyletic relationship. Phylogenetic analysis conducted using nuclear (*Adh1* and *Adh2*) and chloroplast (*Mat*K) genes indicated that both *O. meyeriana* and *O. granulata* are the earliest diverging *Oryza* species (Ge et al. 1999). The evolutionary relationship between *O. meyeriana* with other species as the most ancestral species had been earlier reported by Wang et al. (1992) based on nuclear restriction fragment length polymorphisms (RFLP). Whole chloroplast genomes of these two species have not been reported. No efforts seem to have been made to study the evolutionary relationships between these species using current genomic approaches.

3.3.4 *Oryza ridleyi* Complex

This group comprises of two species, namely *O. ridleyi* and *Oryza longiglumis,* both of which are tetraploid species. A chloroplast-based study showed the two species occupying the basal clade (Kumagai et al. 2010), a finding that contradicts an earlier study where *O. meyeriana* and *O. granulata* were found to be the most basal taxa (Ge et al. 1999). The topological incongruence between different studies can be attributed to gene choice, use of insufficient data as well as choice of out-group. Just like *O. meyeriana* complex, species in *O. ridleyi* complex have received less research attention. Whole chloroplast sequences of these two species have not been reported. It appears that no studies have been conducted to elucidate evolutionary relationships within this complex using the current advances in DNA sequencing and other genomic approaches.

Among the species that were not included in any of the species complex is *O. brachyantha*. As reviewed by Wambugu et al. (2013), the evolutionary relationships between *O. brachyantha* and other species are not very clear due to its inconsistent phylogenetic placement. By placing *O. brachyantha* in the basal clade, Mullins and Hilu (2002) suggested an African origin of *Oryza*. This is, however, not supported by Ge et al. (1999) whose study showed *O. meyeriana* and *O. granulata,* both of the GG genome group to be the most ancestral species. Even more tenuous was the study by Wang et al. (1992) where *O. brachyantha* was nested within the AA genome group. As it usually occurs with single gene phylogenies, the three phylogenies obtained by Ge et al. (1999) after analysing *Adh1*, *Adh2* and *Mat*K genes, all resulted in incongruent placement of *O. brachyantha*. These authors recommended the analysis of additional genes in an attempt at resolving this topological incongruence. The availability of whole genome sequences of *O. brachyantha* provides an opportunity for resolving these inconsistencies in evolutionary relationships. The whole genome sequences of *O. brachyantha* provide one of the most important resources for studying evolution in *Oryza* as it is one of the most diverged and hence contains a lot of signatures of gene and genome evolution (Chen et al. 2013). The chloroplast genome for *O. brachyantha* was recently released and was used to study relationships with 13 other *Oryza* species whose chloroplast genomes have been published (Liu et al. 2016). The analysis indicated that *O. brachyantha* was the most diverged, a finding that is consistent with that of other studies.

3.4 Challenges in Elucidating *Oryza* Evolutionary Relationships

As highlighted earlier in this chapter, evolutionary relationships in the *Oryza* genus have in past studies showed high degree of inconsistency. Among the causes of these inconsistencies is incomplete lineage sorting, rapid diversification and accession misidentification (Wambugu et al. 2013 and the references therein). The use of samples that have been collected from areas that have been disturbed through human activities is also likely to cause contradictory evolutionary relationships. Introgression between various rice populations/species might lead to erosion of important genomic signatures of evolution and is therefore likely to confound evolutionary genetic studies. Gene flow has resulted in a wide array of intermediate phenotypes commonly referred as weedy rice (Bres-Patry et al. 2001) whose identification is usually problematic. With the continued availability of whole genome data, it is becoming increasingly possible to study patterns of genetic admixture across the genome in addition to conducting haplotype structure analysis (Morrell et al. 2012). Using genome scale data, Brozynska et al. (2017) conducted an admixture test that revealed possible introgression between Australian and Asian wild rice populations. Such analysis is likely to enable scientists to address some of the challenges that have faced previous studies on the origin and demographic history of rice. Similarly, accession misidentification especially in genebank collections which is common in rice is also a possible cause of these contradictory theories and conclusions. Efforts to ensure accurate identification prior to making such conclusions are thus imperative.

The majority of studies on the evolutionary and phylogenetic relationships in *Oryza* genus have been conducted using statistical methods that mainly involve clustering of samples and species. Results of such methods, however, can be misleading as genetic groupings do not necessarily reflect geographic origin or evolutionary relationships (Jeffrey and Brandon 2008; Robin et al. 2008). Data concatenation is common in phylogenetic and evolutionary studies. Simulation methods have shown that data concatenation can result in inconsistent phylogenetic estimates (Kubatko and Degnan 2007), thus leading to conflicting conclusions on evolutionary histories. The choice of the right loci and amount of data to use in studies aimed at determining phylogenetic and evolutionary relationships has always been challenging. The use of insufficient amount of data and non-informative loci leads to poor resolution and has been blamed on the inconsistent and contradictory results on phylogenetic and evolutionary relationships in the *Oryza* genus. Increasing the number of loci, for example, helps to increase the accuracy of divergence estimates (Kumar and Hedges 1998). The use of whole chloroplast genome sequences is therefore likely to result in more reliable phylogenetic inference as it involves the analysis of higher number of loci (Wambugu et al. 2015). Moreover, as highlighted earlier, analysis of the same data set using different bioinformatics methods has in some cases resulted in contradictory conclusions (Molina et al. 2011; Civan et al. 2015).

References

Ammiraju JS, Lu F, Sanyal A, Yu Y, Song X, Jiang N, Pontaroli AC, Rambo T, Currie J, Collura K, Talag J, Fan C, Goicoechea JL, Zuccolo A, Chen J, Bennetzen JL, Chen M, Jackson S, Wing RA (2008) Dynamic evolution of *Oryza* genomes is revealed by comparative genomic analysis of a genus-wide vertical data set. Plant Cell 20:3191–3209

Asaf S, Waqas M, Khan AL, Khan MA, Kang S-M, Imran QM, Shahzad R, Bilal S, Yun B-W, Lee I-J (2017) The complete chloroplast genome of wild rice (*Oryza minuta*) and its comparison to related species. Front Plant Sci 8:304

Bres-Patry C, Lorieux M, Clément G, Bangratz M, Ghesquière A (2001) Heredity and genetic mapping of domestication-related traits in a temperate japonica weedy rice. Int J Plant Breed Res 102:118–126

Brozynska M, Omar ES, Furtado A, Crayn D, Simon B, Ishikawa R, Henry RJ (2014) Chloroplast genome of novel rice germplasm identified in Northern Australia. Trop. Plant Biol. 7:111–120

Brozynska M, Copetti D, Furtado A, Wing RA, Crayn D, Fox G, Ishikawa R, Henry RJ (2017) Sequencing of Australian wild rice genomes reveals ancestral relationships with domesticated rice. Plant Biotechnol J. 15:765–774

C-c Yang, Kawahara Y, Mizuno H, Wu J, Matsumoto T, Itoh T (2012) Independent domestication of Asian rice followed by gene flow from japonica to indica. Mol Biol Evol 29:1471–1479

Chen J, Liang C, Chen C, Zhang W, Sun S, Liao Y, Zhang X, Yang L, Song C, Wang M, Shi J, Huang Q, Liu G, Liu J, Zhou H, Zhou W, Yu Q, An N, Chen Y, Cai Q, Wang B, Liu B, Gao D, Min J, Huang Y, Wu H, Li Z, Zhang Y, Yin Y, Song W, Jiang J, Jackson SA, Wing RA, Wang J, Wang J, Chen M, Lang Y, Liu T, Li B, Bai Z, Luis Goicoechea J (2013) Whole-genome sequencing of *Oryza brachyantha* reveals mechanisms underlying *Oryza* genome evolution. Nat. Commun. 4:1595

Cheng C, Tsuchimoto S, Ohtsubo H, Ohtsubo E (2002) Evolutionary relationships among rice species with AA genome based on SINE insertion analysis. Genes Genet Syst 77:323–334

Cheng C, Motohashi R, Tsuchimoto S, Fukuta Y, Ohtsubo H, Ohtsubo E (2003) Polyphyletic origin of cultivated rice: based on the interspersion pattern of SINEs. Mol Biol Evol 20:67–75

Choi JY, Platts AE, Fuller DQ, Hsing Y-I, Wing RA, Purugganan MD (2017) The rice paradox: multiple origins but single domestication in Asian rice. Mol Biol Evol 34:969–979

Civan P, Craig H, Cox CJ, Brown TA (2015) Three geographically separate domestications of Asian rice. Nat Plants 1:15164

Eisen J, Fraser C (2003) Phylogenomics: intersection of evolution and genomics. In: Eisen JA (ed) Science. pp 1706–1707

Frédéric D, Henner B, Hervé P (2005) Phylogenomics and the reconstruction of the tree of life. Nat RevGenet 6:361

Federici MT, Shcherban AB, Capdevielle F, Francis M, Vaughan D (2002) Analysis of genetic diversity in the *Oryza officinalis* complex. Electron J Biotechnol 5:16–17

Fuller D, Sato Y-I, Castillo C, Qin L, Weisskopf A, Kingwell-Banham E, Song J, Ahn S-M, van Etten J (2010) Consilience of genetics and archaeobotany in the entangled history of rice. Archaeol Anthropol Sci 2:115–131

Ge S, Sang T, Lu B-R, Hong D-Y (1999) Phylogeny of rice genomes with emphasis on origins of allotetraploid species. Proc Natl Acad Sci USA 96:14400–14405

Gong Y, Borromeo T, Lu B (2000) A biosystematics study of the *Oryza meyeriana* complex (Poaceae). Plant Syst Evol 224:139–151

Gronau I, Hubisz M, Gulko B, Danko C, Siepel A (2011) Bayesian inference of ancient human demography from individual genome sequences. Nat Genet 43:1031

He Z, Shi S, Zhai W, Wen H, Tang T, Wang Y, Lu X, Greenberg AJ, Hudson RR, Wu C-I (2011) Two evolutionary histories in the genome of rice: the roles of domestication genes. PLoS Genet 7:e1002100

Hu FY, Tao DY, Sacks E, Fu BY, Xu P, Li J, Yang Y, Mcnally K, Khush GS, Paterson AH, Li ZK (2003) Convergent evolution of perenniality in rice and sorghum. Proc Natl Acad Sci USA 100:4050

Huang X, Li W, Guo Y, Lu Y, Zhou C, Fan D, Weng Q, Zhu C, Huang T, Zhang L, Wang Y, Kurata N, Feng L, Furuumi H, Kubo T, Miyabayashi T, Yuan X, Xu Q, Dong G, Zhan Q, Li C, Fujiyama A, Wei X, Toyoda A, Lu T, Feng Q, Qian Q, Li J, Han B, Wang Z-X, Wang A, Zhao Q, Zhao Y, Liu K, Lu H (2012) A map of rice genome variation reveals the origin of cultivated rice. Nature 490:497

IRGSP (2005) The map-based sequence of the rice genome. Nature 436:793–800

Iwamoto M, Nagashima H, Nagamine T, Higo H, Higo K (1999) p-SINE1-like intron of the CatA catalase homologs and phylogenetic relationships among AA-genome *Oryza* and related species. Theor Appl Genet 98:853–861

Jason PL, Yu-Chung C, Kuo-Hsiang H, Tzen-Yuh C, Barbara AS (2006) Phylogeography of Asian wild rice, *Oryza rufipogon*, reveals multiple independent domestications of cultivated rice, *Oryza sativa*. Proc Natl Acad Sci USA 103:9578–9583

Jeffrey R-I, Brandon SG (2008) Multiple domestications do not appear monophyletic. Proc Natl Acad Sci USA 105:E105

Khush GS (1997) Origin, dispersal, cultivation and variation of rice. Plant Mol Biol 35:25–34

Kubatko LS, Degnan JH (2007) Inconsistency of phylogenetic estimates from concatenated data under coalescence. Syst Biol 56:17–24

Kumagai M, Wang L, Ueda S (2010) Genetic diversity and evolutionary relationships in genus *Oryza* revealed by using highly variable regions of chloroplast DNA. Gene 462:44–51

Kumar S, Hedges SB (1998) A molecular timescale for vertebrate evolution. Nature 392:917–920

Li C, Zhou A, Sang T (2006) Rice domestication by reducing shattering. Science 311:1936–1939

Li ZM, Zheng XM, Ge S (2011) Genetic diversity and domestication history of African rice (*Oryza glaberrima*) as inferred from multiple gene sequences. Theor Appl Genet 123:21–31

Liu F, Tembrock RL, Sun C, Han G, Guo C, Wu Z (2016) The complete plastid genome of the wild rice species *Oryza brachyantha* (Poaceae). Mitochondrial DNA B Resour 1:218–219

Lu BR, Zheng KL, Qian HR, Zhuang JY (2002) Genetic differentiation of wild relatives of rice as assessed by RFLP analysis. Theor Appl Genet 106:101–106

Ma J, Bennetzen JL (2004) Rapid recent growth and divergence of rice nuclear genomes. Proc Natl Acad Sci USA 101:12404–12410

Mariac C, Sabot F, Santoni S, Vigouroux Y, Couvreur TLP, Scarcelli N, Pouzadou J, Barnaud A, Billot C, Faye A, Kougbeadjo A, Maillol V, Martin G (2014) Cost-effective enrichment hybridization capture of chloroplast genomes at deep multiplexing levels for population genetics and phylogeography studies. Mol Ecol Resour

Meyer RS, Purugganan MD (2013) Evolution of crop species: genetics of domestication and diversification. Nature Rev Genet 14:840–852

Meyer RS, Choi JY, Sanches M, Plessis A, Flowers JM, Amas J, Dorph K, Barretto A, Gross B, Fuller DQ, Bimpong IK, Ndjiondjop M-N, Hazzouri KM, Gregorio GB, Purugganan MD (2016) Domestication history and geographical adaptation inferred from a SNP map of African rice. Nat Genet 48:1083–1088

Molina J, Sikora M, Garud N, Flowers JM, Rubinstein S, Reynolds A, Huang P, Jackson S, Schaal BA, Bustamante CD, Boyko AR, Purugganan MD (2011) Molecular evidence for a single evolutionary origin of domesticated rice. Proc Natl Acad Sci USA 108:8351–8356

Morrell PL, Buckler ES, Ross-Ibarra J (2012) Crop genomics: advances and applications. Nat Rev Genet 13:85–96

Mullins IM, Hilu KW (2002) Sequence variation in the gene encoding the 10-kDa prolamin in *Oryza* (Poaceae). 1. Phylogenetic implications. Theor Appl Genet 105:841–846

Nabholz B, Sarah G, Sabot F, Ruiz M, Adam H, Nidelet S, Ghesquière A, Santoni S, David J, Glémin S (2014) Transcriptome population genomics reveals severe bottleneck and domestication cost in the African rice (*Oryza glaberrima*). Mol Ecol 23:2210–2227

Nayar NM (1973) Origin and cytogenetics of rice. In: Caspari EW (ed) Advances in genetics. Academic, New York, pp 153–292

Nayar NM (2010) The history and genetic transformation of the African rice, *Oryza glaberrima* Steud. (Gramineae). Curr Sci 99:1681–1689

Nayar NM (2012) Evolution of the African Rice: a historical and biological perspective. Crop Sci 52:505–516

Nayar NM (2014) The origin of African Rice. In: Nayar NM (ed) Origin and phylogeny of rices. Academic, San Diego, pp 117–168

Nock CJ, Waters DLE, Edwards MA, Bowen SG, Rice N, Cordeiro GM, Henry RJ (2011) Chloroplast genome sequences from total DNA for plant identification. Plant Biotechnol J 9:328–333

Oka H-I, Morishima H (1982) Phylogenetic differentiation of cultivated rice, XXIII. Potentiality of wild progenitors to evolve the Indica and Japonica types of rice cultivars. Euphytica 31:41–50

Orjuela J, Sabot F, Chéron S, Vigouroux Y, Adam H, Chrestin H, Sanni K, Lorieux M, Ghesquière A (2014) An extensive analysis of the African rice genetic diversity through a global genotyping. Theor Appl Genet 127:2211–2223

Porteres R (1962) Berceaux Agricoles Primaires Sur le Continent Africain. J Afr Hist 3:195–210

Purugganan MD (2014) An evolutionary genomic tale of two rice species. Nat Genet 46:931–932

Rakshit S, Rakshit A, Matsumura H, Takahashi Y, Hasegawa Y, Ito A, Ishii T, Miyashita NT, Terauchi R (2007) Large-scale DNA polymorphism study of *Oryza sativa* and *O. rufipogon* reveals the origin and divergence of Asian rice. Theor Appl Genet 114:731–743

Robin GA, Dorian QF, Terence AB (2008) The genetic expectations of a protracted model for the origins of domesticated crops. Proc Natl Acad Sci USA 105:13982

Sang T, Ge S (2007a) Genetics and phylogenetics of rice domestication. Curr Opin Genet Dev 17:533–538

Sang T, Ge S (2007b) The puzzle of rice domestication. J Integ Plant Biol 49:760–768

Sanyal A, Ammiraju J, Lu F, Yu Y, Rambo T, Currie J, Kollura K, Kim H-R, Chen J, Ma J, San Miguel P, Mingsheng C, Wing R, Jackson S (2010) Orthologous comparisons of the Hd1 region across genera Reveal Hd1 gene lability within diploid *Oryza* species and disruptions to microsynteny in Sorghum. Mol Biol Evol 27:2487

Shahid Masood M, Nishikawa T, S-i Fukuoka, Njenga PK, Tsudzuki T, K-i Kadowaki (2004) The complete nucleotide sequence of wild rice (*Oryza nivara*) chloroplast genome: first genome wide comparative sequence analysis of wild and cultivated rice. Gene 340:133–139

Sotowa M, Sato YI, Sato T, Crayn D, Simon B, Waters DLE, Henry RJ, Ishikawa R, Ootsuka K, Kobayashi Y, Hao Y, Tanaka K, Ichitani K, Flowers JM, Purugganan MD, Nakamura I (2013) Molecular relationships between Australian annual wild rice, *Oryza meridionalis*, and two related perennial forms. Rice 6

Tan L, Xie D, Sun C, Li X, Liu F, Sun X, Li C, Zhu Z, Fu Y, Cai H, Wang X (2008) Control of a key transition from prostrate to erect growth in rice domestication. Nat Genet 40:1360–1364

Tang J, Xia H, Cao M, Zhang X, Zeng W, Hu S, Tong W, Wang J, Wang J, Yu J, Yang H, Zhu L (2004) A comparison of rice chloroplast genomes. Plant Physiol 135:412–420

Tang L, X-h Zou, Achoundong G, Potgieter C, Second G, D-y Zhang, Ge S (2010) Phylogeny and biogeography of the rice tribe (Oryzeae): evidence from combined analysis of 20 chloroplast fragments. Mol Phylogenet Evol 54:266–277

Thottappilly G, Rossel HW (1993) Evaluation of resistance to rice yellow mottle virus in *Oryza* species. Indian J Virol 9:65–73

Tieyan L, Mingsheng C (2014) Genome evolution of *Oryza*. Biod Sci 22:51–65

Vaughan D (1989) The genus *Oryza* L. Current status of taxonomy. Rice research paper series. International Rice Research Institute (IRRI), Manilla, Phillipines

Vaughan DA, Morishima H, Kadowaki K (2003) Diversity in the *Oryza* genus. Curr Opin Plant Biol 6:139–146

Vaughan DA, Lu B-R, Tomooka N (2008) The evolving story of rice evolution. Plant Sci 174:394–408

Vitte C, Ishii T, Lamy F, Brar D, Panaud O (2004) Genomic paleontology provides evidence for two distinct origins of Asian rice (*Oryza sativa* L.). Mol Genet Genomics 272:504–511

Wambugu P, Furtado A, Waters D, Nyamongo D, Henry R (2013) Conservation and utilization of African *Oryza* genetic resources. Rice 6:29

Wambugu PW, Brozynska M, Furtado A, Waters DL, Henry RJ (2015) Relationships of wild and domesticated rices (*Oryza* AA genome species) based upon whole chloroplast genome sequences. Sci Rep 5:13957

Wang ZY, Second G, Tanksley SD (1992) Polymorphism and phylogenetic relationships among species in the genus *Oryza* as determined by analysis of nuclear RFLPS. Theor Appl Genet 83:565–581

Wang MH, Yu Y, Haberer G, Marri PR, Fan CZ, Goicoechea JL, Zuccolo A, Song X, Kudrna D, Ammiraju JSS, Cossu RM, Maldonado C, Chen J, Lee S, Sisneros N, de Baynast K, Golser W, Wissotski M, Kim W, Sanchez P, Ndjiondjop MN, Sanni K, Long MY, Carney J, Panaud O, Wicker T, Machado CA, Chen MS, Mayer KFX, Rounsley S, Wing RA (2014) The genome sequence of African rice (*Oryza glaberrima*) and evidence for independent domestication. Nat Genet 46:982–988

Wang K, Wambugu PW, Zhang B, Wu AC, Henry RJ, Gilbert RG (2015) The biosynthesis, structure and gelatinization properties of starches from wild and cultivated African rice species (*Oryza barthii* and *Oryza glaberrima*). Carbohydr Polym 129:92–100

Waters DLE, Nock CJ, Ishikawa R, Rice N, Henry RJ (2012) Chloroplast genome sequence confirms distinctness of Australian and Asian wild rice. Ecol Evol 2:211–217

Wei X, Qiao W-H, Chen Y-T, Wang R-S, Cao L-R, Zhang W-X, Yuan N-N, Li Z-C, Zeng H-L, Yang Q-W (2012a) Domestication and geographic origin of *Oryza sativa* in China: insights from multilocus analysis of nucleotide variation of *O. sativa* and *O. rufipogon*. Mol Ecol 21:5073–5087

Wei X, Wang R, Cao L, Yuan N, Huang J, Qiao W, Zhang W, Zeng H, Yang Q (2012b) Origin of *Oryza sativa* in China inferred by nucleotide polymorphisms of organelle DNA. PLoS ONE 7:e49546

Wu W, Liu X, Wang M, Meyer RS, Luo X, Ndjiondjop M-N, Tan L, Zhang J, Wu J, Cai H, Sun C, Wang X, Wing RA, Zhu Z (2017) A single-nucleotide polymorphism causes smaller grain size and loss of seed

shattering during African rice domestication Nat. Plants 3:17064
Xuehui H, Bin H (2015) Rice domestication occurred through single origin and multiple introgressions. Nat Plants
Zang LL, Zou XH, Zhang FM, Yang Z, Ge S (2011) Phylogeny and species delimitation of the C-genome diploid species in *Oryza*. J Syst Evol 49:386–395
Zhang L-B, Ge S (2007) Multilocus analysis of nucleotide variation and speciation in *Oryza officinalis* and its close relatives. Mol Biol Evol 24:769–783
Zhang L-B, Zhu Q, Wu Z-Q, Ross-Ibarra J, Gaut BS, Ge S, Sang T (2009) Selection on grain shattering genes and rates of rice domestication. New Phytol 184:708–720
Zhu Q, Ge S (2005) Phylogenetic relationships among A-genome species of the genus *Oryza* revealed by intron sequences of four nuclear genes. New Phytol 167:249–265
Zhu B-F, Lin H, Qian Q, Sang T, Zhou B, Minobe Y, Han B, Si L, Wang Z, Zhou Y, Zhu J, Shangguan Y, Lu D, Fan D, Li C (2011) Genetic control of a transition from black to straw-white seed hull in rice domestication. Plant Physiol 155:1301–1311
Zhu T, Xu P-Z, Liu J-P, Peng S, Mo X-C, Gao L-Z (2014) Phylogenetic relationships and genome divergence among the AA—genome species of the genus *Oryza* as revealed by 53 nuclear genes and 16 intergenic regions. Mol Phylogenet Evol 70:348–361
Zou XH, Zhang FM, Zhang JG, Zang LL, Tang L, Wang J, Sang T, Ge S (2008) Analysis of 142 genes resolves the rapid diversification of the rice genus. Genome Biol 9:R49–R49

Oryza alta Swollen

4

C. Gireesh

Abstract

Oryza alta Swallen is an allotetraploid perennial wild rice belonging to the *Officinalis* complex and contains a CCDD genome with 2n = 48 chromosomes. It is closely related to other two species of the *Officinalis* complex, namely *O. latifolia* and *O. grandiglumis*. It is widely distributed in South American regions and grows along the river basins, canals in open and sunny areas. The total genome size of *O. alta* is estimated to be 1008 Mbp. Genome sequence analysis reveals that duplicated contigs in the genome are due to orthologous genes, while gene loss and gene rearrangement could be due to polyploidization. The phylogenetic analysis of mitochondrial and chloroplast using SSR markers has revealed that CC genome (*O. eichengeri*) could be the maternal parent. *O. alta* has a higher biomass and resistance to several insects and diseases. Hence, it could serve as important genetic resources for improvement of domesticated rice cultivars.

C. Gireesh (✉)
ICAR—Indian Institute of Rice Research, Ministry of Agriculture, Govt. of India, Rajendranagar, Hyderabad 500 030, Telangana, India
e-mail: giri09@gmail.com

4.1 Economic and Academic Importance

Oryza alta Swallen is a species of allotetraploid wild rice (CCDD, 2n = 48) that possesses a high biomass production and is resistant to several insects and diseases. Resistance to striped stem borer and bacterial grain rot disease was also reported in *O. alta* (Mizobuchi et al. 2016). Eizenga et al. (2009) identified the R genes for blast and sheath blight resistance from several wild species including *O. alta* which indicated the presence of novels genes for disease resistance. Development of an introgression population derived from *O. alta* will enable to utilise the useful traits in improving cultivated rice.

4.2 Botanical Description and Distribution

O. alta Swallen is an American allotetraploid species which consist of CCDD genome with 2n = 4x = 48 and belong to the *O. officinalis* complex (Fig. 4.1). It is a perennial type with plants growing up to 4 m tall as an erect herb with broad leaves (>5 cm) and a long spikelet length (>7 mm) (Vaughan 1994). *O. alta* is closely related to *O. latifolia* and *O. grandiglumis* which also share same genome and chromosome numbers. Earlier researchers have distinguished *O. alta* from *O. latifolia* based on spikelet length

T. K. Mondal and R. J. Henry (eds.), *The Wild Oryza Genomes*,
Compendium of Plant Genomes, https://doi.org/10.1007/978-3-319-71997-9_4

and awn length (Tateoka 1962). Those with a spikelet length of 8–9 mm were grouped in *O. alta,* and a spikelet length above 5 mm were placed in *O. latifolia*. *O. alta* generally possesses a culm of more than 3–15 m, an open and drooping panicle (Tateoka 1962).

The American allotetraploid species is widely distributed in Belize, Brazil, Colombia, Guyana and Paraguay (Fig. 4.2). It occurs in savanna and also in woodland and is found along river basins, streams, lake edges and canals in deep water. It usually grows in open and sunny areas from sea level to 120 m. The plants flower in two seasons in a year from March to July and from September to October (Vaughan 1994).

4.3 Cytogenetic Studies

Although *O. alta* is closely related to *O. latifolia* and *O. grandiglumis*, it possesses variations in morphological and cytological characters which differentiates it from the other two species of the *Officinalis* complex (Oliveira 1992 and Veasey et al. 2008). Tateoka (1962, 1964) grouped the wild rice accessions into *O. alta* based on morphological traits like spikelet length and awn length. Several studies have suggested to group *O. alta, O. latifolia* and *O. grandiglumis* into the same species (Sampath 1962; Kihara 1964; Gopalakrishnan and Sampath 1967). Jena and Kush (1985, 1988) also advocated that these

Fig. 4.1 *O. alta* adult plants

Fig. 4.2 Map showing the geographical distribution of *O. alta*

three species belong to same species in the *O. latifolia* complex as they have no crossability barriers and their genomes are completely homologous.

4.4 Genomics Studies

The genome size of *O. alta* is estimated to be 1008 Mbp (Ammiraju et al. 2006). Ammiraju et al. (2006) analysed the 10 genome types of the genus *Oryza* including *O. alta* (CCDD) using a set of 12 BAC libraries. The coverage of libraries in *O. alta* ranged from 10 to 14 fold. Nine out of eleven probes identified clones assembled into two contigs. These duplicated contigs were derived from orthologous genes while the gene loss and genome rearrangement is attributed to polyploidization during evolution (Ammiraju et al. 2006). Wand et al. (1992) studied the phylogenetic relationship among the *Oryza* species using RFLP probes and revealed that the American tetraploid (CCDD) species, viz. *O. latifolia, O. alta* and *O. grandiglumis,* may be of ancient origin and showed closer proximity with each other than any other diploid species of the genus *Oryza.* The study showed that the closest living diploid relative could be the CC genome *(O. eichingeri)* and EE genome *(O. qustraliensis)* species. RFLP analysis of the progeny derived from hybrids of interspecific crosses *O. sativa* and *O. alta* showed chromosome elimination, sterile triploids and recombination between chromosomes of the two species in F_2 (Long et al. 1994). On the basis gene trees of alcohol dehydrogenase genes (*Adh1* and *Adh2)* and one chloroplast gene (*mat*K), the three species with a CCDD genome form a monophyletic cluster wherein *O. alta* and *O. grandiglumis* are sister line. This infers that three allotetraploid species with a CCDD genome have originated from a single hybridization event (Ge et al. 1999). Based on the high level of sequence conservation between the diploid CC genome species and allotetraploid species (CCDD), phylogenetic study using chloroplast and mitochondrial SSR analyses of CCDD revealed that the CC genome is the maternal parent for both CCBB genome (*O. punctate* and *O. minuta*) and CCDD genome species (*O. latifolia*) (Vaughan et al. 2005). C0t-1 DNA is moderately and highly repetitive sequences of chromosomes which are present in satellite DNA, microsatellite DNA, rDNA, telomeric regions and centromeric regions. The C0t-1 DNA plays significant role in maintaining structure and characteristics of chromosomes. There is significant difference in distribution of moderately and highly repetitive sequences of CC genome between *O. alta* and *O. latifolia* as revealed by study of fluorescence in situ hybridization (FISH) with C0t-1 DNA from C genome as probe (Lan et al. 2006). Further, it revealed that CC and DD genomes are more closely related with each other in *O. alta* than those in *O. latifolia*.

4.5 Future Prospects

Considering the wide diversity of environment in which *O. alta* is grown, it is expected to have useful genetic variability which can be explored to improve the cultivated rice. Although genomes of 16 species of wild rice are being sequenced, the genome of *O. alta* is not yet sequenced. Availability of genome sequence will help in identification of thousands of genes for various traits, understanding the gene functions, the regulation of gene expression, and also helps in identification of underlying factors controls quantitative traits like yield, drought tolerance, cold/heat tolerance. This in turn hastens breeding program for genetic enhancement of cultivated rice.

References

Ammiraju JSS, Meizhong Luo JL, Goicoechea WW et al. (2006) The Oryza bacterial artificial chromosome library resource: construction and analysis of 12 deep-coverage large-insert BAC libraries that represent the 10 genome types of the genus Oryza. Genome Res 16:140–147

Eizenga GS, Agrama HS, Lee FN, Jia Y (2009) Exploring genetic diversity and potential novel disease resistance genes in a collection of rice (*Oryza* spp.) wild relatives. Genet Resour Crop Evol 56:65–76

Ge S, Sang T, Lu B-R, Hong D-Y (1999) Phylogeny of rice genomes with emphasis on origins of allotetraploid species. PNAS 96:14400–14405

Gopalakrishnan R, Sampath S (1967) Taxonomic status and origin of American tetraploid species of the series *Latifoliae* Tateoka in the genus *Oryza*. Indian J Agr Sci 37:465475

Jena KK, Khush GS (1985) Genome analysis of *Oryza officinalis* complex. RGN 2:4445

Jena KK, Khush GS (1988) Cytogenetic relationships among the three species of *Oryza latifolia* complex. RGN 5:7475

Kihara H (1964) Need for standardization of genetic symbols and nomenclature in rice. In: IRRI (ed) Rice genetics and cytogenetics. Elsevier, Amsterdam, London, New York, p 311

Lan WW, He GC, Wu SJ, Qin R (2006) Comparative analysis of *Oryza sativa*, *O. officinalis* and *O. meyeriana* genome with Cot-1 DNA and genomic DNA. Sci Agric Sin 39(6):1083–1090

Long M, Zhou Q, Wang X, Hu H, Zhu L (1994) RFLP analysis of progeny from hybrid of Oryza alta×O. sativa. Acta Botanica Sinica:36(1):11–18

Mizobuchi R, Fukuoka S, Tsushima S, Yano M, Sato H (2016) QTLs for resistance to major rice diseases exacerbated by global warming: brown spot, bacterial seedling rot, and bacterial grain rot. Rice. 9(23):1–12

Oliveira GCX (1992) Padrões de variaçaofenotípica e ecologia de Oryzae (Poaceae) selvagens da Amazônia. MSc Thesis, Escola Superior de Agricultura "Luiz de Queiroz", Universidade de São Paulo, Piracicaba, Brazil

Sampath S (1962) The genus *Oryza*: its taxonomy and species relationships. Oryza 1(1):129

Tateoka T (1962) Taxonomic Studies of *Qryza* I. *O. latifolia* Complex. Bot Mag Tokyo 75:418–427

Tateoka T (1964) Taxonomic studies of the genus *Oryza*. In: IRRI (ed) Rice Genetics and cytogenetics. Elsevier, Amsterdam, London, New York, p 1521

Vaughan AD (1994) The wild relatives of rice: a genetic resource handbook. IRRI publication, Los Baños, p 147

Vaughan DA, Kadowaki K, Kaga A, Tomooka, N (2005) On the phylogeny and biogeography of the genus Oryza. Breed Sci 55:113–122

Veasey EA, da Silva EF, Schammass EA, Oliveira GCX, Ando A (2008) Morphoagronomic genetic diversity in American wild rice species. Braz Arch Biol Technol 51(1):95–104

Wang ZY, Second G, Tanksley SD (1992) Polymorphism and phylogenetic relationships among species in the genus *Oryza*as determined by analysis of nuclear RFLPs. Theor Appl Genet 83:565–581

Oryza australiensis Domin

5

Robert J. Henry

Abstract

Oryza australiensis is a perennial widely distributed across northern Australia. This species has a large and poorly characterized genome with a high content of repetitive elements. The large plants have rhizomes that help them survive the dry season. The grains have starch properties that differ significantly from those of other *Oryza* species. The sequence of the nuclear genome has not been reported, but a chloroplast genome sequence is available. This species represents a potential source of resistance to abiotic and biotic stress and grain traits for rice breeding.

5.1 Economic and Academic Importance

Oryza australiensis is a wild rice in Australia that is a unique representative of the E genome in the genus *Oryza*. The species is illustrated in Fig. 5.1, depicting a plant growing in North Queensland (Australia). It has traits such as a rhizome that allows the plants to survive in the dry season and makes it an important potential source of novel genes for rice improvement (Henry et al. 2010). The relationships to other *Oryza* species suggest that this species may provide important insights into the evolution of the *Oryza* genus. The species also has desirable grain characteristics (Tikapunya et al. 2016) that suggest it may have some value as a crop if domesticated.

5.2 Botanical Descriptions and Distribution

O. australiensis was described more than 100 years ago (Domin 1915). It is perennial and rhizomatous in nature. The description in the Flora of Australia (Kodela 2009) includes the details provided in Table 5.1.

O. australiensis (Fig. 5.1) can usually be readily distinguished from other *Oryza* in the field on the basis of the long (up to 50 cm) open panicles and short awns. The rhizome (Fig. 5.2) is also distinctive.

O. australiensis is widespread across northern Australia (Fig. 5.3). It has not been reported outside Australia. The species is found in seasonally wet locations extending further from sites of permanent water than the other perennial *Oryza* species found in these areas. The rhizome

R. J. Henry (✉)
Queensland Alliance for Agriculture and Food Innovation, University of Queensland, Brisbane, QLD 4072, Australia
e-mail: robert.henry@uq.edu.au

T. K. Mondal and R. J. Henry (eds.), *The Wild Oryza Genomes*,
Compendium of Plant Genomes, https://doi.org/10.1007/978-3-319-71997-9_5

Fig. 5.1 Adult plants of *O. australiensis* growing in northern Australia

Table 5.1 Morphological description of *O. australiensis*

Name	Dimensions
Clums	0.8–2.5 (−3) m
Leaves: ligule	2–8 mm
Blade	2–13 cm long 4–12 (−15) mm wide
Panicles	13–40 (−50) cm long
Spikelets	6–7.5 (−8.5) mm long
Awn	(10–) 15–55 (−65) mm long
Anthers	3–5 mm long
Caryopsis	obloid to ellipsoid 4.5–6.4 mm long

of this species (Fig. 5.2) allows the plants to survive the dry season away from areas of permanent water.

5.3 Cytological Study

Chromosome morphology (2n = 24) was reported by Wu and Li (1964) using an F_1 cross with *O. sativa*. The 12 chromosomes were all longer in *O. australiensis* and had larger centromeres. Data on the relative length of chromosome arms was also presented (Table 5.2).

5.4 Physiological Studies

Analysis of grain characteristics has shown that *O. australiensis* has shorter grains than that other wild species of rice found in northern Australia (Tikapunya et al. 2016). More recent research has investigated the starch properties and has shown the presence of a high amylose content (Tikapunya et al. 2017). The starch properties differed significantly from those of other rice species with more short chain amylose, a low pasting viscosity, and lower retrogradation. The

Fig. 5.2 Rhizomes of *O. australiensis*

Fig. 5.3 Geographical distribution of *O. australiensis*

gelatinization temperature was high, but the time to gelatinization was low. Kasem et al. (2012) compared sequences of GBSS1, SSIIa, and SBEIIB from *O. australiensis* with those from other *Oryza* species revealing very significant differences in gene sequence that suggested divergent properties. Grain morphology of *O. australiensis* was compared to that of other *Oryza* species by light (Kasem et al. 2010) and electron microscopy (Kasem et al. 2011). *O. australiensis* has shorter grains than AA genome species from the same areas.

5.5 Molecular Mapping of Genes and QTLs

There are no maps of the *O. australiensis* genome. However, QTLs for brown planthopper resistance from *O. australiensis* have been mapped in crosses with *O. sativa* (Hu et al. 2015). These indicate the position of insertion of the introgressed genes from *O. australiensis*. A QTL on chromosome 4 had the largest effect on resistance.

Table 5.2 Size or length of chromosomes of *O. australiensis* compared to *O. sativa* (Wu and Li 1964)

Chromosome	*O. sativa*	*O. australiensis*
1	46	65
2	39	56
3	37	46
4	30	44
5	29	42
6	27	34
7	26	30
8	24	27
9	23	25
10	23	24
11	20	24
12	18	22

5.6 Structural and Functional Genomic Resources Developed

Chromosome addition lines have been developed in an *O. sativa* background (Multani et al. 1994). Monosomic additional lines were produced for chromosomes 1, 4, 5, 7, 9, 10, 11, and 12. Phenotypic characterization of these lines should indicate the chromosomal location of traits from *O. australiensis*. Lines from this experiment would be a useful source of novel traits for rice improvement.

5.7 Requirement of Whole-Genome Sequencing

The high content of repetitive elements will make genome sequencing and assembly for this species a significant challenge. Advances in sequencing technology are making the sequencing of this genome more feasible.

5.8 Repetitive Sequences

The genome of *O. australiensis* which is estimated to be 965 Mbp is the largest among the diploid genomes in the *Oryza* genus (Piegu et al. 2006). This is more than twice the size of *O. sativa*. The large size of the genome has been explained by two rounds of genome expansion (Piegu et al. 2006). Expansion of three LTR-retrotransposon families has resulted in more than 90,000 retrotransposons doubling the genome size since the species diverged. The three retrotransposon elements, *RIRE1, Kangourou*, and *Wallabi,* account for about 60% of the genome (Table 5.3). Differences in transposon amplification seem to explain most of the differences in genome size within the *Oryza* (Lu et al. 2009). The repetitive elements of *O. australiensis* have been compared with those of other *Oryza* species (Gill et al. 2010).

Table 5.3 Retrotransposons in the genome of *O. australiensis* (Piegu et al. 2006)

Retrotransposons	Size in bp	Number of copies (full elements)
RIRE1	8300	30,000
Kangourou	9200	9500
Wallabi	9000	27,000

5.9 Organelle Genome

Nock et al. (2011) used a chloroplast genome based on mapping to a *O. sativa* reference to show the relationship of *O. australiensis* to other taxa. Variation in the chloroplast genome within the species has not been explored. The complete chloroplast genome of *O. australiensis* was reported by Wu and Ge (2016). They reported a 135,224 bp genome (39% GC) with a pair of 25,776 bp inverted repeats separated by a large single-copy region (LSC) of 82,212 bp and a small single-copy region (SSC) of 12,470 bp. Gene annotation identified 76 protein-coding genes, 4 rRNA genes, and 30 tRNA genes.

The mitochondrial genome has not been reported.

5.10 Future Prospects

O. australiensis has a large genome with a high frequency of repetitive elements. This will make the generation of a high-quality genome sequence for this species difficult. However because of the unique position within the *Oryza* and some of the unique attributes of this species (Kasem et al. 2010, 2014), a genome sequence will be of high value to both biology and rice improvement and production. It may be a source of unique alleles of important genes (Kasem et al. 2012). This species might also have value if domesticated directly (Henry et al. 2016). Advances in breeding technology make this an important resource for breeding novel rice types (Anacleto et al. 2015). The availability of more data on the genomes and characteristics of *O. australiensis* will increase the evidence available to support both in situ and ex situ conservation (Henry 2014).

References

Anacleto R, Cuevas RP, Jimenez R, Llorente C, Nissila E, Henry R, Sreenivasulu N (2015) Prospects of breeding high-quality rice in the post-genomic era. Theor App Genet 128(8):1449–1466. https://doi.org/10.1007/s00122-015-2537-6

Domin K (1915) Bibliotheca Botanica 20:333

Gill N, San Miguel P, Dhillon BDS, Abernathy B, Kim H, Stein L, Ware D, Wing R, Jackson SA (2010) Dynamic *Oryza* Genomes: repetitive DNA sequences as genome modelling agents. Rice 3:251–269

Henry RJ (2014) Sequencing crop wild relatives to support the conservation and utilization of plant genetic resources. Plant Genetic Res Charact Util 12: S9–S11. https://doi.org/10.1017/S1479262113000439

Henry RJ, Rice N, Waters DLE, Kasem S, Ishikawa R, Dillon SL, Crayn D, Wing R, Vaughan D (2010) Australian Oryza: utility and conservation. Rice 3:235–241

Henry RJ, Rangan P, Furtado A (2016) Functional cereals for production in new and variable climates. Curr Opin Plant Biol 30:11–18

Hu J, Xiao C, Cheng M-X, Gao G-J, Zhang Q-L, He Y-Q (2015) A new finely mapped *Oryza australiensis*-derived QTL in rice confers resistance to brown planthopper. Gene 561:132–137

Kasem S, Waters DLE, Rice N, Shapter FM, Henry RJ (2010) Whole grain morphology of Australian rice species. Plant Genetic Res Charact Util 8:74–81

Kasem S, Waters DLE, Rice N, Shapter FM, Henry RJ (2011) The endosperm morphology of rice and its wild relatives as observed by scanning electron microscopy. Rice 4:12–13

Kasem S, Waters DLE, Henry RJ (2012) Analysis of genetic diversity in starch genes in the wild relatives of rice. Tropical Plant Biol 5:286–308

Kasem S, Waters DLE, Ward RM, Rice N, Henry RJ (2014) Wild *Oryza* grain physico-chemical properties. Tropical Plant Biol 7:13–18

Kodela PG (2009) Oryza. Flora of Australia vol 44, A Poaceae 2. ABRS/CSIRO, Melbourne, Australia, pp 361–368

Lu F, Ammiraju JSS, Sanyal A, Zhang S, Song R, Chena J, Lia G, Sui Y, Song X, Cheng Z, de Oliveira AC, Bennetzen JL, Jackson SA, Wing RA, Chena M (2009) Comparative sequence analysis of MONOCULM1-orthologous regions in 14 *Oryza* genomes. PNAS doi/. https://doi.org/10.1073/pnas.0812798106

Multani DS, Jena KK, Brar DS, de las Reyes BG, Angele ER, Khush GS (1994) Development of monosomic alien addition lines and introgression of genes from *Oryza australiensis* Domin to cultivated rice *O sativa* L. Theor Appl Genet 88:102–109

Nock C, Waters DLE, Edwards MA, Bowen S, Rice N, Cordeiro GM, Henry RJ (2011) Chloroplast genome sequence from total DNA for plant identification. Plant Biotechnol J 9:328–333

Piegu B, Guyot R, Picault N, Roulin A, Saniyal A, Kim H, Collura K, Brar DS, Jackson S, Wing RA, Panaud O (2006) Doubling genome size without polyploidization: dynamics of retrotransposition-driven genomic expansions in *Oryza australiensis*, a wild relative of rice. Genome Res 16:1262–1269

Tikapunya T, Fox G, Furtado A, Henry R (2016) Grain physical characteristic of the Australian wild rice. Plant Genetic Res 15:1–12

Tikapunya T, Zou W, Yu W, Powell P, Fox G, Furtado A, Henry RJ, Gilbert R (2017) Molecular structures and properties of starch of Australian wild rice. Carbohydrate Poly. 172:213–222

Wu H-K, Li SSY (1964) Chromosome morphology of *Oryza sativa* and *Oryza australiensis* and their paining in the F1 hybrid at earlier meiosis. Bot Bull Acad Scinica 5:162–169

Wu Z, Ge S (2016) The whole chloroplast genome of wild rice (*Oryza australiensis*). Mitochondrial DNA Part A 27(2):1062–1063. https://doi.org/10.3109/19401736.2014.928868

Oryza barthii A. Chev

6

Peterson W. Wambugu and Robert J. Henry

Abstract

Oryza barthii (formerly known as *Oryza stapfii or O. breviligulata*) is a part of the *Oryza* primary gene pool and possesses useful genetic diversity which is important for rice improvement. It is however poorly studied, and its genetic potential remains largely untapped. Inadequate understanding of the genetic mechanisms underlying various important traits such as amylose content remains one of the challenges hindering the use of this species in rice improvement. Though *O. barthii* germplasm collections exist, available data indicate collection gaps that need to be addressed before these resources are lost from the habitats where they are currently found. Various genomic resources among them whole genome sequences are now available, and these have the potential for promoting the utility of this species in rice improvement. This chapter has reviewed the latest advances in genetics and genomics for *O. barthii*.

6.1 Academic and Economic Importance

Oryza barthii is the presumed progenitor of African domesticated rice (*O. glaberrima*), and it is found in Africa. *O. barthii* is most closely related to the AA genome of Oryza species of Asia, such as *O. rufipogon,* that were the progenitors of Asian domesticated rice (*O. sativa*) (Wambugu et al. 2015). It has useful genetic diversity which is important for rice improvement and critical to the study of domestication of rice in Africa (Wambugu et al. 2013). The domestication process left behind enormous genetic diversity in *O. barthii* that is important for environmental resilience, adaptive capacity, and resistance to various biotic and abiotic stresses. As will be highlighted later in this chapter, *O. barthii* possesses resistance to a variety of pests and diseases. This diversity is important in broadening the genetic base of domesticated rice species.

It is a diploid (2n = 24) AA genome species. However, this species is not well-studied, and its genetic potential therefore remains poorly understood and largely untapped. Compared to some of the wild *Oryza* taxa such as *O. rufipogon* and *O. nivara,* very little is known about

P. W. Wambugu
Kenya Agricultural and Livestock Research Organization (KALRO), Genetic Resources Research Institute, 781-00902, Kikuyu, Kenya

R. J. Henry (✉)
Queensland Alliance for Agriculture and Food Innovation, University of Queensland, Brisbane, QLD 4072, Australia
e-mail: robert.henry@uq.edu.au

T. K. Mondal and R. J. Henry (eds.), *The Wild Oryza Genomes*,
Compendium of Plant Genomes, https://doi.org/10.1007/978-3-319-71997-9_6

O. barthii. The continued advances in genomics are offering opportunities for increasing our knowledge on its potential genetic value as well as enhancing its deployment in rice improvement. This chapter highlights the advances in the genetics and genomics of *O. barthii*.

6.2 Botanical Descriptions and Geographical Distribution

O. barthii is an erect to semierect annual herb which is primarily self-pollinating (Vaughan 1994). It produces aerial roots from its lower nodes. Its glumes are either absent or obscure. Its inflorescence is an open panicle with primary panicle branches either appressed or ascending. Its fruit is a caryopsis with a sticking pericarp (Clayton et al. 2006). The following details (Table 6.1) are from Flora Zambesiaca (2013). *O. barthii* is depicted in Fig. 6.1. It is widespread in West Africa but is also found in some parts of eastern and southern Africa (Fig. 6.2). *O. barthii* is usually found growing in shallow ponds, stagnant or slowly flowing water and rice fields (Flora Zambesiaca 2013).

6.3 Genomics Resources

6.3.1 Whole Genome Sequences

A reference sequence is perhaps the most valuable genomic resource in promoting the conservation and utility of plant genetic resources. The *O. sativa* Nipponbare reference sequence provided great opportunities for exploring the genetic potential in the *Oryza* genus. It was however acknowledged that due to the enormous genetic diversity found in the *Oryza* genus, one reference sequence was not enough to understand and/or exploit this genetic potential. The International *Oryza* Mapping and Alignment Project (I-OMAP) was therefore formed with the aim of developing reference genome sequences for additional species in the genus (Goicoechea et al. 2010). The genome of *O. barthii* was first sequenced by I-OMAP and though these I-OMAP assembled genome sequences have been deposited in GenBank, they are yet to be published. An integrated approach involving whole genome shot-gun sequencing and physical maps was used in the sequencing and assembly of this genome (Table 6.2). It has an estimated 34,575 protein-coding genes.

A de novo sequence of *O. barthii* was later assembled and reported by Zhang et al. (2014) based upon 18.9 Gbp of shot-gun sequencing data representing an estimated 51-fold coverage of the genome. The genome was assembled using SOAP de novo giving a scaffold N50 of 237,573 bp and a contig N50 of 16,126 bp providing an estimated 89% coverage of the 376 Mbp genome. An estimated 42,283 protein-encoding genes were predicted in this genome assembly. Zhang et al. (2014) estimated that 29.8% of the genome was transposable elements. This was much lower than that of its close relative *O. longistaminata* where almost half of the genome is composed of transposable elements (Zhang et al. 2015). Baldrich et al. (2016) reported mRNA in *Oryza* species including *O. barthii*.

Table 6.1 Description of *O. barthii*

Morphological characteristic	Description
Culms	0.6–1.2 m
Leaves: ligule	3–6 (−9) mm
Leaves: blade	15–45 cm long 4–14 mm wide
Panicles	20–35 cm long
Spikelets	8–10.5 mm long
Awn	(6.5)8–16 (−19) cm long

Fig. 6.1 *O. barthii* plant at flowering stage [*Source* Marie-Noelle Ndjiondjop, Africa Rice Center]

Fig. 6.2 Distribution of *O. barthii* [*Source* NARO Genebank—Illustrated Plant Genetic Resources Database]

6.3.2 Transcriptome

Transcriptome analysis is useful in defining the expression patterns of genes and determining the genetic basis of important traits. RNA-Seq analysis was used to show differential expression of the shattering locus (*Sh4*) in *O. barthii* and *O. glaberrima* (Wang et al. 2014). When assembling the genome of *O. barthii*, RNA-Seq data was used to support gene annotation. This

Table 6.2 Features of *O. barthii* genomes sequenced and assembled by different initiatives

Feature	I-OMAP genome	Genome assembled by Zhang et al. (2014)
Estimated genome size	411 Mb	376 Mb
Assembled genome size	308 Mb	335 Mb
Number of protein-coding genes	34,575	42,283
Contig N50	18,930 bp	16,126 bp
Sequencing platform	454 and Illumina	Illumina
Sequencing strategy	Whole genome shot-gun sequencing and physical map integration	Whole genome shot-gun sequencing

data revealed more functional variation in *O. barthii* as a result of alternative splicing that was observed in a significant portion of genes (Zhang et al. 2014). In the same study, RNA-Seq data was used to confirm the loss of agronomically important genes during the evolutionary process, among them the *S5* locus which is responsible for female sterility. A transcriptome population genomics study revealed extremely low genetic diversity in *O. glaberrima* and *O. barthii*. This low diversity could be attributed to a severe domestication bottleneck in *O. glaberrima,* while in *O. barthii,* it could be due to its high self-pollination (Nabholz et al. 2014).

6.3.3 Organelle Genome

Chloroplast genome sequences for *O. barthii* have been reported by Wambugu et al. (2015) including data for multiple accessions. Its size ranges from 134,596 to 134,674 Kb. These sequences have been used to reconstruct the phylogenetic and evolutionary relationships among the species that constitute the *Oryza* primary gene pool. This analysis showed a close relationship between *O. barthii* and African rice, thus supporting the popular theory that *O. barthii* is the putative progenitor of African rice (Wambugu et al. 2015). Analysis of the chloroplast genome for various *Oryza* species revealed that *O. barthii* has a total of 420 microsatellites and 104 unique genes (Asaf et al. 2017). Mitochondrial sequences have not been reported.

6.3.4 Comparative Genomics

The increased availability of whole genome data for a wide range of species is enabling multi-species comparative analysis. Comparative genomics is supporting many areas of plant science among them structural, functional, and evolutionary analyses. The genome of *O. glaberrima* has been reported (Wang et al. 2014). Re-sequencing (threefold coverage) of 94 *O. barthii* accessions was used to infer that *O. glaberrima* had been domesticated from *O. barthii* populations in Nigeria. Zhang et al. (2014) compared the genomes of *O. barthii* and 4 other AA genome species. This comparative analysis revealed extensive structural variation including segmental duplication between the various genomes. This analysis also reported gene loss and gain between various species and helped shed light on the evolutionary impact of these events on adaptation, speciation, and trait evolution in *Oryza*. Comparative genomics has been used to provide information on the syntenic relationships between genomes which is important in extending our knowledge on evolutionary history between various species. Goicoechea et al. (2010) reported extensive colinearity between the genomes of *O. barthii* and *O. sativa*.

Despite this colinearity, a high frequency of reversals and trans-chromosomal events in terms of translocations and transpositions has been observed between these two species. Genome expansions and contractions in *O. barthii* were also reported in nearly equal proportions.

6.4 Molecular Mapping of Genes and QTLs

Though the genetic potential of *O. barthii* as a source of useful genes is known, the genetic mechanisms underlying these useful traits are not well studied. This poor understanding on the genetic basis of useful traits therefore hinders utility of this species in rice improvement. The transfer of genes from wild species to cultivated rice is usually hindered by hybrid sterility. The *S1* locus has been identified as the main locus responsible for hybrid sterility in crosses between *O. sativa* and O. *glaberrima* as well as *O. sativa* and *O. longistaminata* (Sano 1990; Chen et al. 2009). However, the mechanism of hybrid sterility in *O. barthii* is not well known. Recently, using a cross between *O. sativa* and *O. barthii*, Yang et al. (2016) identified a QTL for hybrid sterility designated as *qHS6-c* on chromosome 6. Similarly, QTLs for hybrid sterility were identified in the same interval in crosses between *O. sativa* and *O. nivara* as well as *O. sativa* and *O. rufipogon*. These suggest that the loci controlling hybrid sterility are present in the same genomic region in AA genome species. Fine mapping of the bacterial blight resistance gene was done using BC_2F_2, BC_2F_3, and BC_2F_4 populations obtained from crosses between *O. barthii* and *O. glaberrima*. Physical mapping was conducted using e-landing approach. The gene was mapped to chromosome 6S at a distance of 19.3 cM from RM197 (Gupta 2010). Chromosome segment substitution lines (CSSLs) developed within the framework of a Generation Challenge project, with *O. barthii* as a donor, will be useful in identifying genomic regions associated with various useful traits (Gutierrez et al. 2010; McCouch et al. 2010). Identification of genomic regions underlying various useful traits will be important for more targeted transfer of these traits from wild to cultivated species.

6.5 Genetic Potential

6.5.1 Physiological Studies

The starch properties of *O. barthii* have been compared with those of domesticated rice such as *O. glaberrima* and *O. sativa* (Wang et al. 2015). The starch was found to contain more amylose and smaller amylose molecules than domesticated rice. The health benefits of high amylose foods are increasingly being recognized, and *O. barthii* seems to have potential for improving starch-related traits, particularly amylose content in domesticated rice. A study on nitrogen use efficiency among wild and cultivated rice genotypes revealed that *O. barthii* has higher dry mater accumulation and higher nitrogen use efficiency than *O. sativa* genotypes under conditions of low nitrogen (Hamaoka et al. 2013). This African wild species therefore seems to be a valuable resource for breeding rice for low input agriculture. In a closely related study, though *O. barthii* was found to have lower silicon uptake than cultivated rice, it has the capacity to optimize silicon accumulation in the shoots, an attribute that is vital for healthy rice production (Mitani-Ueno et al. 2014).

6.5.2 Impact on Plant Breeding Including Pre-breeding

As already highlighted, *O. barthii* has immense genetic potential for various traits which are important for improving domesticated rice (Table 6.3). It is a source of heat or drought resistance traits (Sanchez et al. 2014). Long grain and awnless genotypes of *O. barthii* have been reported (Aladejana and Faluyi 2007). Though resistance to bacterial blight has been reported in a number of *Oryza* cultivated and wild species including *O. barthii*, this resistance has been found to breakdown due to emergence of more virulent forms of *Xanthomonas* bacteria.

Table 6.3 Useful or potentially useful traits found in *O. barthii*

Trait	Reference
Drought resistance/avoidance	Brar and Khush (2002), Sanchez et al. (2014)
Resistance to green leaf hopper	Brar and Khush (2002)
Resistance to bacterial blight	Brar and Khush (2002)
Resistance to bacterial leaf streak	McCouch et al. (2013)
Resistance to blast (*Pyricularia grisea*)	Eizenga et al. (2004)
Moderate resistance to sheath blight (*Rhizoctonia solani*)	Prasad and Eizenga (2008)
Resistance to nematode *Heterodera sacchari/schachtii*	Reversat and Destombes (1998)
Resistance to rice tungro virus	Jain et al. (1989)
Resistance to stem borer attack	AfricaRice (2010)

However, *O. barthii* has recently been found to have resistance to some of the newly evolved pathotypes (Neelam et al. 2016), thereby making it a valuable resource in the management of this highly damaging disease. Interspecific crosses between *O. barthii* and *O. sativa* possessing valuable attributes among them high yield, aroma, and very early maturity have been developed at African Rice Center. With early maturity of less than 90 days, these lines are able to avoid drought which is one of the major challenges currently facing rice production (AfricaRice 2013).

6.6 Future Prospects

More extensive collection, sequencing, and re-sequencing of *O. barthii* will provide a key resource for rice improvement globally. As the progenitor, *O. barthii* is the primary genetic resource for *O. glaberrima* genetic improvement, but it also represents a genetic resource for *O. sativa* as part of the wider AA gene pool. Data on in-country collection and conservation of *O. barthii* genetic resources presented by Wambugu et al. (2013) suggests the existence of collection gaps. Numerical assessment of various collections held in national and international genebanks shows that *O. barthii* is in most cases represented by very few or no accessions. This points to under representation of *O. barthii* in the existing national and international collections. Similarly, the analysis of herbarium data presented by the same authors lends support on the presence of clear collection gaps that need to be addressed. While herbarium specimens have been collected from some countries, there appear to be no collections of *O. barthii* undertaken in these countries. Ex situ and in situ conservation approaches should complement each other, but currently there are no documented in situ conservation sites for *O. barthii.* These collection gaps and the lack of sound conservation programs expose these genetic resources to risks of genetic erosion. There is need for coordinated efforts at the national, regional, and global levels to ensure that these genetic resources are collected and properly conserved. Analysis of genomes can provide information on novel populations that need to be prioritized for in situ conservation and ex situ collection (Henry 2014). Conserved germplasm is poorly characterized, and this arguably acts as the greatest hindrance to the use of these collections. Several initiatives aimed at sequencing the genomes of accessions conserved in various genebanks are currently underway. This genomic data will allow detailed characterization of these resources, thereby providing information on their potential value.

References

Africa Rice Center (AfricaRice) (2010) New breeding directions at AfricaRice: beyond NERICA. Cotonou,

Benin: 24 pp. http://www.africarice.org/publications/Beyond_Nerica.pdf

Africa Rice Center (AfricaRice) (2013) Unlocking the potential of wild *Oryza barthii*. https://africarice.blogspot.co.ke/2013/04/. Accessed on 02 May 2017

Aladejana F, Faluyi JO (2007) Agro botanical characteristics of some West African indigenous species of the a genome complex of the Genus *Oryza* Linn. Int J Botany 3:229–239

Asaf S, Waqas M, Khan AL, Khan MA, Kang S-M, Imran QM, Shahzad R, Bilal S, Yun B-W, Lee I-J (2017) The complete chloroplast genome of wild rice (*Oryza minuta*) and its comparison to related species. Front Plant Sci 8:304. https://doi.org/10.3389/fpls.2017.00304

Baldrich P, Hsing Y-LC, San Segundo B (2016) Genome-wide analysis of polycistronic microRNAs in cultivated and wild rice. Genome Biol Evol 8:1104–1114. https://doi.org/10.1093/gbe/evw062

Brar DS, Khush GS (2002) Transferring genes from wild species into rice. CABI Publishing, Wallingford, UK, pp 197–217

Chen Z, Hu F, Xu P, Li J, Deng X, Zhou J, Li F, Chen S, Tao D (2009) QTL analysis for hybrid sterility and plant height in interspecific populations derived from a wild rice relative, *Oryza longistaminata*. Breed Sci 59:441–445

Clayton WD, Vorontsova MS, Harman KT, Williamson H (2006 onwards). GrassBase—the online world grass flora. http://www.kew.org/data/grasses-db.html. Accessed 09 May 2017

Eizenga GC, Lee FN, Jia Y (2004) Identification of blast resistance genes in wild relatives of rice (*Oryza* spp.) and newly introduced rice (*O. sativa*) lines. Research series—Arkansas agricultural experiment station No 517: 29–36. http://arkansasagnews.uark.edu/517-3.pdf

Flora Zambesiaca (2013) Flora Zambesiaca 10(4). Kew, UK: Kew Royal Botanic Gardens. http://apps.kew.org/efloras/search.do. Accessed on 02 May 2017

Goicoechea JL, Ammiraju JSS, Marri PR, Chen M, Jackson S, Yu Y, Rounsley S, Wing RA (2010) The future of rice genomics: sequencing the collective *Oryza* Genome. Rice 3:89–97

Gupta V (2010) Fine mapping of baterial blight resistance genes from *Oryza glaberrima* (Steud.) and *O. barthii* (A. Chev). Msc thesis. Department of plant breeding and genetics. Punjab Agricultural University, Ludhiana, p 131

Gutierrez AG, Carabali SJ, Giraldo OX, Martinez CP, Correa F, Prado G, Tohme J, Lorieux M (2010) Identification of a rice stripe necrosis virus resistance locus and yield component QTLs using *Oryza sativa* x *O. glaberrima* introgression lines. BMC Plant Biol 10:6

Hamaoka N, Uchida Y, Tomita M, Kumagai E, Araki T, Ueno O (2013) Genetic variations in dry matter production, nitrogen uptake, and nitrogen use efficiency in the AA genome oryza species grown under different nitrogen conditions. Plant Prod Sci 16:107

Henry RJ (2014) Sequencing of wild crop relatives to support the conservation and utilization of plant genetic resources. Plant Genet Resour 12:S9

Jain RK, Mishra MD, Singh J (1989) *Oryza barthii*—a promising resistance source to rice tungro virus. Indian Phytopathol 42:551–552

McCouch SR, Brondani C, Manneh B, Ndjiondjop M, Martínez CP, Diago M (2010) Exploring natural genetic variation: developing genomic resources and introgression lines for four AA genome rice relatives: final technical report. CIAT, Cali, CO, p 18

McCouch S, Wing R, Semon M, Venuprasad R, Atlin GN, Sorrells ME, Jannink JL (2013) Making rice genomics work for Africa. In: Wopereis M (ed) Realizing Africa's rice promise. CABI, Boston, MA, pp 108–127

Mitani-Ueno N, Ogai H, Yamaji N, Ma JF (2014) Physiological and molecular characterization of Si uptake in wild rice species. Physiol Plant 151:200–207

Nabholz B, Sarah G, Sabot F, Ruiz M, Adam H, Nidelet S, Ghesquière A, Santoni S, David J, Glémin S (2014) Transcriptome population genomics reveals severe bottleneck and domestication cost in the African rice (*Oryza glaberrima*). Mol Ecol 23:2210–2227

Neelam K, Lore JS, Kaur K, Pathania S, Kumar K, Sahi G, Mangat GS, Singh K (2016) Identification of resistance sources in wild species of rice against two recently evolved pathotypes of *Xanthomonas oryzae* pv oryzae. Plant Genet Resour 15(6):1–5

Prasad B, Eizenga GC (2008) Rice sheath blight disease resistance identified in *Oryza* spp. accessions. Plant Dis 92:1503–1509

Reversat G, Destombes D (1998) Screening for resistance to *Heterodera sacchari* in the two cultivated rice species, *Oryza sativa* and *O. glaberrima*. Fundam Appl Nematol 21:307–317

Sanchez PL, Wing RA, and Brar DS (2014) The Wild Relative of Rice: Genomes and Genomics In: Zhang Q, Wing RA (eds) Genetics and genomics of rice, plant genetics and genomics: crops and models 5, pp 9–25, Springer, New York. https://doi.org/10.1007/978-1-4614-7903-1_2

Sano Y (1990) The genic nature of gamete eliminator in rice. Genetics 125:183–191

Vaughan D (1994) The wild relatives of rice: a genetic resources handbook. IRRI, Phillipines

Wambugu P, Furtado A, Waters D, Nyamongo D, Henry R (2013) Conservation and utilization of African Oryza genetic resources Rice 6:29

Wambugu PW, Brozynska M, Furtado A, Waters DL, Henry RJ (2015) Relationships of wild and domesticated rice species based on whole chloroplast genome sequences. Sci Rep 5:13957. https://doi.org/10.1038/srep13957

Wang M, Yu Y, Haberer G, Marri PR, Fan C, Goicoechea JL, Zuccolo A, Song X, Kudrna D, Ammiraju JSS, Cossu RM, Maldonado C, Chen J, Lee S, Sisneros N, de Baynast K, Golser W, Wissotski M, Kim W, Sanchez P, Ndjiondjop M-N, Sanni K,

Long M, Carney J, Panaud O, Thomas Wicker T, Machado CA, Chen M, Mayer KFX, Rounsley S, Wing RA (2014) The genome sequence of African rice (*Oryza glaberrima*) and evidence for independent domestication. Nature Genet 46:982–988

Wang K, Wambugu PW, Zhang B, Wu AC, Henry RJ, Gilbert RG (2015) The biosynthesis, structure and gelatinization properties of starches from wild and cultivated African rice species (*Oryza barthii* and *Oryza glaberrima*). Carbohydr Polym 129:92–100. https://doi.org/10.1016/j.carbpol.2015.04.035

Yang Y, Zhou J, Li J, Xu P, Zhang Y, Tao D (2016) Mapping QTLs for hybrid sterility in three AA genome wild species of Oryza. Breed Sci 66:367–371

Zhang Q-J, Zhu T, Xia E-H, Shi C, Liu Y-L, Zhang Y, Liu Y, Jiang W-K, Zhaoa Y-J, Maoa S-Y, Zhang L-P, Huang H, Jiaoa J-Y, Xu P-Z, Yao Q-Y, Zeng F-C, Yang L-L, Gao J, Tao D-Y, Wang Y-J, Bennetzen JL, Gao L-Z (2014) Rapid diversification of five *Oryza* AA genomes associated with rice adaptation. PNAS 111:E4954–E4962

Zhang Y, Zhang S, Liu H, Fu B, Li L, Xie M, Song Y, Li X, Cai J, Wan W, Kui L, Huang H, Lyu J, Dong Y, Wang W, Huang L, Zhang J, Yang Q, Shan Q, Li Q, Huang W, Tao D, Wang M, Chen M, Yu Y, Wing RA, Wang W, Hu F (2015) Genome and comparative transcriptomics of African wild rice *Oryza longistaminata* provide insights into molecular mechanism of Rhizomatousness and self-incompatibility. Mol Plant 8:1683–1686

Oryza brachyantha A. Chev. et Roehr

7

Felipe Klein Ricachenevsky, Giseli Buffon, Joséli Schwambach and Raul Antonio Sperotto

Abstract

The growing interest in the *Oryza* genus comes from the feasibility of studying genome evolution of these closely related species, as well as the direct impact of identifying desirable phenotypes that could be transferred to *Oryza sativa*, one of the world's most important cereals. Among the *Oryza* species, *Oryza brachyantha* is unique: it is highly divergent, has the smallest genome in the genus and is the only FF species, and has several traits that could be useful to improve *Oryza sativa*. However, our understanding of the basic biology of *O. brachyantha* and conservation of its diversity in germplasm are still preliminary. In this chapter, we summarize the current knowledge on *O. brachyantha*, especially on its recently published genomes (nuclear and chloroplastidic), basic genetics, and sequence comparisons with other *Oryza* species. The information gathered here should be useful to guide efforts to conserve and explore *O. brachyantha* diversity, a necessary step in order to achieve both basic and applied science goals in the future.

F. K. Ricachenevsky
Programa de Pós-Graduação em Agrobiologia (PPGAgrobio), Departamento de Biologia, Universidade Federal de Santa Maria (UFSM), Santa Maria, Rio Grande do Sul, Brazil

F. K. Ricachenevsky (✉)
Programa de Pós-Graduação em Biologia Celular e Molecular (PPGBCM), Universidade Federal do Rio Grande do Sul (UFRGS), Porto Alegre, Rio Grande do Sul, Brazil
e-mail: felipecruzalta@gmail.com

G. Buffon · R. A. Sperotto
Programa de Pós-Graduação em Biotecnologia (PPGBiotec), Universidade do Vale do Taquari - UNIVATES, Lajeado, Rio Grande do Sul, Brazil

J. Schwambach
Programa de Pós-Graduação em Biotecnologia (PPGBIO), Universidade de Caxias do Sul (UCS), Caxias do Sul, Rio Grande do Sul, Brazil

R. A. Sperotto
Centro de Ciências Biológicas e da Saúde (CCBS), Universidade do Vale do Taquari - UNIVATES, Lajeado, RS, Brazil

7.1 Academic and Economic Importance

Rice plays an essential role in human nutrition and culture for millennia and has important syntenic relationships with the other cereal species. As such, *Oryza* is among the most well known and important genus for the study of Poaceae evolution and domestication (Lu et al. 2009). The wild relatives of *Oryza* genus have provided invaluable genomic resources for the improvement and breeding of rice (Dong et al. 2013; Menguer et al. 2017) and should become even more important in the following years (Ricachenevsky and Sperotto 2016).

T. K. Mondal and R. J. Henry (eds.), *The Wild Oryza Genomes*,
Compendium of Plant Genomes, https://doi.org/10.1007/978-3-319-71997-9_7

Africa contains a great diversity of both cultivated and wild rice species. The region has eight species representing six out of the ten known genome types (Wambugu et al. 2013). Only a fraction of the genetic variation of these species is conserved in global germplasm collection, and even these genotypes are uncharacterized and therefore underutilized. The lack of in situ conservation programs exposes them to possible genetic erosion or extinction (Wambugu et al. 2013). In order to obtain maximum benefits from these resources, it is imperative that they are collected, efficiently conserved, and utilized. High-throughput molecular approaches such as whole genome sequencing could be employed to more precisely study their genetic diversity and value, and thereby enhance their use in rice improvement (Chen et al. 2013; Wambugu et al. 2013).

O. brachyantha A. Chev. et Roehr. is one of the *Oryza* species endemic to Africa. It has a unique phylogenetic position, since it does not fall into any of the complexes of species within *Oryza* (Joshi et al. 2000). *O. brachyantha* is probably the closest species to the ancestral state of the *Oryza* genomes (Chen et al. 2013). Thus, comparisons of *O. brachyantha* and other *Oryza* genomes will provide us a unique opportunity to explore the genomic changes and the underlying mechanisms of *Oryza* genome evolution (Zhang et al. 2007; Chen et al. 2013). Beyond important reservoir of useful genes, it can be exploited both to broaden the existing narrow genetic base and enrich the existing varieties with desired agronomically important traits (Joshi et al. 2000).

In the literature, there are approximately 30 articles about *O. brachyantha.* Most of these articles are related to phylogenetics and evolution of the *Oryza* genus. The main objective of these studies was understand how genetic diversity affects phenotypic differences between species (Jacquemin et al. 2014). Along with these evolutionary studies of the *Oryza* genus, there are only a few on the genetic diversity of *O. brachyantha* for resistance to biotic stress (Ramachandran and Khan 1991; Brar and Khush 2002; Yamakawa et al. 2008; Ram et al. 2010). It was found that *O. brachyantha* is resistant to rice leaf folder, *Cnaphalocrocis medinalis* (Guenée), an important insect that causes havoc to rice cultivation. It was hypothesized that the high levels of resistance observed in *O. brachyantha* may be due to an additive or synergistic action of the absence of attractants or feeding stimulants, the presence of deterrents, and the physical resistance offered by closer arrangement of silica cells in the epidermal layer (Ramachandran and Khan 1991). These results could be important for breeding programs aiming at transferring resistance factors from this species to cultivated rice.

7.2 Botanical Description and Distribution

The genus *Oryza*, named by Linnaeus in 1753, is part of the botanical family Poaceae and has been placed in the tribe *Oryzeae*, subfamily Pooideae (Tzvelev 1989), supertribe Oryzanae (Watson et al. 1985), and subfamily Oryzoideae (Duistermaat 1987). Within the tribe Oryzeae, the genus *Leersia* is most closely related to *Oryza*. Some species of both genera appear to be intermediate between them. For example, *Leersia stipitata* of Thailand has some anatomical characteristics that resemble *Oryza*. Similarly, *O. brachyantha* has some features that resemble *Leersia* (Vaughan 1989). Further detailed studies are needed for the species at the boundary of *Leersia* and *Oryza* genera to better understand their evolutionary relationships (Smith and Dilday 2003). However, the diploid genome (FF) of *Leersia* is dissimilar to all other species in the *Oryza* genus.

O. brachyantha is an annual or weakly perennial, tufted grass, with slender culms; compact panicles; spikelets small and slender, 7.7–10 mm long and 1.4–1.8 mm wide, with long awns (6–17 cm long) (Vaughan 1989) (Fig. 7.1), anthers 2–3.4 mm long, and chromosome number 2n = 24.

Fig. 7.1 Adult *O. brachyantha* plant and dehusked seeds. Pictures were kindly provided by Dr. Soham Ray, Dr. Lotan K. Bose, and Dr. Kutubuddin A. Molla (National Rice Research Institute, India), and Dr. Peterson W. Wambugu (Kenya Agricultural Research Institute, Kenya)

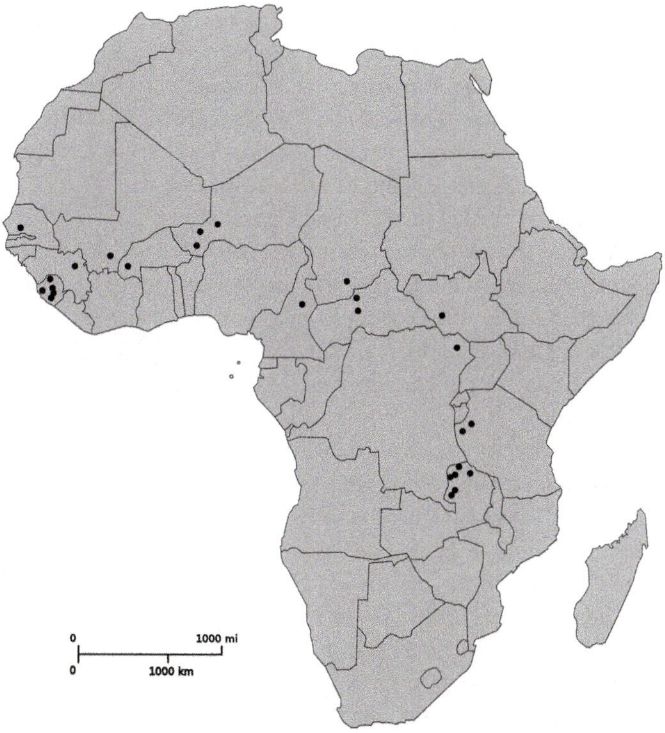

Fig. 7.2 Distribution of *O. brachyantha* on the African continent. Adapted from www.gene.affrc.go.jp—Sengupta and Majumder (2010), Jacquemin et al. (2013)

O. brachyantha is grown in The West and Central Africa (Burkina Faso, Cameroon, Central Africa Republic, Chad, Guinea, Mali, Niger, Senegal, Sierra Leone, Sudan, Tanzania, Democratic Republic of Congo, Zambia—Figure 7.2) and grows in open wetland habitats (Chen et al. 2013).

7.3 Genetic and Genomic Resources

In contrast to the rapid progress of molecular analyses of the cultivated rice genome, careful cytological characterization is still needed for wild rice genomes. Chromosome identification is the most important step for cytogenetic studies. However, karyotype analysis within the *Oryza* genus can be difficult due to small chromosomes sizes and the high morphological similarity between them. The development of a species karyotype, with unambiguous identification of individual chromosomes, is critical for the integration of genetic and physical map data. Fluorescent in situ hybridization (FISH) allowed the localization of genes in rice chromosomes and was used to differentiate *O. brachyantha* from *O. sativa*, *O. glaberrima*, *O. meridionalis* and *O. officinalis* by the distribution pattern of tandem repeated sequences (*TrsA*, *TrsB*, and *TrsC*; Ohmido et al. 1996). In a study comparing species of the *Oryza* genus (*O. sativa*, *O. glaberrima*, *O. rufipogon*, *O. longistaminata*, *O. glumaepatula*, *O. meridionalis*, *O. punctata*, *O. officinalis*, *O. eichingeri*, *O. australiensis*, and *O. brachyantha*) using FISH, the genome of *O. brachyantha* was shown to be the smallest one (Uozu et al. 1997). Also, it was shown that *O. brachyantha* presents a higher number of *TrsB* copies when compared to the other species, and that this explosive amplification of *TrsB* may have caused the speciation of *O. brachyantha* (Uozu et al. 1997).

The hybrids of *O. sativa* x *O. brachyantha* showed high sterility, with limited chromosomal pairing and unequal distribution during anaphase, resulting in abnormal meiosis (Abbasi et al. 2010). However, early work characterizing trait introgression from *O. brachyantha* into *O. sativa* showed that it is possible to introgress chromosome segments even from this distantly related genome into cultivated rice (Aggarwal et al. 1996; Brar et al. 1996). Recently, the development of monosomic alien addition lines (MAALs) from *O. brachyantha* for introgression of yellow stem borer (*Scirpophaga incertulas*) resistance in *O. sativa* was carried out. All MAALS exhibited different morphological features with an overall low or moderate resistance to yellow stemborer (Narain et al. 2016). Considering that *O. brachyantha* is also resistant to bacterial blight (*Xanthomonas oryzae* pv. *oryzae*), leaf folder (*Cnaphalocrocis medinalis*), whorl maggot (*Hydrellia philippina*), and tolerant to laterite soil (Brar and Khush, 2002; Yamakawa et al. 2008; Ram et al. 2010; Sanchez et al. 2013), coupled with the recent development of low cost, easy to use cross-transferable markers to detect *O. brachyantha* DNA in introgressed *O. sativa* (Ray et al. 2016), it is feasible to use *O. brachyantha* traits to improve *O. sativa*.

There are few accessions of *O. brachyantha* that can be obtained from seed stocks. The international seed stock database Genesys (http://www.genesys.org/) has 27 entries for *O. brachyantha* until recently, and only a few other accessions are found in seed distribution banks localized outside Africa (Wambugu et al. 2013). Thus, *O. brachyantha* accessions are clearly underrepresented in germplasm banks, and there is a need to improve seed collection for conservation (Wambugu et al. 2013).

7.4 Genome Sequence and Evolutionary Relationship

Being FF genotype, *O. brachyantha* occupies a unique position in the *Oryza* genus in terms of genome evolution, with the smallest genome size of the genus ($\sim$261 Mb). It is considered the closest genome to the ancestral state *Oryza* (Chen et al. 2013; Jacquemin et al. 2013) and may display a faster evolution rate compared to other *Oryza* (Zou et al. 2008). *O. brachyantha* is estimated to have diverged from the *O. sativa* lineage 15 million years ago (Ammiraju et al. 2008). Recently, the high-quality genomic sequence of *O. brachyantha* (accession IRGC101232) was generated. Sequencing was performed using an Illumina platform, with a coverage of $\sim$104-fold. Small read libraries were assembled into scaffolds, and these anchored to *O. brachyantha* chromosomes using BAC libraries previously generated at the

Arizona Genomics Institute (Ammiraju et al. 2006). Further anchoring was done using information of *Oryza* gene collinearity and confirmed by FISH experiments (Chen et al. 2013).

O. brachyantha genome is only 35% conserved with the *O. sativa* genome. The number of annotated gene models is also lower than *O. sativa*: 32,038, compared to 37,544 (IRGSP 2005) or 49,110 according to recent update (Kawahara et al. 2013). Around 70% of these are located in collinear positions in comparison with the rice genome (Chen et al. 2013). Based on syntenic block analyses, 22,405 and 24,103 genes showed conserved collinearity between *O. brachyantha* and *O. sativa*, respectively. These were grouped into 19,222 gene clusters, 2,468 likely being gene duplication events. There were higher expanded clusters in *O. sativa* than that of *O. brachyantha* (1,363 and 663, respectively). Among the expanded gene families, defense and reproduction process categories were enriched (Chen et al. 2013).

The genome size difference between *O. brachyantha* and *O. sativa* was mainly caused by differences in the lineage-specific evolution of intergenic sequences, especially transposable elements (TEs; Ammiraju et al. 2008; Chen et al. 2013). TEs comprise around 29% of the genome sequence, which is lower than other species of the Poaceae family while consistent with genome size (Paterson et al. 2009; Schnable et al. 2009). This is significantly lower than rice (35%). As expected, the TEs are unevenly distributed on each chromosome with retrotransposons concentrated in pericentromeric or heterochromatic regions. Long Terminal Repeats (LTR) retrotransposons comprise 10% of the *O. brachyantha* genome, contributing to 50% of the size difference (Chen et al. 2013). In *O. brachyantha*, the amplification of LTR retrotransposons occurred over a relatively long period, while in cultivated rice around 40% of the LTR retrotransposons were inserted within the last 0.5 million years. Clearly, non-homologous and unequal recombination played a role in shaping the compact genome of *O. brachyantha*, as shown by the high number of solo and truncated LTR repeats found in the genome sequence and in previous studies (Ammiraju et al. 2008; Chen et al. 2013). Conversely, the *Mutator*-like DNA transposon is more abundant in *O. brachyantha* (18.3 Mb) than in rice (13.4 Mb Chen et al. 2013) and accounts for 7.5% of the genome and 25% of the DNA transposons in total.

More than 30% of the *O. brachyantha* genes are in non-collinear positions compared to rice genome (Chen et al. 2013). Non-collinear genes were enriched in low recombination regions such as pericentromeric or heterochromatic knobs. Double-strand break repair, particularly non-homologous end joining, was shown to have an important role in gene movement and erosion of collinearity when comparing *O. brachyantha*, *O. sativa*, and *O. glaberrima* genomes. Duplicated genes were also enriched in the heterochromatic regions, which is also likely to be linked to genome expansion of these regions, since redundancy leads to tolerance to mutations, insertions, and rearrangements (Chen et al. 2013). Still, analysis of gene families revealed that 95% (17,076 out of 18,020) are clustered with rice genes and more than 80% of the gene families shared by *O. brachyantha* and rice have a one-to-one orthologous relationship.

Centromere sequences are commonly associated with a histone variant CenH3 (De Rop et al. 2012). Despite conservation in eukaryotes, sequences to which CenH3 binds can be highly divergent. In rice, centromeres are populated by arrays of a 155 bp repeat called *CentO* and dispersed copies of *CRR* retrotransposons, commonly found in other grasses and in many *Oryza* species (Lee et al. 2005). However, CC, GG, and FF species showed distinct sequences in their centromeres. Remarkably, *O. brachyantha* (FF) has a species-specific 154 bp single repeat, named *CentO-F*, which composes 90% of its centromeres, and is not found in other *Oryza* genomes (Lee et al. 2005). *CentO-F* is present in all 12 chromosomes only in centromeres and varies in size among chromosomes. Later, it was shown that CenH3-binding sequences can be classified in three distinct classes, with little sequence variation, and confirmed that they are FF genome-specific (Yi et al. 2013). Moreover, the canonical *CRR* retrotransposons, found in centromeres of all other *Oryza* species, are not found in *O. brachyantha*. Instead, a *Ty3-gypsy* retroelement was shown to have colonized *O. brachyantha* centromeres (Lee et al. 2005; Gao et al. 2009).

Recombination is usually suppressed in centromeric regions (Shi et al. 2010), and thus deleterious mutations can easily accumulate and cause degeneration of genes and genomes. However, the centromere of chromosome 8 (*Cen8*) of rice contains several transcribed genes (Fan et al. 2011). Analysis of the *Cen8* region of *O. sativa*, *O. glaberrima*, and *O. brachyantha* showed that despite containing a different sets of repetitive DNA sequences compared with cultivated rice (Gao et al. 2009), a set of seven orthologous transcribed *Cen8* genes have undergone purifying selection and are highly conserved, showing a striking example of active gene survival over a long evolutionary time in a low recombination, TE, and repetitive DNA-rich genomic region (Fan et al. 2011).

The plastome sequence of *O. brachyantha* (accession IRGC101232) was recently fully sequenced. It has $\sim$134 Mb and is circularly structured, showing a pair of inverted repeats of $\sim$20 Mb separated by a large ($\sim$80 Mb) and a small ($\sim$12 Mb) single copy regions (Liu et al. 2016). The genome is AT rich, with only 39% GC content, and shows 110 unique annotated genes, including 76 protein-coding, 30 tRNA genes, and four ribosomal genes, which represent around 44, 2, and 7% of the whole sequence, respectively. From these, 20 genes are duplicated in the inverted repeats (Liu et al. 2016). Phylogenetic analysis of *Oryza* plastome sequences placed *O. brachyantha* as an early diverging species, confirming results obtained with nuclear markers (Tang et al. 2015).

7.5 Comparison of Gene Families

The *Oryza* genus has experienced 15 million years of evolution, which resulted in more than 24 species and ten genome types. However, major disagreements still exist on the systematics and genome evolution of this genus (Mullins and Hilu 2002; Jacquemin et al. 2013). Elucidating the evolutionary history of *Oryza* species is very important to understand how genetic diversity affects phenotypic differences between species (Jacquemin et al. 2014; Sui et al. 2014). This section is dedicated to the recent efforts that have elucidated the diversity and relatedness among wild and domesticated species at the DNA and protein levels for the *Oryza* genus using several different phylogenetic approaches.

Prolamin, a seed storage protein and site of nitrogen and sulfur accumulation, is sequestered in the subaleurone layer of the starchy endosperm for use during seedling germination (Mullins and Hilu 2004). Sequence variation in the gene that encodes the 10-kDa prolamin polypeptide was used to determine phylogenetic relationships and evaluate current systematics for 19 *Oryza* species (Mullins and Hilu 2002). Later, Mullins and Hilu (2004) analyzed the deduced amino acid variability for the same 10-kDa prolamin protein in 16 *Oryza* species, including rice. Examination of nucleotide sequences demonstrated systematic utility and also showed that *O. brachyantha* (FF) is sister to all *Oryza* species examined, since a strict consensus tree shows *O. brachyantha* (FF) as the most basal species, followed by a polytomy of three clades: (i) the GG clade: *O. granulata* and *O. meyeriana*; (ii) the EE clade: *O. australiensis*; and (iii) the ABCD clade: the remaining *Oryza* species. This finding has important genomic and biogeographic implications, pointing to the FF genome as the ancestral type and suggesting an African origin for *Oryza* (Mullins and Hilu 2002). Based on amino acid sequences, *O. brachyantha* exhibited one of the greatest residue variability within the signal peptide region, compared to *O. sativa*. As this region is thought to be removed prior to prolamin packaging within protein bodies that reside in the subaleurone layer of the starchy endosperm (Coffman and Juliano 1987), *O. brachyantha* may exhibit low levels of prolamin synthesis, thereby naturally optimizing the production of the glutelin fraction and not allowing the crop improvement by standard molecular breeding and manipulation of the prolamin fraction itself (Mullins and Hilu 2004).

In the genus *Oryza*, 45 S rDNA loci are polymorphic in both number and location. From one to four rDNA loci have been detected in *Oryza* chromosomes. *O. brachyantha* has only one 45 S rDNA locus at the end of the short arm of chromosome 9 (9 S), which is the most conserved position of 45 S rDNAs in *Oryza* genus (Chung et al. 2008). Chang et al. (2010) revealed that *O. brachyantha* is different from other *Oryza* species. The intergenic spacer (IGS), which separates two adjacent rDNAs in the tandem arrays and consists of a nontranscribed region, contains four tandem repeats and three transcriptional initiation sites in *O. brachyantha.* Also, the typical 254 bp repeat, which commonly exists in the IGS of most *Oryza* species, is absent in *O. brachyantha.* It is important to highlight that 45 S rDNA organization in *O. brachyantha* is different even from *O. australiensis*, which also has only one 45 S rDNA locus, suggesting different evolution processes of orthologous rDNA loci in the genus *Oryza* (Chang et al. 2010). The mutations presented in the 45 S rDNA locus of *O. brachyantha* could reduce transcriptional activity and alter the site of transcription initiation, since sub-repeats in IGS regions may carry promoter or potential transcription initiation sequences (Suzuki et al. 1996; Chang et al. 2010).

Some disease resistance-related gene families such as NB-ARC, leucine-rich repeat (LRR), and F-box are overrepresented in rice relative to *O. brachyantha*, reflecting its role in adaptation to various environments (Jones and Dangl 2006; Chen et al. 2013). Similar results were obtained by Jacquemin et al. (2014), who observed large size expansions in the rice genome compared to *O. brachyantha* for the superfamilies F-box and NB-ARC, and five additional families: the aspartic proteases, BTB/POZ proteins (BTB), glutaredoxins, trypsin α-amylase inhibitor proteins, and Zf-Dof proteins. These seven gene families have all been shown to play critical roles in plant stress responses or development (Jacquemin et al. 2014). Such expansions resulted from amplification, largely by tandem duplications, and contraction by gene losses. Previously, Ammiraju et al. (2008) also detected dynamic patterns of gene gain and loss for the F-box gene family, which displayed significant copy number variation throughout the evolution of the *Oryza* genus, with a tendency to expand in the more recently diverged genomes. Only two F-box genes were identified in the *Adh1–Adh2* locus of *O. brachyantha*, while 12 family members were identified in the orthologous region of rice.

Calcium (Ca) permeable channels that function as osmosensors are important to receive and respond to exogenous as well as endogenous stimuli that lead to osmotic changes, sustaining plant growth, and development under non-optimal conditions (Shavrukov 2013). Osmotic stress along with various other stimuli trigger increases in the cytosolic/intracellular free Ca concentration $[Ca^{2+}]_i$ in plants (McAinsh and Pittman 2009). Hyperosmolality-gated Ca-permeable channels (OSCA) is one of these sensors involved in the increase of $[Ca^{2+}]_i$ induced by hyperosmolarity. Li et al. (2015) identified 11 *OSCA*s in the *O. brachyantha* genome. Phylogenetic analysis revealed that members of the *OSCA* family were separated into four distinct clades, designated I, II, III, and IV. Compared with rice, *OSCA* members from *O. brachyantha* shared almost identical intron/exon structures and intron phases, suggesting that *OSCA*s structures were formed prior to the split between wild and cultivated rice. However, further studies with *O. brachyantha* are needed to check whether a functional divergence may occur among the *OSCA* genes.

The resistance (*R*) and defense response (*DR*)-genes have become very important resources for the development of disease-resistant cultivars (Singh et al. 2015). *O. brachyantha* has 251 *R*-genes and 86 *DR*-genes, approximately three- and two-fold less than rice, respectively. Out of 251 predicted *R*-genes, 45% (114) belongs to LRR category. The 86 *DR*-genes were identified and categorized in three classes such as chitinases, glucanases, and thaumatin-like proteins (Singh et al. 2015). Analysis of paralogs across rice species indicated that *O. brachyantha* has three times less duplicated genes than rice. Interestingly, phylogenetic tree reconstruction for all *DR*-genes showed that they are closely related clusters, which were divided on the basis of *DR*-genes category and not on the basis of rice species, suggesting that *DR*-genes of rice species are closely related (Singh et al. 2015). The paralogs of *R*- and *DR*-genes were studied separately for each chromosome. Most of the paralogs in *japonica* and *indica* were found in the parent chromosome, showing few shifts of genes across the chromosomes, whereas in *O. brachyantha* the result is just opposite showing more gene shift across chromosomes. In *O. brachyantha*, out of 251 *R*-genes, only 38.5% paralogs were found to be located on the same chromosome and 61.5% in different chromosomes. On the other hand, out of 86 *DR*-genes, 70% of paralogs were found in different chromosomes, whereas 30% were found to be on the same chromosome (Singh et al. 2015).

Cultivated rice, *Oryza sativa*, produces antimicrobial diterpene phytoalexins represented by phytocassanes and momilactones, and the majority of their biosynthetic genes are clustered on chromosomes 2 and 4, respectively, termed the Os02g and Os04g clusters (Okada 2011). Recently, Miyamoto et al. (2016) compared the rice gene clusters Os02g and Os04g with wild *Oryza* species. It was demonstrated that the *O. brachyantha* loci corresponding to the Os02g and Os04g clusters have only two genes involved with phytocassanes biosynthesis, CYP76 M-related and CYP99A-related homologs, respectively, suggesting that the phytocassane biosynthetic gene clusters were present in the common ancestor of the *Oryza* species despite the different locations, directions, and numbers of their member genes (Miyamoto et al. 2016). No gene related to momilactones was detected in *O. brachyantha* genome. Also, no class of phytoalexin was detected in *O. brachyantha*, which contains neither an Ob02g nor an Ob04g functional cluster (Miyamoto et al. 2016). Genes for the biosynthesis of gibberellins (GA), which are labdane-related phytohormones (Toyomasu 2008), were investigated by Miyamoto et al. (2016). The location and numbers of these genes in *O. brachyantha* are highly similar to those of *O. sativa*. Thus, GA biosynthetic genes have been conserved throughout the genus *Oryza* during evolution as part of the primary metabolism.

A major quantitative trait locus called *SUBMERGENCE-1* (*SUB1*) confers submergence tolerance to rice plants. The locus encodes three genes such as *Ethylene-Responsive Factors* (*ERF*): *SUB1A*, *SUB1B*, and *SUB1C* (Xu et al. 2006). Recently, dos Santos et al. (2017) showed that *SUB1* locus is not present on *O. brachyantha* genome. Instead, a gene similar to *SUB1* was found on chromosome 9, but the absence of a serine at position 13 and other differences along their sequence prevented its classification as a *SUB1A*-like gene. On the contrary, *Leersia perrieri*, a grass tolerant to deep flooding, presents three *ERF* genes in the *SUB1* locus, similar to that found in flooding tolerant rice (dos Santos et al. 2017).

7.6 Future Perspectives

O. brachyantha has a unique position within the *Oryza* genus: it is the only extant FF genome species, is highly divergent, and has the smallest genome size (Chen et al. 2013). Sequence information of *O. brachyantha* is indispensable to understand the evolution of the *Oryza* genus and to understand the mechanisms of genome compaction/expansion in these models for plant genome evolution. Therefore, it is not surprising that *O. brachyantha* was the first non-AA genome to be fully sequenced (Chen et al. 2013) Other genomes outside the AA cluster have already been sequenced. Future exploration of the full genome sequence and more in-depth comparative analyses with other species of *Oryza* should uncover interesting new aspects of the genus evolution, as well as details about genes and gene families that have potential to be used for introgression in *O. sativa*.

O. brachyantha is also a valuable source for biotic resistance genes, especially considering the limited number of accessions collected so far (Wambugu et al. 2013; Sanchez et al. 2013). As we progress in sampling the species diversity and search for specific traits, it is likely that new interesting phenotypes will emerge. Although analyses based on the environment where the current germplasm samples are found does not indicate *O. brachyantha* as a good candidate for abiotic stress tolerance (Atwell et al. 2014), we are probably only starting to uncover the potential of this species to provide useful traits to crops. Thus, we believe that conservation of genetic diversity of *O. brachyantha* is key for both basic, evolutionary understanding of the genus, as well as for breeding/engineering improved crops plants in the future.

References

Abbasi FM, Shah AH, Perveen F, Afzal M, Sajid M, Masood R, Nawaz F (2010) Genomic affinity between *Oryza sativa* and *Oryza brachyantha* as revealed by in situ hybridization and chromosome pairing. Afr J Biotech 9:3068–3072

Aggarwal RK, Brar DS, Huang N, Khush GS (1996) Molecular analysis of introgression in *Oryza sativa/O. brachyantha* and *O. sativa/O. granulata* derivatives. Int Rice Res. Notes 21:14

Ammiraju JS, Luo M, Goicoechea JL, Wang W, Kudrna D, Mueller C, Talag J, Kim H, Sisneros NB, Blackmon B, Fang E, Tomkins JB, Brar D, MacKill D, McCouch S, Kurata N, Lambert G, Galbraith DW, Arumuganathan K, Rao K, Walling JG, Gill N, Yu Y, SanMiguel P, Soderlund C, Jackson S, Wing RA (2006) The *Oryza* bacterial artificial chromosome library resource: construction and analysis of 12 deep-coverage large-insert BAC libraries that represent the 10 genome types of the genus *Oryza*. Genome Res 16:140–147

Ammiraju JS, Lu F, Sanyal A, Yu Y, Song X, Jiang N, Pontaroli AC, Rambo T, Currie J, Collura K, Talag J, Fan C, Goicoechea JL, Zuccolo A, Chen J, Bennetzen JL, Chen M, Jackson S, Wing RA (2008) Dynamic evolution of *Oryza* genomes is revealed by comparative genomic analysis of a genus-wide vertical data set. Plant Cell 20:3191–3209

Atwell BJ, Wang H, Scafaro AP (2014) Could abiotic stress tolerance in wild relatives of rice be used to improve *Oryza sativa*? Plant Sci 215–216:48–58

Brar DS, Dalmacio R, Elloran R, Aggarwal R, Angeles R, Khush GS (1996) Gene transfer and molecular characterization of introgression from wild *Oryza* species into rice. In: Khush GS (ed) Rice Genetics III. International Rice Research Institute, Manila, pp 477–485

Brar DS, Khush GS (2002) Transferring genes from wild species into rice. In: Kang MS (ed) Quantitative genetics, genomics and plant breeding. CABI, Wallingford, pp 197–217

Chang KD, Fang SA, Chang FC, Chung MC (2010) Chromosomal conservation and sequence diversity of ribosomal RNA genes of two distant *Oryza* species. Genomics 96:181–190

Chen J, Huang Q, Gao D, Wang J, Lang Y, Liu T, Li B, Bai Z, Goicochea JL, Liang C, Chen C, Zhang W, Sun A, Liao Y, Zhang X, Yang L, Song C, Wang M, Shi J, Liu G, Liu J, Zhou H, Zhou W, Yu Q, An N, Chen Y, Cai Q, Wnag B, Liu B, Min J, Huang Y, Wu H, Li Z, Zhang Y, Yin Y, Song W, Jiang J, Jackson SA, Wing RA, Wang J, Chen M (2013) Whole genome sequencing of *Oryza brachyantha* reveals mechanisms underlying Oryza genome evolution. Nat Commun 4:1595

Chung MC, Lee YI, Cheng YY, Chou YJ, Lu CF (2008) Chromosomal polymorphism of ribosomal genes in the genus *Oryza*. Theor Appl Genet 116:745–753

Coffman WR, Juliano BO (1987) Rice. Nutritional quality of cereal grains: genetic and agronomic improvement. ASACSSA-SSSA, Madison, WI, pp 101–131

De Rop V, Padeganeh A, Maddox PS (2012) CENP-A: the key player behind centromere identity, propagation, and kinetochore assembly. Chromosoma 121:527–538

Dong Z, Wang H, Dong Y, Wang Y, Liu W, Miao G, Xiuyun L, Daqing W, Liu B (2013) Extensive microsatellite variation in rice induced by introgression from wild rice (*Zizania latifolia* Griseb.). PLoS ONE 8:e62317

dos Santos RS, Farias DR, Pegoraro C, Rombaldi CV, Fukao T, Wing RA, de Oliveira AC (2017) Evolutionary analysis of the *SUB1* locus across the *Oryza* genomes. Rice 10:4

Duistermaat H (1987) A revision of *Oryza* (Gramineae) in Malesia and Australia. Blumea 32:157–193

Fan C, Walling JG, Zhang J, Hirsch CD, Jiang J, Wing RA (2011) Conservation and purifying selection of transcribed genes located in a rice centromere. Plant Cell 23:2821–2830

Gao D, Gill N, Kim HR, Walling JG, Zhang W, Fan C, Yu Y, Ma J, SanMiguel P, Jiang N, Cheng Z, Wing RA, Jiang J, Jackson SA (2009) A lineage-specific centromere retrotransposon in *Oryza brachyantha*. Plant J 60:820–831

International Rice Genome Sequencing P (2005) The map-based sequence of the rice genome. Nature 436:793–800

Jacquemin J, Bhatia D, Singh K, Wing RA (2013) The International *Oryza* Map Alignment Project: development of a genus-wide comparative genomics platform to help solve the 9 billion-people question. Curr Opin Plant Biol 16:147–156

Jacquemin J, Ammiraju JS, Haberer G, Billheimer DD, Yu Y, Liu LC, Rivera LF, Mayer K, Chen M, Wing RA (2014) Fifteen million years of evolution in the *Oryza* genus shows extensive gene family expansion. Mol Plant 7:642–656

Jones JD, Dangl JL (2006) The plant immune system. Nature 444:323–329

Joshi SP, Gupta VS, Aggarwal RK, Ranjekar PK, Brar DS (2000) Genetic diversity and phylogenetic relationship as revealed by inter simple sequence repeat (ISSR) polymorphism in the genus *Oryza*. Theor Appl Genet 100:1311–1320

Kawahara Y, de la Bastide M, Hamilton JP, Kanamori H, McCombie WR, Ouyang S, Schwartz DC, Tanaka T, Wu J, Zhou S, Childs KL, Davidson RM, Lin H, Quesada-Ocampo L, Vaillancourt B, Sakai H, Lee SS, Kim J, Numa H, Itoh T, Buell CR, Matsumoto T (2013) Improvement of the *Oryza sativa* Nipponbare reference genome using next generation sequence and optical map data. Rice 6:4

Lee HR, Zhang W, Langdon T, Jin W, Yan H, Cheng Z, Jiang J (2005) Chromatin immunoprecipitation cloning reveals rapid evolutionary patterns of centromeric DNA in *Oryza* species. Proc Natl Acad Sci U S A 102:11793–11798

Li Y, Yuan F, Wen Z, Li Y, Wang F, Zhu T, Zhuo W, Jin X, Wang Y, Zhao H, Pei ZM, Han S (2015) Genome-wide survey and expression analysis of the *OSCA* gene family in rice. BMC Plant Biol 15:261

Liu F, Tembrock LR, Sun C, Han G, Guo C, Wu Z (2016) The complete plastid genome of the wild rice species *Oryza brachyantha* (Poaceae). Mitochondrial DNA 1:218–219

Lu F, Ammiraju JS, Sanyal A, Zhang S, Song R, Chen J, Liu G, Sui Y, Song J, Cheng Z, De Oliveira AC, Bennetzen JL, Jackson AS, Wing RA, Chen M (2009) Comparative sequence analysis of MONOCULM1-orthologous regions in 14 *Oryza* genomes. Proc Natl Acad Sci U S A 106:2071–2076

McAinsh MR, Pittman JK (2009) Shaping the calcium signature. New Phytol 181:275–294

Menguer PK, Sperotto RA, Ricachenevsky FK (2017) A walk on the wild side: *Oryza* species as source for rice abiotic stress tolerance. Genet Mol Biol 40:238–252 https://doi.org/10.1590/1678-4685-GMB-2016-0093

Miyamoto K, Fujita M, Shenton MR, Akashi S, Sugawara C, Sakai A, Horie K, Hasegawa M, Kawaide H, Mitsuhashi W, Nojiri H, Yamane H, Kurata N, Okada K, Toyomasu T (2016) Evolutionary trajectory of phytoalexin biosynthetic gene clusters in rice. Plant J 87:293–304

Mullins IM, Hilu KW (2002) Sequence variation in the gene encoding the 10-kDa prolamin in *Oryza* (Poaceae). I. Phylogenetic implications. Theor Appl Genet 105:841–846

Mullins IM, Hilu KW (2004) Amino acid variation in the 10 kDa *Oryza* prolamin seed storage protein. J Agric Food Chem 52:2242–2246

Narain A, Kar MK, Kaliaperumal V, Sen P (2016) Development of monosomic alien addition lines from the wild rice (*Oryza brachyantha* A. Chev. et Roehr.) for introgression of yellow stem borer (*Scirpophaga incertulas* Walker.) resistance into cultivated rice (*Oryza sativa* L.). Euphytica 209:603–613

Ohmido N, Ohtsubo H, Ohtsubo E, Fukui K (1996) Physical mapping of several genes in rice using fluorescent in situ hybridization. In: Khush GS (ed) Rice Genetics III, Manila, pp 467–470

Okada K (2011) The biosynthesis of isoprenoids and the mechanisms regulating it in plants. Biosci Biotechnol Biochem 75:1219–1225

Paterson AH, Bowers JE, Bruggmann R, Dubchak I, Grimwood J, Gundlach H, Haberer G, Hellsten U, Mitros T, Poliakov A, Schmutz J, Spannagl M, Tang H, Wang X, Wicker T, Bharti AK, Chapman J, Feltus FA, Gowik U, Grigoriev IV, Lyons E, Maher CA, Martis M, Narechania A, Otillar RP, Penning BW, Salamov AA, Wang Y, Zhang L, Carpita NC, Freeling M, Gingle AR, Hash CT, Keller B, Klein P, Kresovich S, McCann MC, Ming R, Peterson DG, Mehboob-ur-Rahman, Ware D, Westhoff P, Mayer KF, Messing J, Rokhsar DS (2009) The Sorghum bicolor genome and the diversification of grasses. Nature 457:551–556

Ram T, Laha GS, Gautam SK, Deen R, Madhav MS, Brar DS, Viraktamath BC (2010) Identification of a new gene introgressed from *Oryza brachyantha* with broad-spectrum resistance to bacterial blight of rice in India. RGN 25:57–58

Ramachandran R, Khan ZR (1991) Mechanisms of resistance in wild rice *Oryza brachyantha* to rice leaffolder *Cnaphalocrocis medinalis* (Guenée) (*Lepidoptera: Pyralidae*). J Chem Ecol 17:41–65

Ray S, Bose LK, Ray J, Ngangkham U, Katara JL, Samantaray S, Behera L, Anumalla M, Singh ON, Chen M, Wing RA, Mohapatra T (2016) Development and validation of cross-transferable and polymorphic DNA markers for detecting alien genome introgression in *Oryza sativa* from *Oryza brachyantha*. Mol Genet Genomics 291:1783–1794

Ricachenevsky FK, Sperotto RA (2016) Into the wild: *Oryza* species as sources for enhanced nutrient accumulation and metal tolerance in rice. Front Plant Sci 7:974

Sanchez PL, Wing RA, Brar DS (2013) The wild relatives of rice: genomes and genomics. In: Zhang Q, Wing RA (eds) Genetics and genomics of rice. Plant genetics and genomics. Springer, New York, pp 9–26

Schnable PS, Ware D, Fulton RS, Stein JC, Wei F, Pasternak S, Liang C, Zhang J, Fulton L, Graves TA, Minx P, Reily AD, Courtney L, Kruchowski SS, Tomlinson C, Strong C, Delehaunty K, Fronick C, Courtney B, Rock SM, Belter E, Du F, Kim K, Abbott RM, Cotton M, Levy A, Marchetto P, Ochoa K, Jackson SM, Gillam B, Chen W, Yan L, Higginbotham J, Cardenas M, Waligorski J, Applebaum E, Phelps L, Falcone J, Kanchi K, Thane T, Scimone A, Thane N, Henke J, Wang T, Ruppert J, Shah N, Rotter K, Hodges J, Ingenthron E, Cordes M, Kohlberg S, Sgro J, Delgado B, Mead K, Chinwalla A, Leonard S, Crouse K, Collura K, Kudrna D, Currie J, He R, Angelova A, Rajasekar S, Mueller T, Lomeli R, Scara G, Ko A, Delaney K, Wissotski M, Lopez G, Campos D, Braidotti M, Ashley E, Golser W, Kim H, Lee S,

Lin J, Dujmic Z, Kim W, Talag J, Zuccolo A, Fan C, Sebastian A, Kramer M, Spiegel L, Nascimento L, Zutavern T, Miller B, Ambroise C, Muller S, Spooner W, Narechania A, Ren L, Wei S, Kumari S, Faga B, Levy MJ, McMahan L, Van Buren P, Vaughn MW, Ying K, Yeh CT, Emrich SJ, Jia Y, Kalyanaraman A, Hsia AP, Barbazuk WB, Baucom RS, Brutnell TP, Carpita NC, Chaparro C, Chia JM, Deragon JM, Estill JC, Fu Y, Jeddeloh JA, Han Y, Lee H, Li P, Lisch DR, Liu S, Liu Z, Nagel DH, McCann MC, SanMiguel P, Myers AM, Nettleton D, Nguyen J, Penning BW, Ponnala L, Schneider KL, Schwartz DC, Sharma A, Soderlund C, Springer NM, Sun Q, Wang H, Waterman M, Westerman R, Wolfgruber TK, Yang L, Yu Y, Zhang L, Zhou S, Zhu Q, Bennetzen JL, Dawe RK, Jiang J, Jiang N, Presting GG, Wessler SR, Aluru S, Martienssen RA, Clifton SW, McCombie WR, Wing RA, Wilson RK (2009) The B73 maize genome: complexity, diversity, and dynamics. Science 326:1112–1115

Sengupta S, Majumder AL (2010) *Porteresia coarctata* (Roxb.) Tateoka, a wild rice: a potential model for studying salt-stress biology in rice. Plant Cell Environ 33:526–542

Shavrukov Y (2013) Salt stress or salt shock: which genes are we studying? J Exp Bot 64:119–127

Shi J, Wolf SE, Burke JM, Presting GG, Ross-Ibarra J, Dawe RK (2010) Widespread gene conversion in centromere cores. PLoS Biol 8:e1000327

Singh S, Chand S, Singh NK, Sharma TR (2015) Genome-wide distribution, organisation and functional characterization of disease resistance and defence response genes across rice species. PLoS ONE 10:e0125964

Smith CW, Dilday RH (2003) Rice: Origin, history, technology, andproduction, 1st edn. Wiley, Hoboken, NewJersey, p 656

Sui Y, Li B, Shi J, Chen M (2014) Genomic, regulatory and epigenetic mechanisms underlying duplicated gene evolution in the natural allotetraploid *Oryza minuta*. BMC Genom 15:11

Suzuki A, Tanifuji S, Komeda Y, Kato A (1996) Structural and functional characterization of the intergenic spacer region of the rDNA in *Daucus carota*. Plant Cell Physiol 37:233–238

Tang L, Zou X, Zhang L, Ge S (2015) Multilocus species tree analyses resolve the ancient radiation of the subtribe Zizaniinae (Poaceae). Mol Phylogenet Evol 84:232–239

Toyomasu T (2008) Recent advances regarding diterpene cyclase genes in higher plants and fungi. Biosci Biotechnol Biochem 72:1168–1175

Tzvelev N (1989) The system of grasses (Poaceae) and their evolution. Bot Rev 55:141–203

Uozu S, Ikehashi H, Ohmido N, Ohtsubo H, Ohtsubo E, Fukui K (1997) Repetitive sequences: cause for variation in genome size and chromosome morphology in the genus *Oryza*. Plant Mol Biol 35:791–799

Vaughan D (1989) The genus *Oryza L.* current status of taxonomy. IRRI research paper series number 138

Wambugu PW, Furtado A, Waters DL, Nyamongo DO, Henry RJ (2013) Conservation and utilization of African *Oryza* genetic resources. Rice 6:29

Watson L, Clifford HT, Dallwitz MJ (1985) The classification of Poaceae: subfamilies and supertribes. Austral J Bot 33:433–484

Xu K, Xu X, Fukao T, Canlas P, Maghirang-Rodriguez R, Heuer S, Ismail AM, Bailey-Serres J, Ronald PC, Mackill DJ (2006) Sub1A is an ethylene-response-factor like gene that confers submergence tolerance to rice. Nature 442:705–708

Yamakawa H, Ebitani T, Terao T (2008) Comparison between locations of QTLs for grain chalkiness and genes responsive to high temperature during grain filling on the rice chromosome map. Breed Sci 58:337–343

Yi C, Zhang W, Dai X, Li X, Gong Z, Zhou Y, Liang G, Gu M (2013) Identification and diversity of functional centromere satellites in the wild rice species *Oryza brachyantha*. Chromosome Res 21:725–737

Zhang S, Gu YQ, Singh J, Coleman-Derr D, Brar DS, Jiang N, Lemaux PG (2007) New insights into *Oryza* genome evolution: high gene colinearity and differential retrotransposon amplification. Plant Mol Biol 64:589–600

Zou XH, Zhang FM, Zhang JG, Zang LL, Tang L, Wang J, Sang T, Ge S (2008) Analysis of 142 genes resolves the rapid diversification of the rice genus. Genome Biol 9:R49

Oryza coarctata Roxb

8

Soni Chowrasia, Hukam Chand Rawal,
Abhishek Mazumder, Kishor Gaikwad, Tilak Raj Sharma,
Nagendra Kumar Singh and Tapan K. Mondal

Abstract

Climate change-induced abiotic stresses are major limitations to crop growth and development. Among the various stresses, soil salinity is a major concern, as percentage of soil salinization has increased due to the increase in the level of ocean water and increase in irrigated area. Biotechnology and precision breeding techniques can be efficiently utilized to cope up with this abiotic stress. However, the prerequisite of the utilization of such technique requires suitable genetic resources consisting salt stress responsive genes can be deployed against this stress. Wild relatives are known to be the excellent source of such favorable alleles. *Oryza coarctata* is the only wild halophyte in the genus *Oryza*, which can withstand salinity up to 40 ds/m due to presence of distinct anatomical, morphological and physiological characteristics. Several metabolites and their genes had been elucidated in this plant for their role in imparting salt tolerance. In this chapter, we have compiled all the relevant information to understand the mechanism for salinity and waterlogging tolerance of this species. Additionally, we also identified the research gaps that need to be addressed to harness the beneficial genes/QTLs from this important halophyte.

8.1 Introduction

High salt concentration in soils is considered as an important abiotic stress, which limits the crop production worldwide. It contributes about 5% of the total crop loss, globally (Tabatabaei 2006). Salinity stress causes three major physiological responses in plants, namely osmotic adjustment, ion exclusion and tissue tolerance to accumulated ions. These imbalances lead to the reduction of the plant growth and could also lead to cell death under extreme conditions (Munns and Tester 2008). This is because salinity stress reduces plant growth by impairing its important metabolic processes, cell division, ion accumulation, and necrosis of cells, which lead to cell death. Halophytes are salt-loving plants which can remove the excess salt through salt glands by coordinating action of various genes (Askari et al. 2006; Yu et al. 2011). Depending upon the degree of salt tolerance, halophytes are distinguished as facultative and obligate halophytes. The facultative halophytes are those which can

T. R. Sharma
National Agri-Food Biotechnology Institute, Mohali, Punjab, India

S. Chowrasia · H. C. Rawal · A. Mazumder · K. Gaikwad · N. K. Singh · T. K. Mondal (✉)
ICAR-National Research Centre on Plant Biotechnology, Pusa 110012, New Delhi, India
e-mail: mondaltk@rediffmail.com

T. K. Mondal and R. J. Henry (eds.), *The Wild Oryza Genomes*,
Compendium of Plant Genomes, https://doi.org/10.1007/978-3-319-71997-9_8

Fig. 8.1 *O. coarctata*. **a** Vegetatively grown plant in the greenhouse of NRCPB, New Delhi; **b** flowers on the same plant

survive under both sweet water as well as saline water, they did not need saline water for their survival, whereas obligate halophytes are those which require salt for their normal growth and development throughout the entire life. Again, halophytes are divided into two groups, depending upon their habitats, i.e., xero-halophytes and hydro-halophytes. The xero-halophytes are those plants which can survive in the arid saline soil. The hydro-halophytes are usually found in wet and marsh land areas. One of model species of hydro-halophyte is *Oryza coarctata* which has the ability to survive under salinity stress (Sengupta and Majumder 2010). It is a facultative halophyte but requires saline environment for initial establishment. It is a perennial herb that belongs to family Poaceae (Fig. 8.1). It is the only species in the genus *Oryza* which is a halophyte by nature. Therefore, it can serve as an excellent source of genes and QTLs for salinity and submergence tolerance. Different areas of biotechnological research progress done in the species are given in Fig. 8.2.

8.2 Academic Importance

O. coarctata known as Asian wild rice grows naturally in the coastal region, where plants experience the lunar tide and are exposed to submerge saline sea water in every alternative 12 h. It is locally called as 'Uri-dhan' in India (Vaughan et al. 2005). It can withstand saline water as high as 40 dS m^{-1} ECe (electrical conductivity) (Bal and Dutt 1986). It had been found that this species had the ability to bind with peripheral soils in mangrove forests. Thus, it is considered as one of the important species in ecological succession. The analysis based on the carbon dioxide compensation points and phosphoenol pyruvate carboxylase activity revealed that it is a C3 species (Garcia 1992). Although, this species had many ecological importance but its distribution pattern decrease in abundance in several mangrove areas because of both natural as well as anthropogenic activities. The pharmaceutical potential antifungal compound

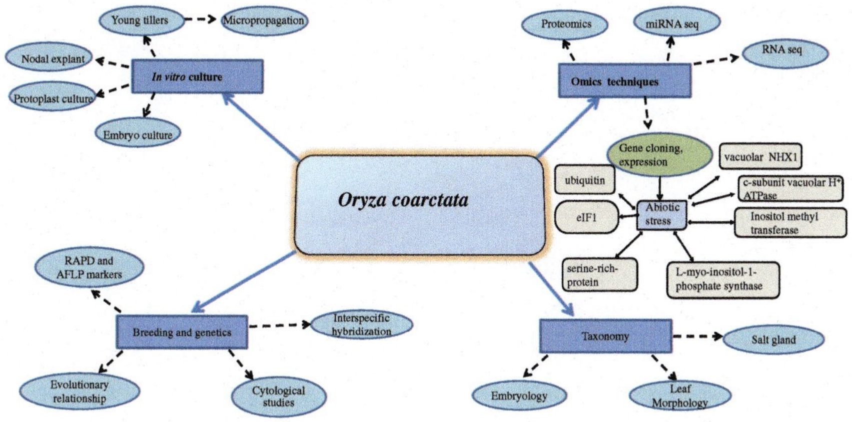

Fig. 8.2 Schematic explanation of *O. coarctata.* improvement. The bold arrows are the major area of research achievements. Dotted arrows are the subareas. Thin arrows are the different applications with a major or submajor area

Cyclosporine-A had been extracted from marine fungi *Microdochium nivale* which were associated with the *O. coarctata* leaves (Bhosale et al. 2011). It had been reported that the biomass of the *O. coarctata* was increased when soil is enriched with nitrite and nitrate (Jagtap et al. 2006). The increased growth of *O. coarctata* that occurs during monsoon is due to high humidity with less sunlight (Takemura et al. 2000; Jagtap et al. 2006). It had been found from the C:N ratio in *O. coarctata* that it could be used as compost, organic mannure and in aquaculture (Bauersfeld et al. 1969). Based on morphological characteristics such as leaf anatomy and embryo morphology, *O. coarctata* taxa was transferred to another monotypic genus *Porteresia coarctata* (Tateoka 1964). But later it was found that this morphological difference was due to their adaptation in saline habitat. Additionally, since *O. coarctata* showed morphological similarities with other species of *Oryza,* such as *O. brachyantha, O. granulata* and *O. schlectheri,* it was strongly recommended for retention of the *O. coarctata* in the genus *Oryza* (Lu and Ge 2003).

8.3 Botanical Descriptions and Geographical Distribution

The morphological characteristics of the *O. coarctata* are so unique in the genus *Oryza* that it can be easily identified. The height of the plant is found to be maximum 1 m under favorable growth conditions. It is a perennial herb with thick leathery leaves. The stems are prostrate, soft, green in color, covered with hair and stem solid and round. Presence of distinct nodes and internodes length varies between 1 and 10 cm. The shape of the leaves is linear with short petiole and leathery. The leaves have many ridges and furrows and each ridge contains one small vascular bundle. The epidermal cells are composed of short and long cells in rows. The vascular bundles are prominent and marginally tuberculated. The leaves are without midrib, distinct from rice leaves. The leaves are succulent and waxy which help in maintaining relative water content of the plant by checking transpiration rate, similar to the morphological adaptation seen in the halophyte *Tecticornia medusa* (Flowers and Colmer 2015). The leaf blades are coriaceous and two types of salt hairs i.e., peg shaped present on lower surface of leaf and finger shaped present on upper surface of leaf, as observed under scanning electron microscopy. The salt hairs had been found to be associated with many plasmodesmatal connections with surrounding epidermal cells as observed through the transmission electron microscope (Flowers et al. 1990). The adaxial surface hairs for secretion of the salt crystal under high salinity stress but the abaxial surface hairs were rupture at high salt concentration and regrow under low salt concentration (Sengupta and Majumder 2009; Flowers et al.

Fig. 8.3 Diagramatic representation of salinity tolerance mechanism of *O. coarctata* under salinity stress

1990). The lower surface hairs are peg-shaped, small and arranged all over the surface in groups. It had been reported that 2 hairs are present on each of the stomatal guard cells. The salt hairs are unicellular, without cuticle and highly vacuolated. Rhizomes are present in each nodes of the root which are highly branched and present on the surface of soil. To overcome intertidal strong flow, it often formed pseudo-taproots up to a depth of 1 m and fibrous roots develop from the tip of those pseudo-tap roots and internodes of a widespreading underground stem, called a sobole. Numerous plantlets emerged vegetatively from these soboles (Latha et al. 2004). The emergence of inflorescence is found to be more under low saline condition. The inflorescence is spikelets and mature flowers are present on the upper end of the spikelet. It has been found that glumes consisted of stomata. The flowers of *O. coarctata* are seasonal and small in size, mainly found in the month of August–October and persist only for one week. The anthers are yellow in color, oblong-shaped, six in number and remain attached with filaments. It was observed that the anthers of mature flowers protruded out, i.e., synchronized with flower opening and dehisce longitudinally. The ovaries are white in color and fruits are caryopsis type with sizes varying from 0.5 to 1 cm. The embryo is large in size, short-lived, and recalcitrant (Probert and Longley 1989). Principle method of propagation is vegetative under ex situ condition. Although, there was no report on stem and root anatomy till today, but due to the presence of aerenchymatous cell and sunken stomata, it can thrive well in submerged as well as in saline soil. It is mainly found in abundance along the Eastern and Western coasts of India, Pakistan, and Bangladesh and in the intertidal meadow on the river banks (Fig. 8.3).

8.4 Cytology and Phylogenetic Relationship

The chromosome number in *O. coarctata* is $2n = 4x = 48$ chromosomes (Parthasarathy 1938) with genome size 665 Mb (Mondal et al. 2017). The chromosomes are small with 4 nucleoli, i.e., 4 satellite chromosomes are present in the mitotic cell division (in metaphase plate). The nucleolus usually disappears before the dissolution of the nuclear membrane, i.e., on the prior phase of the metaphase. It was found that sometimes nucleoli dividing at the equatorial plane migrate to polar end and remain in the cytosol, but do not form a part of the daughter nuclei. Similar, kind of nucleoli division was reported in the lower as well as higher plants (Frew and Bowen 1929). This analysis showed that the *O. coarctata* is an ancient plant. It represented the evolutionary transition of aquatic plant to land plant because of the presence of some ancient as well as advanced characteristics of terrestrial habitat. *O. coarctata* is an allotetraploid. An allotetraploid consisted of two distinct diploid genomes. It has four sets of chromosomes which had been originated through hybridization of diverged diploid species. Cytogenetics analyses distinguished *O. coarctata* as KKLL type genome (Ge et al. 1999; Lu et al. 2009) based on chromosome pairing behavior during the zygotene stage of meiotic in interspecific hybrids. Previously, genome of *O. coarctata* was denoted as HHKK type based on a phylogenetic study of two nuclear genes, alcohol dehydrogenase gene 1 (*Adh1*) as well as 2 (*Adh2*) and the chloroplast gene *matK* (maturase K) sequences in rice species. This study revealed that HHKK genomes occupied the basal positions relative to those of the BBCC and CCDD allotetraploid genomes which suggested the most ancient origins and allowed genomic rearrangement events like deletions of some homologous loci. Similar type of phylogenetic tree was predicted with advance technology (Ammiraju et al. 2010; Lu et al. 2009; Ge et al. 1999), where genotype of the *O. coarctata* was depicted as KKLL and position of KKLL type is between KK and HHLL type. However, there was no reports on chromosome complementation in *O. coarctata*.

8.5 Breeding Approaches

Plant breeding is a powerful technique to produce desired genotypes with a specific trait. It is one of the ancient techniques that includes crossing of closely related species. The crossability of *O. sativa* and its wild species can be categorized as the primary, secondary and tertiary gene pools (Khush 1997). The cultivar genotype of *Oryza* species (*O. sativa* and *O. glaberrima*) had AA genome, which were easily crossable among AA genome and regarded as the primary gene pool. The hybridization between BB and EE genomes is regarded as the secondary gene pool, and the remaining genomes (KK × FF) constitute the tertiary gene pool. The crossability in *Oryza* species is related to their phylogenetic relationships. Incase of *O. coarctata*, although attempt has been made for a successful hybridization with salinity susceptible rice genotype but with little success. For example, crosses were made between *O. coarctata* and different salt-sensitive rice cultivars (IR28, IR36, and Tellashamsa) using *O. coarctata* as pollen donor. The hybrid of *O. coarctata* and rice zygotes was produced in vitro from 6-day-old fertilized ovule which subsequently developed to embryo on MS media, devoid of hormone, but failed to develop further. Within the limited success, the percentage of viable hybrid was more in cross between *O. coarctata* and IR28 compared to IR36 and no hybrid was found in the cross with Tellashamsa. The hybrids were male sterile, triploid and possess all dominant traits of *O. coarctata* (Jena 1994), although details of their growth are not reported in the literature.

The genetic analysis through RAPD and AFLP markers revealed that *O. coarctata* was closely related to *O. australiensis* (Rangan et al. 2002). In order to make a connecting bridge between *O. coarctata* and *Oryza* sp., it was hypothesized that the F_1 of *O. coarctata* with recipient *O. australiensis* (diploid) may be successful and repeat backcrossing with the recipient may produce the diploid hybrid progeny (Latha et al. 2004). Nevertheless, till today no commercially successful cross between this species and rice is reported.

8.6 In Vitro Studies

Plant tissue culture is used for growth of plant cells, tissues and its organs under sterile conditions in a well-defined nutrient culture medium. In general, this technique is suitable for multiplication of perennial plants or genetic manipulation of the any plants. Nevertheless, there are some attempts of in vitro culture of this species mainly to provide salinity stress uniformly under in vitro conditions. Additionally, micropropagation of *O. coarctata* was done using nodal segment as explants (Latha et al. 1998). The major problem with *O. coarctata* in vitro culture is its association with an endophyte fungus, *Acremonium* sp., which caused a major problem in establishing axenic cultures (Ramanan et al. 1996). The explant was cultured on WPM (Woody Plant Medium, Quoirin and Lepoivre 1977) and MS (Murashige and Skoog 1962) media containing BA (2.2–5.5 mM) and kinetin (0.4–2.3 mM). The new shoots were produced after two weeks of the culture. But the percentage of multiplication of nodal segments was found to be high on WPM provided with 5.5 mM BA and 2.3 mM kinetin. Further, trimming of the apical meristem enhanced the lateral tillers growth. These multiplied shoots were transferred to basal rooting media followed by supplementing with 10.7 mM NAA (Naphthaleneacetic acid). It had been found that the root length decreased in media supplemented with auxin after it was immediately transferred from shooting media. Overall, the best response was found on WPM in compared to MS media (Latha et al. 1998). In a separate work, it had been found that the anther culture was unsuccessful because the frequency of callusing from anthers was extremely low (Subramaniam et al., unpublished). Somatic hybridization of *O. coarctata* and *O. sativa* L. cv 'Taipei 309' was made in which cell suspension-derived protoplasts of *O. sativa* cv 'Taipei 309' ($2n = 2x = 24$) were fused with non-dividing leaf mesophyll protoplasts of *O. coarctata* ($2n = 4x = 48$). Majority of the hybrid were diploid with $2n = 24$ except for few lines that were found to be tetraploid or hexaploid ($2n = 6x = 72$). The cytological analysis revealed that tetraploids were amphidiploid and contain chromosome sets of both parental genomes and hence was regarded as a symmetric somatic hybrid, which showed slow growth (Jelodar et al. 1999). Further, the growth of the somatic hybrids was not reported in their study.

8.7 Strategies and Tools for Genome Sequencing

The genome of the *O. coarctata* has been sequenced recently. A *de novo* sequence of *O. coarctata* genome was done for the first time (Mondal et al. 2017) based upon illumina shot gun, mate-pair as well as oxford nanopore platform representing an estimated 372.48× depth. The genome was assembled with various assembly software giving a scaffold N50 of 1858627 providing an estimated 87.71% coverage of the 665 Mb genome but previously its genome size was estimated to be 1283 Mb (Zuccolo et al. 2007). An estimated 34469 protein-encoding genes were predicted in this genome assembly (Table 8.1).

But several important genes were sequenced by PCR-based cloning. Only 498 expressed sequence tags are available in the NCBI database till today (Table 8.2).

In general, these genes are cloned, characterized for academic purpose to demonstrate their function for salinity stress tolerance, though no systematic study has been conducted to develop transgenic rice plants for salinity stress tolerance.

Table 8.1 Statistics of *O. coarctata* genomes sequenced and assembled (Mondal et al. 2017)

Feature	Data
Estimated genome size	665 Mb
Assembled genome size	Approx. 560 Mb
Number of protein-coding genes	34,469
Contig N50	15,132 bp
Sequencing platform	Illumina and oxford nanopore
Sequencing strategy	Whole genome sequencing (WGS)

Table 8.2 Sequenced genes, their function, and approaches used for cloning

Name	Genbank ID	Function	Approaches used	Reference
Serine-rich protein	AF110148	Accumulated under salinity stress especially in root	Partial cDNA clone and identified with heterologous probe	Mahalakshmi et al. (2006)
L-myo-inositol-1-phosphate Synthase	AF412340	Convert glucose-6-phosphate to L-myo-inositol-1-phosphate to form inositol	PCR-based cloning	Majee et al. (2004)
Inositol methyl transferase	EU240449	Pinitol synthesis under salinity stress	PCR-based cloning	Sengupta et al. (2008)
NHX1	JQ782416	Antiporters (Na^+/H^+ and K^+/H^+ antiport)	Identified through heterologous probe and 5′ end sequenced through RACE	Kizhakkedath et al. (2015)
Ubiquitin	HQ340170	Degrade superfluous protein	PCR-based cloning and 5′ end cloned through RAGE PCR	Philip et al. (2013)
Phosphoenolpyruvate carboxylase	EU371116	Convert phosphoenolpyruvate into oxaloacetic acid in presence of carbonic acid	Partial clone with heterologous probe	
Metallothionein	AF257465	Homeostasis of essential metals	PCR-based cloning	
CatA	AB014455	Catalase	PCR-based cloning	Iwamoto et al. (1999)
Adh	EU371995	Convert ethanol to its acetaldehyde and acetic acid and generate NAD^+	PCR-based cloning	Sengupta et al. (2016)
V-ATPase subunit c	AF286464	Proton transport in the vacuole	PCR-based cloning	
eIF 1	AF380357	Translational initiation factor	PCR-based cloning	Latha et al. (2004)
Homeobox protein	AF384375	Induce cellular differentiation and signal transduction	PCR-based cloning	Latha et al. (2004)
NADH dehydrogenase	AY507935	Dehydrogenase(remove H^+ atom)	PCR-based cloning	
Fructose-1,6-bisphosphatase	EU371109	Photosynthesis (Calvin cycle) and gluconeogenesis	PCR-based cloning	Chatterjee et al. (2013)
Maturase (*matK*)	AF148669	Spliced intron	PCR-based cloning using heterologous primer information	Ge et al. (1999)
G protein alpha subunit	AY792545	Cell signaling	PCR-based cloning using heterologus primer information	Guo and Ge (2005)

(continued)

Table 8.2 (continued)

Name	Genbank ID	Function	Approaches used	Reference
tRNA-Leu	AY792522	Translation	PCR-based cloning using heterologus primer information	Guo and Ge (2005)
Triosephosphate isomerase	EU371994	Convert triose phosphate isomers dihydroxyacetone phosphate (glycolysis)	PCR-based cloning	Sengupta et al. (2008)
HKT1	KU994892	Sodium transporter	PCR-based cloning	–
SOS2	KP330207	Serine/threonine-type protein kinase	PCR-based cloning	–
SOS3	KP330206	Calcium-binding protein	PCR-based cloning	–
eIF5a	KP330205	Translation elongation and regulates mRNA turnover, actin dynamics and cell cycle	PCR-based cloning	–
E3 ubiquitin-protein ligase	KM349311	Transfer ubiquitin from *E2* to the target protein	PCR-based cloning	–
ASR	KM349310	Induced abscisic acid formation	PCR-based cloning	–
salT	KM349309	Binds to lactin to activate abiotic-responsive genes	PCR-based cloning	–
Histone H3	AF109910	Structural component of nucleosome	PCR-based cloning	–
DREB2	KM349308	Transcriptional activator and binds to DRE element to enhance transcription of dehydration-responsive genes	PCR-based cloning	–
Plastidal Ribosomal protein (*rpl33*)	FJ908552	Plastid translation under cold stress	PCR-based cloning using heterologus primer information	Tang et al. (2010)
Ycf3	FJ908683	Thylakoid protein involves in the stable accumulation of PSI	PCR-based cloning using heterologus primer information	Tang et al. (2010)
PetB	FJ908378	Involves in electron transport and the generation of ATP	PCR-based cloning using heterologus primer information	Tang et al. (2010)
Ribosomal protein S19 (*rpl19*)	FJ908444	Acts as secondary rRNA binding protein	PCR-based cloning using heterologus primer information	Tang et al. (2010)
Ribosomal protein L20 (*rpl20*)	FJ908172	Structural component of large ribosome	PCR-based cloning using heterologus primer information	Tang et al. (2010)
NADH dehydrogenase subunit 3	FJ908313	Help in formation of a ferredoxin binding site of NDH	PCR-based cloning using heterologus primer information	Tang et al. (2010)

(continued)

Table 8.2 (continued)

Name	Genbank ID	Function	Approaches used	Reference
Ribosomal protein S16	FJ908411	Structural component of small ribosome	PCR-based cloning using heterologus primer information	Tang et al. (2010)
ATP synthase beta chain	FJ908106	Form the catalytic core of F_1 subunit	PCR-based cloning using heterologus primer information	Tang et al. (2010)
F-box family-1	GQ203300	Mediates degradation of protein, cell signaling and regulate cell cycle	PCR-based cloning	Ammiraju et al. (2010)
Monoculm1	FJ032635	Regulates tiller number	Sequence identified through probe *MOC1* from BAC clones library and gaps filled through primer walking	Lu et al. (2009)
Ribulose bisphosphate carboxylase	HE577876	Help in carbon dioxide fixation in the Calvin cycle	PCR based with pleurocarp mosses sequence as template for primers design	Edwards et al. (2011)
NBS-LRR disease resistance protein family-2	GQ203301	Disease resistance	PCR-based cloning	Ammiraju et al. (2010)
PolA1	AB366710	Mediate initiation of DNA replication	PCR-based cloning	

8.8 Physiological Basis of Salt Tolerance

Many physiological as well as morphological adaptations have been reported in *O. coarctata*. The important physiological adaptation included maintenance of low Na:K ratio, high photosynthetic efficiency, maintenance of relative water content and efficient function of vacuolar H^+-ATPase under salinity stress. It had been found that physiological adaptations of *O. coarctata* have been an excellent example, among the halophytes.

Under high salinity stress, the sodium ions shed off from the salt hairs instead of getting compartmentalized in the mesophyll cells. It was also reported that the calcium concentration in root increased with the increase in the salinity (Sengupta and Majumder 2009). The calcium concentration was also found to be lower in the leaf than stem (Bal and Dutt 1986). The ion compartmentalization occurs in different parts of *O. coarctata* under salinity stress condition (Garcia 1992). The Na^+ accumulation was maximum in the roots than in the floral parts. The ion accumulation in the leaves helped in maintaining the low Na:K ratio within shoots (Latha et al. 2004), which decrease the deleterious impact of the salinity. There was no change in osmotic potential of leaf sap under saline condition except at 31.7 dS m^{-1} ECe, which may be due to the increased concentration of organic solutes in the tissues. It is noteworthy to mention that with the increase in soil salinity up to ECe −8.1 dSm^{-1}, the total biomass production increased 65.0% over the control (Bal and Dutt 1986). The relative water content of *O. coarctata* was retained 90% under both normal and salinity stress. However, the positive correlation had been found between leaf sodium and water content because of the presence of efficient membrane channels (Latha et al. 2004). The vacuolar Na^+/H^+ antiporters which had been found to be regulated diurnally in leaves under salinity stress in the morning, i.e., after 12 h of treatment and decreased after 24 h of the treatment. Although there is high similarity between Na^+/H^+ (NHX) antiporters of rice and *O. coarctata*, but it performed more efficiently in the later (Kizhakkedath et al. 2015). Besides, these ions and water homeostasis capacity under salinity stress, this halophyte had light-harvesting complex with the ability to absorb excess light energy during abiotic stress. There was no loss of photosynthetic system I and II in function under high hypertonic condition. The electron transport and energy trapping system remain 88% functionally active under saline condition. It was also found that there were differences in the sodium, potassium, calcium and chloride content in the tissue of *O. coarctata* under salinity stress. Besides, they are able to maintain the mineral concentration in their tissues according to their requirement, such as they exclusion of sodium ions under salinity stress but retention of potassium ions. These physiological adaptations provided us a valuable information regarding salinity-responsive metabolites. Moreover, further exploration of the stress biology of this halophyte may provide unique glimpses for salinity tolerance mechanism.

8.9 Transcriptomic Resources

These genomic resources are very important for studying the gene expression. The RNA-seq data (transcriptomic sequences) generated from *O. coarctata*, subjected to high salinity stress treatments such as 450 and 700 mM NaCl, submerged and submerged in 450 mM NaCl concentration (Garg et al. 2013). In this experiment, both salinity and submerged tolerance-related transcripts were identified. It was found that in total, 15,158 genes were differentially expressed. Out of which 3020 and 2553 genes were up-regulated under 400 and 700 mM NaCl salinity stress regimes, respectively. Under salinity with the submerged condition, more genes were expressed in compared to the submerged condition only; this showed that *O. coarctata* had specific genes for submergence as well as salinity tolerance. Several transcription factors had been found to be expressed differentially under stress condition such as *Aux/IAA*, *bHLH*, bromodomain, *bZIP*, *C2C2-Yabby*, *C3H*, *CCAAT*, *CCHC*, *HB*, *HSF*, *PHD*, *SET*, *SNF2* and *WRKY*. The GO analysis of these genes revealed that they are involved in several biological events such as photosynthesis, photorespiration, rRNA processing, and secondary metabolites synthesis (such as oxoacid, ketone, amino acid, oxylipin, flavonoid, phenylpropanoid and chorismic acid). Under submergence condition, ABA, signaling gibberellin biosynthesis and several anaerobic metabolic pathways have been found to be up-regulated. Several biochemicals metabolism pathways such as suberin, ABA, fatty acid, jasmonates, carbohydrate, sesquiterpenoid, and ammonia assimilation were induced in this plant under salinity stress. On the other hand, under the anoxic condition, biosynthetic pathways, such as cellulose, cell structure, chlorophyll, ethylene, sugar, tryptophan, suberin, and ammonia assimilation have been found to be induced.

The miRNAs are non-coding RNAs that play pivotal roles in every domains of life including stress response. There are very few reports on the discovery of salt-responsive miRNAs from halophytes. In a study, small RNA-seq libraries were made from the leaves of *O. coarctata* which yielded 338 known and 95 novel miRNAs (Mondal et al. 2015). Additionally, 48 conserved miRNAs along with their pre-miRNA sequences were discovered through in silico analysis. In total, 36 known and 7 novel miRNAs were found to be up-regulated, whereas 12 known and 7 novel miRNAs were down-regulated under salt treatment. Further, 233 and 154 target genes were predicted for 48 known and 14 novel differentially regulated miRNAs, respectively. These targets with the help of gene ontology analysis were found to be involved in several important biological processes that could be involved in salinity tolerance. Relative expression trends of majority of the miRNAs as detected by real-time PCR as well as predicted by Illumina sequencing were found to be coherent. Additionally, expression of most of the target genes was negatively correlated with their corresponding miRNAs. The GO studies revealed that several metabolic processes such as cellular homeostasis, cell death, regulation of transcription and transportation were significantly repressed under salinity condition. From the target prediction analysis, it was also revealed that the differentially expressed salt-responsive miRNAs had many target genes such as transcription factors, SPB-like proteins, mybdomainproteins, auxin response factor, APETELA 2, NAC domain-containing proteins, WRKY transcription factors (WRKY-30, 35, 47, 52, 61, 90), and nuclear factor Y subunit. These transcriptional factors had been found to activate many stress-related genes which included stresses related proteins belonging to cell elongation and carbohydrate metabolism. Several pathways in *O. coarctata* were found to be exclusively induced under salinity stress, such as phenylpropanoid biosynthesis, phenylalanine metabolism, sucrose metabolism, phenyl propanoid metabolism, ubiquitin biosynthesis pathways, and other terpenoid pathways. The major aim of deep sequencing of miRNAs was depiction of the networking between miRNAs and its targeted genes in salinity tolerance.

8.10 Proteomics Studies

The proteomics study of the *O. coarctata* revealed that severity of loss of functional proteins under hypertonic state was low as compared to other rice varieties such as Pokkali (salt tolerant) and IR64 (salt sensitive). This result showed that intrinsic salinity tolerance property in *O. coarctata* was high. The 2D-PAGE image analysis depicted that around 59 and 74 was highly up-regulated spots under 200 and 400 mM of NaCl concentration, respectively. Out of these spots, the 16 abundant spots were analyzed by laser-desorption ionization time-of-flight mass spectrometry (MALDI-TOF). The photosynthetic proteins such as oxygen-evolving complex (OEC), CP47 (maintain PSII centre functionally active), PSI reaction centre subunit IV protein and RuBisCO had been found to be up-regulated under increased salinity stress. Moreover, other important proteins such as chloroplastic glutamine synthase, CRTDRE, Hsp70 chaperones, cellulose synthase, alcohol dehydrogenase, and sucrose synthase have been found to be enriched under salinity stress. Although, rice is taxonomically related to *O. coarctata* but the proteomics studies revealed that less than 50% of the expressed proteins in *O. coarctata* matches with rice (Sengupta and Majumder 2009). These identified proteins were functionally related to the inherent physiological properties of the halophytic character of *O. coarctata*.

8.11 Molecular Basis of Salt Tolerance

Different approaches which are known to the salinity tolerance mechanism of *O. coarctata* are depicted in Fig. 8.4. The major pathways that have been discovered so far have been discussed below.

8.11.1 Serine-Rich Protein

The root of *O. coarctata* when subjected to high salinity stress (427 mM NaCl, for 12 h) showed high level of expression of the protein which was rich in serine followed by leucine, phenylalanine, and proline. This protein (PcSrp) showed >97% of similarity with heterologous probe from *Medicago sativa*. The PcSrp protein consisted of 288 amino acids with the molecular mass of about 32 kDa. The PcSrp protein showed ~50% homology to the rice protein in the N-terminal region only. This protein had specific sequence stretch SPSPSPSS and SSSSS amino acid, with isoelectric point (PI)-8.84. In vitro localization of this PcSrp mRNA through northern blot showed hybridized signals in NaCl stressed root and rhizome. However, no such signal was observed in the stressed leaf and unstressed leaf as well as root indicating that it accumulated only in root under salinity stress. The PcSrp protein highly expressed in the epidermal and cortex cells of the salt-stressed root which probably provide stress tolerance. In rhizome, the vascular tissues and in

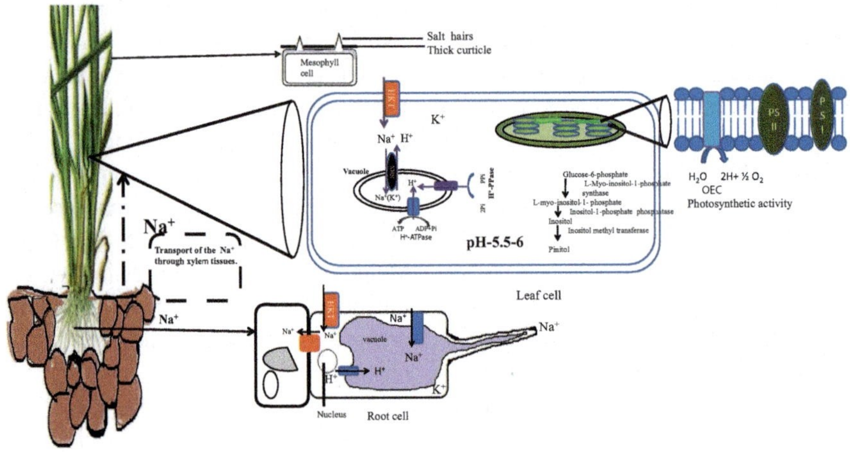

Fig. 8.4 Diagramatic representation of salinity tolerance mechanism of *O. coarctata* under salinity stress

certain cortex cells showed high expression of this protein indicating its salinity-responsive nature. However, under salinity stress, in yeast encoding PcSrp protein showed better growth and tolerance when compared to control suggesting that this gene rendered salt tolerance. This experiment depicted that post-translational modification was essential for the functional activity of this protein. Over expression of *PcSrp* in finger millet under promoter of rice *Actin1* showed a high germination rate under salinity stress. The transgenic plant (T_1) showed good tolerance to salinity stress and maintained low Na^+ and K^+ ion content (Mahalakshmi et al. 2006).

8.11.2 L-myo-Inositol-1-Phosphate Synthase

Inositol is a sugar alcohol with six carbon atoms (cyclohexane). Inositol played an important role as secondary messengers in eukaryotic cells. The enzyme required for the synthesis of the inositol, is known as L-myo-inositol-1-phosphate synthase, which converts glucose-6-phosphate to L-myo-inositol-1-phosphate. This L-myo-inositol 1-phosphate is further converted to inositol by the enzyme, inositol-1-phosphate phosphatase. It has been reported that L-myo-inositol-1-phosphate synthase from *Archaeoglobus fulgidis* provided thermo-tolerance in *Escherichia coli* (Chen et al. 2000).

Comparative nucleotide sequence analysis of *O. coarctata* L-myo-inositol-1-phosphate synthase (PINO1) and its rice orthologue showed that there was difference in the organization between amino acids. Two deletion mutants, i.e., PINO1-1 (deletion between Asn-342 and Lys-361) and PINO1-2 (deletion between Trp-174 to Ser-210), were synthesized by in vitro mutagenesis. The expression analysis in yeast and bacteria showed that RINO1 and PINO1-2 were functionally active under normal conditions, but with the increase in NaCl concentration these enzymes became inactive (Majee et al. 2004). However, PINO1-1 retained its functional activity with the increased salt concentration. It had been found that mutant PINO1-1 is functionally active in every condition depicting that this portion of the enzyme is the catalytic active domain. Furthermore, hybrid gene of RINO1 and PINO1-1 showed salt tolerance properties (Dastidar et al. 2006). The structure of PINO1 protein was found to be stable with the addition of salts because of electrostatic interactions were less, i.e., there was considerable difference in the exposition of the charged residues on the outer surface of the protein. The transgenic tobacco line with PINO1 showed better growth and photosynthetic efficiency, with an increased in the synthesis of inositol under salinity stress (Majee et al. 2004).

8.11.3 Inositol Methyl Transferase

Pinitol (O-Methylated inositol), a cytosolic sugar alcohol, is accumulated under salinity stress in leaves of this plant (Sengupta et al. 2008). Pinitol synthetic pathway is considered as a halophyte specific pathway. Pinitol acts as an osmoprotectant because it maintains the enzymes and cell membrane integrity under osmotic stress (Ishitani et al. 1996). Pinitol synthesis depends upon the methylation of the inositol in the presence of the enzyme inositol methyl transferase (EC 2.1.1.4). Inositol methyl transferase is encoded by an *IMT1* gene in an S-adenosyl methionine (SAM)-dependent reaction. The activity of the *O. coarctata* IMT1 (*PcIMT1*) is depended upon the intracellular ratio of SAM to SAH (S-adenosyl homocysteine) (Vernon and Bohnert 1992). The methyl cycle became activated under salinity stress because of the upregulation of photorespiration, since the carbon dioxide level decreases due to closure of the stomata (Sheveleva et al. 1997). The inverse relationship of the content of inositol and pinitol had been reported. Under salinity stress condition, the pinitol content increased more than inositol. Nucleotide sequence of the *PcIMT1* showed 99% of similarity with the ice halophyte *Mesembryanthemum crystallinum*, whereas no significant match was found within any member of *Oryza* methyl transferase. *PcIMT1* had two O-methyl transferase domains, SAM-binding motifs and dimerization domain. The N-glycosylation site and transmembrane domain have also been reported in PcIMT1 protein with amino acid length of 365. It was reported that the both transcript as well as protein level increased under salinity stress, oxidative stress (paraquat) and ABA treatment in *O. coarctata*. But no accumulation of pinitol has been found in rice as analyzed through semi-quantitative RT-PCR. However, under cold stress, no upregulation of the *PcIMT1* transcript had been found, whereas under heat stress *PcIMT1* transcript level decreased to some extend because of the wilting of the *O. coarctata*. In phylogenetic analysis, *PcIMT1* was found to be distantly related with rice orthologue indicating that the *O. coarctata* is more primitive, evolutionary distantly related to rice (Ge et al. 1999; Sengupta et al. 2008).

8.11.4 Vacuolar Membrane Transporter

The vacuolar *NHX1* is a type of Na^+/H^+ and K^+/H^+ antiporter that mediates the vacuolar compartmentalization of Na^+ ions under salinity condition. It prevents the toxic effects of the sodium ions and maintains the Na^+/K^+ ratio in the cytosol. There are four vacuolar NHX isoforms (NHX1–NHX4) reported in Arabidopsis (McCubbin et al. 2014). Under the normal growth condition, NHX antiporters accumulated the K^+ ions and nutrition (Leidi et al. 2010). The amino acid sequence of PcNHX1 showed 96% similarity with rice NHX1 (*OsNHX1*). The expression of *PcNHX1* was found to be up-regulated under salt treatment of 6 h, but decrease at 12 h and further increased again at 24 h of the treatment. Maximum expression is detectable at 48 h of NaCl application (Kizhakkedath et al. 2015). Such expression level was found to be an adaptation in their habitat. The sequestering of sodium ions in the salt hairs was thought to be performed by the most efficient channels such as NHX1 and V-type H^+-ATPase. The transient expression level of *PcNHX1* in tobacco showed tissues specific expression. The expression of the *PcNHX1* in transgenic tobacco was found to be high in external and internal phloem of leaf petiole, stem, petiole as well as in epidermal, sub-epidermal layer-mesodermal cells guard cells, and trichomes. Under salinity stress condition, the expression of *PcNHX1* was up-regulated in root vascular bundle, cortex and root tips. Similarly, the type of tissues-specific expression *LmNHX1* was found in the halophyte *Lobularia maritima* (Popova and Golldack 2007). Other membrane channel such as *HKT1* also played important role maintaining the cytosolic homeostasis under osmotic stress in halophyte such as *M. crystallinum* and *Suaeda salsa* (Barkla et al. 2002; Shao et al. 2014). Likewise, in the *O. coarctata* such type of transporter also helped in maintaining K^+ ions homeostasis under stress condition.

The vacuolar H^+-ATPase (V-ATPase) is mainly involved in the acidification of the vacuole. The H^+-ATPase was found to be induced under chilling and salinity stress (Senthilkumar et al. 2005). This transporter acidification helps in maintaining Na^+ under higher concentration. The H^+-ATPase has two domains V_1 and V_0. It has been reported that the transcript of the V_0 domain (c-subunit) of the *O. coarctata* enhanced with the increase of the duration of the salinity stress. But the expression of the c-subunit was found to be more in the roots than in the leaves under salinity stress. It is present as multigenic family. The efficiency of the V-ATPase was found to be higher under salinity stress and it was thought that V-ATPase played important role in the tolerance mechanism of the *O. coarctata.*

8.11.5 Ubiquitin

The promoter region of a gene is very important under regular as well as in the stress condition. The stronger the induction of promoter higher will be the gene expression. The ubiquitin was found to be constitutively expressed. The transient expression of the promoter region of the ubiquitin from *O. coarctata* (Port Ubi2.3) in sugarcane was found to be stronger than the commonly used promoters such as maize Ubil, CaMV 35S. The expression of PortUbi2.3 was stronger in the monocot than in dicot. This difference level was due to heterologous intron processing between monocots and dicots (Philip et al. 2013). The ubiquitin gene of this species consisted of two exons and introns. The promoter sequence along with the proximal exon (I) and intron (I) was sufficient to drive higher levels of the expression of *GUS* (β-glucuronidase) reporter gene in monocots than Port *Ubi2.3*. The same constructed fragment of the Port *Ubi2.3* had higher expression of GUS in tobacco than *CaMV 35S* promoter. The 5′ UTR RNA folding was analyzed and it was observed that the folding was weak and unstable, which reflected the higher rate of the translation of the protein. Many cis-elements for both biotic and abiotic stresses were observed in the promoter region of the Port *Ubi2.3*. The tissue-specific and abiotic stress-responsive elements were found around the transcription start site (TSS), whereas biotic stress-responsive elements were clustered away from the TSS. The cis-elements such as *MYB*, *MYC* for dehydration stress, *WBOX* for wound-responsive, *WRKY* for pathogen-responsive, *DOFCOREZM* for signal-responsive and tissue-specific gene expression. Port*Ubi2.3* sequence consisted of the Matrix Attachment Regions (MARs) which act as nuclear-anchoring sites and found in many AT-rich regions. There were many long poly (A) or poly (T) sequence stretches, direct repeats and inverted repeats or palindromes found to be associated with the *O. coarctata* 3′ regulatory region of ubiquitin. The multiple alignment of the Port *Ubi2.3* sequence with the ubiquitin sequences from maize, sugarcane and rice showed the sequence similarity ranged from 27 to 47%, whereas the similarity among their respective promoter regions (−1 to −640 bp) ranged from 61 to 64% (Philip et al. 2013). It was thought that because of the presence of such stronger promoter as well as *cis*-elements the expression levels of stress-responsive genes were higher in *O. coarctata.* This higher expression of the stress-responsive genes helped *O. coarctata* to survive under such unfavorable conditions.

8.11.6 Eukaryotic Translation Initiation Factor 1 (*eIF1*)

The translation factors play vital role in the translation and gene expression. The translation factor, *eIF1* had ability to bind with 40S ribosome subunit-mRNA complex to open the mRNA conformation for activated tRNA. The expression analysis of the *eIF1* in *O. coarctata* (*PceIF1*) from leaf under different abiotic stresses was studied. The expression of the *eIF1* was higher under salinity stress for 5 and 10 h of treatment in compared to control. However, after 10th day of treatment, the transcript level of the *eIF1* decreased. It was analyzed that the expression of the *PceIF1* was higher under ABA and mannitol treatments which suggested the transient increase in the expression of the *eIF1*. In this study, it was observed that *PceIF1* was induced under both ABA dependent as well as independent pathway. The *PceIF1* consisted of 154 amino acids with a molecular weight of 12.7 kDa (Latha et al. 2004). This study showed the transient expression of the *PceIF1* under an abiotic condition in the leaves which predicted the existence of a salt tolerance pathway including *eIF1*.

8.11.7 Chloroplast Fructose-1, 6-Bisphosphatase

Chloroplast fructose-1,6-bisphosphatase is a regulatory enzyme in the Calvin cycle and gluconeogenesis pathway. It converts fructose-1,6-bisphosphate to fructose 6-phosphate. The fructose 6-phosphate was found to be converted into its isomer glucose 6-phosphate, which played an important role in several metabolic processes. The fructose-1,6-bisphosphatase of *O. coarctata* (*PcCFR*) was found to be active under saline condition. The sequence analysis of fructose-1,6-bisphosphatase showed that this enzyme had ability to withstand high salinity condition due to the presence of three serine amino acid residues in specific positions (S^{14}, S^{48}, S^{168}) and mutation of S^{168} reduced its efficiency under salinity stress in compared to mutations in other serine residues. The introgression of *PcCFR* gene in tobacco enhanced its salt tolerance. It has been reported that transgenic tobacco showed higher photosynthetic efficiency and growth rate under saline conditions. The *PcCFR* played important role in enzymatic activation of RuBisCO.

There were several reports on introgression of genes from halophyte into model plant which enhances the tolerance mechanism toward several abiotic stresses such as transfer of inositol polyphosphate kinase from *Thellungiella halophila* into soybean which increased its tolerance to water deficit, salt and oxidative stress (Liu et al. 2012). The photosynthetic enzymes as well as reaction centres mainly PSI and PSII were found to be active under the high Na^+ ions concentration because of the presence of osmolytes. The ability to withstand under high saline condition was due to the presence of several efficient membrane transporters, osmolytes, and salt exclusion glands. These properties were lacking in cultivar variety, so they cannot withstand such adverse conditions.

8.12 Future Prospects

Wild species is a good source of agronomically important genes. Though, this species could serve as good source of favorable alleles, non-coding RNA and regulatory sequences for salt and submergence stresses yet it remains unexplored. In future, attempts will be made to transfer the salt-tolerant characters of this species into rice cultivar. It was considered as the best halophytic monocot model to study the mechanism of salinity tolerance property. From phylogenetic analysis, it was observed that the *O. coarctata* origin was basal, i.e., in the course of evolution, this plant came earlier than other *Oryza* species. The habitat of the *O. coarctata* reflected the water to land transition of the plants in the mean time period. It was also assumed that many abiotic-responsive genes were absent in rice due to this transition. In general, due to the presence of abiotic stress-responsive genes and distinct morphological adaptation, *O. coarctata* is considered as a gold mine to study the salt-tolerance mechanism in plant and identification of genes which can be used for rice improvement against salinity stress.

References

Ammiraju JS, Fan C, Yu Y, Song X, Cranston KA, Pontaroli AC, Lu F, Sanyal A, Jiang N, Rambo T, Currie J (2010) Spatio-temporal patterns of genome evolution in allotetraploid species of the genus *Oryza*. Plant J 63:430–442

Askari H, Edqvist J, Hajheidari M, Kafi M, Salekdeh GH (2006) Effects of salinity levels on proteome of *Suaeda aegyptiaca* leaves. Proteomics 6:2542–2554

Bal AR, Dutt SK (1986) Mechanism of salt tolerance in wild rice (*Oryza coarctata* Roxb). Plant Soil 92: 399–404

Barkla BJ, Vera-Estrella R, Camacho-Emiterio J, Pantoja O (2002) Na^+/H^+ exchange in the halophyte *Mesembryanthemum crystallinum* is associated with cellular sites of Na^+ storage. Funct Plant Biol 29:1017–1024

Bauersfeld P, Kifer RR, Durrant NW, Sykes J (1969) Nutrient contents of turtle grass (*Thalassia testudinum*). Proc Int Seaweed Symp 6:637–645

Bhosale SH, Patil KB, Parameswaran PS, Naik CG, Jagtap TG (2011) Active pharmaceutical ingredient (api) from an estuarine fungus, *Microdochium nivale* (Fr.). J Environ Biol 32:653–658

Chatterjee J, Patra B, Mukherjee R, Basak P, Mukherjee S, Ray S, Bhattacharyya S, Maitra S, Ghosh D, Ghosh S, Sengupta S (2013) Cloning, characterization and expression of a chloroplastic fructose-1, 6-bisphosphatase from *Porteresia coarctata* conferring salt-tolerance in transgenic tobacco. J Plant Biotech 114:395–409

Chen L, Zhou C, Yang H, Roberts MF (2000) Inositol-1-phosphate synthase from *Archaeoglobus fulgidus* is a class II aldolase. Biochem 39: 12415–12423

Dastidar KG, Maitra S, Goswami L, Roy D, Das KP, Majumder AL (2006) An insight into the molecular basis of salt tolerance of L-myo-inositol 1-P synthase (PcINO1) from *Porteresia coarctata* (Roxb.) Tateoka, a halophytic wild rice. Plant Physiol 140:1279–1296

Edwards E, Aliscioni S, Bell H (2011) New grass phylogeny resolves deep evolutionary relationships and discovers C4 origins. New Phytol 193:304–312

Flowers TJ, Flowers SA, Hajibagheri MA, Yeo AR (1990) Salt tolerance in the halophytic wild rice,Porteresia coarctata Tateoka. New Phytol 114:675–684

Flowers TJ, Colmer TD (2015) Plant salt tolerance: adaptations in halophytes. Ann Bot 115:327–331

Frew PE, Bowen RH (1929) Memoirs: nucleolar behaviour in the mitosis of plant cells. J Cell Sci 2:197–212

Garcia A (1992) Salt tolerance in the halophytic wild rice, *Porteresia coarctata* Tateoka. Ph.D. Thesis, University of Sussex

Garg R, Verma M, Agrawal S, Shankar R, Majee M, Jain M (2013) Deep transcriptome sequencing of wild halophyte rice, *Porteresia coarctata*, provides novel insights into the salinity and submergence tolerance factors. DNA Res 8:1–16

Ge S, Sang T, Lu BR, Hong DY (1999) Phylogeny of rice genomes with emphasis on origins of allotetraploid species. Proc Nat Acad Sci 96:14400–14405

Guo YL, Ge S (2005) Molecular phylogeny of Oryzeae (Poaceae) based on DNA sequences from chloroplast, mitochondrial, and nuclear genomes. Am J Bot 92:1548–1558

Ishitani M, Majumder AL, Bornhouser A, Michalowski CB, Jensen RG, Bohnert HJ (1996) Coordinate transcriptional induction of myo-inositol metabolism during environmental stress. Plant J 9:537–548

Iwamoto M, Nagashima H, Nagamine T, Higo H, Higo K (1999) p-SINE1-like intron of the CatA catalase homologs and phylogenetic relationships among AA-genome *Oryza* and related species. Theor Appl Genet 98:853–861

Jagtap TG, Bhosale S, Charulata S (2006) Characterization of *Porteresia coarctata* beds along the Goa coast, India. Aquat Bot 84:37–44

Jelodar NB, Blackhall NW, Hartman TP, Brar DS, Khush G, Davey MR, Cocking EC, Power JB (1999) Intergeneric somatic hybrids of rice [*Oryza sativa* L. (+) *Porteresia coarctata* (Roxb.) Tateoka]. Theor Appl Genet 99:570–577

Jena KK (1994) Production of intergeneric hybrid between *Orzya sativa* L. and *Porteresia coarctata* T. Curr Sci 67:744–746

Khush GS (1997) Origin, dispersal, cultivation and variation of rice. In *Oryza*: from molecule to plant, vol 35. Springer Netherlands, pp 25–34

Kizhakkedath P, Jegadeeson V, Venkataraman G, Parida A (2015) A vacuolar antiporter is differentially regulated in leaves and roots of the halophytic wild rice *Porteresia coarctata* (Roxb.) Tateoka. Mol Biol Rep 42:1091–1105

Latha R, Ajith A, Srinivasa RC, Eganathan P, Balakrishna P (1998) In vitro propagation of salt-tolerant wild rice relative, *Porteresia coarctata* Tateoka. J Plant Growth Regul 17:231–235

Latha R, Salekdeh GH, Bennett J, Swaminathan MS (2004) Molecular analysis of a stress-induced cDNA encoding the translation initiation factor, eIF1, from the salt-tolerant wild relative of rice, *Porteresia coarctata*. Funct Plant Biol 31:1035–1042

Leidi EO, Barragán V, Rubio L, ElHamdaoui A, Ruiz MT, Cubero B, Fernández JA, Bressan RA, Hasegawa PM, Quintero FJ, Pardo JM (2010) The AtNHX1 exchanger mediates potassium compartmentation in vacuoles of transgenic tomato. Plant J 61:495–506

Liu M, Li D, Wang Z, Meng F, Li Y, Wu X, Teng W, Han Y, Li W (2012) Transgenic expression of *ThIPK2* gene in soybean improves stress tolerance, oleic acid content and seed size. Plant Cell Tissue Organ Cult 111:277–288

Lu BR, Ge S (2003) *Oryza coarctata*: the name that best reflects the relationships of *Porteresia coarctata* (Poaceae: Oryzeae). Nordic J Bot 23:555–558

Lu F, Ammiraju JS, Sanyal A, Zhang S, Song R, Chen J, Li G, Sui Y, Song X, Cheng Z, De Oliveira AC (2009) Comparative sequence analysis of MONOCULM1-orthologous regions in 14 *Oryza* genomes. Proc Nat Acad Sci 106:2071–2076

Mahalakshmi S, Christopher GS, Reddy TP, Rao KV, Reddy VD (2006) Isolation of a cDNA clone (PcSrp) encoding serine-rich-protein from *Porteresia coarctata* T. and its expression in yeast and finger millet (*Eleusine coracana* L.) affording salt tolerance. Planta 224:347–358

Majee M, Maitra S, Dastidar KG, Pattnaik S, Chatterjee A, Hait NC, Das KP, Majumder AL (2004) A novel salt-tolerant L-myo-inositol-1-phosphate synthase from *Porteresia coarctata* (Roxb.) Tateoka, a halophytic wild rice molecular cloning, bacterial overexpression, characterization, and functional introgression into tobacco-conferring salt tolerance phenotype. J Biol Chem 279:28539–28552

McCubbin T, Bassil E, Zhang S, Blumwald E (2014) Vacuolar Na+/H+ NHX-type antiporters are required for cellular K^+ homeostasis, microtubule organization and directional root growth. Plants 3:409–426

Mondal TK, Ganie SA, Debnath AB (2015) Identification of novel and conserved miRNAs from extreme halophyte, *Oryza coarctata*, a wild relative of rice. PLoS ONE 10:1–27

Mondal TK, Rawal HC, Gaikwad K et al (2017) First de novo draft genome sequence of *Oryza coarctata*, the only halophytic species in the genus Oryza. F1000Research 6:1750. (https://doi.org/10.12688/f1000research.12414.1)

Munns R, Tester M (2008) Mechanisms of salinity tolerance. Annu Rev Plant Biol 59:651–681

Murashige T, Skoog F (1962) A revised medium for rapid growth and bioassays with tobacco tissue cultures. Physiol Plant 15:473–497

Parthasarathy N (1938) Cytological studies in Oryzeae and Phalarideae. Cytologia 9:307–318

Philip A, Syamaladevi DP, Chakravarthi M, Gopinath K, Subramonian N (2013) 5' Regulatory region of ubiquitin 2 gene from *Porteresia coarctata* makes efficient promoters for transgene expression in monocots and dicots. Plant Cell Rep 32:1199–1210

Popova OV, Golldack D (2007) In the halotolerant *Lobularia maritima* (Brassicaceae) salt adaptation correlates with activation of the vacuolar H^+-ATPase and the vacuolar Na^+/H^+ antiporter. J Plant Physiol 164:1278–1288

Probert RJ, Longley PL (1989) Recalcitrant seed storage physiology in three aquatic grasses (*Zizania palustris, Spartina anglica* and *Porteresia coarctata*). Ann Bot 63:53–63

Quoirin M, Lepoivre P (1977) Improved media for in vitro culture of Prunus sp. Acta Hort 78:437–442

Ramanan BV, Balakrishna P, Suryanarayanan TS (1996) Search for seed borne endophytes in rice (*Oryza sativa*) and wild rice (*Porteresia coarctata*). Rice Biotechnol Q 27:7–8

Rangan L, Sankararamasubramanian HM, Radha R, Swaminathan MS (2002) Genetic relationship of *Porteresia coarctata* Tateoka using molecular markers. Plant Biosyst 136:339–348

Sengupta S, Patra B, Ray S, Majumder AL (2008) Inositol methyl tranferase from a halophytic wild rice, *Porteresia coarctata* Roxb. (Tateoka): regulation of pinitol synthesis under abiotic stress. Plant Cell Environ 31:1442–1458

Senthilkumar P, Jithesh MN, Parani M, Rajalakshmi S, Praseetha K, Parida A (2005) Salt stress effects on the accumulation of vacuolar H^+-ATPase subunit c transcripts in wild rice, *Porteresia coarctata* (Roxb.). Tateoka Curr Sci 89:1386–1393

Sengupta S, Majumder AL (2009) Insight into the salt tolerance factors of a wild halophytic rice, *Porteresia coarctata*: a physiological and proteomic approach. Planta 229:911–928

Sengupta S, Majumder AL (2010) Porteresia coarctata (Roxb.) Tateoka, a wild rice: apotential model for studying salt-stress biology in rice. Plant Cell Environ 33:526–542

Shao Q, Han N, Ding T, Zhou F, Wang B (2014) SsHKT1; 1 is a potassium transporter of the C3 halophyte *Suaeda salsa* that is involved in salt tolerance. Funct Plant Biol 41:790–802

Sheveleva E, Chmara W, Bohnert HJ, Jensen RG (1997) Increased salt and drought tolerance by D-ononitol production in transgenic *Nicotiana tabacum*. Plant Physiol 115:1211–1218

Tabatabaei SJ (2006) Effects of salinity and N on the growth, photosynthesis and N status of olive (*Olea europaea* L.) trees. Sci Horticult 108:432–438

Tang L, Zou XH, Achoundong G, Potgieter C, Second G, Zhang DY, Ge S (2010) Phylogeny and biogeography of the rice tribe (Oryzeae): evidence from combined analysis of 20 chloroplast fragments. Mol Phylogen Evol 54:266–277

Takemura T, Hanagata N, Suighara K, Baba S, Karube I, Dubinsky Z (2000) Physiological and biochemical response to salt stress in the mangrove *Bruguiera gymnorrhiza*. Aquat Bot 68:15–28

Tateoka T (1964) Notes on some grasses. XVI. Embryo structure of the genus Oryza in relation to the systematics. Am J Bot 1:539–543

Vaughan D, Kadowaki KI, Kaga A, Tomooka N (2005) Eco-genetic diversification in the genus *Oryza*:

implications for sustainable rice production. Inter Rice Res Inst 2005:44–46
Vernon DM, Bohnert HJ (1992) A novel methyl transferase induced by osmotic stress in the facultative halophyte *Mesembryanthemum crystallinum.* EMBO J 6:2077–2085
Yu J, Chen S, Zhao Q, Wang T, Yang C, Diaz C, Sun G, Dai S (2011) Physiological and proteomic analysis of salinity tolerance in *Puccinellia tenuiflora.* J Proteom Res 10:3852–3870
Zuccolo A, Sebastian A, Talag J, Yu Y, Kim H, Collura K, Kudrna D, Wing RA (2007) Transposable element distribution, abundance and role in genome size variation in the genus *Oryza.* BMC Evol Biol 7:152–167

Oryza glaberrima Steud. 9

Marie Noelle Ndjiondjop, Peterson Wambugu, Jean Rodrigue Sangare, Tia Dro, Bienvenu Kpeki and Karlin Gnikoua

Abstract

Oryza glaberrima is the African cultivated rice species, domesticated from its wild ancestor by farmers living in Inland Delta of Niger River. Several studies indicated that it has extremely narrow genetic diversity compared to both its wild progenitor, *Oryza barthii* and the Asian rice, *Oryza sativa* which can mainly be attributed to a severe domestication bottleneck. Despite its scarcity in farmer's field due to its low yield potential, high shattering and lodging susceptibility, *O. glaberrima* is of great value not only to Africa but also globally. Perhaps its greatest contribution to regional and global food security is as a source of genes, as it possesses resistance/tolerance to various biotic and abiotic stresses. It also has unique starch-related traits which give it good cooking and eating properties. Advances in DNA sequencing have provided useful genomic resources for African rice, key among them being whole genome sequences. Genomic tools are enabling greater understanding of the useful functional diversity found in this species. These advances have potential of addressing some of the undesirable attributes found in this species which have led to its continued replacement by Asian rice. Development of new generation of rice varieties for African farmers will therefore require the adoption of advanced molecular breeding tools as these will allow efficient utilization of the wealth and resilience found in African rice in rice improvement.

9.1 Introduction

Rice belongs to the genus *Oryza*, classified under the tribe Oryzeae, subfamily Oryzoideae of the grass family, Poaceae. There are 24 species belonging to the genus *Oryza* (Kellogg 2009; Jacquemin et al. 2013). *Oryza glaberrima*, commonly known as African rice, is one of the two independently domesticated rice species, having been domesticated in West Africa more than 3500 years ago (Portères 1956; Linares 2002). Though intensively produced in the past, *O. glaberrima* is being replaced by *Oryza sativa*

M. N. Ndjiondjop (✉) · J. R. Sangare · T. Dro · B. Kpeki · K. Gnikoua
Africa Rice Center (AfricaRice), 01 B.P. 2031, Cotonou, Benin
e-mail: m.ndjiondjop@cgiar.org

P. Wambugu
Kenya Agricultural and Livestock Research Organization (KALRO), Genetic Resources Research Institute, P. O. Box 781-00902, Kikuyu, Kenya

T. K. Mondal and R. J. Henry (eds.), *The Wild Oryza Genomes*, Compendium of Plant Genomes, https://doi.org/10.1007/978-3-319-71997-9_9

in African farmer's fields mainly due to its low yield potential, high shattering, lodging susceptibility, and also the pressure of widespread introduction of high-yielding improved cultivars (Ghesquière et al. 1997). Despite its decline, it is reported that it is still grown in various environments by smallholder rice farmers in several countries such as Nigeria, Mali, Sierra Leone, Senegal, and Togo (Vaughan 1994; Teeken 2015). Its resilience is mainly due to its hardiness, pest and disease resistance, and good response to low-input farming system (Wang et al. 2014). *O. glaberrima* is not only a cereal used in native diet but it also has sociocultural importance making it unique to Africa (Mohapatra 2010). It is used in ritual ceremonies by Jola people in Casamance in Senegal and in Danyi, a region in the hills of Togo, to appease the souls of the ancestors (Linares 2002).

As will be highlighted later in this chapter, the genetic potential of this African indigenous cultivated rice as a rich reservoir of genes, particularly for resistance to several biotic and abiotic stresses, is well known (Jones et al. 1997b; WARDA 1997; Sarla and Swamy 2005; Futakuchi and Sié 2009; Wambugu et al. 2013). African rice therefore plays a critical role in regional and global food security both as source of food for many households, particularly in West Africa and also as source of genes for rice improvement. Despite this potential, African rice remains grossly understudied and therefore unutilized in rice improvement (Wambugu et al. 2013) because crossing between *O. sativa* a cultivated species largely grown in Africa is complicated due to incompatibility that causes hybrid mortality, hindering heterogenic recombination, and progeny sterility. The problem was mainly overcome through backcrossings with the *O. sativa* parent coupled with anther culture, permitting the development of interspecific lines well adapted to African conditions which are trademarked as New Rice for Africa (NERICA). The NERICAs are adapted to ecology of both upland and lowland. The first generation of NERICA rice is interspecific inbred progeny derived from crosses between *O. sativa japonica* varieties WAB 56-104, WAB56-50, and WAB181-18 as the recurrent parents and an *O. glaberrima* variety (CG 14) as the donor parent. Three thousand lines were obtained from these crosses. After two backcrosses and anther culture to break infertility barrier were conducted, 70 fertile and fixed lines were selected and evaluated throughout several African countries (Jones et al. 1997a; Semagn et al. 2007; Somado et al. 2008). Among these 70 lines, 18 were nominated by AfricaRice as NERICAs in which NERICA 1 to 11 were derived from crosses between WAB 56-104 and CG14 while NERCA 12 to 18 were derived from crosses between WAB 56-50 or WAB 181-18 and CG 14. Actually, 18 NERICA varieties are available for the upland rice ecology of SSA. Some of the NERICA varieties have a yield advantage over their *O. glaberrima* and *O. sativa* parents, either through superior weed competitiveness, drought tolerance, and pest or disease resistance or higher yielding potential. NERICA rice also has more protein in the grain, about 10–12% compared to 8–10% in other varieties. All these traits can significantly contribute to food security and improved nutrition in SSA. NERICA lines have been tested in 31 SSA countries, and about 150,000 ha are under upland NERICA production in Africa.

Concerning NERICA for lowland ecology, several lines have also been obtained after crosses between *O. sativa* and *O. glaberrima* with different levels of backcross among which over 500 fixed lines were selected and evaluated in lowland conditions in several West African country for phenotypic adaptability, yield potential, and resistance to Rice yellow mottle virus (RYMV). Out of these, 60 promising lines were selected and named NERICA-L. These lines were obtained from crosses between three *O. glaberrima* varieties (TOG5681, TOG5674, and TOG5675) and four *O. sativa* varieties (IR64, IR31785-58-1-2-3-3, IR31851-96-2-3-2-1, and IR1529-680-3-2). The backcross level of the 60 NERICA-L varieties was 4 BC1, 22 BC2, 19 BC3, and 15 BC4 (Ndjiondjop et al. 2008; Moukoumbi et al. 2015).

The continued advances in genomics are offering opportunities for extending our knowledge on the genetic value of this species and how its utility can be enhanced. Analysis of the genomes of African rice is enabling the identification of useful functional diversity.

This chapter discusses molecular and phenotypic diversity (Fig. 9.1) found in *O. glaberrima*. The advances in knowledge that have been witnessed in the last few years in the area of genetics and genomics particularly on genome sequencing of this important crop are also reviewed. The genetic potential of African rice in rice improvement, as well as the future perspectives on the utilization of this species, is presented.

9.2 History

Rice is an indigenous staple food in Africa. According to Buddenhagen et al. (1978), "rice is not only Asian, rice is also African." Just like in *O. sativa*, the origin and domestication history of African rice have been contentious, with various conflicting theories being advanced. Analysis of the genome of African rice is providing useful insights on the evolutionary patterns of this cultivated species (Sakai et al. 2011). Africa is the only continent where the two species, both of which are diploid (genome AA, $2n = 24$), are cultivated (Vaughan et al. 2003). *O. sativa* was domesticated in Asia, probably in China about 7000 years ago (Fuller et al. 2009). It was introduced in Africa through the Eastern region of the continent, probably in the first century AD by traders from India (Linares 2002) and subsequently to West Africa by the Portuguese in about 1500 AD (Portères 1962). The domestication of African rice occurred much later than that of Asian rice, but certainly prior to the introduction of Asian rice to Africa (Sarla and Swamy 2005; Sweeney and McCouch 2007). *O. glaberrima* was described and named as a new species by the botanist Steudel in 1855 after observation of rice plants collected by Jardin, a French naval officer in 1845–1848. He distinguished *O. glaberrima* from *O. sativa* on the basis of the smoothness of spikelets and leaves (Portères 1956), but he did not mention its origin. French botanists investigated the origin of this new species and in 1914 recognized that this species is indigenous to West Africa (Carney 2000). By 1970s, *O. glaberrima* became fully recognized by the international scientific

Fig. 9.1 Phenotypic diversity within *O. glaberrima* in term of **a** panicle and **b** grains **c** seeds conservation containers and packs for distribution to users in AfricaRice genetic resource unit

community as a new species, with its unique domestication history in West Africa (Carney 2000). Two hypotheses have been proposed regarding the domestication process of *O. glaberrima*. According to Portères (1962, 1976), African rice was first domesticated from its wild relative *Oryza barthii* (formerly known as *O. breviligulata*) by people living in the Inland Delta of the Upper Niger River, which is located within what is known today as the Republic of Mali about 3500 years ago. *O. glaberrima* is diploid ($2n = 24$) and shares the same genome type (AA) with its progenitor *O. barthii* and *O. longistaminata* from which *O. barthii* is derived. *O. barthii* is more widely distributed in Africa than *O. glaberrima*. However, *O. longistaminata* is the most widely distributed in sub-Saharan Africa and Madagascar (Khush 1997; Brar and Khush 2003). From its primary center of domestication, *O. glaberrima* diffused to two secondary centers, one along the Senegambian coast and Guinea Bissau where it was grown under irrigated conditions and the second in the Guinea forest between Sierra Leone and the western part of Côte d'Ivoire where it was produced under rainfed (upland) conditions (Traore 2005). Based on fragmented linguistic and archaeo-botanical evidence, Harlan and Stemler (1976) proposed an alternative theory, stating that *O. glaberrima* was domesticated in different localities by ancient hunting–gathering human populations who harvested and selected *O. barthii* forms found in vast forest and savanna areas.

Recently, studies using genomic data confirmed the hypothesis of Portères (1962) saying that *O. glaberrima* was domesticated from its annual ancestor *O. barthii* in a single region along the Niger River in Mali (Li et al. 2011; Wang 2014). Thereafter, the cultivation areas of African rice were extended to several African countries, from the delta of the river Senegal in the West to Lake Chad in the East (Fig. 9.2). *O. glaberrima* has also been found outside West African regions, in the islands of Pemba and Zanzibar in Tanzania (Protadatabase record, Bezançon 1993; Vaughan 1994). It was introduced into American countries, particularly in the Caribbean, Brazil, Guyana, El Salvador, Panama, and Carolina in USA during the slave trade era, by enslaved West African rice farmers (Vaughan 1994; Carney 2004). This idea is supported by results reported by van Andel et al. (2016) who, by analyzing 1,649,769 SNPs of 109 *O. glaberrima* accessions representing diversity across West Africa and African rice from a Maroon market in Paramaribo, Suriname, found no evidence of introgression from Asian rice.

Progress has been made in knowledge concerning the evolution of morphological traits and associated gene networks in relation to rice domestication of African rice. It is known today that several morphological traits were selected during domestication (Sweeney and McCouch 2007). Among them, the inflorescence architecture and large seeds are some of the main morphological traits modified during rice domestications.

It has been recently demonstrating that the higher complexity of the *O. glaberrima* panicle compared to that of its wild relative *O. barthii* is associated with a wider rachis meristem and resulted from early alterations to branching-related genes expression during reproductive

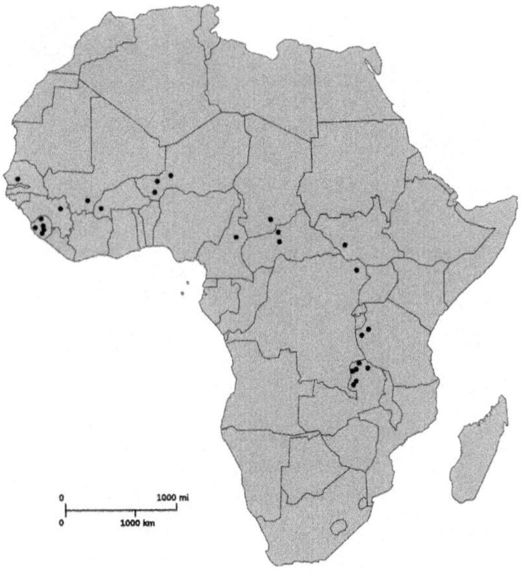

Fig. 9.2 Map of *O. glaberrima* distribution. The blue dots indicate countries of origin of *O. glaberrima* accessions conserved in AfricaRice genebank

development (Ta et al. 2017). The grain of African cultivated rice typically is smaller than that of its progenitor, *O. barthii*. Cloning and characterization of a QTL, GL4, that controls the grain length on chromosome 4 in African rice, and that regulates longitudinal cell elongation of the outer and inner glumes, revealed that a single-nucleotide polymorphism (SNP) mutation in the GL4 gene is the basis of a premature stop codon and leading to small seeds and loss of seed shattering during African rice domestication (Wu et al. 2017). These finding supported the independent origin of domestication in the one hand, brings a new light on the domestication and selection processes of African rice.

9.3 Genetic and Genomic Resources

9.3.1 Collection and Conservation of Genetic Resources

Rice is a highly diverse crop with a variety of genetic resources among them wild accessions, landraces, new and old improved varieties, genetic stocks, and breeding lines (Jackson 1997). Collection of *O. glaberrima* germplasm began in the 1950s when collecting trips were undertaken across various West African countries. This was mainly organized and conducted by national and international research institutes working on rice in Africa, such as the Office de la Recherche Scientifique et Technique d'Outre-Mer (ORSTOM) (France), Institut de Recherches Agronomiques Tropicales (IRAT) (France), and International Institute for Tropical Agriculture (IITA) (Nigeria) (Abifarin 1988). Some *O. glaberrima accessions* are also conserved at "Institut de recherche pour le développement" (IRD), International Rice Research Institute (IRRI) (Philippines), and Bangladesh Rice Research Institute. The AfricaRice genebank has been involved in rice germplasm collection, ex situ conservation, multiplication, and distribution to the global community. Currently, the AfricaRice genebank holds 21,300 registered rice samples, which includes the two cultivated species (*O. sativa* and *O. glaberrima*) and five African wild species (*O. longistaminata, O. barthii, O. punctata, O. brachyantha,* and *O. eichingeri*). Approximately, 88% of the registered rice samples at the AfricaRice genebank originated within Africa, of which African rice represents approximately 14% of the collection (Fig. 9.2). This germplasm is conserved in medium-term storage at 5 °C at the temporary headquarter of AfricaRice in Cotonou and in a long-term conservation at −18 °C at IITA in Ibadan. Two safety duplications are also maintained at the National Center for Genetic Resources Preservation (NCGRP), Fort Collins, USA and Svalbard Global Seed Vault (SGSV), Norway. The duplicate collections at SGSV are maintained in collaboration with Global Crop Diversity Trust as part of efforts aimed at ensuring global security of crop diversity in the world. The entire collection of AfricaRice's genebank is kept in trust for humanity under the Multi-Lateral System (MLS) of access and benefit sharing within the purview of the International Treaty on Plant Genetic Resources for Food and Agriculture as part of the global ex situ collections. This will enhance the facilitated access to these genetic resources by breeders and other interested users.

Compared to other areas of plant science, plant genetic resources conservation is lagging behind in embracing advances in molecular biology, especially in genome sequencing (FAO 2010). These advances have the potential to aid in addressing fundamental biological questions and greatly impact many aspects of the conservation and utilization of plant genetic resources. The rice collection at AfricaRice genebank, in Benin, is benefiting from tools and opportunities offered by the current genomic revolution. A total of 2927 accessions of the African rice collection have been genotyped with 31,739 diversity arrays technology (DArT)-based single-nucleotide polymorphic (SNP) markers (MN Ndjiondjop unpublished results). The results of this molecular analysis have been valuable in supporting the management of the AfricaRice collection. Specifically, the preliminary SNP data has enabled (i) increased understanding of allele frequencies, the extent of genetic variation (genetic distance), and relationships within and among accessions; (ii) identifying unique germplasm and duplicates as well as defining core and mini-core collections for further detailed phenotypic evaluation; (iii) exploring possible human errors during sample collection, processing, conservation, and multiplication/regeneration. In addition to supporting critical genebank management decisions, the data will also be highly useful in conducting genome-wide association studies (GWAS), genomic selection, and parental selection for new pedigree crosses. The data will also be highly valuable to genebank users for selecting accessions/varieties of their preference for seed request. The analysis will provide the largest molecular data set so far on the *O. glaberrima* collection, and it is expected that it will greatly help to promote the conservation and utility of this germplasm.

9.3.2 Genome Sequences and Genomic Resources

A reference genome sequence is arguably the most important genomic resource that helps promote utility of a species. The history of rice genome sequencing started with the International Rice Genome Sequencing Project (IRGSP) established in 1997, with *O. sativa* being the first crop species to have its genome sequenced to high quality. This reference has served as a useful tool for functional, structural, and comparative genomics. It was however realized that due to the wide diversity in the *Oryza* genus, one reference sequence is not enough to enable proper understanding and comprehensive exploitation of this diversity. The International Oryza Mapping and Alignment Project (I-OMAP), an international collaborative effort, was therefore formed with the aim of developing reference sequences for additional species in the *Oryza* genus. The aim of the initiative was to produce RefSeqs through de novo approaches rather than through read mapping approaches. This is based on the fact that there exists substantial structural divergence between the various species and mapping reads to the Nipponbare RefSeq would lead to an incorrect reference as the resultant assembly would be biased toward the Nipponbare RefSeq (Brozynska et al. 2016).

The variety CG14, the donor parent of the first generation of NERICA, was the first *O. glaberrima* to be sequenced by the International *Oryza* Map Alignment Project (I-OMAP) consortium (Wang et al. 2014). This publicly available genome has a size of 316 Mb and a total of 33,164 genes, far much less than the *O. sativa* Nipponbare reference (Table 9.1). Massive gene losses have been reported in African rice during process of evolution (Zhang et al. 2014). Despite the good overall quality of these genome sequences, recent studies have identified various gaps. For example, analysis of the publicly available sequence of *O. glaberrima* revealed no duplication of the OsSWEET14 (the major bacterial blight susceptibility S gene) promoter (Hutin et al. 2015). However, this duplication was confirmed by re-sequencing CG14 and obtaining whole genome sequences for an additional variety of African rice, TOG5681, using the Pacific Biosciences (PacBio) platform. Similarly, Pariasca-Tanaka et al. (2014) noted that the *PSTOL1* locus that is associated with tolerance to phosphorus deficiency is missing from the assembled genome sequence but is present in an unanchored scaffold. This situation is likely to negatively affect the successful mapping of a QTL in that, as reported by Wenger et al. (2010), there is a likelihood that the causal loci lie in the missing region. There is therefore need for concerted efforts with the rice scientific community to address these inaccuracies so as to improve the quality of the assembly. The sequencing of the genome of African rice lays a strong foundation for enhancing its utility in rice improvement. This foundation is further reinforced by the availability of diverse germplasm resources of both wild and cultivated species held in various national, regional, and international genebanks which makes it possible to effectively dissect genotype–phenotype relationships.

In an effort at improving the quality of the publicly available CG14 reference, Monat et al. (2016) re-sequenced the genome of CG14 and sequenced additional varieties among them TOG5681 and G22. TOG5681 is the parent of

Table 9.1 Gene feature statistics of various Oryza genomes [modified from Wang et al. 2014]

Feature	*O. glaberrima*	*O. sativa* sp. Japonica (MSU V. 6.1)
Gene count	33,164	41,620
Exon count	142,095	171,163
Intron count	108,931	129,543
Gene size (bp)	2393	2306
Exon size (bp)	255	254

the second generation of New Rice for Africa (NERICA). These varieties were chosen on the basis of their genetic diversity, with CG14 and TOG5681 each representing the extreme ends of the variability found in African rice while G22 takes a median position in this variability spectrum. Though these de novo assembled genomes were smaller and more fragmented than the publicly available sequences, they provided better sequences and resolution at some loci such as Rice yellow mottle virus (RYMV) loci. These sequences allowed the identification of the SWEET14 gene which appears in the publicly available CG14 sequences as a single copy. This re-sequencing effort identified significantly higher number of genes than identified initially, with 50,000 genes being identified for CG14 as compared to 33,164 identified by Wang et al. (2014). A total of 51,262 and 49,662 genes were identified for TOG5681 and G22, respectively. Though these genomes also have gaps based on their small size and fragmented nature, they nevertheless provide an important contribution in efforts aimed at advancing the genetics and genomics of African rice.

Over the last one decade, there have been tremendous advances in genomics and especially in DNA sequencing which have expanded the frontiers of genomic and genetic research in African rice. The sequencing of African rice has been followed by the development of large-scale genomic and genetic resources. These resources include bacterial artificial chromosome libraries (BAC), genetic and physical maps, complete reference sequence, novel mapping populations (Table 9.2), and numerous high-throughput molecular markers, among them SSR and SNP markers. Among the high-throughput markers, SNP markers are the most important because they are hugely abundant in the rice genome and promise to be cost-effective due to their potential for multiplexing and automation (McCouch et al. 2010; Thomson et al. 2012).

Oryza glaberrima possesses genes for tolerance to both biotic and abiotic stress. Interspecific crosses between African rice and *O. sativa* were made to combine the superior yield potential of *O. sativa* and the high resistance to biotic and abiotic stresses found in African rice. The development of NERICA varieties is considered as a success story in rice breeding efforts in Africa. In order to enhance the utility of African rice in developing interspecific crosses, it is important to develop a high-throughput SNP system to specifically detect *O. glaberrima* introgressions in an *O. sativa* or another species background (Pariasca-Tanaka et al. 2015). The introduction of next-generation sequencing (NGS) technology has dramatically changed the landscape for detecting and monitoring genome-wide polymorphism. Coupled with this is the increased availability of *Oryza* species reference genomes and genomic variation information which has become available in recent years (Metzker 2005; Craig et al. 2008; Huang et al. 2009; Ohyanagi et al. 2016). Over the last decade, significant progress has been made in identification of SNPs in *O. glaberrima* genome. By re-sequencing 20 accessions of African rice, Wang et al. (2014) identified a total of 4,447,424 SNPs which were used to measure genome-wide levels of nucleotide diversity. In the same study, a total 8,374,445 SNPs which were identified between African rice and *O. barthii*, its putative progenitor, were used to successfully study the genetic relationships between the two species (Wang et al. 2014). Additional SNPs were recently identified by sequencing of three

Table 9.2 Mapping populations developed from crosses between *O. glaberrima* and *O. sativa*

Population type	Donor parent	Recurrent parent	Trait	References
F_2	*O. glaberrima* (Tog5675)	Indica (cv. IR64)	BPH resistance (Bph1)	Ram et al. (2010)
BC_1	*O. glaberrima* (Tog5681)	Indica (cv. IR64)	Rice yellow mottle virus resistance	Ghesquière et al. (1997)
Chromosome segment substitution lines (CSSLs)	*O. glaberrima*	Japonica (cv. Koshihikari)	Glabrous gene	Angeles-Shim et al. (2009)
Advanced backcross	*O. glaberrima* (IRGC103544)	Indica (cv. Milyang 23)	Yield and yield components	Kang et al. (2008)
Chromosome segment substitution lines (CSSLs)	*O. glaberrima* (IRGC104038)	Japonica (cv. Taichung 65)	Pollen fertility, days to heading	Doi et al. (1997)
Back cross inbred lines (BC2F4)	*O. glaberrima* (CG14/IRGC96717)	Japonica (WAB56-104)	Drought resistance, early vigor, amylose content, P deficiency tolerance	Ndjiondjop et al. (2010), Wambugu et al. (2017), Saito (personal communication)
Chromosome segment substitution lines (CSSLs)	*O. glaberrima* (MG12/IRGC103544)	Tropical japonica (cv. Caiapo)	Rice stripe necrosis virus resistance	Gutiérrez et al. (2010)
Introgression lines	TOG 5681 *Oryza glaberrima* (TOG 5681)	Indica (cv. IR64)	Drought tolerance	Bocco et al. (2012)
BC_3F_1Doubled-haploid (DH) lines	*O. glaberrima* (MG12)	Tropical japonica (cv. Caiapo)	Iron toxicity	Dufey et al. (2015)
Near-isogenic line (NIL)	*O. glaberrima* (IRGC102375)	Japonica (Dianjingyou1-DJY1)	Pollen and embryo sac sterility	Shen et al. (2015)

O. glaberrima varieties (Monat et al. 2016). Recent molecular characterization of the 2179 *O. glaberrima* accessions held in AfricaRice genebank using DArTseq has led to the identification of 3834 polymorphic SNPs markers within these accessions (MN Ndjiondjop unpublished results). These available SNPs need to be converted into KASP markers so that they can be used in the cost-effective and flexible throughput Competitive Allele-Specific PCR (KASPar) assays. These KASP markers can then be used by African breeders more efficiently to evaluate and utilize the wealth of natural variation that exists in both *O. glaberrima* and its introgression lines.

A recent study conducted a comparative analysis of various lines obtained from crosses between CG 14 and WAB 56-104, generated a set of 9523 polymorphic SNPs. A set of 1540 polymorphic SNPs was also identified between *O. glaberrima* versus *O. sativa.* The study successfully converted 2015 SNPs into KASP markers providing 745 polymorphic SNPs for the parents CG 14/WAB 56-104 and 752 for TOG 5681/IR 64 (Pariasca-Tanaka et al. 2015). These new SNP resources will be useful in modern breeding applications such as QTL mapping, association, mapping, positional cloning pedigree analysis, variety identification, marker-assisted selection, and genomic selection

(McCouch et al. 2010). These advances will help improve rice productivity and sustainability of agriculture which is important in feeding the growing population in Africa.

9.3.3 Genetic Diversity in *O. glaberrima* Genome

The genetic diversity and population structure of African rice have been the subject of numerous studies, although not as highly studied as *O. sativa*. It has long been known that Asian rice possesses more genetic diversity than African rice. In fact, African rice is reported to have only one-sixth of the nucleotide diversity found in Asian rice and about 70% less diversity compared to its wild progenitor. This extremely low diversity in African rice which is suggestive of a severe bottleneck during its domestication could also be due to its self-pollinating nature. Using isozymes, the genetic diversity of a wide collection of *O. glaberrima* was studied, revealing a decline in genetic diversity from 0.14 to 0.03 compared to *O. barthii*, its putative progenitor. The genetic diversity of *O. glaberrima* was found to be about half of that of *O. sativa* in terms of number of alleles per locus (Second 1982). This low diversity was also reported in a study that analyzed 14 unlinked nuclear genes (Li et al. 2011) and recently confirmed by a whole genome-based population genomic-based study which analyzed 20 *O. glaberrima* and 19 *O. barthii* accessions. This analysis showed that the levels of nucleotide variability in *O. glaberrima* (π per kb, 2.281) were significantly lower than those in *O. barthii* (π per kb, 5.324). Similarly, the levels of nucleotide variability of intergenic regions in *O. glaberrima* (π per kb, 4.799) were also significantly lower than those in *O. barthii* (π per kb, 8.506), resulting in a significant reduction in genetic diversity of African rice as a consequence of domestication (Wang et al. 2014). It has been suggested that this could be the lowest diversity ever recorded for any grass crop species (Nabholz et al. 2014) or indeed for any crop species (Li et al. 2011). It appears intriguing that a species with such low genetic diversity could possess such wide genetic potential as demonstrated by its resistance to biotic and abiotic stresses. The current advances in genomics allow identification of those genomic regions that have lost genetic diversity during domestication and provide opportunities for restoration of such diversity in a well-targeted approach.

9.4 Advances in the Genetics

9.4.1 Genetic Potential of *O. glaberrima* in Rice Improvement

Though the genetic potential of African rice is generally well known, effective utilization of these important traits in rice improvement is hindered by the inadequate understanding of the genetic basis underlying these traits. Efforts aimed at identifying valuable traits in African rice have resulted in the development of several mapping populations, both primary and secondary (advanced) populations, ranging from introgression lines (ILs), recombination inbred lines (RILs), chromosomal segment substitution lines (CSSLs), near-isogenic lines (NILs), and doubled haploid lines (DH) (Table 9.2). These populations have continued to be a valuable genetic resource for genetic mapping studies in African rice. Majority of these populations have been used in QTL mapping with the goal of identifying QTLs underlying important traits. Most of the QTL mapping studies involving African rice have been conducted using interspecific crosses between *O. sativa* and *O. glaberrima* (Table 9.2).

In addition to QTL mapping, the genetic mechanisms controlling important traits can also be studied through association studies. Though association studies are widespread in *O. sativa*, it appears that no such studies have been conducted for African rice. Association studies require a lot of genomic information such as SNPs for the organism being studied and hence may not be suitable for organisms without well-annotated genomes. This might explain the lack of genome-wide association studies for African rice as compared to *O. sativa*. However, with the continued release of well-annotated genome sequences for African rice, the situation might change and we are likely to witness an increase in these studies. Meanwhile, it is likely that QTL mapping will continue to be an important tool in the near future for African rice for dissecting genotype–phenotype relationships.

A study conducted to assess the genetic diversity of 106 introgression lines (IL) from *O. glaberrima* and *O. sativa* revealed large phenotypic and genetic diversity (Chen et al. 2016). The estimation of effect of the proportion of glaberrima genome revealed significant positive effect on several agronomic traits including grain yield per plant, plant height, panicle length, and spikelet number per panicle. Pyramiding more favorable genes for developing more adapted and productive rice could be achieved by developing lines with a good level of proportion of glaberrima genome to boost rice yield in Africa.

9.4.1.1 Resistance to Biotic and Abiotic Stresses

Oryza glaberrima is a part of the primary gene pool of cultivated common rice and is therefore a source of readily available diversity for rice improvement. Several studies have demonstrated that *O. glaberrima* possesses immense genetic value especially agronomically interesting traits such as resistance to biotic and abiotic stresses, as well as ability to respond to low-input farming system (Table 9.3) (Ghesquière et al. 1997; Linares 2002; Sarla and Swamy 2005). Despite this immense genetic potential of African rice being well known and documented, the genetic and molecular basis of some of the traits has only recently been deciphered. Aided by the increased availability of genomic resources, the last couple of years have seen concerted efforts in linking genotypes and phenotypes. These have led to discovery of more locus or causal mutations associated with various traits particularly tolerance to various biotic and abiotic stresses. For example, recently a novel *PSTOL1* allele for phosphorus deficiency tolerance was identified in *O. glaberrima* and NERICA varieties (Pariasca-Tanaka et al. 2014). A second major gene, *RYMV2,* controlling resistance to RYMV which is one of the most devastating rice-infecting viruses in Africa has been identified in *O. glaberrima* (Thiémélé et al. 2010). Efforts to fine map the *RYMV2* gene led to the identification of a putative loss-of-function one base deletion mutation in one of the candidate genes for *RYMV2*. This low-frequency mutation was highly associated with RYMV resistance and affected a gene homologous to the CPR5 defense gene in *Arabidopsis thaliana* (Orjuela et al. 2013). Using *O. sativa* and *O. glaberrima* introgression lines, Gutiérrez et al. (2010) for the first time identified a major QTL controlling rice stripe necrotic virus located on chromosome 11. Transgressive segregation was observed in these interspecific crosses, with *O. glaberrima* contributing the superior alleles resistance to rice stripe necrosis virus (Gutiérrez et al. 2010), thereby showing the potential of *O. glaberrima* in improving *O. sativa*. These authors also identified a host of other QTLs for various traits, demonstrating the power of chromosome segment substitution lines (CSSL) as a genetic mapping tool. Novel QTLs for resistance to iron toxicity have recently been identified in *O. glaberrima* (Dufey et al. 2015). The continued identification of such locus and alleles is important in rice improvement as it assists in marker-assisted selection.

One of the good agronomic characteristics of African rice is its tolerance to many biotic stresses including Rice yellow mottle virus (RYMV) (Albar et al. 2003), root-knot nematodes (Soriano et al. 1999) and bacteria (Linares 2002; Djedatin et al. 2011). Previous studies reported high tolerance of some *O. glaberrima* accessions to *Mycosphaerella graminicola* (Soriano et al. 1999; Plowright et al. 1999; Bimpong et al. 2010). Though for a long time, several *O. glaberrima* varieties have been identified as highly resistant to nematodes, it is only now that the mechanisms underlying resistance are being understood. Recently, Petitot et al. (2017) provided new insights into root-knot nematode resistance. Histological analysis and

Table 9.3 Resistance or tolerance trait found in African rice to various biotic and abiotic stresses

Trait	References
Weed competitiveness	Jones et al. (1997b), Johnson et al. (1998), Fofana and Rauber (2000), Moukoumbi et al. (2011)
Resistance to drought	(IRRI (1974), Sano et al. (1984), Jones et al. (1997a), Maji et al. (2001), Ndjiondjop et al. (2012)
Resistance to nematodes	Reversat and Destombes (1998), Plowright et al. (1999)
Resistance to iron toxicity	Sikirou et al. (2016), Sahrawat and Sika (2002)
Resistance to African gall midge	Williams et al. (1999), Nwilene et al. (2002).
Resistance to Rice yellow mottle virus	IITA (1986), Thottappilly and Rossel (1993), Ndjiondjop et al. (1999), Albar et al. (2003), Thiémélé et al. (2010)
Resistance to bacterial leaf blight	Djedatin et al. (2011)
Tolerance to lodging	Futakuchi et al. (2008)
Resistance to green rice leafhopper (*Nephotettix cincticeps* Uhler)	Fujita et al. (2010)
Tolerance to salinity	Akbar et al. (1987), Linares (2002), Platten et al. (2013)
Tolerance to soil acidity	Sano et al. (1984)
Tolerance to submergence	Watarai and Inouye (1998)
Resistance to Blast	Bidaux (1978), Silué and Nottéghem (1991)
Resistance to grain shattering	Montcho et al. (2013)

transcriptome profiling of the resistant variety TOG 5681 and the susceptible "Nipponbare" (*O. sativa*) whose roots had been infected by *M. graminicola* revealed that resistance in African rice involves different complex mechanisms and pathways. Early resistance involves restriction of juvenile penetration and later resistance requires degeneration of giant cells. Both of these resistance mechanisms involve strong activation of defense genes. Furthermore, that investigation identified several candidate genes for *M. graminicola* resistance in rice (Petitot et al. 2017).

Drought is a major constraint of rice production in Africa. It has been intensively studied, but identifying tolerant lines is a challenge for scientists due to the complexity and specificity of this stress over environment. Ndjiondjop et al. (2012) reported that some *O. glaberrima* cultivars, namely TOG 5691, TOG 6679, and TOG 5591 are more resistant to drought and produce higher yield than the resistant checks. Though African rice is generally known to be susceptible to grain shattering, two cultivars namely as TOG 12303 and TOG 9300 have been found to be highly resistant to grain shattering compared to *O. sativa* checks (Montcho et al. 2013). The yields of 21 *O. glaberrima* accessions were evaluated alongside four *O. sativa* check varieties. The results revealed yield reduction ranging from 57.88 to 66.76% in *O. glaberrima* due to its grain shattering. This implied that *O. glaberrima* yield potential would not be inferior to *O. sativa* if the grain shattering problems are overcome.

Among crop species, rice is considered as the most sensitive to salt stress. Salinity tolerance has been extensively studied in *O. sativa*, but little is known about the salt tolerance levels in *O. glaberrima* (Awala et al. 2010). Platten et al. (2013) were the first to report significant salinity tolerance from *O. glaberrima.* They identified 12 *O. glaberrima* accessions from diverse geographic regions with moderate to high tolerance to salinity. These accessions were also genetically and phenotypically characterized. These results demonstrated that major tolerance mechanisms found in these species are ones responsible for limiting sodium uptake and accumulation in active leaves. However, a recent study suggested that the relative concentrations of sodium in leaves is not the only tolerance mechanism. Several others strategies are involved in salt tolerance. Salinity tolerance is a trait that is likely associated with geographical adaptation (Carney 1996; Temudo 2011). Screening of 100 African rice landraces from upland and coastal West African environments identified 20 African rice genotypes tolerant to salinity (Meyer 2014). An evaluation of 121 diverse West African genotypes for salinity tolerance showed phenotypic variability regarding diverse traits associated with salinity. Furthermore, variability in salinity tolerance is likely associated with geographical adaptation (Meyer 2014; Meyer et al. 2016). These results open the door to the understanding of the mechanism of salinity tolerance in African rice. Such results are promising for developing rice genotypes with high level of salt tolerance.

In addition to the problem of grain shattering, low yield potential in African rice is also a concern which is caused by lodging, low spikelet number, and leaf senescence. Senescence is a normal event in the life cycle of plants, and it represents an endogenously controlled degenerative process that is initiated at full maturity and may ultimately lead to leaf death. Several studies have reported that *O. glaberrima* appeared to senescence faster than *O. sativa* in the maturing stage (Sumi and Katayama 1994; Takasaki et al. 1994). However, some genotypes from AfricaRice genebank showed variability in regard to senescence (Fig. 9.3). Some varieties are very susceptible to senescence since early in their life cycle while others do not show any symptoms until full maturity.

Lodging is an undesirable trait that causes yield reduction in African rice. An evaluation of six genotypes for lodging tolerance identified the variety TOG 7235 and CG 14 as fairly tolerant to lodging (Futakuchi et al. 2008). Jones et al. (1997a, b) had previously identified some of the short-statured and thick-leaved accessions of *O. glaberrima* such as Old Ayoma, Fufore-yola, UG 75 (1), UG 75 (2), NG 26 (3), CG 66, Katasina ala, and Shendam as resistant cultivars to lodging.

9.4.1.2 Physicochemical and Organoleptic Properties

Rice eating and cooking qualities are highly related to some easily measurable physicochemical properties including gelatinization temperature and pasting viscosity which reflect the starch functionality of the grain and apparent amylose content (Singh et al. 2005; Bao 2012). Though African rice has potential for contributing genes to improve rice quality (Wambugu et al. 2013), only few studies have been reported on African rice in the aspects of physicochemical as well as organoleptic proprieties and the genetic mechanisms controlling them. Interest in studying physicochemical properties in African rice is however growing (Gayin et al. 2015; Wang et al. 2015; Gayin et al. 2016a, b). African rice generally has higher amylose content and gelatinization temperature than Asian rice (Wang et al. 2015). High amylose content has been associated with an increase in slowly digestible and resistant starch, thereby leading to low digestibility (Chung et al. 2011; Cai et al. 2015). The high amylose content perhaps explains why African rice has slow digestibility and provides increased post-meal satiety (Nuijten et al. 2013). Though the genetic basis of amylose content is well studied in Asian rice, it remains grossly understudied in African rice. Recently, Wambugu et al. (2017) conducted whole genome-based bulk segregant analysis of progenies of an interspecific cross between *O. sativa japonica* (WAB56104), an Asian rice variety and *O. glaberrima* (CG14), an African rice. The aim was to identify markers that may be linked with amylose content in African rice and which can potentially find application in marker-assisted selection. This study identified a G/A functional SNP associated with granule-bound starch synthase (GBSS1) located on chromosome 6. This SNP may influence GBSS1 activity, thereby controlling amylose content. Analysis of natural variation found in African rice identified novel non-synonymous SNPs in GBSS1 whose functional importance especially in relation to amylose content is unknown. This study provides a foundation which future studies can build on in dissecting the genetic architecture of amylose content in African rice.

Studies have shown that generally *O. glaberrima* tends to break easily during milling; thus, head rice recovery is lower compared with *O. sativa*. In contrast to Asian domesticated rice, most cultivars of African domesticated rice have red pericarp and so do not differ in grain color from their wild relative. This makes African rice less attractive when polished. *O. glaberrima* is preferred by consumers, particularly in regions where it is still cultivated, because of its distinct taste and flavor. It is reported that African rice has higher protein content than Asian rice (Juliano and Villareal 1993; Traore et al. 2011) although CG 14, a *O. glaberrima* genotype showed lower protein content than WAB 56-104, an Asian rice variety (Watanabe et al. 2006). Increased protein content is particularly desirable for poor regions such as Asia and Africa where malnutrition is widespread. Data on grain quality of the *O. glaberrima* collection conserved at the AfricaRice Genetic Resources Unit revealed that African rice has good characteristics such as firmer texture, perhaps due to their higher apparent amylose content (AAC) and unique pasting properties (low peak, low breakdown, and higher setback viscosities) compared to *O. sativa*. Asian rice however has greater diversity in AAC than African rice (Gayin et al. 2015; Manful and Graham-Acquaah 2016). These studies also showed a large variability within *O. glaberrima* accessions for each of these traits.

9.4.2 *O. glaberrima* as Donor to Improve *O. sativa*

9.4.2.1 The NERICA Varieties

Though the African rice accessions have adaptive or protective mechanisms for different abiotic and biotic stresses in the continent, they are generally characterized by a wide range of undesirable agronomic traits (Jones et al. 1997a; Linares 2002; Vikal et al. 2007). As highlighted earlier, in order to combine traits of economic importance from both the Asian rice and African rice, interspecific breeding programs were initiated by the AfricaRice in the early 1990s. These breeding efforts led to the development of NERICA varieties which clearly demonstrated the usefulness of the African rice germplasm in developing modern improved varieties that combined the high yield potential from the *O. sativa* parents and the adaptability to different abiotic and biotic stresses from the *O. glaberrima* parents. These varieties were created using conventional breeding coupled with tissue culture for the upland varieties. These high-yielding varieties have already had great impacts on farmer's livelihoods and poverty reduction in SSA (Obilana and Okumu 2005). Although their impact on rice production in SSA remains undeniable, the upland NERICAs were created using only one *O. glaberrima* genotype as donor. Additionally, the assessment of the genome content of those varieties using molecular markers showed a weak proportion of *O. glaberrima's* genome (Semagn et al. 2006, 2007) and the incorporation of a known sterility locus S1 on chromosome 6. This was confirmed by Ndjiondjop et al. (2008) who undertook molecular profiling of interspecific lowland rice populations derived from a cross between IR 64 and TOG 5681 using microsatellites markers. The results revealed that the estimated average rate of introgression of *O. glaberrima* genome varies from 7.2 (83.5 cM) to 8.5% (99.3 cM) and 8.7 to 13.2%, respectively. It therefore appears that the introgression of genes initially targeted for transfer from *O. glaberrima* to *O. sativa* during the development of NERICA varieties was still impaired by residual sterility problem. In order to further take advantage of the immense genetic potential of African rice more efforts aimed at increasing the proportion of *O. glaberrima* genome in the interspecific crosses were made by AfricaRice upland breeders. As a result new interspecific lines namely Advanced Rice for Africa (ARICA 4) were generated by crossing (WAB 56-104 × CG 14) × Moroberekan. The level of *O. glaberrima* genome in ARICA 4 may reach 25% (Semon, personal communication).

9.4.2.2 Interspecific Bridges

A major drawback in the use of *O. glaberrima* is the very strong and remnant sterility observed in the interspecific crosses. A more systematic approach of introgressions has been developed by construction of two sets of chromosome substitution segment lines (CSSLs) with 2 *O. glaberrima* accessions TOG 5681, MG 12/Ac 103544 in *O. sativa* background which are more suitable to reveal alien variation (Ali et al. 2010). The S1 locus responsible for a major effect on interspecific sterility and giving a strong segregation distortion was finely mapped and served to propose a genetic model of reproductive barrier (Garavito et al. 2010). The complete genomic sequence of the S1 region was compared with *O. sativa* to characterize structural differences and rearrangements between the two species (Guyot et al. 2012). Based on this work, Institut de Recherche pour le développement (IRD) has developed a research project named Ibridges project. Interspecific bridges give full access to the African rice allele pool for enhancing drought tolerance of Asian rice funded by the Generation Challenge Program (GCP). This project aims at overcoming the interspecific sterility barrier separating the two cultivated rice species, *O. sativa* and *O. glaberrima*. After interspecific hybridization, a marker-assisted selection was carried out on backcross progenies and focused on the S1 locus, which is the key factor of the interspecific sterility. In collaboration with IRD, AfricaRice has used the power of molecular markers tightly linked to the sterility gene S1 to select 400 BC1F2 fertile progenies each containing at least 25% of *O. glaberrima's* genome out of 7000 interspecific lines. Moreover, these new interspecific progenies are free from the major sterility loci, thus constitute bridges that capture the best of the two cultivated rice in one. Consequently, the newly developed lines look more like *O. glaberrima* for resistance to the harsh rice growing ecologies of Africa and more like *O. sativa* in term of yield and other agronomically important traits.

9.4.2.3 Other Interspecific and Intraspecific Lines Developed Using *O. glaberrima* as Genetic Resources

As already highlighted, the development of NERICA varieties is hailed as huge success in rice improvement in Africa. However, despite the current success of NERICA, *O. glaberrima* is still considered as underexploited (Futakuchi and Sié 2009). In order to build on the gains of NERICA and continue the momentum of these breeding efforts, AfricaRice genetic improvement program has put in place a strategy to better exploit the reservoir of *O. glaberrima* germplasm. As a result, the *O. glaberrima* collection was screened for several abiotic stresses including drought, iron toxicity, submergence, anaerobic germination, and early flowering. New donors for drought tolerance and anaerobic germination were found (Venuprasad Ramaiah, unpublished results). These results pave the way for the development of a new project named Rapid Alleles Mobilization (RAM) funded by Bill and Melinda Gates Foundation. This project focuses on identification of novel genes associated with abiotic stress tolerance, present in lowland ecology in Africa using African rice as genetic resources. The project is also developing new interspecific lines derived from crosses between several of these new *O. glaberrima* donors and popular high-yielding *O. sativa* varieties that are available in Africa. The MENERGEP (Methodologies and new resources for genotyping and phenotyping of African rice species and their pathogens for developing strategic disease resistance breeding programs) is another initiative which was undertaken by IRD and AfricaRice. It focuses on identifying donors from *O. glaberrima* accessions associated with certain diseases, identifying new resistance genes, and characterizing rice—pathogen interactions in search of more durable resistance strategies.

To better exploit the genetic assets of *O. glaberrima* without being hampered by the sterility problems of hybridization with the Asian rice species, AfricaRice scientists have begun working on the intraspecific *O. glaberrima* breeding lines and are taking steps to develop plants that are less prone to lodging, shattering and resistant to Rice yellow mottle virus (RYMV, Sobemovirus), a highly damaging disease of rice in Africa. The best hope in reducing yield losses lies in the development of resistant varieties. A single gene of high resistance, *RYMV1*, has been identified so far in *O. sativa* (Ndjiondjop et al. 1999). On the other hand, high resistance appears to be more frequent and more diverse in *O. glaberrima*. Indeed until now two genes, resistance to RYMV (*RYMV1*, *RYMV2*) have been identified in *O. glaberrima* collection. Twenty-nine *O. glaberrima* accessions have been described as resistant to RYMV (Thiémélé et al. 2010). Crosses were made between *O. glaberrima and O. glaberrima* accessions by IRD (France) in collaboration with AfricaRice breeder. The new intra-specific lines are under evaluation in African environments by AfricaRice breeders. The promising lines which are expected to combine the genes of resistance from different ecotypes as well as optimize the yield potential of African rice will then be disseminated as varieties to farmers through participatory varietal selection (PVS).

More recently, Pidon et al. (2017) reported a new resistance gene to Rice yellow mottle virus named *RYMV3* in TOG 5307 (*O. glaberrima*) which has none of the two previously identified resistance genes to RYMV. The R*YMV3* would also be involved in TOG 5672 resistance. This is an important finding as the emergence of resistance-breaking virus variants is frequent and the durability of resistance genes is still a challenge. Mutations in the virus gene which can break down the resistance have been noticed by Kobayashi et al. (2014). For example, some accessions which carry the resistance alleles at *RYMV1* and *RYMV2* genes showed resistance breakdown in accessions carrying resistance in experimental conditions (Pinel-Galzi et al. 2007, 2016; Traoré et al. 2010).

9.5 Future Perspectives

The major challenge in the achievement of sustainable food self-sufficiency in SSA is the huge gap between actual yields and potential yields. The challenge in the hands of the rice scientific community is therefore on how to reduce this gap so as to enhance food and nutritional security in the region. This objective could be achieved mainly by better exploitation of indigenous African rice germplasm. AfricaRice which holds the largest collection of African rice is currently undergoing major changes in order to position itself to better face these challenges which are at the core of its mandate. The AfricaRice genebank now based in Cotonou (Benin) will be relocated back to M'be (Bouake-Cote d'Ivoire) as part of the multi-phase return of AfricaRice to its permanent headquarters in Côte d'Ivoire. With financial support from The Africa Development Bank (ADB) through the SARDC project and major extra funding through the CGIAR, world class facilities and laboratories are being established at the AfricaRice genebank in M'be. The new genebank will become the AfricaRice Biodiversity Center. The design and operation of the new genebank are expected to meet FAO/IBPGR standards and will accommodate up to 60,000 rice germplasm accessions by 2050. Based on a collection gap analysis that will be conducted using DArTseq data that will be obtained from genotyping the entire *O. glaberrima* collection, more germplasm collection will be undertaken. We also hope to build a core and mini-core collection of *O. glaberrima* which will enhance further in-depth evaluation by breeders and scientists.

AfricaRice scientists and their partners are currently investigating the African rice gene pools. They are integrating phenotypic screening (physical characteristics) with molecular analysis (genetic composition) to unravel the genetic value of local germplasm. AfricaRice is screening the entire *O. glaberrima* collection on the basis of their agro-morphological characters. Through molecular analysis, scientists are identifying the genes and/or the genetic regions (quantitative trait loci or QTLs) that possess local stress tolerance traits. After identifying these specific genes for major diseases and environmental stresses such as acidity, iron toxicity, cold, and salinity, scientists can then transfer them into improved rice varieties. A 3-pronged approach is being used to improve rice varieties' tolerance to West and Central Africa constraints. This involves the characterization of biotic and abiotic stresses profiles of rainfed, irrigated, and lowland rice production systems using GIS, conventional breeding, and marker-assisted selection. Since *O. glaberrima* had been considered to have generally low yield potential, interspecific hybridization with *O. sativa*, which possess high yield potential, was a major method to better exploit the genetic potential of the two cultivated species. However, AfricaRice breeders now think that *O. glaberrima* can potentially produce yields of about 5–6 t/ha, that is sufficient for rainfed and lowland rice ecosystems in Africa. Initial results from crossing between different *O. glaberrima* genotypes also showed that completely different sets of genes are responsible for tolerance to submergence, Rice yellow mottle virus, and phosphorus deficiency in soils from those found in *O. sativa*. By characterizing the *O. glaberrima* collection available in AfricaRice genebank, new *O. glaberrima* accessions with better traits than the current donor's parents of NERICA may be identified and used as heads of lines in the new breeding program. This will help to exploit the treasure existing within the African rice germplasm following the new concept of AfricaRice whose aim is to "combine the adaptability of *O. glaberrima* to local environments with the optimal conjunction of the best traits of the two species in relation to yielding ability" (Futakuchi and Sié 2009). Finally, in collaboration with partners, a worldwide project is being developed through an International African Rice Improvement Consortium (IAfRIC) aimed at improving *O. glaberrima* for traits such as yield.

9.6 Conclusion

Despite the great potential of African rice, this potential is not yet fully exploited for the benefit of rice consumers in Africa and especially in sub-Saharan Africa (SSA) where only few countries have attained self-sufficiency in rice production. To address this challenge, it is imperative that research programs should focus on the development of more intraspecific and interspecific breeding lines possessing the interesting traits of African rice. The African rice scientific community should take advantage of the current genomic revolution as it is providing tools that allow better conservation and utilization of available diversity. Greater technical, human, and infrastructural capacity will be required in order to take advantage of the current advances in genomics. This also calls for greater

Fig. 9.3 Diversity within *O. glaberrima* in term of leaf senescence. **a** Some *O. glaberrima* accessions highly susceptible to senescence. 80% of leaves and tillers are yellow and dead before the maturity of the panicles. **b** An *O. glaberrima* accession tolerant to senescence

collaboration between scientists working on African rice in Africa with those in developed countries in order to share genetic resources, experiences, and the available scientific capacity. Well-coordinated efforts should be put in place to address the various gaps that are present in the publicly available reference sequence of African rice so as to improve its quality.

References

Abifarin A (1988) West Africa Rice Development Association (WARDA) activities in rice germplasm collection, conservation and utilization. In: Proceedings of an international conference on crop genetic resources of Africa, Ibadan, Nigeria, pp 35–40

Akbar M, Gunawardena I, Ponnamperuma N (1987) Breeding for soil stresses: Progress in rainfed lowland rice. International Rice Research Institute, Los Banos, The Philippines

Albar L, Ndjiondjop M-N, Esshak Z, Berger A, Pinel A, Jones M, Fargette D, Ghesquière A (2003) Fine genetic mapping of a gene required for Rice yellow mottle virus cell-to-cell movement. TAG Theor Appl Genet Theor Angew Genet 107:371–378. https://doi.org/10.1007/s00122-003-1258-4

Ali ML, Sanchez PL, Yu S, Lorieux M, Eizenga GC (2010) Chromosome segment substitution lines: a powerful tool for the introgression of valuable genes from oryza wild species into cultivated rice (*O. sativa*). Rice 3:218–234. https://doi.org/10.1007/s12284-010-9058-3

Angeles-Shim RB, Asano K, Takashi T, Kitano H, Ashikari M (2009) Mapping of the glabrous gene in rice using CSSLs derived from the cross Oryza sativa subsp. Japonica cv. Koshihikari × O. glaberrima. Paper presented at the 6th International Rice Genetics Symposium, Manilla, Phillipines.

Awala SK, Nanhapo I, Sakagami J-I, Kanyomeka L, Iijima M (2010) Differential salinity tolerance among *Oryza glaberrima*, *Oryza sativa* and their interspecies including NERICA. Plant Prod Sci 13:3–10. https://doi.org/10.1626/pps.13.3

Bao JS (2012) Toward understanding the genetic and molecular bases of the eating and cooking qualities of rice. Cereal Foods World 57:148–156. https://doi.org/10.1094/CFW-57-4-0148

Bezançon G (1993) Le riz cultivé d'origine africaine *Oryza glaberrima* Steud et les formes sauvages et adventrices apparentées: diversité, relations génétiques et domestication

Bidaux J (1978) Screening for horizontal resistance to rice blast (*Pyricularia oryzae*) in Africa. In: Buddenhagen IW, Persley GJ (eds) Rice in Africa. Academic Press, United kingdom, London, pp 159–174

Bimpong K, Carpena A, Mendioro M, Fernandez L, Ramos J, Reversat G, Brar D (2010) Evaluation of Oryza sativa x O. glaberrima derived progenies for resistance to rootknot nematode and identification of introgressed alien chromosome segments using SSR markers. Afr J Biotechnol 9(26):3988–3997. https://doi.org/10.5897/AJB2010.000-3275

Bocco R, Lorieux M, Seck PA, Futakuchi K, Manneh B, Baimey H, Ndjiondjop MN (2012) Agromorphological characterization of a population of introgression lines derived from crosses between IR 64 (*Oryza sativa indica*) and TOG 5681 (*Oryza glaberrima*) for drought tolerance. Plant sci 183: 65–76. https://doi.org/10.1016/j.plantsci.2011.09.010

Brar DS, Khush GS (2003) Utilization of wild species in rice. In: Nanda JS, Sharma SD (eds) Monograph on genus Oryza. Science Publishers, Enfield, NH, pp 283–309

Brozynska M, Furtado A, Henry RJ (2016) Genomics of crop wild relatives: expanding the gene pool for crop improvement. Plant Biotechnol J 14:1070–1085. https://doi.org/10.1111/pbi.12454

Buddenhagen IW, Persley GJ et al (1978) Rice in Africa. In: Proceedings of a conference held at the International Institute of Tropical Agriculture Ibadan, Nigeria, 7–11 March 1977. Academic Press

Cai J, Man J, Huang J, Liu Q, Wei W, Wei C (2015) Relationship between structure and functional properties of normal rice starches with different amylose contents. Carbohydr Polym 125:35–44. https://doi.org/10.1016/j.carbpol.2015.02.067

Carney J (1996) Landscapes of technology transfer: rice cultivation and African continuities. Technol Cult 37:5–35. https://doi.org/10.2307/3107200

Carney J (2000) The African origins of Carolina rice culture. Ecumene 7:125–149

Carney JA (2004) "With grains in her hair": rice in colonial Brazil. Slavery Abolit 25:1–27. https://doi.org/10.1080/0144039042000220900

Chen C, He W, Nassirou TY, Zhou W, Yin Y, Dong X, Rao Q, Shi H, Zhao W, Efisue A, Jin D (2016) Genetic diversity and phenotypic variation in an introgression line population derived from an interspecific cross between *Oryza glaberrima* and *Oryza sativa*. PLoS ONE 11:e0161746. https://doi.org/10.1371/journal.pone.0161746

Chung H-J, Liu Q, Lee L, Wei D (2011) Relationship between the structure, physicochemical properties and in vitro digestibility of rice starches with different amylose contents. Food Hydrocoll 25:968–975. https://doi.org/10.1016/j.foodhyd.2010.09.011

Craig DW, Pearson JV, Szelinger S, Sekar A, Redman M, Corneveaux JJ, Pawlowski TL, Laub T, Nunn G, Stephan DA, Homer N, Huentelman MJ (2008) Identification of genetic variants using bar-coded multiplexed sequencing. Nat Methods 5:887–893. https://doi.org/10.1038/nmeth.1251

Djedatin G, Ndjiondjop M-N, Mathieu T, Cruz CMV, Sanni A, Ghesquière A, Verdier V (2011) Evaluation of African cultivated rice *Oryza glaberrima* for resistance to bacterial blight. Plant Dis 95:441–447. https://doi.org/10.1094/PDIS-08-10-0558

Doi K, Iwata N, Yosimura A (1997) The construction of chromosome substitution lines of African rice (*Oryza glaberrima Steud.*) in the background of Japonica rice (*O. sativa L.*). Rice Genet Newsl 14:39–41

Dufey I, Draye X, Lutts S, Lorieux M, Martinez C, Bertin P (2015) Novel QTLs in an interspecific backcross *Oryza sativa* × *Oryza glaberrima* for resistance to iron toxicity in rice. Euphytica 204:609–625. https://doi.org/10.1007/s10681-014-1342-7

FAO FAAOOTU (2010) Second Report on the World's Plant Genetic Resources for Food and Agriculture

Fofana B, Rauber R (2000) Weed suppression ability of upland rice under low-input conditions in West Africa. Weed Res 40:271–280. https://doi.org/10.1046/j.1365-3180.2000.00185.x

Fujita D, Doi K, Yoshimura A, Yasui H (2010) A major QTL for resistance to green rice leafhopper (*Nephotettix cincticeps* Uhler) derived from African rice (*Oryza glaberrima* Steud.). Breed Sci 60:336–341. https://doi.org/10.1270/jsbbs.60.336

Fuller DQ, Qin L, Zheng Y, Zhao Z, Chen X, Hosoya LA, Sun G-P (2009) The domestication process and domestication rate in rice: spikelet bases from the lower Yangtze. Science 323:1607–1610. https://doi.org/10.1126/science.1166605

Futakuchi K, Sié M (2009) Better exploitation of African rice (*Oryza glaberrima* Steud.) in varietal development for resource-poor farmers in West and Central Africa. Agric J 4:96–102

Futakuchi K, Fofana M, Sie M (2008) Varietal differences in lodging resistance of African rice (*Oryza glaberrima* Steud.). Asian J Plant Sci 7:569–573. https://doi.org/10.3923/ajps.2008.569.573

Garavito A, Guyot R, Lozano J, Gavory F, Samain S, Panaud O, Tohme J, Ghesquière A, Lorieux M (2010) A genetic model for the female sterility barrier between Asian and African cultivated rice species. Genetics 185:1425–1440. https://doi.org/10.1534/genetics.110.116772

Gayin J, Chandi GK, Manful J, Seetharaman K (2015) Classification of rice based on statistical analysis of pasting properties and apparent amylose content: the case of *Oryza glaberrima* accessions from Africa. Cereal Chem J 92:22–28. https://doi.org/10.1094/CCHEM-04-14-0087-R

Gayin J, Abdel-Aal E-SM, Manful J, Bertoft E (2016a) Unit and internal chain profile of African rice (*Oryza glaberrima*) amylopectin. Carbohydr Polym 137:466–472. https://doi.org/10.1016/j.carbpol.2015.11.008

Gayin J, Bertoft E, Manful J, Yada RY, Abdel-Aal E-SM (2016b) Molecular and thermal characterization of starches isolated from African rice (*Oryza glaberrima*). Starch Stärke 68:9–19. https://doi.org/10.1002/star.201500145

Ghesquière A, Séquier J, Second G, Lorieux M (1997) First steps towards a rational use of African rice, *Oryza glaberrima*, in rice breeding through a "contig line" concept. Euphytica 96:31–39. https://doi.org/10.1023/A:1003045518236

Gutiérrez AG, Carabalí SJ, Giraldo OX, Martínez CP, Correa F, Prado G, Tohme J, Lorieux M (2010) Identification of a rice stripe necrosis virus resistance locus and yield component QTLs using *Oryza sativa* × *O. glaberrima* introgression lines. BMC Plant Biol 10:6. https://doi.org/10.1186/1471-2229-10-6

Guyot G, Scoffoni C, Sack L (2012) Combined impacts of irradiance and dehydration on leaf hydraulic conductance: insights into vulnerability and stomatal control. Plant Cell Environ 35:857–871. https://doi.org/10.1111/j.1365-3040.2011.02458.x

Harlan JR, Stemler A (1976) The races of sorghum in Africa. In: Harlan JR, De Wet JM, Stemler AB (eds) Origins of African plant domestication, Reprint 2011 edition. The Hague, Mouton, Netherlands

Huang X, Feng Q, Qian Q, Zhao Q, Wang L, Wang A, Guan J, Fan D, Weng Q, Huang T, Dong G, Sang T, Han B (2009) High-throughput genotyping by whole-genome resequencing. Genome Res 19:1068–1076. https://doi.org/10.1101/gr.089516.108

Hutin M, Sabot F, Ghesquière A, Koebnik R, Szurek B (2015) A knowledge-based molecular screen uncovers a broad-spectrum OsSWEET14 resistance allele to bacterial blight from wild rice. Plant J 84:694–703. https://doi.org/10.1111/tpj.13042

IITA (1986) *Oryza glaberrima* as a source of resistance to Rice yellow mottle virus. Annual Report and Research Highlights, International Institute for Tropical Agriculture, Ibadan, Nigeria

IRRI (1974) Annual report for 1973. Los Bafios, Philippines

Jackson MT (1997) Conservation of rice genetic resources: the role of the International Rice Genebank at IRRI. Plant Mol Biol 35:61–67. https://doi.org/10.1023/A:1005709332130

Jacquemin J, Bhatia D, Singh K, Wing RA (2013) The International Oryza Map Alignment Project: development of a genus-wide comparative genomics platform to help solve the 9 billion-people question. Curr Opin Plant Biol 16:147–156. https://doi.org/10.1016/j.pbi.2013.02.014

Johnson DE, Dingkuhn M, Jones M, Mahamane M (1998) The influence of rice plant type on the effect of weed competition on *Oryza sativa* and *Oryza glaberrima*. Weed Res 38:207–216. https://doi.org/10.1046/j.1365-3180.1998.00092.x

Jones MP, Dingkuhn M, Aluko GK, Semon M (1997a) Interspecific *Oryza Sativa* L. × *O. Glaberrima* Steud. progenies in upland rice improvement. Euphytica 94:237–246. https://doi.org/10.1023/A:1002969932224

Jones MP, Mande S, Aluko K (1997b) Diversity and potential of *Oryza glaberrima* Steud in upland rice breeding. Jpn J Breed 47:395–398. https://doi.org/10.1270/jsbbs1951.47.395

Juliano BO, Villareal CP (1993) Grain quality evaluation of world rices. International Rice Research Institute, Manila

Kang JW, Suh JP, Kim DM, Oh CS, Oh JM, Ahn SN (2008) QTL Mapping of Agronomic Traits in an Advanced Backcross Population from a Cross between *Oryza sativa L.* cv. Milyang 23 and *O. glaberrima*. Korean J Breed Sci 40:243–249

Kellogg EA (2009) The evolutionary history of Ehrhartoideae, Oryzeae, and Oryza. Rice 2:1–14. https://doi.org/10.1007/s12284-009-9022-2

Khush GS (1997) Origin, dispersal, cultivation and variation of rice. In: Oryza: from molecule to plant. Springer, The Netherlands, pp 25–34

Kobayashi K, Sekine K-T, Nishiguchi M (2014) Breakdown of plant virus resistance: can we predict and extend the durability of virus resistance? J Gen Plant Pathol 80:327–336. https://doi.org/10.1007/s10327-014-0527-1

Li Z-M, Zheng X-M, Ge S (2011) Genetic diversity and domestication history of African rice (*Oryza glaberrima*) as inferred from multiple gene sequences. Theor Appl Genet 123:21–31. https://doi.org/10.1007/s00122-011-1563-2

Linares OF (2002) African rice (*Oryza glaberrima*): history and future potential. Proc Natl Acad Sci 99:16360–16365. https://doi.org/10.1073/pnas.252604599

Maji T, Singh B, Akenova M (2001) Vegetative stage drought tolerance in *Oryza glaberrima* Steud and *Oryza sativa* L. and relationship between drought parameters. Oryza Cuttack 38:17–23

Manful JT, Graham-Acquaah S (2016) African rice (*Oryza glaberrima*): a brief history and its growing importance in current rice breeding efforts. In: Colin WW, Harold C, Koushik S, Jonathan F (eds) Encyclopedia of food grains, 2nd edn. Elsevier, Reference Module in Food Science

McCouch SR, Zhao K, Wright M, Tung C-W, Ebana K, Thomson M, Reynolds A, Wang D, DeClerck G, Ali ML, McClung A, Eizenga G, Bustamante C (2010) Development of genome-wide SNP assays for rice. Breed Sci 60:524–535. https://doi.org/10.1270/jsbbs.60.524

Metzker ML (2005) Emerging technologies in DNA sequencing. Genome Res 15:1767–1776. https://doi.org/10.1101/gr.3770505

Meyer RS (2014) African rice (*Oryza glaberrima*) salinity tolerance and genetic variation across coastal West Africa. Plant and Animal Genome Conference XXII 2014

Meyer RS, Choi JY, Sanches M, Plessis A, Flowers JM, Amas J, Dorph K, Barretto A, Gross B, Fuller DQ, Bimpong IK, Ndjiondjop M-N, Hazzouri KM, Gregorio GB, Purugganan MD (2016) Domestication history and geographical adaptation inferred from a SNP map of African rice. Nat Genet 48:1083–1088. https://doi.org/10.1038/ng.3633

Mohapatra S (2010) Pockets of gold. Rice Today 9:32–33

Monat C, Pera B, Ndjiondjop M-N, Sow M, Tranchant-Dubreuil C, Bastianelli L, Ghesquière A, Sabot F (2016) de novo assemblies of three *Oryza glaberrima* accessions provide first insights about pan-genome of African rices. Genome Biol Evol evw253. https://doi.org/10.1093/gbe/evw253

Montcho D, Futakuchi K, Agbangla C, Fofana M, Dieng I (2013) Yield loss of *Oryza glaberrima* caused by grain shattering under rainfed upland conditions. Int J Biol Chem Sci 7:535–543. https://doi.org/10.4314/ijbcs.v7i2.10

Moukoumbi YD, Sié M, Vodouhe R, Toulou B, Ahanchede A (2011) Screening of rice varieties for their weed competitiveness. Afr J Agric Res 6:5446–5456. https://doi.org/10.5897/AJAR11.1162

Moukoumbi YD, Kolade O, Drame KN, Sie M, Ndjiondjop MN (2015) Genetic relationships between interspecific lines derived from *Oryza glaberrima* and *Oryza sativa* crosses using microsatellites and agro-morphological markers. Span J Agric Res 13:701

Nabholz B, Sarah G, Sabot F, Ruiz M, Adam H, Nidelet S, Ghesquière A, Santoni S, David J, Glémin S (2014) Transcriptome population genomics reveals severe bottleneck and domestication cost in the African rice (*Oryza glaberrima*). Mol Ecol 23:2210–2227. https://doi.org/10.1111/mec.12738

Ndjiondjop MN, Albar L, Fargette D, Fauquet C, Ghesquière A (1999) The genetic basis of high Resistance to Rice yellow mottle virus (RYMV) in cultivars of two cultivated rice species. Plant Dis 83:931–935. https://doi.org/10.1094/PDIS.1999.83.10.931

Ndjiondjop MN, Semagn K, Sie M, Cissoko M, Fatondji B, Jones M (2008) Molecular profiling of interspecific lowland rice populations derived from IR64 (*Oryza sativa*) and Tog5681 (*Oryza glaberrima*). Afr J Biotechnol 7(23):4219–4229

Ndjiondjop MN, Manneh B, Cissoko M, Drame NK, Kakai RG, Bocco R, … Wopereis M (2010) Drought resistance in an interspecific backcross population of rice (*Oryza spp.*) derived from the cross WAB56-104 (*O. sativa*) × CG14 (*O. glaberrima*). Plant Sci 179 (4):364–373. https://doi.org/10.1016/j.plantsci.2010.06.006

Ndjiondjop MN, Seck PA, Lorieux M, Futakuchi K, Yao KN, Djedatin G, Sow ME, Bocco R, Cisse F, Fatondji B (2012) Effect of drought on *Oryza glaberrima* rice accessions and *Oryza glaberrima* derived-lines. Asian J Agric Res 6:144–157. https://doi.org/10.3923/ajar.2012.144.157

Nuijten E, Temudo M, Richards P, Okry F, Teeken B, Mokuwa A, Struik PC (2013) Towards a new approach for understanding interactions of technology with environment and society in small-scale rice farming. In: Wopereis MCS, Johnson DE, Ahmadi N (eds) Realizing Africa's Rice Promise. CABI

Nwilene FE, Williams CT, Ukwungwu MN, Dakouo D, Nacro S, Hamadoun A, Kamara SI, Okhidievbie O, Abamu FJ, Adam A (2002) Reactions of differential rice genotypes to African rice gall midge in West Africa. Int J Pest Manag 48:195–201. https://doi.org/10.1080/09670870110103890

Obilana AB, Okumu BN (2005) Evaluation study report for interspecific hybridization between African and Asian Rice Species. UNDP and WARDA. Available at: http://www.africarice.org/publications/UNDP%20Report/UNDP%20report.pdf

Ohyanagi H, Ebata T, Huang X, Gong H, Fujita M, Mochizuki T, Toyoda A, Fujiyama A, Kaminuma E, Nakamura Y, Feng Q, Wang Z-X, Han B, Kurata N (2016) OryzaGenome: genome diversity database of Wild Oryza species. Plant Cell Physiol 57:e1. https://doi.org/10.1093/pcp/pcv171

Orjuela J, Deless EFT, Kolade O, Chéron S, Ghesquière A, Albar L (2013) A recessive resistance to Rice yellow mottle virus is associated with a rice homolog of the CPR5 gene, a regulator of active defense mechanisms. Mol Plant Microbe Interact 26:1455–1463. https://doi.org/10.1094/MPMI-05-13-0127-R

Pariasca-Tanaka J, Chin JH, Dramé KN, Dalid C, Heuer S, Wissuwa M (2014) A novel allele of the P-starvation tolerance gene OsPSTOL1 from African rice (*Oryza glaberrima* Steud) and its distribution in the genus Oryza. TAG Theor Appl Genet Theor Angew Genet 127:1387–1398. https://doi.org/10.1007/s00122-014-2306-y

Pariasca-Tanaka J, Lorieux M, He C, McCouch S, Thomson MJ, Wissuwa M (2015) Development of a SNP genotyping panel for detecting polymorphisms in *Oryza glaberrima/O. sativa* interspecific crosses. Euphytica 201:67–78. https://doi.org/10.1007/s10681-014-1183-4

Petitot A-S, Kyndt T, Haidar R, Dereeper A, Collin M, de Almeida Engler J, Gheysen G, Fernandez D (2017) Transcriptomic and histological responses of African rice (*Oryza glaberrima*) to Meloidogyne graminicola provide new insights into root-knot nematode resistance in monocots. Ann Bot 119:885–899. https://doi.org/10.1093/aob/mcw256

Pidon H, Ghesquière A, Chéron S, Issaka S, Hébrard E, Sabot F, Kolade O, Silué D, Albar L (2017) Fine mapping of RYMV3: a new resistance gene to Rice yellow mottle virus from *Oryza glaberrima*. Theor Appl Genet 130:807–818. https://doi.org/10.1007/s00122-017-2853-0

Pinel-Galzi A, Rakotomalala M, Sangu E, Sorho F, Kanyeka Z, Traoré O, Sérémé D, Poulicard N, Rabenantoandro Y, Séré Y, Konaté G, Ghesquière A, Hébrard E, Fargette D (2007) Theme and variations in the evolutionary pathways to virulence of an RNA plant virus species. PLoS Pathog 3:e180. https://doi.org/10.1371/journal.ppat.0030180

Pinel-Galzi A, Dubreuil-Tranchant C, Hébrard E, Mariac C, Ghesquière A, Albar L (2016) Mutations in Rice yellow mottle virus polyprotein P2a Involved in RYMV2 gene resistance breakdown. Front Plant Sci. https://doi.org/10.3389/fpls.2016.01779

Platten JD, Egdane JA, Ismail AM (2013) Salinity tolerance, Na+ exclusion and allele mining of HKT1;5 in *Oryza sativa* and *O. glaberrima*: many sources, many genes, one mechanism? BMC Plant Biol 13:32. https://doi.org/10.1186/1471-2229-13-32

Plowright RA, Coyne DL, Nash P, Jones MP (1999) Resistance to the rice nematodes *Heterodera sacchari*, Meloidogyne graminicola and M. incognita in *Oryza glaberrima* and *O. glaberrima* × *O. sativa* interspecific hybrids. Nematology 1:745–751. https://doi.org/10.1163/156854199508775

Portères R (1956) Taxonomie Agrobotanique des Riz cultivés *O. sativa* L. et *O. glaberrima* Steudel. J Agric Trop Bot Appliquée 3:341–384. https://doi.org/10.3406/jatba.1956.2318

Portères R (1962) Berceaux Agricoles Primaires Sur Le Continent Africain. J Afr Hist 3:195–210. https://doi.org/10.1017/S0021853700003030

Portères R (1976) African cereals: eleusine, fonio, black fonio, teff, Brachiaria, Paspalum, Pennisetum, and African rice. In: Harlan JR, de Wet JMJ, Stemler ABL (eds) Origins of African Plant domestication. Mouton, The Hague, Netherlands, pp 409–452

Reversat G, Destombes D (1998) Screening for resistance to *Heterodera sacchari* in the two cultivated rice species, *Oryza sativa* and *O. glaberrima*. Fundam Appl Nematol 21:307–317

Sahrawat KL, Sika M (2002) Comparative tolerance of *O. sativa* and *O. glaberrima* rice cultivars for iron toxicity in West Africa. Int Rice Res Notes 27:30–31

Sakai H, Ikawa H, Tanaka T, Numa H, Minami H, Fujisawa M, Shibata M, Kurita K, Kikuta A, Hamada M, Kanamori H, Namiki N, Wu J, Itoh T, Matsumoto T, Sasaki T (2011) Distinct evolutionary patterns of *Oryza glaberrima* deciphered by genome sequencing and comparative analysis. Plant J Cell Mol Biol 66:796–805. https://doi.org/10.1111/j.1365-313X.2011.04539.x

Sano Y, Sano R, Morishima H (1984) Neighbour effects between two co-occurring rice species, *Oryza sativa* and *O. Glaberrima*. J Appl Ecol 21:245–254. https://doi.org/10.2307/2403050

Sarla N, Swamy BPM (2005) *Oryza glaberrima*: a source for the improvement of *Oryza sativa*. Curr Sci 89:955–963

Second G (1982) Origin of the genic diversity of cultivated rice (Oryza spp.): study of the polymorphism scored at 40 isozyme loci. Jpn J Genet 57:25–57

Semagn K, Bjørnstad A, Ndjiondjop MN (2006) An overview of molecular marker methods for plants. Afr J Biotechnol 5:2540–2568

Semagn K, Ndjiondjop M-N, Lorieux M, Cissoko M, Jones MP, McCouch S (2007) Molecular profiling of an interspecific rice population derived from a cross between WAB 56-104 (*Oryza sativa*) and CG 14 (*Oryza glaberrima*). Afr J Biotechnol 6:2014–2022. https://doi.org/10.5897/AJB2007.000-2310

Shen Y, Zhao Z, Ma H, Bian X, Yu Y, Yu X, Chen H, Liu L, Zhang W, Jiang L, Zhou J, Tao D, Wan J (2015) Fine mapping of S37, a locus responsible for pollen and embryo sac sterility in hybrids between *Oryza sativa L.* and *O. glaberrima Steud.* Plant Cell Rep 34(11):1885–1897

Sikirou M, Shittu A, Konaté KA, Maji AT, Ngaujah AS, Sanni KA, Ogunbayo SA, Akintayo I, Saito K, Dramé KN, Ahanchédé A, Venuprasad R (2016) Screening African rice (*Oryza glaberrima*) for tolerance to abiotic stresses: I. Fe toxicity. Field Crops Res. https://doi.org/10.1016/j.fcr.2016.04.016

Silué D, Nottéghem J-L (1991) Resistance of 99 *Oryza glaberrima* Steud. varieties to blast. Int Rice Res Newsl 16:13–14

Singh N, Kaur L, Sodhi NS, Sekhon KS (2005) Physicochemical, cooking and textural properties of milled rice from different Indian rice cultivars. Food Chem 89:253–259

Somado EA, Guei RG, Keya SO (2008) NERICA the new rice for Africa: a compendium. Africa Rice Center (WARDA), Cotonou

Soriano IR, Schmit V, Brar DS, Prot J-C, Reversat G (1999) Resistance to rice root-knot nematode Meloidogyne graminicola identified in *Oryza longistaminata* and *O. glaberrima*. Nematology 1:395–398. https://doi.org/10.1163/156854199508397

Sumi A, Katayama TC (1994) Studies on Agronomic Traits of African Rice (*Oryza glaberrima* Steud.) I. Growth, yielding ability and water consumption. Jpn J Crop Sci 63:96–104. https://doi.org/10.1626/jcs.63.96

Sweeney M, McCouch S (2007) The complex history of the domestication of rice. Ann Bot 100:951–957. https://doi.org/10.1093/aob/mcm128

Ta KN, Adam H, Staedler YM, Schönenberger J, Harrop T, Tregear J, Do NV, Gantet P, Ghesquière A, Jouannic S (2017) Differences in meristem size and expression of branching genes are associated with variation in panicle phenotype in wild and domesticated African rice. EvoDevo. https://doi.org/10.1186/s13227-017-0065-y

Takasaki Y, Seki Y, Nojima H, Isoda A (1994) Growth of an annual strain of *Oryza glaberrima* Steud. and a perennial cultivar of *Oryza sativa* L. after heading. Jpn J Crop Sci 63:632–637. https://doi.org/10.1626/jcs.63.632

Teeken B (2015) African rice (*Oryza glaberrima*) cultivation in the Togo Hills: ecological and socio-cultural cues in farmer seed selection and development. Ph.D. Dissertation, Wageningen University & Research, Centre for Crop Systems Analysis, Wageningen School of Social Sciences (WASS).

Temudo MP (2011) Planting knowledge, harvesting agro-biodiversity: a case study of Southern Guinea-Bissau rice farming. Hum Ecol 39:309–321. https://doi.org/10.1007/s10745-011-9404-0

Thiémélé D, Boisnard A, Ndjiondjop M-N, Chéron S, Séré Y, Aké S, Ghesquière A, Albar L (2010) Identification of a second major resistance gene to Rice yellow mottle virus, RYMV2, in the African cultivated rice species, *O. glaberrima*. Theor Appl Genet 121:169–179. https://doi.org/10.1007/s00122-010-1300-2

Thomson MJ, Zhao K, Wright M, McNally KL, Rey J, Tung C-W, Reynolds A, Scheffler B, Eizenga G, McClung A, Kim H, Ismail AM, de Ocampo M, Mojica C, Reveche MY, Dilla-Ermita CJ, Mauleon R, Leung H, Bustamante C, McCouch SR (2012) High-throughput single nucleotide polymorphism genotyping for breeding applications in rice using the BeadXpress platform. Mol Breed 29:875–886. https://doi.org/10.1007/s11032-011-9663-x

Thottappilly G, Rossel H (1993) Evaluation of resistance to Rice yellow mottle virus in Oryza species. Indian J Virol 9:65–73

Traore K (2005) Characterization of novel rice germplasm from West Africa and genetic marker associations with rice cooking quality. A&M University, Texas

Traore K, McClung AM, Fjellstrom R, Futakuchi K (2011) Diversity in grain physico-chemical characteristics of West African rice, including NERICA genotypes, as compared to cultivars from the United States of America. Int Res J Agric Sci Soil Sci 1(10):435–448

Traoré O, Pinel-Galzi A, Issaka S, Poulicard N, Aribi J, Aké S, Ghesquière A, Séré Y, Konaté G, Hébrard E, Fargette D (2010) The adaptation of Rice yellow mottle virus to the eIF(iso)4G-mediated rice resistance. Virology 408:103–108. https://doi.org/10.1016/j.virol.2010.09.007

van Andel TR, Meyer RS, Aflitos SA, Carney JA, Veltman MA, Copetti D, Flowers JM, Havinga RM, Maat H, Purugganan MD, Wing RA, Schranz ME (2016) Tracing ancestor rice of Suriname Maroons back to its African origin. Nat Plants 2:16149. https://doi.org/10.1038/nplants.2016.149

Vaughan DA (1994) The wild relatives of rice: a genetic resources handbook. International Rice Research Institute, Los Banos

Vaughan DA, Morishima H, Kadowaki K (2003) Diversity in the Oryza genus. Curr Opin Plant Biol 6:139–146. https://doi.org/10.1016/S1369-5266(03)00009-8

Vikal Y, Das A, Patra B, Goel RK, Sidhu JS, Singh K (2007) Identification of new sources of bacterial blight (*Xanthomonas oryzae* pv. oryzae) resistance in wild Oryza species and *O. glaberrima*. Plant Genet Resour 5:108–112. https://doi.org/10.1017/S147926210777661X

Wambugu PW, Furtado A, Waters DL, Nyamongo DO, Henry RJ (2013) Conservation and utilization of African Oryza genetic resources. Rice. https://doi.org/10.1186/1939-8433-6-29

Wambugu PW, Ndjiondjop MN, Furtado A, Henry R (2017) Sequencing of bulks of segregants allows dissection of genetic control of amylose content in rice. https://doi.org/10.1111/pbi.12752

Wang M (2014) The origin and domestication history of African Rice *Oryza glaberrima*. Plant and animal genome

Wang M, Yu Y, Haberer G, Marri PR, Fan C, Goicoechea JL, Zuccolo A, Song X, Kudrna D, Ammiraju JSS, Cossu RM, Maldonado C, Chen J, Lee S, Sisneros N, de Baynast K, Golser W, Wissotski M, Kim W, Sanchez P, Ndjiondjop M-N, Sanni K, Long M, Carney J, Panaud O, Wicker T, Machado CA, Chen M, Mayer KFX, Rounsley S,

Wing RA (2014) The genome sequence of African rice (*Oryza glaberrima*) and evidence for independent domestication. Nat Genet 46:982–988. https://doi.org/10.1038/ng.3044

Wang K, Wambugu PW, Zhang B, Wu AC, Henry RJ, Gilbert RG (2015) The biosynthesis, structure and gelatinization properties of starches from wild and cultivated African rice species (*Oryza barthii* and *Oryza glaberrima*). Carbohydr Polym 129:92–100. https://doi.org/10.1016/j.carbpol.2015.04.035

WARDA (1997) West Africa rice development association. Annual report 1996. Mbé, Côte d'Ivoire

Watanabe H, Futakuchi K, Jones MP, Sobambo BA (2006) Grain protein content of interspecific progenies derived from the cross of African rice (*Oryza glaberrima* Steud.) and Asian Rice (*Oryza sativa* L.). Plant Prod Sci 9:287–293. https://doi.org/10.1626/pps.9.287

Watarai M, Inouye J (1998) Internode elongation under different rising water conditions in African floating rice (*Oryza glaberrima* Steud.). J Faculty Agric Kyushu Univ 42:301–307

Wenger JW, Schwartz K, Sherlock G (2010) Bulk segregant analysis by high-throughput sequencing reveals a novel xylose utilization gene from *Saccharomyces cerevisiae*. PLoS Genet. https://doi.org/10.1371/journal.pgen.1000942

Williams CT, Okhidievbie O, Harris KM, Ukwungwu MN (1999) The host range, annual cycle and parasitoids of the African rice gall midge *Orseolia oryzivora* (Diptera: Cecidomyiidae) in central and southeast Nigeria. Bull Entomol Res 89:589–597. https://doi.org/10.1017/S0007485399000747

Wu W, Liu X, Wang M, Meyer RS, Luo X, Ndjiondjop M-N, Tan L, Zhang J, Wu J, Cai H, Sun C, Wang X, Wing RA, Zhu Z (2017) A single-nucleotide polymorphism causes smaller grain size and loss of seed shattering during African rice domestication. Nat Plants 3:17064. https://doi.org/10.1038/nplants.2017.64

Zhang Q-J, Zhu T, Xia E-H, Shi C, Liu Y-L, Zhang Y, Liu Y, Jiang W-K, Zhao Y-J, Mao S-Y, Zhang L-P, Huang H, Jiao J-Y, Xu P-Z, Yao Q-Y, Zeng F-C, Yang L-L, Gao J, Tao D-Y, Wang Y-J, Bennetzen JL, Gao L-Z (2014) Rapid diversification of five Oryza AA genomes associated with rice adaptation. Proc Natl Acad Sci 111:E4954–E4962. https://doi.org/10.1073/pnas.1418307111

Oryza glumaepatula Steud.

10

Camila Pegoraro, Daniel da Rosa Farias
and Antonio Costa de Oliveira

Abstract

Oryza glumaepatula is the only native wild rice from the Americas that has a diploid genome ($2n = 24$). Thus, this species constitutes a source of genetic variability for the improvement of *Oryza sativa* cultivated in the New World. Wild rice species are important sources of commercially important characters, such as tolerance to acid soils and drought, yield potential, cytoplasmic male-sterility, heat tolerance, and elongation ability. The transference of alleles associated to target characters from *O. glumaepatula* into *O. sativa* could be done via introgression or through the use of genetic engineering. *Oryza glumaepatula* occurs in Central and South America and Caribe, growing in flooded areas, swamps, rivers, and humid areas with clay soils and that present invasive or colonizing behavior. It presents populations with perennial, annual, or biannual cycle, depending of geographical localization, with bushy growth, fragile stems near the base of plants, which can detach and fluctuate, forming new populations. Flowering occurs between October and November, presenting self- and outcrossing. Recently, *O. glumaepatula* had its genome sequenced, enabling comparative studies among different *Oryza* species. These studies suggest the expansion and contraction of gene families, diversity, and variation in the number of noncoding genes and divergence in the sequences among the species are associated with the adaptation of each species to different environmental conditions. These results demonstrate the importance of species from the *Oryza* genus in the development of new *O. sativa* cultivars.

10.1 Introduction

The genus *Oryza* is composed of 24 species and ten genomes (Ge et al. 2001). Within this genus there are two cultivated species, *O. sativa* L. and *O. glaberrima* Steud., which were domesticated independently from *O. rufipogon* in Asia and *O. barthii* in Africa, respectively (Win et al. 2016). Asian rice (*O. sativa*), besides its importance as a staple food for over half of the world's population (Xiong et al. 2016), is a model species, since it was the second plant species and the first cultivated species to have its genome sequenced (Jackson 2016).

The growth of the world's population added by intensification of biotic and abiotic stresses,

C. Pegoraro · D. da Rosa Farias · A. C. de Oliveira (✉)
Federal University of Pelotas, Plant Genomics and Breeding Center, FAEM 3rd Floor, Campus UFPel, 354, Pelotas-RS, Brazil
e-mail: acostol@gmail.com

T. K. Mondal and R. J. Henry (eds.), *The Wild Oryza Genomes*,
Compendium of Plant Genomes, https://doi.org/10.1007/978-3-319-71997-9_10

climate changes, and the scarcity of new available agricultural areas, has generated a continuous demand of new and more productive genotypes, more resistant to adverse environmental conditions, with a lower environmental impact (less water, fertilizer and chemical demands). This scenario is observed for all cultivated species, including rice. Rice can grow in different environments, however presents severe losses in yield when grown under adverse conditions. Wild genotypes, better adapted to adverse environments, are potential sources of genetic diversity due to the presence of useful genes that can be transferred to cultivated species. The identification and transference of target genes between different genotypes and species are enabled through classical tools of hybridization and introgression. These tools are now improved by more precise assessments made using genomics tools, such as molecular markers, bioinformatics, and genetic engineering/genome editing.

Wild species belonging to the genus *Oryza* are valuable genetic resources for plant breeding. There are 22 wild species of rice identified, being diploid and tetraploid. *O. glumaepatula* (AA) is the only diploid ($2n = 24$) species native of South America (Fuchs et al. 2016). Therefore, *O. glumaepatula* is a potential source of genetic diversity for rice breeding programs in South America, since it is adapted to soil and climate conditions. However, few studies have addressed the potential of wild relatives of rice, including *O. glumaepatula*. In this sense, the aim of this chapter is to collect and summarize different studies involving this species, including characterization, genetic diversity aiming to improve conservation of gene pools and its genome sequence, in order to enable the development of high genetic potential genotypes.

10.2 Anatomy and Morphology

In South and Central America, there are four wild species of the *Oryza* genus described. One diploid, *O. glumaepatula* (AA genome), and three tetraploid species, *O. alta* Swallen, *O. latifolia* Desv., and *O. grandiglumis* Doell (CCDD genome). Early reports considered *O. glumaepatula* as an American form of *O. rufipogon* (Vaughan 1994). However, since the hybrid originated from the cross between *O. glumaepatula* and *O. rufipogon* is sterile, a new classification was created, and *O. glumepatula* was described as AA. This new classification was later confirmed by other studies (Juliano et al.1998; Ge et al. 1999; Sánchez and Espinoza 2005).

A comparison of 26 *O. glumaepatula* accessions from five countries in South America (Brazil, Colombia, French Guiana, Suriname, and Venezuela) and one from Cuba with *O. rufipogon* and *O. nivara* was performed (Juliano et al. 1998). The morphological traits of 19 out of 26 accessions were clearly distinct from *O. rufipogon* and *O. nivara*. The majority of accessions from Brazil, two from Suriname and one from the French Guiana (France), presented long and plain grains, pair of sterile long and wide lemmas, long second leaf and wide flag leaf. Two accessions originating from Cuba and Colombia presented a pair of short sterile long and narrow lemmas and narrow flag leaf. Two accessions originating from Venezuela presented short anthers and small grains. Two accessions originating from Brazil and Venezuela presented short and wide anthers and short grains. One access from Brazil was clearly distinct from the other accesses, presenting long, wide, thick grains and sterile long and wide lemmas. The *O. rufipogon* accesses were characterized by narrow grains, sterile lemmas, long anthers, and narrow second leaves.

With the objective of analyzing *O. glumaepatula* taxonomical characters that are distinct from *O. sativa* and *O. rufipogon*, electron microscopy analyses were performed (Sánchez and Espinoza 2005). All characters observed for *O. glumaepatula* from Costa Rica were common to the ones found in Brazilian accesses. Leaf blade in *O. glumepatula* is covered with oblong and spheroid epicutilar wax in the shape of tombstones; the edge of leaf blades exhibits voluminous spiky trichomes; stomata of diamond shape surrounded by silicon spheric papillae, with voluminous and small trichomes; and the central vein is covered with large globular papillae, a trait that distinguishes *O. glumaepatula* from *O. sativa* and *O. rufipogon*. Other traits were also present in *O. glumaepatula* but not in *O. sativa* and *O. rufipogon*, such is the case of auricles that present attenuated trichomes on the edges and bicellular trichomes in the surface; ligule that presents many short attenuated trichomes and sterile lemmas, which are in the shape of wings with rough edges. Spikelets have a characteristic morphology of the genus *Oryza*, and the fertile lemmas have abundant thorns, similar to *O. rufipogon*, but not to *O. sativa*. *O. glumaepatula* plant can be visualized in Fig. 10.1.

10.3 Genetic Variability

O. glumaepatula occurs in South, Central America and Caribbean, growing in flooded areas, swamps, rivers, and humid areas with clay soils. It is a species that presents populations with perennial cycle, annual or biannual, depending of its geographical location (Karasawa et al. 2007a), with bushy growth, fragile stem near the plant base, which can detach and fluctuate, forming new populations. Flowering occurs between October and November, presenting self- as well as cross-pollination habit (Fuchs et al. 2016). Plants grow in the river margins present behavior of invasive or colonizing plants (Karasawa et al. 2012).

The evaluation of genetic variability and the understanding of the mating system is key to determine in situ and ex situ *O. glumaepatula* conservation strategies, since this species is a potential source of genetic variability for *O. sativa* breeding programs. Thus, different studies were developed aiming to characterize *O. glumaepatula* populations originating from diverse geographic origins.

In order to evaluate the genetic variability of *O. glumaepatula* populations, thirty populations originating from three different Brazilian biomes (Cerrado, Amazonia, and Pantanal) were analyzed (Brondani et al. 2005). A high index of genetic variability was observed, being two populations much more variable than the others. Part of the variability found in *O. glumaepatula* was a consequence of proximity with commercial areas, suggesting that cross-pollination between cultivated and wild accesses could be responsible for some introgressions of cultivated into wild genotypes. This could cause a future loss of genetic identity of wild populations (Brondani et al. 2005).

Eleven *O. glumaepatula* populations originating from two Brazilian biomes, Pantanal and Amazonia, genetic diversity (H_e) among the populations was detected, as well as unique

Fig. 10.1 *O. glumaepatula* plants. Top corner: Adult one growing in the State of Tocantins, TO, Brazil. *Source* Paulo Hideo Nakano Rangel and Claudio Brondani). Close up view of an adult plant. *Photograph courtesy* Dr. Gireesh C, DRR, India

alleles in each population. However, low observed heterozygosity (H_o) were found, indicating an excess of homozygous genotypes. Also, deviations in the Hardy–Weinberg equilibrium were significant for all populations. It was suggested that differences between H_e and H_o could be caused by differences in cross-pollination rates (Karasawa et al. 2007b). In a complementary study, evaluating three populations of *O. glumaepatula*, two originating from the Amazonia and one from Pantanal, they verified that one population is predominantly annual, while a high perenniality degree was detected in the other two (Karasawa et al. 2007a). Also, the reproductive system is variable among the populations, ranging from self-pollination to an intermediate system with predominance of selfing. On the other hand, when six populations from the Amazonia biome were compared to one population of the Pantanal biome, no population showed the Hardy–Weinberg equilibrium, due to self-crossing predominance and consequent absence of gene flow, contributing for a high degree of interpopulacional divergence (Silva et al. 2007).

When *O. glumaepatula* accesses from the Amazonia and Cerrado Biomes were compared, Amazonian accesses showed higher diversity than those from the Cerrado (Abreu et al. 2015). The Amazonian accesses occur in swampy valleys and are isolated from other water sources, while the Cerrado populations were collected near small creeks. Considering these features, it would be expected that the Amazonian accesses would have lower H_e, but a higher intrapopulacional genetic variability was observed, suggesting a large flow of plants in wider areas, either through seed or pollen dispersal.

The genetic diversity of *O. glumaepatula* populations from Costa Rica was recently reported (Fuchs et al. 2016). All accesses of *O. glumaepatula* known in the country were studied, originating from two regions from northwest and northeast Costa Rica. Intermediate genetic diversity levels, significant evidences of inbreeding, small differences between populations separated by more than 100 km, and indirect evidences of gene flow between cultivated rice and *O. glumaepatula* were seen. Hybrids from the cross *O. glumaepatula* × *O. sativa* are male-sterile. However, estigmas are receptive and can produce viable seeds. Besides, these hybrids are capable of backcrossing with its wild parent and can result in the introgression of genome segments from *O. sativa* into *O. glumaepatula* (Fuchs et al. 2016).

Besides Brazil and Costa Rica, *O. glumaepatula* has been reported in Bolivia, Colombia, Cuba, Dominican Republic, French Guiana, Guiana, Honduras, Mexico, Panama, Suriname, and Venezuela (Veasey et al. 2008). However, genetic diversity studies were found only for Brazilian (Brondani et al. 2005; Karasawa et al. 2007a; Silva et al. 2007; Abreu et al. 2015; Karasawa et al. 2007b) and Costa Rica (Fuchs et al. 2016) populations (Fig. 10.2).

10.4 *O. glumaepatula* as a Source of Genetic Diversity for *O. sativa*

Modern rice varieties result from a process of intense artificial selection, which led to the fixation of important target traits. However, this directional selection reduced the genetic variability of cultivated rice due to the loss of alleles and the fixation of others, resulting in genetic erosion. Besides, most breeding programs are based on methods that favor inbreeding, leading to reduced genetic recombination. Also, the use of elite × elite crosses makes use of narrow genetic basis, resulting in small genetic gains (Rangel et al. 2008). In this sense, additional sources of variability are necessary for rice breeding programs.

Wild rice species have been shown to be important sources of new and economically important characters, such as tolerance to acid soils, drought and yield increases (Fuchs et al. 2016), cytoplasm male-sterility, heat tolerance and elongation ability (Sanchez et al. 2014). For rice breeding programs in South America, it is convenient to use *O. glumaepatula* as a source of genetic diversity and adaptability of this species to local environments.

Fig. 10.2 Distribution of *O. glumaepatula*

The transference of alleles associated to target traits from *O. glumaepatula* into *O. sativa* can be mediated by introgression or through the use of genetic engineering tools. The introgression of wild alleles into commercial genotypes can be performed through the development of introgression lines, i.e., through interspecific backcrossing (*O. sativa*—recurrent × *O. glumaepatula*—donor) followed by a few generations of selfing. Molecular markers associated with target wild alleles can be used to select plants among the progeny that bear specific genomic regions containing the target allele, therefore increasing the efficiency. Once they present small introgressed wild segments in an elite recurrent genome, these lines can be used as allele reservoirs for the development of new varieties with higher diversity and higher resistance to biotic and abiotic stresses. Also, introgression lines can be used for functional genomics and genetic mapping (Rangel et al. 2008). Introgression and Chromosomal Segment Substitution (CSSL) lines are synonyms (Yoshimura et al. 2010).

Erect panicle is a trait that is highly correlated with yield potential in rice (Gao et al. 1999). However, the molecular mechanism responsible for this trait has not been elucidated yet. *O. glumaepatula* with erect panicles is an excellent gene pool for the introduction of this trait into Asian rice (Zhang et al. 2015). One BC_5F_2 introgression line derived from a cross between *O. sativa* spp. *japonica* (recurrent) and *O. glumaepatula* (donor), enabled the mapping of a gene responsible for erect panicle, *EP4*, in the short arm of chromosome 6. This information can be used for marker-assisted selection (MAS) strategies aiming at erect panicle or for the cloning of *EP4* (Zhang et al. 2015).

10.5 Genome and Genomics

The genus *Oryza* is composed by 22 wild and 2 domesticated species, with 10 distinct genomes (AA, BB, CC, BBCC, CCDD, EE, FF, GG, KKLL, HHJJ). Asian rice (*O. sativa)* was the first species to have its genome sequenced by the International Rice Genome Sequencing Project (IRGSP). Data from *Oryza* wild species have been obtained in recent years with the development of *Oryza* Map Alignment (OMAP) and

Oryza Genome Evolution (OGEP) projects. These projects were funded to produce and make available a large collection of genomic data, specially physical maps that represent 17 out of 24 species, covering eight AA species and one each of the other nine genomes plus *Leersia perrieri* as an outgroup (Kim et al. 2008; Ammiraju et al. 2006, 2010; Jacquemin et al. 2013), which are available to the scientific community.

Comparative genomic studies have enabled the identification of key points related to genome structure and evolution. Also, these studies indicate that only one reference genome for the *Oryza* genus is insufficient for the understanding of the allelic diversity within the genus, which is important for the development of superior genotypes in order to supply enough for future demands (Jacquemin et al. 2013). To complement the OMAP, the International Oryza Map Alignment Project (I-OMAP) was developed, which goals are the generation of transcriptome and reference sequence for the species of the genus *Oryza*; generation of AA genome maps and chromosome segment substitution lines for breeding programs; identification and collection of naturally occurring wild rice populations for diversity, conservation, and evolution analyses (Sanchez et al. 2014).

In I-OMAP, the sequencing and assembly of the *O. glumaepatula* (AA) genome was coordinated by the Federal University of Pelotas, Brazil, with the collaboration of University of Perpignan Via Domitia (UPDV), France, and University of Arizona (UA), USA. For the sequence, the DNA was obtained from leaves and seedlings of *O. glumaepatula* genotype 1233, obtained from Embrapa and duplicated at International Center for Tropical Agriculture (CIAT)—Colombia, being sequenced by the National Center for Genome Resource (NCGR)—USA, using the Illumina Genome Analyzer IIx platform.

After sequencing, bioinformatic tools were used to identify adaptors present in the reads to check quality of the bases (Farias 2013). Genome assembly was performed using ABySS (Simpson et al. 2009); RAY (Boisvert et al. 2010); SOAPdenovo (Li et al. 2009); Velvet (Zerbino and Birney 2008); Gossamer (Conway et al. 2012); ZORRO (http://lge.ibi.unicamp.br/zorro/) and SSPACE (Boetzer et al. 2011). Sequence alignments of *O. glumaeopatula* and *O. sativa* were performed using Bowtie (Langmead and Salzberg 2012), and for scaffold alignment on *O. sativa* sequences, BLAST was used (Altschul et al. 1990).

Many assemblies of the *O. glumaepatula* genome were performed, and two contig groups, originating from two independent assemblers were mixed to generate the scaffolds. A total of 292,860 scaffolds covering 87% of the genome were obtained. After the assembly, the scaffolds were compared with the *O. sativa* genome in order to evaluate the quality of generated sequences. A total of 97% of transcribed sequences of *O. sativa* were aligned to the scaffolds, demonstrating that this preliminary version had good quality. On the other hand, 23% of the *O. sativa* genome aligned with *O. glumaepatula* sequences (Farias 2013). The current assembly can be improved by checking the physical map and ordering sequences according to this physical order, enabling the closing of gaps still present.

A comparative analysis between the five wild rice species from the AA genome, including *O. glumaepatula*, was performed. Structural rearrangements, including segmental duplication and gene family turnover with high instability of genes associated to plant defense, were observed. The effect of gene family expansions and contractions on the morphological and reproductive characters, indicating a possible path for evolution and speciation at a genomic scale was reported. The diversity of genes among the species has a key role in the adaptation to the different environmental conditions (Asia, South America, Africa, and Australia) of each species. Also, the variation in the number of noncoding RNA genes and sequence divergence is associated with the adaptation process in these species (Zhang et al. 2014).

Based on the genome of *O. glumaepatula*, zinc finger family WRKY members were identified (Santos et al. 2016). The majority of WRKY genes from *O. sativa* are present in *O. glumaepatula*. However, some genes are located in distinct chromosomes. The evolutionary analysis of WRKY proteins between the *O. sativa* and *O. glumaepatula* species indicated the formation of nine distinct groups. This was the first study aiming to associate WRKY transcription factors between two species, contributing to functional studies. The obtained information can be useful for breeding, enabling the transference of genes associated to stress tolerance.

The evolutionary fate of the *Submergence 1 (SUB1)* locus was studied comparing the *Oryza* genomes. *O. glumaepatula*, as well as *O. rufipogon*, *O. nivara*, *O. glaberrima*, *O. sativa* and *O. punctata* present the genes *SUB1B-like* and *SUB1C-like* on chromosome 9. The location of the *SUB1* locus in chromosome 9 is highly conserved in *O. nivara*, *O. rufipogon*, *O. glumaepatula*, and *O. glaberrima*, with a deviation smaller than 2.0 Mb. Also, an intron was present in the coding region of *SUB1B-like*, in all but *O. glumaepatula*, *O. glaberrima*, and *O. rufipogon*. The gene *SUB1A-like* was not detected in *O. glumaepatula* and the remaining wild species, being found only in *O. nivara* (Santos et al. 2017).

Recently, the distribution of transposable elements (TEs) in the cultivated (*O. sativa* spp. *japonica*, *O. sativa* spp. *indica*, and *O. glaberrima*) and wild (*O. nivara*, *O. rufipogon*, *O. barthii*, *O. glumaepatula*, and *O. meridionalis*) species from the genus *Oryza* which present the AA genome were reported. In general, TEs were deleted from genomic regions which corresponded to genes in cultivated rice species when compared to the wild regions, and when present in expressed genes, are preferentially located in introns, both for cultivated and wild species. Specifically, 37.5% of the *O. sativa* spp. *japonica* genome consists of TEs, while in *O. glumaepatula* the TE fraction corresponds to 21.9% of the genome. When expression data were evaluated in leaves, roots, and panicles, 33,069 expressed genes were detected in *O. sativa*, from which 11,480 presented TEs associated to the genic regions, and 29,442 expressed genes in *O. glumaepatula*, from which 13,522 presented TEs (Li et al. 2017).

Long terminal repeat (LTR) retrotransposons are the most abundant TEs in plant genomes. A new LTR retrotransposon called *RTPOSON* was identified in the genus *Oryza*, and its presence was investigated in the different species of this genus (Hsu et al. 2016). LTR retrotransposons can have an impact on genome evolution, gene creation, and genetic variation. *O. glumaepatula* presents 83 *RTPOSON* copies being 10 intact, 32 fragments, and 41 intact LTR sequences, while in *O. sativa* spp. *japonica* 93 *RTPOSONs* were found, where 15 were intact, 37 were fragments, and 41 were intact LTR sequences.

10.6 Conclusion and Future Direction

Rice has an important role to supply most of the world's food demand and to guarantee food security of a constantly increasing population. However, in order to achieve this objective, it is necessary to develop supervarieties, with less water and input requirements; more resistant to biotic and abiotic stresses; and with threefold higher yields. In order to achieve this gain, new sources of genetic variability have to be exploited, such as wild *Oryza* species. In South America, especially in Brazil, a promising source of variability is *O. glumaepatula*. However, in order to exploit this genetic resource, better understanding of its genome and genetic diversity are needed. Also, one of the most important points for this success is the conservation of *O. glumaepatula*, to avoid genetic erosion and the loss of promising genes, some of them lost due to the high selective pressure during the domestication of *O. sativa*.

Reports available in the literature, covered by this chapter, supply the initial support for the pre-breeding of *O. sativa* from *O. glumaepatula*. However, larger efforts are needed in order to achieve the major goal of plant scientists for the next 30 years, that is to feed 9 billion people with as low as possible environmental impacts.

References

Abreu AG, Rosa TM, Borba TCO, Vianello RP, Rangel PHN, Brondani C (2015) SSR characterization of *Oryza glumaepatula* populations from the Brazilian Amazon and Cerrado biomes. Genetica 143:413–423

Altschul SF, Gish W, Miller W, Myers EW, Lipman DJ (1990) Basic local alignment search tool. J Mol Biol 215:403–410

Ammiraju JS, Luo M, Goicoechea JL, Wang W, Kudrna D, Mueller C, Talag J, Kim H, Sisneros NB, Blackmon B, Fang E, Tomkins JB, Brar D, MacKill D, McCouch S, Kurata N, Lambert G, Galbraith DW, Arumuganathan K, Rao K, Walling JG, Gill N, Yu Y, SanMiguel P, Soderlund C, Jackson S, Wing RA (2006) The *Oryza* bacterial artificial chromosome library resource: construction and analysis of 12 deep-coverage large-insert BAC libraries that represent the 10 genome types of the genus *Oryza*. Genome Res 16:140–147

Ammiraju JSS, Song X, Luo M, Sisneros N, Angelova A, Kudrna D, Kim HR, Yu Y, Goicoechea JL, Lorieux M, Kurata N, Brar D, Ware D, Jackson S, Wing RA (2010) The *Oryza* BAC resource: a genus-wide and genome scale tool for exploring rice genome evolution and leveraging useful genetic diversity from wild relatives. Breed Sci 60:536–543

Boetzer M, Henkel CV, Jansen HJ, Butler D, Pirovano W (2011) Scaffolding pre-assembled contigs using SSPACE. Bioinformatics 27:578–579

Boisvert S, Laviolette F, Corbeil J (2010) Ray: simultaneous assembly of reads from a mix of high-throughput sequencing technologies. J Comp Biol 17:1519–1533

Brondani RPV, Zucchi MI, Brondani C, Rangel PHN, Borba TCO, Rangel PN, Magalhães MR, Vencovsky R (2005) Genetic structure of wild rice *Oryza glumaepatula* populations in three Brazilian biomes using microsatellite markers. Genetica 125:115–123

Conway T, Wazny J, Bromage A, Zobel J, Beresford-Smith B (2012) Gossamer—a resource-efficient de novo assembler. Bioinformatics 28:1937–1938

Farias DR (2013) Desenvolvimento de ferramentas de bioinformática para montagem e prospecção de genomas selvagens visando o melhoramento de arroz. Programa de Pós-Graduação em Agronomia da Universidade Federal de Pelotas, Tese, p 81p

Fuchs EJ, Martínez AM, Calvo A, Muñoz M, Arrieta-Espinoza G (2016) Genetic diversity in *Oryza glumaepatula* wild rice populations in Costa Rica and possible gene flow from *O. sativa*. Peer J 4:e1875

Gao SJ, Chen WF, Zhang BL (1999) Studies of erect panicle in rice. J Jilin Agric Sci 24:12–15

Ge S, Oliveira GCX, Schaal BA, Gao LZ, Hong DY (1999) RAPD variation within and between natural populations of the wild rice *Oryza rufipogon* from China and Brazil. Heredity 82:638–644

Ge S, Sang T, Lu B-R, Hong D-Y (2001) Rapid and reliable identification of rice genomes by RFLP analysis of PCR-amplified *Adh* genes. Genome 44:1136–1142

Hsu T-C, Wang C-S, Lin Y-R, Wu Y-P (2016) Structural Diversity of a Novel LTR Retrotransposon, RTPOSON, in the Genus *Oryza*. Evol Bioinform 12:29–40

Jackson SA (2016) Rice: the first crop genome. Rice 9:14

Jacquemin J, Bhatia D, Singh K, Wing RA (2013) The International Oryza Map Alignment Project: development of a genus-wide comparative genomics platform to help solve the 9 billion-people question. Curr Opin Plant Biol 16:147–156

Juliano AB, Naredo MEB, Jackson M (1998) Taxonomic status of *Oryza glumaepatula* Steud. I. Comparative morphological studies of New World diploids and Asian AA genome species. Genet Resour Crop Evol 45:197–203

Karasawa MMG, Vencovsky R, Silva CM, Zucchi MI, Oliveira GCX, Veasey EA (2007a) Mating system of Brazilian *Oryza glumaepatula* populations studied with microsatellite markers. Ann Bot 99:245–253

Karasawa MMG, Vencovsky R, Silva CM, Zucchi MI, Oliveira GCX, Veasey EA (2007b) Genetic structure of Brazilian wild rice (*Oryza glumaepatula* Steud., Poaceae) populations analyzed using microsatellite markers. Genet Mol Biol 30:400–410

Karasawa MMG, Vencovsky R, Silva CM, Cardim DC, Bressan EA, Oliveira GCX, Veasey EA (2012) Comparison of microsatellites and isozymes in genetic diversity studies of *Oryza glumaepatula* (Poaceae) populations. Rev Biol Trop 60:1463–1478

Kim H, Hurwitz B, Yu Y, Collura K, Gill N, SanMiguel P, Mullikin JC, Maher C, Nelson W, Wissotski M, Braidotti M, Kudrna D, Goicoechea JL, Stein L, Ware D, Jackson SA, Soderlund C, Wing RA (2008) Construction, alignment and analysis of twelve framework physical maps that represent the ten genome types of the genus *Oryza*. Genome Biol 9:R45

Langmead B, Salzberg SL (2012) Fast gapped-read alignment with Bowtie 2. Nat Methods 9:357–359

Li H, Handsaker B, Wysoker A, Fennell T, Ruan J, Homer N, Marth G, Abecasis G, Durbin R (2009) 1000 genome project data processing subgroup. The sequence alignment/map format and SAMtools. Bioinformatics 25:2078–2079

Li X, Guo K, Zhu X, Chen P, Li Y, Xie G, Wang L, Wang Y, Persson S, Peng L (2017) Domestication of rice has reduced the occurrence of transposable elements within gene coding regions. BMC Genom 18:55

Rangel PN, Brondani RPV, Rangel PHN, Brondani C (2008) Agronomic and molecular characterization of introgression lines from the interspecific cross *Oryza sativa* (BG90-2) × *Oryza glumaepatula* (RS-16). Genet Mol Res 7:184–195

Sánchez E, Espinoza AM (2005) Ultrastructure of *Oryza glumaepatula*, a wild rice species endemic of tropical America. Revista de Biología Tropical 53:15–22

Sanchez PL, Wing RA, Brar DS (2014) The wild relative of rice: genomes and genomics. In: Zhang Q, Wing RA (eds) Genetics and genomics of rice. Plant Genetics and Genomics, Crops and Models

Santos RS, Viana VE, Farias DR, Junior ATA, Costa de Oliveira A (2016) Identificação e filogenia da família WRKY em *Oryza glumaepatula*. Revista Congrega

Santos RS, Farias DR, Pegoraro C, Rombaldi CV, Fukao T, Wing RA, de Oliveira Costa (2017) Evolutionary analysis of the SUB1 locus across the *Oryza* genomes. Rice 10:4

Silva CM, Karasawa MMG, Vencovsky R, Veasey EA (2007) Elevada diversidade genética interpopulacional em *Oryza glumaepatula* Steud. (Poaceae) avaliada com microssatélites. Biota Neotropica 7: bn04607022007

Simpson JT, Wong K, Jackman SD, Schein JE, Jones SJ, Birol I (2009) ABySS: a parallel assembler for short read sequence data. Genome Res 19:1117–1123

Vaughan DA (1994) The wild relatives of rice. International Rice Research Institute, Los Baños, Philippines

Veasey EA, Cardin D, Silva RM, Bressan EA, Vencovsky R (2008) Assessing the genetic structure of *Oryza glumaepatula* Populations with isozyme markers. Braz Arch Biol Technol 51:873–882

Win KT, Yamagata Y, Doi K, Uyama K, Nagai Y, Toda Y, Kani T, Ashikari M, Yasui H, Yoshimura A (2016) A single base change explains the independent origin of and selection for the nonshattering gene in African rice domestication. New Phytol 213:1925–1935

Xiong Y, Zhang K, Cheng Z, Wang G-L, Liu W (2016) Data for global lysine-acetylation analysis in rice (*Oryza sativa*). Data Brief 7:411–417

Zhang Q-J, Zhu T, Xia E-H, Shi C, Liu Y-L, Zhang Y, Liu Y, Jiang W-K, Zhao Y-J, Mao S-Y, Zhang L-P, Huan H, Jiao J-Y, Xu P-Z, Yao Q-Y, Zeng F-C, Yang, L-L, Gao J, Tao D-Y, Wang Y-J, Bennetzen JL, Gao L-Z (2014) Rapid diversification of five *Oryza* AA genomes associated with rice adaptation. PNAS E4954–E4962

Yoshimura A, Nagayama H, Sobrizal Kurakazu T, Sanchez PL, Doi K, Yamagata Y, Yasui H (2010) Introgression lines of rice (*Oryza sativa* L.) carrying a donor genome from the wild species, *O. glumaepatula* Steud. and *O. meridionalis* Ng. Breeding Sci 60:597–603

Zerbino DR, Birney E (2008) Velvet: algorithms for de novo short read assembly using de Bruijn graphs. Genome Res 18:821–829

Zhang Y, Zhou J, Yang Y, Li J, Xu P, Deng X, Deng W, Wu Z, Tao D (2015) A novel gene responsible for erect panicle from *Oryza glumaepatula*. Euphytica 205:739–745

11 *Oryza grandiglumis* (Doell) Prod.

Abubakar Mohammad Gumi and Adamu Aliyu Aliero

Abstract

Oryza grandiglumis (Doell) Prod. is a wild species of the genus *Oryza* and is endemic to tropical America. It is a perennial tall plant that grows in areas subject to seasonal flooding due to its characteristic culm elongation that may reach up to 760 cm in height. *Oryza grandiglumis* has an allotetraploid (CCDD) genome type (along with *Oryza alta* and *Oxalis latifolia*) consisting of $2n = 4x = 48$ with 891 Mbp genome size. Wild species of *Oryza* serve as untapped reservoir of important agronomic traits which can be used to improve cultivated rice. *O. grandiglumis* is tolerant to submergence by an unknown mechanism different from the *Sub1* QTL of rice and its useful traits were successfully introgressed into *O. sativa* populations using the *Japonica* subspecies cv. Hwaseongbyeo as recurrent parent. In this chapter, we describe its distribution, cytogenetics, progress on genomics research, submergence tolerance mechanism, and introgressions of genes into cultivated rice species for improvement.

A. M. Gumi (✉) · A. A. Aliero
Department of Biological Sciences, Faculty of Science, Usmanu Danfodiyo University, PMB 2346, Sokoto, Nigeria
e-mail: muhammadag@yahoo.co.uk

11.1 Economic and Academic Interest

The species *O. grandiglumis* has a unique geographical distribution occurring only in South America and can be distinguished easily from the rest of *Oryza* species by the characteristic larger plants and broad leaves (up to 2 cm). It shows a wide range of resistance to biotic and abiotic stresses such as physical wounding, fungal elicitor, and flooding (Kim et al. 2005; Niroula et al. 2012) which are important traits for molecular breeding of cultivated rice species. Of agronomic traits, it has large and open panicles, long ligules and higher plant biomass which are important for yield, flood resistance, and weed competition, respectively.

11.2 Botanical Descriptions and Distribution

It is a perennial tall plant with a big and open panicle (Nayar 1973). However, Vaughan (1989) inferred that species with CCDD genome types can easily be distinguished from the rest of the *Oryza* species by their larger plants and wider leaves (3–5 cm). *Oryza grandiglumis* is endemic to Latin America and reports on its origins and phylogenetic relationships with members of the *officinallis* complex (*O. alta* and *O. latifolia*) have long been in dispute and are still ambiguous

T. K. Mondal and R. J. Henry (eds.), *The Wild Oryza Genomes*,
Compendium of Plant Genomes, https://doi.org/10.1007/978-3-319-71997-9_11

due to their homogeneous genome type, similar morphological characteristics and overlapping distribution (Bao and Ge 2004). The leaf blade has a characteristic cuticular wax pattern with dense rod-like structures and is surrounded by papillae, zipper-like silica cells, abundant bulky prickle trichomes, and hooked trichomes. The ligule is a blunt membrane covered by short prickly trichomes. Spikelet morphology is characteristic of the Poaceae family, but the sterile lemmas are nearly as long as the fertile lemmas, and they have an unique crown-like structure of lignified spines between the rachilla and the fertile lemmas (Sánchez et al. 2006).

The species is endemic to the Neotropics (Morishima and Martins 1993) and found in open (savanna) as well as shaded (woodlands) habitats with clay and alluvial soils. It is found up to an altitude of 230 m above sea level in Argentina, Bolivia, Brazil, Colombia, Ecuador, French Guyana, Paraguay, Venezuela, and Peru (Morishima 1994). Zamora et al. (2003) reported that *O. grandiglumis* in Costa Rica has a blunt and pubescent ligule and may reach up to 760 cm in height as a result of culm elongation due to changes in water level. The auricles embrace the culm and are completely white, unlike most *Oryza* species that have purple-colored auricles (Figs. 11.1 and 11.2).

11.3 Cytogenetics

Oryza grandiglumis (Döell) Prod. belongs to the genus *Oryza* and is very closely related to *O. alta* and *O. latifolia.* It is tetraploid with CCDD genome type ($2n = 4x$, 48) with 891 Mb genome size and along with *O. alta* and *O. latifolia* forming the *officinallis* complex (Vaughan 1994; Akimoto 1998). Like all species of *Oryza,* it has small sized chromosomes that are difficult to analyze for karyotype even with the use of a conventional metaphase plate of its root tips. Generally, pachytene analysis in the cytology of rice was first introduced in 1958 and ever since it was used in Indian Agricultural Research Institute (IARI) to investigate many species of *Oryza* such as *sativa, perrenis, australiensis, officinallis, barthii, granulate,* and *glaberrima* (Yao et al. 1958).

Fig. 11.1 A growing plant of *O. grandiglumis. Photograph courtesy* Dr. Gireesh C, DRR, India. Top right: Close-up view of panicles

Fig. 11.2 Map of America showing the distribution of *O. grandiglumis*. *Source* NARO database Japan

11.4 Biotic Stress Tolerance in *O. grandiglumis*

Wild species of plants are reported to be diverse genetically and display a wide range of resistance to various stresses. The wild species of the genus *Oryza* exhibit tremendous diversity in morphology, agronomic traits, and adaptations to different biotic and abiotic stress (Sanchez et al. 2013). The rice blast disease caused by the *Magnaportheqrisae* fungus is one of the devastating crop pathogens leading to severe loss of yields in rice fields. Wounds are the regions of pathogen entry in plants though they are mostly caused by insects; they also cause severe damage to plants leading to yield losses (Reymond et al. 2000; Cheong et al. 2002). A study by Kim et al. (2005) revealed that *O. grandiglumis* subjected to a fungal elicitor and physical wounding expressed the ogfw library threefold higher than blast infected *O. sativa* through the use of a subtracted library (ogfw) constructed by suppression subtractive hybridization. These subtracted ESTs can be presumed to be related to the defense/resistance system and will be used to investigate the defense mechanisms of wild rice and to provide new insights into the genome of wild rice, which in turn will assist molecular breeding strategies of cultivated rice.

11.5 Submergence Tolerance in *O. grandiglumis*

In Asia (mostly south and southeast), seasonal flooding is a common occurrence annually and results in huge economic losses. *Oryza grandiglumis* mostly grows in Amazonian floodplains, where water levels can reach up to 10 meters, and thus is expected to show some degree of submergence tolerance (Okishio et al. 2014, 2015). Depending on the flooding conditions, *O. grandiglumis* showed distinct responses: when progressively submerged, the internodes elongated, resembling the escape strategy of *O. sativa.* However, when plants were completely submerged, growth was reduced, as in the quiescent strategy (Okishio et al. 2014).

11.5.1 Characterization of Submergence-Related QTLs

In rice, SUB1 is the major QTL associated with submergence tolerance using quiescent strategy and is mapped to chromosome 9 with a cluster of 3 ethylene-responsive factor (ERF) genes, namely SUB1A, 1B, and 1C (Fukao et al. 2009). Different rice genotypes have two genes in the cluster, SUB1B and SUB1C, whereas a third gene, SUB1A, is present only in a subset of them. Interestingly, tolerance to submergence is linked to a specific allele of the SUB1A gene, named SUB1A-1 (Xu et al. 2006). Accessions that lack the SUB1A gene, or carry the SUB1A-2 allele, are sensitive to submergence. Introgression of functional copies of SUB1A-1 in sensitive genotypes is sufficient to generate submergence tolerant plants (Xu et al. 2006). The tolerant SUB1A-1 allele is derived from the aus subgroup of indica rice (Xu et al. 2006). Wild species from the *Oryza* genus commonly grow in constantly or seasonally wet habitats (Vaughan et al. 2003), and thus submergence tolerance could be found in other species.

Niroula et al. (2012) tested 109 accessions of rice and wild relatives, including 12 species, for submergence tolerance, and found *O. rufipogon* and *O. nivara* (AA genome) tolerant accessions that carry the SUB1A-1 allele, showing that SUB1 locus architecture determines submergence tolerance in these species, as in *O. sativa* (Niroula et al. 2012). Strikingly, accessions of *O. rhizomatis* and *O. eichingeri* (CC genome) were also found to be submergence tolerant, but SUB1A sequences were absent from genomes of tested accessions, indicating that a novel, SUB1A-independent mechanism is responsible for submergence tolerance, at least in these two CC genome species (Table 11.1). Interestingly, SUB1A is absent in *O. grandiglumis*, indicating that a SUB1A-independent mechanism for a quiescent strategy is present, as observed for *O. rhizomatis* and *O. eichingeri* (Table 11.1). Genes similar to SNORKEL1 and SNORKEL2, responsible for the escape strategy in deepwater-adapted *O. sativa*, as well as *O. rufipogon* and *O. glumeapatula* (AA genome; Table 11.1), are absent in *O. grandiglumis* (Hattori et al. 2009; Okishio et al. 2015). These results indicate that *O. grandiglumis* is tolerant to both gradual and full submergence by an unknown mechanisms (Okishio et al. 2014), indicating that CC genome *Oryza* species might provide new molecular mechanisms to improve cultivated rice.

11.6 Genome Sequence

Advancement in sequencing technology has led to the completion of a high-quality map-based genome sequence of *O. sativa* ssp. *japonica* cv. Nipponbare which today serves as an indispensable tool in plant genomic research. The draft sequences of the two subspecies of Asian rice (*O. sativa* ssp. *japonica* and *indica*) were published fifteen years ago (Goff et al. 2002; Yu et al. 2002). However, many rice projects and programs have been created with the aim to produce high-quality sequence of rice germplasm and its wild relatives for breeding, species characterization, and evolutionary studies (Sanchez et al. 2013).

To the best of our knowledge, the genome of *O. grandiglumis* has not been decoded. Efforts

Table 11.1 Allele combinations of SUB1 locus and submergence tolerance in *O. grandiglumis* and other *Oryza* species

Species/genome	SUB1B Locus (Allele combination)		
	SUB1A-1, SUB1B, SUB1C	*SUB1A-2, SUB1B, SUB1C*	*SUB1B–SUB1C*
O. sativa (AA)	Tolerant	Sensitive	Sensitive
O. nivara (AA)	Tolerant	Sensitive	Sensitive
O. rufipogon (AA)	Tolerant	Sensitive	Sensitive
O. eichingeri (CC)[a]	Absent	Absent	Tolerant
O. rhizomatis (CC)[a]	Absent	Absent	Tolerant
O. grandiglumis (CCDD)[a]	Absent	Absent	Tolerant

[a]The tetraploid CCDD and diploid CC species lack the *SUB1A* allele and are tolerant to submergence, indicating that the SUB1 locus does not contribute to tolerance in these species

by OMAP and OGEP have led to the establishment of a platform for comparative genomics at the genus level which paved the way for creation of large publicly available sets of manually edited BAC-based physical maps of 18 out of the 24 recognized *Oryza* species, i.e., all 8 AA genome species and one each of the other 9 genome types; BB, CC, EE, FF, GG, BBCC, CCDD, KKLL, HHJJ (Ammiraju et al. 2006, 2010a, b; Kim et al. 2008). This will help in future for genome sequencing of this species.

11.7 Introgression of Genes from *O. grandiglumis*

Genes from wild species have provided crops with resistance to diseases, pest, and abiotic stresses and even improved quality traits such as protein content. In the genus *Oryza,* the wild species of the genus represent an important untapped genetic reservoir that can be used to improve the two cultivated species (*O. sativa* and *O. glaberrima*) though *O. sativa* has received more attention than its African counterpart due to its high yielding potentials. Introgression of useful genes from wild relatives to cultivated species of rice has led to the improvement of rice in recent time, thereby increasing yield and tolerance (Yoon et al. 2006; Eizenga et al. 2009).

A number of breeders have produced interspecific hybrids between rice and CCDD genome species which are close relatives of *O. glaberrima* in terms of genome type (Sitch 1990; Brar et al. 1991). Some introgression lines derived from *O. sativa* × *O. grandiglumis* have been evaluated for introgression of useful traits. Yoon et al. (2006) reported a successful introgression from *O. grandiglumis* (Accession Number: 101154) into *O. sativa* subsp. *japonica* cv. Hwaseongbyeo as a recurrent parent. It has also been seen that the species *O. alta* and *O. grandiglumis* can be hybridized, and that the F_1 is moderately fertile. In addition, alleles associated with resistances to bacterial blight, brown planthoppers, and white-backed planthoppers from *O. latifolia* Desv. (CCDD genome), *O. officinalis* Wall ex Watt (CC genome), and *O. australiensis* Domin. (EE genome) have been successfully introgressed into *O. sativa* populations (Brar and Khush 1997; Multani et al. 1994, 2003).

11.8 Conclusion and Future Direction

In conclusion, *O. grandiglumis* can be distinguished from cultivated rice and other wild species of *Oryza* by its unique morphology such as robust size and big open panicles. Though it was reported not so long ago, work on stress tolerance, morphology, and gene introgression has been initiated with the prospect of improving cultivated rice. Studies on genomics of this species are gaining interest though not much has been done in areas like genome sequence, large-scale marker generation, transcriptomes and linkage mapping. Despite these genomics setbacks, the advent of sequencing technology and unflinching consortium efforts (such as I-OMAP, OGEP, IRGSP, BGI) that constantly supply vast amount of data in the field of *Oryza* genus genomics offer hope and possibilities of sequencing this species in future. Lastly to achieve the global objectives of improving cultivated rice, wild relatives hold the key as they are the reservoirs of so many useful agronomic traits needed to improve cultivated rice.

Acknowledgements The authors are thankful Department of Biological Sciences, Usmanu Danfodiyo University, Sokoto-Nigeria.

References

Akimoto M (1998) Population genetic structure of the wild rice *Oryza glumaepatula* Steud. distributed in the Amazon flood area influenced by-history traits. Mol Ecol 7:1371–1381

Ammiraju JS, Luo M, Goicoechea JL, Wang W, Kudrna D, Mueller C, Talag J, Kim H, Sisneros NB, Blackmon B (2006) The *Oryza* bacterial artificial chromosome library resource: construction and analysis of 12 deep-coverage large-insert BAC libraries that represent the 10 genome types of the genus *Oryza*. Genome Res 16:140–147

Ammiraju JS, Fan C, Yu Y, Song X, Cranston KA, Pontaroli AC, Lu F, Sanyal A, Jiang N, Rambo T (2010a) Spatio-temporal patterns of genome evolution in allotetraploid species of the genus *Oryza*. Plant J 63:430–442

Ammiraju JS, Song X, Luo M, Sisneros N, Angelova A, Kudrna D, Kim HR, Yu Y, Goicoechea JL, Lorieux M (2010b) The *Oryza* BAC resource: a genus-wide and genome scale tool for exploring rice genome evolution and leveraging useful genetic diversity from wild relatives. Breed Sci 60:536–543

Bao Y, Ge S (2004) Origin and phylogeny of *Oryza* species with the CD genomes based on multiple-gene sequence data. Plant Syst Evol 249:55–66

Brar DS, Khush GS (1997) Alien introgression in rice. Plant Mol Biol 35:35–47

Brar DS, Elloran R, Khush GS (1991) Interspecific hybrids produced through embryo rescue between cultivated and eight wild species of rice. Rice Genet Newsl 8:91–93

Cheong YH, Chang HS, Gupta R, Wang X, Zhu T, Luan S (2002) Transcriptional profiling reveals novel interactions between wound, pathogen, abiotic stress, and hormonal responses in *Arabidopsis*. Plant Physiol 129:661–677

Eizenga GC, Agrama HA, Lee FN, Jia Y (2009) Exploring genetic diversity and potential novel disease resistance genes in a collection of rice (*Oryza* spp.) wild relatives. Genet Res Crop Evol 56:65–76

Fukao T, Harris T, Bailey-Serres J (2009) Evolutionary analysis of the *Sub1* gene cluster that confers submergence tolerance to domesticated rice. Annal Bot 103:143–150

Goff SA, Ricke D, Lan TH, Presting G, Wang R, Dunn M (2002) A draft sequence of the rice genome (*Oryza sativa* L. ssp. *japonica*). Sci 296:92–100

Hattori Y, Nagai K, Furukawa S, Song XJ, Kawano R, Sakakibara H, Wu J, Matsumoto T, Yoshimura A, Kitano H (2009) The ethylene response factors *SNORKEL1* and *SNORKEL2* allow rice to adapt to deep water. Nat. 460:1026–1030

Kim KM, Cho SK, Shin SH, Kim GT, Lee JH, Oh BJ, Kang KH, Hong JC, Choi JY, Shin JS, Chung YS (2005) Analysis of differentially expressed transcripts of fungal elicitor- and wound-treated wild rice (*Oryza grandiglumis*). J Plant Res 118(5):347–354

Kim H, Hurwitz B, Yu Y, Collura K, Gill N, SanMiguel P, Mullikin JC, Maher C, Nelson W, Wissotski M (2008) Construction, alignment and analysis of twelve framework physical maps that represent the ten genome types of the genus *Oryza*. Genome Biol 9(2):R45

Morishima H (1994) Background information about *Oryza* species in tropical America. In: Investigations of plant genetic resources in the American basin with emphasis on the genus *Oryza*: Report of 1992/93 Amazon Project. Monbusho International Science Research Program, pp 102

Morishima H, Martins P (1993) Investigations of plant genetic resources in the Amazon Basin with the emphasis on the genus *Oryza*. Report of 1992/93 Amazon Project. The Mombusho International Science Research Program, Japan Research Support Foundation of the State of Sao Paulo, Brazil, pp 96

Multani DS, Jena KK, Brar DS, Delos Reyes BC, Angeles ER, Khush GS (1994) Development of monosomic alien addition lines and introgression of genes from *Oryza australiensis* Domin. to cultivated rice *O. sativa* L. Theor Appl Genet 88:102–109

Multani DS, Khush GS, Delos Reyes BG, Brar DS (2003) Alien genes introgression and development of monosomic alien additional lines from *Oryza latifolia* Desv. to rice. Theor Appl Genet 107:395–405

Nayar NM (1973) Origin and cytogenetics of rice. Adv Genet 17:153–292

Niroula RK, Pucciariello C, Ho VT, Novi G, Fukao T, Perata P (2012) *SUB1A*-dependent and -independent mechanisms are involved in the flooding tolerance of wild rice species. Plant J. 72:282–293

Okishio T, Sasayama D, Hirano T, Akimoto M, Itoh K, AzumaT (2014) Growth promotion and inhibition of the Amazonian wild rice species *Oryza grandiglumis* to survive flooding. Planta 240:459–469

Okishio T, Sasayama D, Hirano T, Akimoto M, Itoh K, Azuma T (2015) Ethylene is not involved in adaptive responses to flooding in the Amazonian wild rice species *Oryza grandiglumis*. J Plant Physiol 174: 49–54

Reymond P, Weber H, Damond M, Farmer EE (2000) Differential gene expression in response to mechanical wound and insect feeding in *Arabidopsis*. Plant Cell 12:707–719

Sanchez PL, Wing RA, Brar DS (2013) The wild relatives of rice: genome and genomics. In: Zhang Q, Wing RA (eds) Genetics and genomics of rice, plant genetics and genomics: crops and models 5. Springer science + Business media, New York, pp 193. https://doi.org/10.1007/978-1-4614-7903-1_13

Sánchez E, Quesada T, Espinoza AM (2006) Ultrastructure of the wild rice *Oryza grandiglumis* (Gramineae) in Costa Rica. Rev Biol Trop 54(2):377–385

Sitch LA (1990) Incompatibility barriers operating in crosses of *Oryza sativa* with related species and genera. In: Gustafson JP (ed) Genetic manipulation in plant improvement II. Plenum Press, New York, pp 77–94

Vaughan DA(1989) The genus *Oryza* L. Current status of taxonomy. IRPS 138:1–21, IRRI, Manila, Philippines

Vaughan DA (1994) The wild relatives of rice: a genetic resources handbook. International Rice Research Institute, Manila, Philippines

Vaughan DA, Morishima H, Kadowaki K (2003) Diversity in the *Oryza* genus. Curr Opin Plant Biol 6: 139–146

Xu K, Xu X, Fukao T, Canlas P, Maghirang-Rodriguez R, HeuerS Ismail AM, Bailey-Serres J, Ronald PC, Mackill DJ (2006) Sub1A is an ethylene-response-factor- like gene that confers submergence tolerance to rice. Nat 442:705–708

Yao SY, Henderson MT, Jodon NE (1958) Cryptic structural hybridity as a probable cause of sterility in intervarietal hybrids of cultivated rice, Oryza sativa L. Cytofogia 23(1):46–55

Yoon DB, Kang KH, Kim HJ, Ju HG, Kwon SJ, Suh JP, Jeong OY, Ahn SN (2006) Mapping quantitative trait loci for yield components and morphological traits in an advanced backcross population between *Oryza grandiglumis* and the *O. sativa* japonica cultivar Hwaseongbyeo. Theo Appl Genet 112(6):1052–1062

Yu J, Hu S, Wang J, Wong GK-S, Li S, Liu B, Deng Y, Dai L, Zhou Y, Zhang X (2002) A draft sequence of the rice genome (*Oryza sativa* L. ssp. *indica*). Sci 296:79–92

Zamora A, Barboza C, Lobo J, Espinoza AM (2003) Diversity of native rice (*Oryza*: Poaceae) species of Costa Rica. Gen Res Crop Evol 50:855–870

Oryza granulata Nees et Arn. ex Watt 12

Blanca E. Barrera-Figueroa and Julián M. Peña-Castro

Abstract

Oryza granulata Nees et Arn. ex Watt is a wild rice species with importance as a resource for the study of genetic diversity and evolution of the *Oryza* genus, and for the identification of genes and regulatory networks with a potential use in the improvement of tolerance to biotic and abiotic stress in cultivated rice. In this chapter, we reviewed the biology, distribution, genetic diversity, genome resources, and advances on the study of *O. granulata* genome.

12.1 Academic and Economic Importance

O. granulata Nees et Arn. ex Watt, belonging to the Meyeriana complex, is one of the 24 species of the genus *Oryza*. It occupies the most basal phylogenic position in the *Oryza* genus, which makes it a unique source of germplasm for the study of evolution, diversity, genetics, and functional genomics in wild rice (Gee et al. 1999). *O. granulata* could be useful for improvement of agricultural traits in cultivated rice, such as to produce tolerant lines against bacterial blight and brown plant hopper, tolerance to shade and drought, adaptation to aerobic soils (Sanchez et al. 2014), and plasticity to temperature and moisture (Atwell et al. 2014). In spite of its importance, studies on *O. granulata* are scarce compared to other wild rice species.

12.2 Botanical Description and Geographic Distribution

O. granulata is a short perennial herb with dark green leaves and awnless spikelets (<6.4 mm in length), with granulate texture to palea and lemma (Vaughan 1994) (Fig. 12.1). It is distributed in South Asia including Cambodia, China, India, Indonesia, Laos, Myanmar, Nepal, Philippines, Sri Lanka, and Thailand. All the natural populations have been localized to the west of the biogeographical boundary called the Wallace Line, where *O. granulata* grows and associated with vegetation in tropical deciduous forest, tropical uplands with low requirements of water and partial to full shade (Vaughan et al. 2008) (Fig. 12.2).

12.3 Genome Size

According to cytological, morphological, and molecular studies, *O. granulata* has been classified as a diploid GG genome ($2n = 24$), having

B. E. Barrera-Figueroa (✉) · J. M. Peña-Castro
Laboratorio de Biotecnología Vegetal, Instituto de Biotecnología, Universidad Del Papaloapan. Tuxtepec, Oaxaca, Mexico
e-mail: bbarrera@unpa.edu.mx

T. K. Mondal and R. J. Henry (eds.), *The Wild Oryza Genomes*,
Compendium of Plant Genomes, https://doi.org/10.1007/978-3-319-71997-9_12

Fig. 12.1 *O. granulata* (Accession 102118). Adult plants grown at the International Rice Research Institute (IRRI). Courtesy of Dr. Rod A. Wing

Fig. 12.2 Distribution of *O. granulata* according to Vaughan (1994, 2005), and collection sites registered in the Gateway to Genetic Resources (http://www.genesis-pgr.org) and the NARO Database in Japan (https://www.gene.affrc.go.jp/databases-plant_images_en.php)

the second largest genome (1.83 pg/2C) among *Oryza* diploid species, only preceded by *O. australiensis* as estimated by flow cytometry (Ammiraju et al. 2006). Later, also based on flow cytometry and chromosome analysis, Miyamabashi et al. (2007) reported the *O. granulata* genome size being 2.46 pg/2C, larger than previously reported, and with the largest chromosomes (157.8 μm in length) among several wild rice species studied. *O. granulata* genome size is 862 Mbp.

12.4 Genomic Resources

A comparative genomics project, the *Oryza* Map Alignment Project (OMAP) (Wing et al. 2005), was aimed to develop a comparative bacterial artificial chromosome (BAC)-based physical map of 10 genomes of the *Oryza* genus and create a resource useful for the study of genome organization, diversity, and evolution and to facilitate further studies aimed to exploiting the full genetic potential of wild rice species. According to this, Ammiraju et al. (2006) reported the construction of 12 deep-coverage large-insert BAC libraries representing 10 *Oryza* genomes. *O. granulata* (accession number 102118) was included in the study due to its important agronomic traits. The theoretical genome coverage of the *O. granulata* library was generated and calculated to be 10.8X, however, the coverage shown by hybridization/contig analysis resulted in only 6.3 fold genome coverage, which was low but still in the range of 5–10X coverage useful for further studies like genome sequencing and positional cloning. Derivated from this project, 138,171 sequences from *O. granulata* BAC libraries were deposited in the GenBank at the National Center for Biotechnology Information (NCBI, taxid: 110450). In addition, *O. granulata* has been studied at the transcriptomic level. Genome-wide sequencing of transcripts from roots, panicle, and leaves was performed, and the sequences were deposited at the GenBank.

12.5 Population and Conservation Genetics

In spite of its importance as a rich germplasm with high potential to improve rice, *O. granulata* has become a vulnerable species due to habitat disturbance. A survey developed in China revealed that 12.9% of the *O. granulata* populations gets depleted, and 83.9% are under endangered category (Qian et al. 2001b). In order to facilitate the conservation management of *O. granulata*, it is required to investigate the population genetics with the use of genetic markers to explore the genetic diversity between and within natural populations of the species.

Qian et al. (2001a) reported the use of RAPD and ISSR markers to estimate the genetic variation in five populations of *O. granulata* in Yunnan and Hainan, two Provinces in China where this species is naturally distributed. This study showed that there exists low genetic variation within and among populations of *O. granulata*. However, a high amount of genetic differentiation occurred between geographical regions. A similar study was made to include 23 accessions of *O. granulata* sampled from different locations in South and Southeast Asia (Qian et al. 2006). Concurrently with the former study, a high level of genetic variation in *O. granulata* was found between accessions from different geographical regions, indicating that gene flow among populations is limited. Isolation by ecological or geographical factors, selection pressures present in different habitats, are the main driving force for genetic diversity in *O. granulata*.

If genetic variation is lost at the population level, it could result a decrease in the ability to adapt to adverse environmental conditions, which together with habitat disturbance caused by human activities, places *O. granulata* in the focus as an endangered wild rice species.

12.6 Repetitive Sequences

Polyploidy and transposable elements (TEs) are the main factors for genome size increase in higher eukaryotes. *Oryza* species are diverse in ploidy level, with 17 diploid species and seven tetraploid, where the tetraploid have the largest genomes. Among the diploid species, there is also a high degree of variation in genome size, ranging from 357 Mbp (for *O. glaberrima*) to 965 Mbp (for *O. australiensis*) (Ammiraju et al. 2006). This degree of variation indicates that not only ploidy but also other factors, like the flux of repetitive elements such as TEs, are responsible for structural changes.

Ammiraju et al. (2006) reported a positive correlation between long terminal repeat (LTR) retrotransposon content and genome size in *Oryza*. In support of this, Piegu et al. (2006) used *O. australiensis* as a model species and provided evidence that three transposons, i.e., *Wallabi*, *RIRE1*, and *Kangaroo*, contributed to the increase in genome size (genomic obesity) in this species. *O. granulata*, possesses the second largest genome among diploid species, and is another case of genomic obesity in *Oryza*, which is caused by an LTR retrotransposon family named RWG, Gran3, that is closely related to *Wallabi* and *RIRE1*, and constitutes nearly one quarter of the *O. granulata* genome (Ammiraju et al. 2007). In addition, using in silico and experimental approaches, Roulin et al. (2008) found *RIRE1* homologs in *O. granulata* with high sequence identity to that of *O. australiensis* and suggested that *RIRE1* jumped several times and has been transferred horizontally from *O. australiensis* between 60,000–70,000 years ago, representing a strong evolutionary factor for diversification of the *Oryza* genome.

12.7 Impact on Plant Breeding

At difference of other wild rice relatives, *O. granulata* is difficult to cross with cultivated rice. Embryo rescue has been used as a technique to obtain hybrids in the *Oryza* genus. Several hybrid lines have been generated, but there is a low level of introgression of *O. granulata* into the hybrids (Brar et al. 1991). In some cases, the cross between *O. sativa* and *O. granulata* was unsuccessful due to a low percentage of embryo germination and/or inability to survive after germination, indicating the existence of strong crossability barriers between these species (Niroula et al. 2005). Other alternative approaches of molecular breeding might be used to transfer agricultural traits from *O. granulata* into cultivated rice.

12.8 Other Studies

O. granulata has been useful as a model system for the study of endophytic mycobiota in wild rice. Yuan et al. (2010) revealed that *O. granulata* roots bear a high diversity of fungi, with 34.5% of all taxa potentially representing undescribed species. This indicates that *O. granulata* roots are a rich resource of complex endophytic fungi.

12.9 Future Prospects

Because of its agronomical traits and phylogenetic position in the *Oryza* genus, *O. granulata* is unique with a high potential as a genetic resource for the improvement of agricultural traits in cultivated rice. Future directions on the study of this species should lead to the reinforcement of programs to investigate the genetic diversity and variation that are directed to its conservation. In situ and ex situ conservation strategies should be continued for its conservation. On the other hand, *O. granulata* can be regenerated from protoplasts, producing diploid, fertile plants (Baset et al. 1993). It is relevant to develop methods for in vitro culture and propagation of *O. granulata* plants.

A phylogenomics project reported the sequence of 124 single-copy genes from *O. granulata*, which together with the sequences of other wild rice species, were used to resolve the phylogenetic relationships among different genome types in *Oryza* species (Zou et al. 2008). As more wild rice gene sequences are available, it is expected that phylogenomics will become an important tool useful to reconstruct with high resolution the rapid evolutionary diversification that occurred in the *Oryza* genus.

Even though classical hybridization methods have failed to transfer *O. granulata* traits into *O. sativa*, it is needed to take advantage of the natural variation present in *O. granulata* to identify genetic markers for tolerance to biotic and abiotic stress, in order to develop strategies for their future use in marker-assisted selection (MAS) improvement programs.

Since *O. granulata* is a model species with a broad adaptation to abiotic stress such as drought, temperature changes, full shade, and tolerance against bacterial blight and brown plant hopper, it is prominent to broaden the knowledge of the molecular basis for tolerance to biotic and abiotic stress through the study of transcriptome and/or proteome in *O. granulata* accessions under diverse environmental conditions. This will lead to the identification of genes and regulatory networks that are key for tolerance and with a potential use for improvement of rice through modern biotechnology approaches.

References

Ammiraju JSS, Luo M, Goicoechea JL, Wang W, Kudrna D, Mueller C, Talag J, Kim HR, Sisneros NB, Blackmon B, Fang E, Tomkins JB, Brar D, MacKill D, McCouch S, Kurata N, Lambert G, Galbraith DW, Arumuganathan K, Rao K, Walling JG, Gill N, Yu Y, SanMiguel P, Soderlund C, Jackson S, Wing RA (2006) The *Oryza* bacterial artificial chromosome library resource: construction and analysis of 12 deep-coverage large-insert BAC libraries that represent the 10 genome types of the genus *Oryza*. Genome Res 16:140–147

Ammiraju JSS, Zuccolo A, Yu Y, Song X, Piegu B, Chevalier F, Walling JG, Ma J, Talag J, Brar DS, SanMiguel PJ, Jiang N, Jackson SA, Panaud O, Wing RA (2007) Evolutionary dynamics of an ancient retrotransposon family provides insight into evolution of genome size in the genus *Oryza*. Plant J 52:342–351

Atwell BJ, Wang H, Scafaro AP (2014) Could abiotic stress tolerance in wild relatives of rice be used to improve *Oryza sativa*? Plant Sci 215–216:48–58

Baset A, Cocking ED, Finch RP (1993) Regeneration of fertile plants from protoplasts of the wild rice species *Oryza granulata*. J Plant Physiol 141:245–247

Brar DS, Elloran R, Khush GS (1991) Interspecific hybrids produced through embryo rescue between cultivated and eight wild species of rice. RGN 8:91–93

Gee S, Sang T, Lu B-R, Hong D-Y (1999) Phylogeny of rice genomes with emphasis on origins of allotetraploid species. Proc Natl Acad Sci U S A 96:14400–14405

Miyamabashi T, Nonomura K-I, Morishima H, Kurata N (2007) Genome size of twenty wild species of *Oryza* determined by flow cytometric and chromosome analyses. Breed Sci 57:73–78

Niroula RK, Subedi LP, Sharma RC, Upadhyay MP (2005) Interspecific hybrid plants recovered from in vitro embryo rescue in rice. Sci World 3:90–94

Piegu B, Guyot R, Picault N, Roulin A, Saniyal A, Kim H, Collura K, Brar DS, Jackson S, Wing RA, Panaud O (2006) Doubling genome size without polyploidization: dynamics of retotransposition-driven genomic expansions in *Oryza australiensis*, a wild relative of rice. Genome Res 16:1262–1269

Qian W, Ge S, Hong D-Y (2001a) Genetic variation within and among populations of a wild rice *Oryza granulata* from China detected by RAPD and ISSR markers. Theor Appl Genet 102:440–449

Qian W, Xie Z-W, Ge S, Hong D-Y (2001b) Distribution and conservation of an endangered wild rice *Oryza granulata* in China. Acta Bot Sin 43:1279–1287

Qian W, Ge S, Hong D-Y (2006) Genetic diversity in accessions of wild rice *Oryza granulata* from South and Southeast Asia. Genet Resour Crop Evol 53:197–204

Roulin A, Piegu B, Wing RA, Panaud O (2008) Evidence of multiple horizontal transfers of the long terminal repeat retrotransposon RIRE1 within the genus Oryza. Plant J 53:950–959

Sanchez PL, Wing RA, Brar DS (2014) The wild relative of rice: genomes and genomics. In: Zhang Q, Wing RA (eds) Genetics and genomics of rice, plant genetics and genomics: crops and models 5. Springer, New York, pp 9–25

Vaughan DA (1994) The wild relatives of rice. IRRI, Manila

Vaughan DA, Ge S, Kaga A, Tomooka N (2008) Phylogeny and biogeography of the genus *Oryza*. In: Hirano H-Y, Hirai A, Sano Y, Sasaki T (eds) Rice biology in the genomic era. Springer, Heidelberg, pp 219–232

Wing RA, Ammiraju JSS, Luo M (2005) The *Oryza* map alignment project: the golden path to unlocking the genetic potential of wild rice species. Plant Mol Biol 59:53–62

Yuan Z-L, Zhang C-L, Lin F-C, Kubicek CP (2010) Identity, diversity, and molecular phylogeny of the endophytic mycobiota in the roots of rare wild rice (*Oryza granulate*) from a nature reserve in Yunnan, China. Appl Environ Microb 76:1642–1652

Zou X-H, Zhang F-M, Zhang J-G, Zang L-L, Tang L, Wang J, Sang T, Ge S (2008) Analysis of 142 genes resolves the rapid diversification of the rice genus. Genome Biol 9:R49

Oryza latifolia Desv

13

C. Gireesh

Abstract

Oryza latifolia Desv is an aquatic plant belonging to the *Officinalis* complex of *Oryza*. It is an allotetraploid species contains a CCDD genome with $2n = 48$ chromosomes. It grows widely in aquatic ecosystems of Latin American region. *Oryza latifolia* is closely related to two other allotetraploid species of the *Officinalis* complex, namely *O. alta* and *O. grandiglumis*. Morphologically and cytologically, these three species resemble each other with few distinct features. However, genomic studies have demonstrated clear distinctions among these three species of the *Officinalis* complex. Multiple sequence-based phylogenetic studies have revealed that the CD genome of *O. latifolia* might have originated from a single hybridization event between C genome species (*O. officinalis* or *O. rhizomatis*) as maternal parent and an E genome species (*O. australiansis*) as paternal parent during evolution. *O.latifolia* possesses important genes/QTLs for yield-enhancing traits, biotic, and biotic stress. Through interspecific hybridization and embryo rescue, synthetic amphidiploids of *O. sativa* and *O. latifolia* are successfully developed. Several QTLs for plant height, primary branches per panicles, grains per panicle, biomass, and lodging resistance have been identified from *O. latifolia*. Monosomic alien addition lines derived from *O. latifolia* were developed for introgression of beneficial QTLs. Brown planthopper resistance gene *Bph12* was identified and introgressed from *O. latifolia*. Therefore, *O. latifolia* is an important genetic resource for genetic improvement of domesticated cultivars.

13.1 Botanical Description and Distribution

Oryza latifolia (*latis*-wide and *folium*-leaf) is a perennial aquatic grass belonging to the *Officinalis* complex of *Oryza*. It is an allotetraploid species possessing a CCDD genome with 48 chromosomes ($2n = 48$). The plants are characterized by gigantic growth habit, erect tall, hard stems, broad leaves, long panicles, high biomass, and large number of spikelet (De-bin et al. 2010; Bertazzoni et al. 2011). The growth of *O. latifolia* depends on seasonal variations and water levels. An increase in water level is known to increase the rate of cell division and induces higher growth rates of plants (Metraux and Kende 1984). Depending upon the water level, it grows up to 150 cm in flooded condition.

C. Gireesh (✉)
ICAR—Indian Institute of Rice Research,
Rajendranagar, Hyderabad, Telangana 500 030,
India
e-mail: giri09@gmail.com

T. K. Mondal and R. J. Henry (eds.), *The Wild Oryza Genomes*,
Compendium of Plant Genomes, https://doi.org/10.1007/978-3-319-71997-9_13

However, some plants have been reported 6 m tall under flooded condition. The plants are propagated by seeds which often exhibit seed dormancy. Temperature treatment (45 °C for 48 h) is recommended to break the seed dormancy.

The plant possesses a ligule with a stiff membrane, blunt, and rigid pubescent. It flowers throughout the years. The spikelet is characterized by presence of two sterile lemmas (known as glumes) and two fertile lemmas. The floret is enclosed with lemma and palea and contains an oblong ovary with two feathery stigma and six anthers. Anthers consist of filament and four lobes (Pohl 1978, 1980). Generally, *O. latifolia* has 1–2 m culms, panicles are not drooping and verticillate branches will be as much as 25 cm long (Tateoka 1962). Barbosa (2007) carried out chemical analysis of seeds of *O. latifolia* which revealed 9.6% protein, 2.2% lipids, 10.5% moisture (Barbosa 2007).

The *Oryza* species, *O. latifolia,* is endemic to Latin America and grows in ecosystems like swamps, savannas, grassland and woodlands, along the shores of rivers, hill slopes, and coasts (Pohl 1978; Ying and Song 2003). The largest population of *O. latifolia* was found at Verde National Park, Tempisque Conservation Area, Guanacaste, along the Tempisque River in places that are subjected to frequent flooding (Sanchez et al. 2003). In the low lands of Costa Rica, it grows from tropical dry forest (1200 mm annual precipitation) to humid tropical dry forest (4000 mm annual precipitation), while in altitude it grows from zero to 650 m above mean sea level.

13.2 Cytogenetic Studies

Three allotetraploid species ($2n = 48$) of wild rice, viz. *O. latifolia* Desv., *O. alta* Swallen and *O. grandiglumis* Prod possess a CCDD genome and belong to the *Officinallis* complex. Due to their similarity in morphological attributes and chromosomal homology, many earlier workers have advocated for these three species to be considered the same (Sampath (1962), Kihara (1964) and Gopalakrishnan and Sampath (1967). The study of Oka (1961) based on habitats of the three species indicates that the three taxa are con-specific. However, Chatterjee (1948) and Tateoka (1964) considered these three CD genome species to be independent.

Interspecific hybridizations studies showed that successful F_1 hybrids can be produced from hybridization among these three species indicating no crossability barriers (Jena and Kush, IRRI website). *Oryza latifolia* can be easily crossed with the other two allotetraploid species of the *latifolia* complex. However, pollen and seeds from F_1 plants will be usually sterile. Studies of Jena and Kush have shown that the formation of 22II and 1IV in chromosomal arrangement in F_1 plants. Quadrivalent formation could be due to reciprocal translocation. More univalents (22–32) and few bivalents (2–6), trivalents (0–2), quadriavlents (0–1) were observed in an F_1 derived from a cross between *O. sativa* and *O. latifolia* (Multani et al. 2003). On the basis of cytological studies, it was concluded that due to the similarity in morphology and chromosomal homology, the three tetraploid species be included in the *O. latifolia* complex as there is little cytological difference among them to establish their independence. Although F_1 sterility occurs in crosses among them, the sterility alone not sufficient to prove a species-level difference. Therefore, molecular and genomics studies will help in elucidating the difference among the three tetraploid species of Latin America.

Multani et al. (2003) developed monosomic alien addition lines having a 2n chromosome complement of *O. sativa* with one additional chromosome of *O. latifolia*. Alien chromosome transmission through female gametes was 4.4–35.5%, while it ranged from 7 to 11.9% through male gametes. These monosomics were further characterized for morphological traits, isozyme banding pattern, and biotic stress tolerance. Disomic progenies in BC_3 and BC_4 generations showed complete resemblance to the *O. sativa* parent.

13.3 Genomics Studies

Genomics studies using the sequence of nuclear genes, chloroplast gene, RFLP sequence, AFLP sequence, and ITS have clearly demonstrated significant differences between *O. alta* and *O. latifolia* (Ge et al. 1999; Aggarwal et al. 1999; Bao and Ge 2003). Molecular characterization of 21 species of wild species of the genus *Oryza* using RFLP and AFLP has revealed that tetraploid species (CCDD), viz. *O. latifolia, O. alta,* and *O. grandiglumis* have closer proximity among themselves than with any other species of *Oryza*. In addition, the study also revealed that these three species possess the closest proximity to two diploid species, namely O. *eichingeri* (CC genome) and O. *australiensis* (EE genome) diploid species, this indicates that these three species might have originated from common ancient diploid species. Among these three species, O. *alta* and *O. grandiglu*mis are more closely related to each other than to O. *latifolia* (Wang et al. 1992; Aggarwal et al. 1999). Existence of significant divergence between *O. latifolia* and the other two species (*O. alta* and *O. grandiglumis*) was further supported by phylogenetic analysis using sequence information of two chloroplast fragments (*mat*K and *trn*L*trn*F) and three nuclear genes (*Adh*1, *Adh*2 and *GPA*1) (Bao and Ge 2004).

Phylogenetic analysis using multiple sequence data (two chloroplast fragments (*mat*K, *trn*L, and *trn*F) and three nuclear genes (*Adh*1, *Adh*2, and *GPA*1) carried out by Bao and Ge (2004) showed that CD genome originated from a single hybridization events in which the C genome species (*O. officinalis* or *O. rhizomatis*) was the maternal donor parent and the E genome species (*O. australiensis*) was the paternal donor parent during the evolution of CD genome species. However, there is no diploid DD genome species is known and yet to be discovered (Vaughan et al. 2003).

The repetitive sequence in eukaryotic genomes is an important characteristics which elucidate the dismutation of species, origin of ancestors in turn will help in determining of evolution of species and genomes. *C0t*-1 DNA is moderately and highly repetitive DNA of eukaryotic chromosomes which includes satellite DNA, microsatellite DNA, ribosomal RNA, and repetitive sequence of telomere and centromere. FISH (fish fluorescent in situ hybridization) probed with *C0t*-1 DNA determines relationships between different genomes and different species. Karyotype analysis using FISH with *C0t*-1 DNA from *O. alta* as a probe showed high homology and close proximity between *O. alta* and *O. latifolia*, though, there was very clear distinction between the hybridization signals (Wang De-bin et al. 2010).

The study of Bao and Ge (2004) revealed that chloroplast genome of *O. latifolia* is different from that of *O. alta,* and the CC genome is the female parent genome of *O. latifolia.* Lan et al. (2006) had also demonstrated that there was significant difference in the distribution of moderately and highly repetitive sequence of CC genome between the two rice species by using FISH (fluorescence in situ hybridization) with *C0t*-1 DNA from C genome as probe. The CC and DD genomes of *O. alta* had a closer relationship than in *O. latifolia.*

13.4 Pre-breeding

Oryza latifolia is considered to be an importance source of genes for yield-enhancing traits and biotic and abiotic stress tolerance for improvement of cultivated rice (Vaughan 1989, 1994). *O. latifolia* is resistance to many pests and diseases that affect cultivated rice, and therefore, *O. latifolia* is a useful source of genes for improvement of cultivated rice. Introgression of these agronomically important traits/genes through interspecific hybridization could enhance yield and confer resistance to pest and disease.

Resistance to bacterial blight, brown planthopper, and the white-backed planthopper have been reported in *O. latifolia* (Multani et al. 2003). Angeles-Shim et al. (2014) identified genes/QTLs derived from *O. latifolia* for various agronomically important traits like heading date, plant height, number of primary branches per panicle, number of grains per panicle, grain width and length, high biomass production, and lodging resistance.

Molecular characterization of 27 introgression lines derived from *O. latifolia* MAAL (Monosomic Alien Addition Lines) using 168 (38 SSRs, 17 STS, 85 SNPs, and 29 InDels) polymorphic markers between *O. sativa* and *O. latifolia* identified beneficial QTLs for agronomically important traits introgressed from *O. latifolia* (Angeles-Shim et al. 2014). The molecular markers that were identified from the study by Angeles-Shim et al. (2014) will help in differentiation between *O. sativa* and *O. latifolia* genomes, molecular mapping of traits/genes/QTLs mapping and cloning of the genes identified from *O. latifolia,* in addition to genomic and genetic studies on breeding resources derived from wild *Oryza* species that share the same genome as *O. latifolia*.

O. latifolia germplasm is considered to be important source of resistance to BPH (WU et al. 1986). Yang et al. (2002) identified single dominant BPH resistance, *Bph12* in B14 an introgression line derived from wild rice *O. latifolia*. They mapped the resistance genes on to short arm of chromosome number 4 and identified RM261 was closely linked to the resistance gene at a distance of 1.8 cm. Sarao et al. (2016) have screened wild species of rice for BPH (brown planthopper) resistance and have identified nine accessions of *O. latifolia* resistance to BPH. *O. latifolia* is also reported to be an important source of resistance to bacterial leaf blight (Sanchez et al. 2014).

13.4.1 *bph12* (Yang et al. 2002)

Oryza latifolia germplasm is considered to be good source of BPH resistance. Cultivated rice, *O. sativa*, was crossed with *O. latifolia* on a large scale and fertile progenies were recovered through somatic cell culture (SHU et al. 1994) to further develop introgression lines through backcross breeding and to screen for BPH resistance.

An introgression line, B14, derived from *O. latifolia* showed resistance to BPH (Yang et al. 2002). B14 was crossed with a BPH susceptible parent Taichung Native 1 to develop an F_2 mapping population. Genetic analysis in the F_2 population showed BPH resistance in B14 was conferred by a single dominant gene. Bulked segregant analysis carried out using 302 SSR markers identified three putative SSR markers (RM335, RM261, and RM185) linked to BPH resistance gene on chromosome 4 of rice genome. The SSR marker RM261 was closely linked (1.6 cm) to *bph12* to the resistance gene and can be used for marker assisted introgression (Yang et al. 2002). The schematic representation of identification and introgression of *bph12* gene from *O. latifolia* into *O. sativa* is given in Fig. 13.1.

13.4.2 Synthetic Amphiploids of *O. sativa* and *O. latifolia*

Interspecific hybrids between *O. sativa* ($2n = 2x = 24$, AA genome) and *O. latifolia* ($2n = 4x =$ CCDD genome) were obtained through sexual hybridization and embryo rescue (Yi et al. 2008). GISH analysis was employed to identify an allotriploid with an ACD genomic constitution in interspecific hybrids ($2n = 3x = 26$ ACD genome). The abnormal meiosis of the hybrid pollen mother cell and poor chromosome pairing were reported and most of the univalent chromosomes resulted in complete male sterility in the hybrid. Therefore, the hybrids were preserved and propagated as tube seedlings. To ensure that every chromosome had the appropriate homologous chromosome during meiosis, the tube seedlings of the allotriploid hybrids ($2n = 6x = 52$, AACCDD) were treated with colchicine for chromosome doubling. A total of 211 SSR markers evenly distributed across the 12 rice chromosomes was used to identify polymorphic markers and a further 31 polymorphic markers were used for genotyping of allotetraploids lines. Development of introgression lines from *O. sativa* and *O. latifolia* is detailed in Fig. 13.2.

Germplasm of *O. latifolia* were screened for BPH resistance

↓

BPH resistant lines of *O. latifolia* was crossed with *O. sativa*

↓

Fertile progenies were recovered through embryo rescue

↓

Progenies were screened for BPH to identify the resistant lines

↓

One progeny line B14 showed highly tolerant to BPH biotype 1 and 2

↓

B14 line was crossed with susceptible line Taichung Native 1

↓

F_1 plants were obtained and selfed to produce F_2 population

↓

Genetic analysis and bulked segregant analysis in F_2 population was carried out

↓

bph12 gene was mapped to short arm of chromosome 4 and marker RM261

found linked to the resistance gene at a distance of 1.8 cM

Fig. 13.1 Identification of *bph12* gene from *O. latifolia. O. sativa* was crossed with *O. latifolia* and fertile progenies were recovered through somatic cell culture (SHU et al. 1994). An introgression line B14 showed resistance to BPH (Yang et al., 2002) and was crossed with a BPH susceptible parent TN1. Genetic analysis in the F_2 population revealed that BPH resistance is conferred by a single dominant gene. BSA identified three putative SSR markers (RM335, RM261, and RM185) linked to BPH resistance gene on chromosome 4 of rice genome

O. sativa AA (2n=24) was crossed with *O. latifolia* CCDD (2n=48) and F_1 were obtained through sexual hybridization and embryo rescue

↓

F_1 (allotriploid) ACD (2n=36) obtained were confirmed by GISH and FISH. F_1 plants further were treated with colchicine to develop amphidiploids

↓

Amphidiploids (Allotetraploid, AACCDD, 2n=72) plants were identified by GISH and FISH

↓

Five introgression lines were identified

Fig. 13.2 Development of synthetic amphiploids of *O. sativa* and *O. latifolia*. Interspecific hybrids between *O. sativa* and *O. latifolia* were obtained through sexual hybridization and embryo rescue (Yi et al. 2008). GISH analysis was employed to identify an allotriploid with an ACD genomic constitution in interspecific hybrids. Tube seedlings of the allotriploid hybrids ($2n = 6x = 52$, AACCDD) were treated with colchicine for chromosome doubling. A total of 211 SSR were used to identify polymorphic markers and a further 31 polymorphic markers were used for genotyping of allotetraploid lines. The five introgression lines were lines were identified

Fig. 13.3 Adult *Oryza latifolia* plants are growing at ICAR-Indian Institute of Rice Research, Hyderabad, India

13.5 Future Prospects

Oryza latifolia is an aquatic wild rice species and is known to possess many useful genes for various biotic and abiotic tolerance and agronomically important traits useful for improvement of cultivated rice. There has been little research carried out to utilize this species, and therefore, there is a need to have significant consortia research effort to explore *O. latifolia* for improvement of cultivated rice. There have been very little cytological works are carried to understand the structural and functional features of the chromosomes of *O. latifolia*. Advanced cytological tools like FISH will help in understanding the chromosomal variations that exist in *O. latifolia*. This will help in understanding the evolution of the CD genomes of the wild rice species. Recent advances in molecular approaches like genome sequencing, resequencing, and next-generation sequencing technology will help in comparative analysis of different genomes and evolution of *O. latifolia* (Figs. 13.3 and 13.4).

(a) (b)

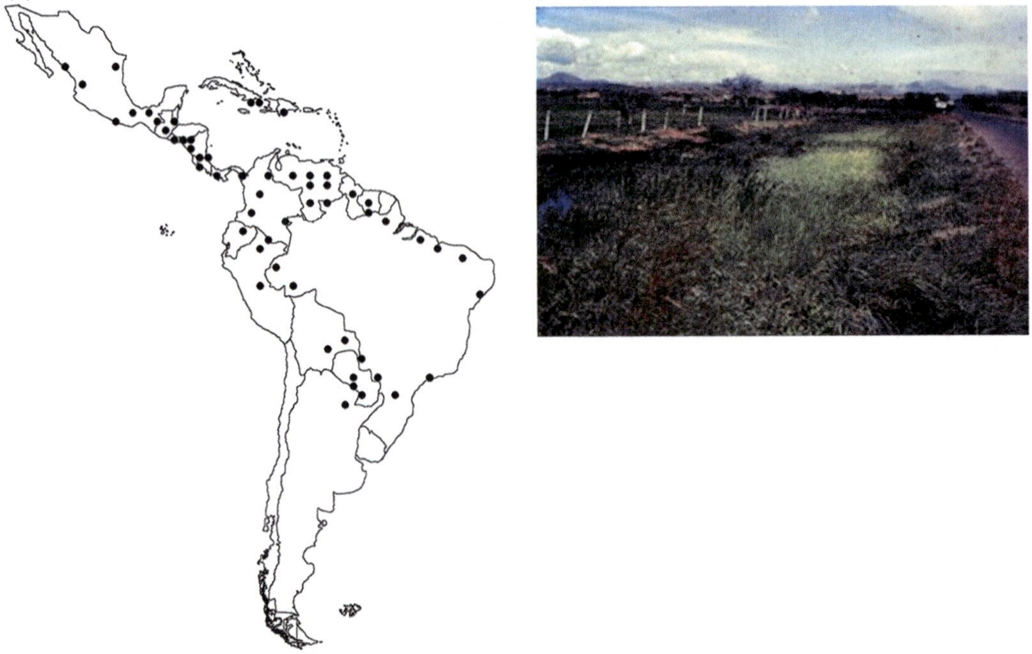

Fig. 13.4 Distribution of *O. latifolia* in **a** Central and South America, **b** A naturela Habitat, Costa Rica. Photograph credit Gerard Second

References

Aggarwal RK, Brar DS, Huang N, KhushGS (1999) Differentiation within CCDD genome species in the genus *Oryza*as revealed by total genomic hybridization and RFLP analysis: II. Wild species and evolution 8. Differentiation within CCDD genome. http://archive.gramene.org/newsletters/rice_genetics/rgn13/v13p54.html. Accessed 18 May 2017

Aggarwal RK, Brar DS, Nandi S, Huang N, Khush GS (1999) Phylogenetic relationships among *Oryza* species revealed by AFLP markers. Theor Appl Genet 98:1320–1328

Angeles-Shim RB, Ricky B, Vinarao BM, Jena KK (2014) Molecular analysis of *Oryza latifolia* Desv. (CCDD genome)-derived introgression lines and identification of value-added traits for rice (*O. sativa* L.) improvement. J Hered 105(5):676–689

Bao Y, Ge S (2003) Identification of *Oryza* species with the CD genome based on RFLP analysis of nuclear ribosomal ITS sequences. Acta Bot Sin 45(7):762–765

Bao Y, Ge S (2004) Origin and phylogeny of *Oryza* species with the CD genome based on multiple gene sequence data. Plant Syst Evol 249(1):5566

Barbosa MM (2007) Development and utilization of food of vegetal origin by three communities of the Pantanal and Cerrado. (CNPQ, Technical Report)

Bertazzoni EC, Damasceno Junior GA (2011) Aspects of the biology and phenology of *Oryza latifolia* Desv. (Poaceae) in the Pantanal wetland in Mato Grosso do Sul, Brazil. Acta Bot Bras 25(2):476–486

Chatterjee D (1948) A modified key and enumeration of the species of *Oryza Linn.* Indian J Agr Sci 18:185–192

De-bin Wang, Wang Yang WU, Qi ZHAO Hou-ming, Gang LI, Rui QIN, Chun-tai Wang, Hong Liu (2010) Comparative analysis of genomes from *Oryza alta* and *Oryza latifolia*by *C0t*-1 DNA. Rice Sci 17(3):131–136

Ge S, Sang T, Lu BR, Hong DY (1999) Phylogeny of rice genomes with emphasis on origins of allotetraploid species. Proc Natl Acad Sci USA 96:14400–14405

Gopalakrishnan R, Sampath S (1967) Taxonomic status and origin of American tetraploid species of the series *latifoliae tateoka* in the genus *Oryza*. Indian J Agr Sci 37:465–475

Kihara H (1964) Need for standardization of genetic symbols and nomenclature in rice. In IRRI (ed.) rice genetics and cytogenetics, Elsevier, Amsterdam, London, New York, pp 3–11

Lan WZ, He GC, Wu SJ, Qin R (2006) Comparative analysis of *Oryza sativa, O. officinalis and O. meyeriana* genomes with Cot-1 DNA and genomic DNA. Sci Agric Sin 39(6):1083–1090

Metraux JP, Kende H (1984) The cellular basis of the elongation response in submerged deep-water rice. Planta. 160:73–77

Multani DS, Khush GS, Reyes BG, Brar DS (2003) Alien genes introgression and development of monosomic alien addition lines from *Oryza latifolia* Desv. to rice, *Oryza sativa* L. Theor Appl Genet 107:395–405

Oka HI (1961) Report of trip for investigation of rice in Latin American countries. National Institute of Genetics, Japan, pp 140

Pohl WR (1978) How to know the grasses. 3rd edn. Wm C Brown, Dubuque, Iowa

Sampath S (1962) The genus *Oryza*: its taxonomy and species relationships. Oryza 1(1):129

Sanchez E, Montiel M, Espinoza AM (2003) Ultrastructural morphologic description of the wild rice species *Oryza latifolia* (Poaceae) in Costa Rica. Rev Biol Trop 51:345–353

Sanchez PL, Wing RA, Brar DS (2014) The wild relative of rice: Genomes and genomics. In: Zhang Q, Wing RA (eds) Genetics and genomics of rice, plant genetics and genomics: crops and models. https://doi.org/10.1007/978-1-4614-7903-1_2, © Springer Science Business Media New York 2013

Sarao RS, Sahi GK, KumariNeelam Mangat GS, Patra BC, Singh Kuldeep (2016) Donors for resistance to brown Planthopper *Nilaparvatalugens* (Stål) from wild rice species. Rice Sci 23(4):21–224

Tateoka T (1962) Taxonomic studies of *Oryza I. O. latifolia* complex. Bot Mag Tokyo 75:418–427

Tateoka T (1964) Taxonomic studies of the genus *Oryza*. In: IRRI (ed) rice genetics and cytogenetics. Elsevier, Amsterdam, London, New York, p 1521

Vaughan DA (1989) The genus *Oryza L.*: current status of taxonomy. In: IRRI research paper series, 138:1–21. Manila, IRRI

Vaughan DA, Morishimay H, Kadowaki K (2003) Diversity in the *Oryza* genus. Curr Opin Plant Biol 6:139–146

Vaughan DA (1994) The wild relatives of rice. A Genetic handbook. IRRI, Manila

Wang Z, Second YG, Tanksley SD (1992) Polymorphism and phylogenetic relationships among species in the genus *Oryza* as determined by analysis of nuclear RFLPs. Theor Appl Genet 83:565–581

Yang H, Ren X, Weng Q, Zhu L, He G (2002) Molecular mapping and genetic analysis of a rice brown plant hopper (*Nilaparvatalugens*) resistance gene. Hereditas 136:39–43

Yi CD, Tang SZ, Zhou Y, Liang GH, Gong ZY, Gu MH (2008) Development and characterization of inter-specific hybrids between *Oryza sativa* and *O. latifolia* by in situ hybridization. Chin Sci Bull 53(19):2973–2980

Ying B, Song GE (2003) Identification of *Oryza* species with the CD genome using RFLP analysis on nuclear ribosomal ITS sequences. Acta Bot Sin 45:762–765

Oryza longiglumis Jansen

14

Mrinmoy Sarker, Dipti Ranjan Pani and Tapan K. Mondal

Abstract

Rice is probably the oldest food crop cultivated by farmers for thousands of years. To obtain better yield, traditionally people use breeding techniques which generate improved varieties of rice. Due to continuous focus on selection for a few major traits such as yield, many other traits may have been lost. On the other hand, wild species are a good source of important alleles. *Oryza longiglumis* is one such species from the genus *Oryza* that has several important traits. In this chapter, we have discussed in detail the research carried out so far on this species.

14.1 Academic and Economic Importance

Among biotic stresses, bacterial and fungal pathogens are most devastating. Every year around 50% of rice yield is damaged due to infectious diseases. Although some genotypes of rice may have some resistance allele to these diseases, wild species are better sources of resistance alleles. It is important to note that *O. longiglumis* shows resistance against two major rice diseases, i.e., bacterial blight (BB) and blast (Sanchez et al. 2013). *O. longiglumis* has been known to grow in the geographical region where the weather conditions vary from tropical to extreme cold and warm temperate (temperature variation 5–37 °C). Although not domesticated and cultivated commercially, plants with these kinds of traits are useful for varietal improvement of rice.

14.2 Botanical Description and Geographical Distribution

O. longiglumis is a species of the *Oryza ridleyi* complex and is closely related to *O. ridleyi.* The plant height can be up to maximum of 2 m under favorable conditions (Fig. 14.1). It is a perennial herb with long leaves. The stem is prostrate, soft, green in color, and at the ground the stem is hollow and round. Distinct nodes and internodes are prominent. The shape of the leaves is linear with a green basal leaf sheath. The leaves have ridges and furrows, and each ridge contains a small vascular bundle (Fig. 14.2). The plant has flexuous sterile lemma as long or longer than the spikelet. The spikelet is 7–8 mm long and 1.8–

M. Sarker (✉)
Molecular Biology Laboratory, Department of Biochemistry and Molecular Biology, University of Dhaka, Dhaka 1000, Bangladesh
e-mail: mrinmoy1048@gmail.com

D. R. Pani
ICAR-National Bureau of Plant Genetic Resources, Base Centre, Cuttack 753006, Odisha, India

T. K. Mondal
ICAR-National Research Centre on Plant Biotechnology, Pusa, New Delhi 110012, India

T. K. Mondal and R. J. Henry (eds.), *The Wild Oryza Genomes*,
Compendium of Plant Genomes, https://doi.org/10.1007/978-3-319-71997-9_14

Fig. 14.1 *O. longiglumis* full plant. *Image courtesy* ICAR-National Bureau of Plant Genetic Resources, India

Fig. 14.2 *O. longiglumis* herbarium specimen. *Image courtesy* ICAR-National Bureau of Plant Genetic Resources, India

2.2 mm wide with trichromes in rows, down its length (Fig. 14.3). The palea and lemma are papery. The awn is about 1 cm in length. The ligule is truncate in shape. The auricle is glabrous to hairy. The lower leaf surface is slightly pubescent to pubescent. The panicle is compact or open (Vaughan 1994; Naredo et al. 2003). The habitat of the plant is shaded forests close to swamps or rivers that may seasonally inundate the forest floor. These are perennials that occur as scattered plants in organic loamy soils in the forest. They are mainly found in abundance in the Komba River, Iranian Jaya, Indonesia, and Papua New Guinea (Fig. 14.4) (Vaughan 1994; Sanchez et al. 2013). The plants flower during January and November in situ (Vaughan 1994) and during April–May in moist humid subtropical climates.

14.3 Cytology

Oryza longiglumis is an allotetraploid having a HHJJ genome with a chromosome number of $2n = 4x = 48$ (Sanchez et al. 2013). The genome size was determined by flow cytometer and was found to be 1330 Mb (Miyabayashi et al. 2007).

14.3.1 Breeding Approaches

Plant breeding is a powerful method to produce desired genotypes with specific traits. The technique is prehistoric and includes crossing of closely related or distantly related species. The crossability of rice with its wild relatives can be categorized into crosses with three gene pools,

Fig. 14.3 *O. longiglumis* spikelet. *Image courtesy* ICAR-National Bureau of Plant Genetic Resources, India

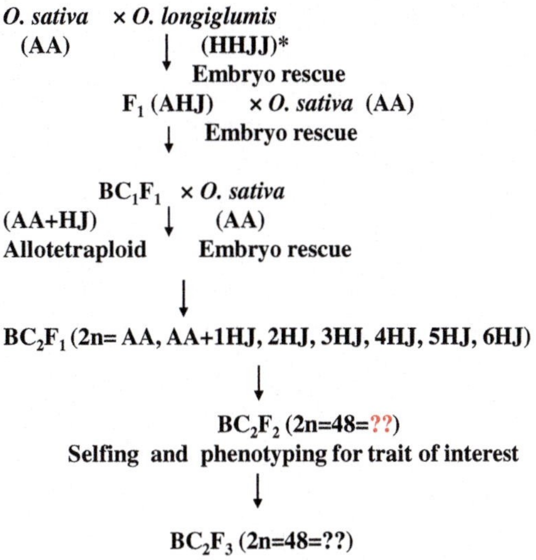

Fig. 14.5 Scheme for gene transfer from *O. longiglumis* into cultivated rice. Adopted and modified from Jena 2010

i.e., the primary, secondary, and tertiary (Khush 1997). The genotype of edible rice, i.e., *Oryza sativa* and *Oryza glaberrima* has an AA genome. AA genomes are easily crossable among themselves and are considered as the primary gene pool. Crossing between BB and EE genomes can be considered as accessing the secondary gene pool and the remaining genomes (crossing between KK and FF) constitute the use of the tertiary gene pool. The possibility of crossing of *Oryza* species depends on their phylogenetic relationships. A diverse set of beneficial qualities such as high yield, disease and pest resistance, and tolerance to several environmental stresses has been incorporated into rice cultivars from its wild relatives. The production of distant hybrids from crosses between cultivated and wild species of rice belongs to the secondary and tertiary pools which is extremely difficult because of the premature death of the embryo. The early abortion of hybrid embryo at different developmental stage is a characteristic feature in rice (Brar and Khush 1997). Nevertheless, resistance alleles which are present in this genus can be introgressed through breeding approaches along with in vitro techniques such as embryo rescue. A schematic representation of this breeding approach is depicted in Fig. 14.5.

Fig. 14.4 Geographical distribution of *O. longiglumis*. *Image courtesy* Genebank Project, NARO

Table 14.1 List of the gene that has been cloned from *O. longiglumis*

NCBI accession	Name	Biological Role	Approach for sequence elucidation	References
AF148625	Alcohol dehydrogenase II (*Adh2*)	Alcohol metabolism	PCR-based approach	Ge et al. (1999)
AF148595	J-type alcohol dehydrogenase I (*Adh1*)	Alcohol metabolism	PCR-based approach	Ge et al. (1999)
AF148591	H-type alcohol dehydrogenase I (*Adh1*)	Alcohol metabolism	PCR-based approach	Ge et al. (1999)
AF148672	Maturase (*matK*)	Intron splicing	PCR-based approach	Ge et al. (1999)
AB018073	5′ flanking region of the catalase gene	Catalyzes the decomposition of hydrogen peroxide into water and oxygen	PCR-based approach	Iwamoto et al. (1999)
AB436272	Chloroplast DNA, including interspecific variant sites, ndhC-tRNA Val intergenic region	Chloroplast-based gene expression and regulation	PCR-based approach	Kumagai et al. (2010)
AB436213	Chloroplast DNA, including interspecific variant sites, tRNA-Gly flanking region	Chloroplast-based gene expression and regulation	PCR-based approach	Kumagai et al. (2010)
AB436331	Chloroplast DNA, including interspecific variant sites, ORF133-psa1 intergenic region	Chloroplast-based gene expression and regulation	PCR-based approach	Kumagai et al. (2010)
AB128732	Chloroplast DNA, containing SSR region (RCt10)	Chloroplast-based gene expression and regulation	PCR-based approach	Nishikawa et al. (2005)
AY507933	NADH dehydrogenase subunit 1 (*nad1*)	Transfer of electrons from NADH to the respiratory chain	PCR-based approach	Song (2004)

14.4 Genomics

Intensive literature search indicates that there is very little information about gene cloning and genetic engineering of this species. The genome of this species is not decoded yet maybe due to the fact that it has largest genome size among the *Oryza* species. Nevertheless, a few genes have been cloned, mainly for academic interest, and are tabulated below (Table 14.1)

14.4.1 Blast and Bacterial Blight Resistance in Rice

Among rice pathogens bacterial blight (BB) caused by *Xanthomonas oryzae* pv. *Oryzae* (*Xoo*) is the most devastating bacterial disease of rice. It occurs in epidemics and can cause a yield loss of up to 50% (Ou 1985). Traditional methods like cultivation strategies, chemical control, and biological control are useful tool to fight BB. However, these methods can be tedious, costly and are not eco-friendly. Another serious biotic stress of rice is blast disease caused by the fungal pathogen *Magnaporthe oryzae* and is the most harmful threat to high productivity of rice (Kwon and Lee 2002; Li et al. 2007). Due to this disease, yield losses range from 1 to 50% and economic losses over $70 billion (Scheuermann et al. 2012). This loss in rice yield should be minimized in order to help the marginal and poor farmers of developing countries (Latif et al. 2011).

A few defense-responsive genes have been reported to positively or negatively regulate partial resistance to *Xoo*, such as *NH1*, *XB3*, *TGA2.1*, *Spl11*, *WRKY*62, *WRKY71*, *WRKY76*,

MPK5, *MPK12*, and *Rac1*. However, their link with resistance QTLs to *Xoo* has not been reported (Kou and Wang 2013, 2010).

14.5 Future Prospects

This species is almost an orphan plant despite the fact that it has many economic traits. The genomic and genetic resources of this species are also very limited. With the advancement of sequencing technology, it is possible to generate sequence-driven information including decoding its genome. Specifically, *O. longiglumis* harbors resistance gene cascades for two of the most devastating rice pathogens. Most studies conducted on BB and blast for rice have been conducted in several species of rice. It is high time that this wild variety of rice is scrutinized using modern sophisticated techniques of molecular biology to elucidate the mechanism of disease resistance in *O. longiglumis*. This will enable the introduction of possible new resistant varieties to farmers, using tools of genetic engineering and biotechnology. It is also a possibility that studying this wild relative of rice might open new wider horizons to disease resistance in plants.

References

Brar DS, Khush GS (1997) Alien introgression in rice. Plant Mol Biol 35:35–47

Ge Song, Sang Tao, Bao-Rong Lu, Hong De-Yuan (1999) Phylogeny of rice genomes with emphasis on origins of allotetraploid species. Proc Nat Acad Sci 96:14400–14405

Iwamoto M, Nagashima H, Nagamine T, Higo H, Higo K (1999) A Tourist element in the 5′-flanking region of the catalase gene CatA reveals evolutionary relationships among Oryza species with various genome types. Mole Gen Genet 262:493–500

Jena KK (2010) The species of the genus Oryza and transfer of useful genes from wild species into cultivated rice, O. sativa. Breed Sci 60:518–523

Khush GS (1997) Origin, dispersal, cultivation and variation of rice. In: *Oryza*: from molecule to plant (Springer)

Kou and Wang (2013) Bacterial blight resistance in rice. Trans Genomics Crop Breed Biotic Stress 1:11–30

Kou Y, Wang S (2010) Broad-spectrum and durability: understanding of quantitative disease resistance. Curr Opin Plant Biol 13:181–185

Kumagai M, Wang L, Ueda Shintaroh (2010) Genetic diversity and evolutionary relationships in genus Oryza revealed by using highly variable regions of chloroplast DNA. Gene 462:44–51

Kwon J-O, Lee S-G (2002) Real-time micro-weather factors of growing field to the epidemics of rice blast. Res Plant Dis 8:199–206

Latif MA, Badsha MA, Tajul MI, Kabir MS, Rafii MY, Mia MAT (2011) Identification of genotypes resistant to blast, bacterial leaf blight, sheath blight and tungro and efficacy of seed treating fungicides against blast disease of rice. Sci Res Essays 6:2804–2811

Li YB, Wu CJ, Jiang GH, Wang LQ, He YQ (2007) Dynamic analyses of rice blast resistance for the assessment of genetic and environmental effects. Plant Breeding 126:541–547

Miyabayashi T, Nonomura K-I, Morishima H, Kurata N (2007) Genome size of twenty wild species of Oryza determined by flow cytometric and chromosome analyses. Breed Sci 57(1):73–78

Naredo MEB, Juliano AB, Lu B-R, Jackson MT (2003) The taxonomic status of the wild rice species Oryza ridleyi Hook. f. and O. longiglumis Jansen (Ser. Ridleyanae Sharma et Shastry) from Southeast Asia. Genet Resour Crop Evol 50:477–488

Nishikawa Tomotaro, Vaughan Duncan A, Kadowaki Koh-ichi (2005) Phylogenetic analysis of Oryza species, based on simple sequence repeats and their flanking nucleotide sequences from the mitochondrial and chloroplast genomes. Theor Appl Genet 110:696–705

Ou SH (1985) Rice diseases (IRRI)

Sanchez PL, Wing RA, Brar DS (2013) The wild relative of rice: genomes and genomics. In: Genetics and genomics of rice, Springer

Scheuermann KK, de Andrade A, Wickert E, Raimondi JV, Marschalek R (2012) Magnaporthe oryzae genetic diversity and its outcomes on the search for durable resistance. INTECH Open Access Publisher

Song, GYLG (2004) The utility of mitochondrial nad1 intron in phylogenetic study of Oryzeae with reference to the systematic position of Porteresia. Acta Phytotaxonomica Sinica 4:001

Vaughan AD (1994) The wild relative of rice: a genetic resources handbook

Oryza longistaminata A. Chev. and Röhr

15

Marie Noelle Ndjiondjop, Peterson Wambugu, Tia Dro, Raphael Mufumbo, Jean Sangare and Karlin Gnikoua

Abstract

Oryza longistaminata is a distantly related wild rice relative found only in Africa. It is agronomically inferior to cultivated rice but possesses several latent useful traits that can be used to improve agronomically important traits in cultivated rice. These useful traits include strong rhizomes, a vigorous biomass, drought avoidance mechanisms, good weed suppression ability via allelopathy, and high nitrogen-use efficiency. It also possesses resistance to several biotic stresses such as brown plant hopper, nematodes, yellow stem borer, rice tungro bacilliform virus, blast, and bacterial blight. Recent studies have revealed that it is a good source of new alleles that can be used to improve yield-related traits in cultivated rice varieties. However, this potential is not well studied and is therefore not optimally exploited. Advances in genomics such as the release of whole genome reference sequence are offering opportunities for enhanced use of this species and its genetic resources. In this chapter, efforts were made to review current knowledge on *O. longistaminata* by addressing phenotypic and molecular studies conducted on this species. The use of advances in DNA sequencing in understanding the potential of this species and dissecting the molecular mechanism underlying various useful traits is highlighted.

M. N. Ndjiondjop (✉) · T. Dro · J. Sangare
K. Gnikoua
Africa Rice Center (AfricaRice), 01 B.P. 2031, Cotonou, Benin
e-mail: m.ndjiondjop@cgiar.org

P. Wambugu
Kenya Agricultural and Livestock Research Organization (KALRO), Genetic Resources Research Institute, P.O. Box 781, Kikuyu 00902, Kenya

R. Mufumbo
National Plant Genetic Resources Center/Uganda National Gene Bank, Biodiversity and Biotechnology Program. National Agricultural Research Organization (NARO), P.O. Box 40, Entebbe, Uganda

15.1 Academic and Economic Importance

Rice, the staple diet of half of the world's population, is the product of the domestication process of wild species in the genus *Oryza* (Callaway 2014). The genus *Oryza*, named by Linnaeus (1753), belongs to the tribe *Oryzeae*, sub-family *Oryzoideae* of the grass family *Poaceae* (Gramineae). According to Kellogg (2009) and Jacquemin et al. (2013), the genus *Oryza* includes 24 species. These 24 species are grouped into four complexes named the *Oryza sativa*, *Oryza officinalis*, *Oryza ridleyi,* and *Oryza granulata* complexes (Banaticla-Hilario 2012). The complex *O. sativa* or serie *sativae*,

T. K. Mondal and R. J. Henry (eds.), *The Wild Oryza Genomes*,
Compendium of Plant Genomes, https://doi.org/10.1007/978-3-319-71997-9_15

also known as AA-genome group, contains two cultigens (*O. sativa*, Asian rice and *Oryza glaberrima*, African rice) and six wild species, namely *O. rufipogon*, *Oryza nivara*, *Oryza meridionalis*, *Oryza longistaminata*, *Oryza barthii*, and *Oryza glumaepatula* (Vaughan et al. 2003). The evolution of rice from wild to cultivated species was done independently in Africa and Asia (Londo et al. 2006; Vaughan et al. 2008). In Asia, there is clear archeo-botanical evidence that *O. sativa* was domesticated from *O. rufipogon* in the lower Yangtze region of Zhejiang in China, between 6900 and 6600 years ago (Fuller et al. 2009). Another hypothesis postulates that *O. rufipogon*, the perennial species and *O. nivara*, the annual species, are both the direct ancestors of *O. sativa* (Sweeney and McCouch 2007).

The African continent is home to species representing five of the ten known Oryza genome types. These include AA (*O. longistaminata*, *O. barthii*, *O. glaberrima*, and *O. sativa*), BB (*O. punctata*), BBCC (*O. schweinfurthiana*), CC (*O. eichingeri*), and FF (*O. brachyantha*) (Vaughan et al. 2003). According to Portères (1962, 1976), *O. glaberrima* was first domesticated from the wild ancestor *O. barthii* (formerly known as *O. breviligulata*) by people living in the Inland Delta of the Upper Niger River about 3500 years ago. However, Harlan and Stemler (1976) suggested that *O. glaberrima* was domesticated from *O. barthii* in several different localities within the vast forest and savanna areas. Recent studies have confirmed that *O. barthii* is the direct progenitor of *O. glaberrima* and they are both restricted to West Africa (Duan et al. 2007; Wang et al. 2014). Though it was initially thought that *O. barthii* was derived from *O. longistaminata* (Khush 1997), this has been discounted by a recent study using whole chloroplast genome sequences (Wambugu et al. 2015). *Oryza longistaminata* is native to most sub-Saharan African countries and Madagascar (Vaughan 1994). It is genetically diverse and together with the other wild relatives of cultivated rice represent an important gene pool that has significantly contributed to rice improvement (Kiambi et al. 2008). Evaluation of these wild species populations continues to reveal new sources of resistance to diseases and pests as well as for enhancing yield and yield-related traits (Khush et al. 2003; Goicoechea et al. 2010; Ramos et al. 2016). Interest in this species has increased as techniques to transfer genes from wild species to cultivated rice have improved (Heinrichs et al. 1985; Khush et al. 1990; Brar and Khush 1995). *Oryza longistaminata* is one of the wild rice species that is already making significant contributions to rice improvement programs, particularly in Asia and Africa (Khush et al. 1990; Cissé and Khouma 2016). In this paper, we review current knowledge on *O. longistaminata* by addressing phenotypic and molecular studies conducted on the species. The use of advances in DNA sequencing in understanding the potential of this species and dissecting the molecular mechanism underlying various useful traits is highlighted.

15.2 Distribution and Morphological Characteristics

Oryza longistaminata is widely distributed in Africa (Fig. 15.1) (Khush 1997; Brar and Khush 2003). It is known as a robust, perennial, and allogamous herb with an erect, soft, and spongy stem (Fig. 15.2) (Phillips 1995). It is also characterized by long stigma and anthers (>3 mm), acute or 2-cleft ligule (>15 mm), strong, extensive, and branched rhizomes. It can propagate by both seeds and rhizomes, but mainly by rhizomes. This is due to its high level of seed sterility, explained by the relatively longer phenological cycle, partial self-incompatibility, and variability of the mature seeds (Jones et al. 1996; Kiambi et al. 2005). *Oryza longistaminata* has a high production of biomass and larger populations than other *Oryza* species in Africa (Melaku et al. 2013). It is tall (≥ 2 m) and grows in stagnant or swampy areas with water flowing through them, flood plains, edges of rivers and lakes. It can also grow along canals in cultivated rice fields and in permanently wet or seasonally dry areas. It can also survive in water depth of 4 m (Kanya et al. 2012). *Oryza longistaminata*

Fig. 15.1 Map of *O. longistaminata* distribution. The blue dots indicate country of origin of *O. longistaminata* accessions conserved in AfricaRice gene bank

displays long awns (Fig. 15.2) with varying pigmentation and severe shattering for seed dispersal (Sweeney and McCouch 2007).

15.3 Genomic Resources

15.3.1 Whole Genome Sequences

The whole genome shotgun assembly of *O. longistaminata* was recently released through the collaborative efforts of Kunming Institute of Zoology, Chinese Academy of Sciences, Kunming, China and BGI-Shenzhen, China (Zhang et al. 2015). This de novo assembled genome with a size of 347 Mb was generated through a combination of Illumina short-read data and Roche GS FLX long-read data. About half of the genome is composed of transposable elements, a proportion which is significantly higher than that found in *O. sativa, O. glaberrima,* and *O. brachyantha*. Key statistics of the genome sequence are presented in Table 15.1. These genome sequences lay the foundation for enhancing the deployment of the valuable traits found in *O. longistaminata* in improving cultivated rice. The availability of whole genome sequences will help usher in resequencing efforts which are important in the identification of various structural and functional variation. These genomic resources also provide a valuable resource for studying evolution in the *Oryza* genus. They also provide tools for promoting the efficient conservation and utilization of *O. longistaminata* genetic resources.

15.3.2 Transcriptome

The transcriptome is a collection of all RNA present in a cell or a population of cells at any given moment. Transcriptome analysis is able to reveal genes that are being actively expressed in specific tissue and species of interest and also

Fig. 15.2 Plant and grain of *O. longistaminata*: **a** *O. longistaminata* plant at flowering stage; **b** *O. longistaminata* seeds with typical long awns and big grains

Table 15.1 Key statistics of *longistaminata* genome sequence

Attribute	Size/number	Proportion of total genome
Genome size	347,790,983 (bp)	–
Repeats	184,380,075 (bp)	50.93
TEs	176,428,297 (bp)	48.73
Predicted protein-coding genes	32,502	–
Non-protein-coding genes	5377	–

The total number of transposable elements (TEs) has been assessed using Tandem Repeat Finder (TRF), RepeatMasker, Proteinmask, and de novo. Non-protein-coding RNA genes include 3954 putative miRNA, 720 tRNA, 34 rRNA, and 669 snRNA genes

facilitate the discovery of potential molecular markers (Waiho et al. 2017). Analysis of the transcriptome has been used to gain increased understanding on the genetic basis of rhizomatousness and self-incompatibility (Hu et al. 2011; Zhang et al. 2015). However, just like genomic data, transcriptomic data for *O. longistaminata* is limited. The first large transcriptomic data was generated by Yang et al. (2010) who conducted a global characterization of the root transcriptome by deep sequencing. This study detected novel expressed sequence tags (ESTs) and provided valuable data that is useful for gene discovery as well as rice functional studies. For example, this publicly available data is vital in the identification of genes responsible for adapting *O. longistaminata* to environments with limited nutrients. A recent investigation using Illumina-based next-generation sequencing technology provided new transcriptomic data from rhizome tissues and other sample tissues including root, stem, and leaf. This study identified various rhizome-specific and preferentially expressed genes including transcription factors, hormone metabolism, and stress response-related genes. Several of the identified genes displayed higher expression levels in the rhizome. In addition, this study reported the presence of high levels of *Magnaporthe oryzae* (a hemibiotrophic fungal pathogen that causes blast in cultivated rice) genes that were preferentially expressed especially in the healthy *O. longistaminata* rhizome tissues (He et al. 2014). This suggests that the red rice is tolerant to this pathogen.

Oryza longistaminata has been used as a model species for dissecting the genetic and molecular mechanisms underlying the rhizomatous trait. This has been conducted through transcriptome sequencing (Zhang et al. 2015) and genome-wide differential analysis (Hu et al. 2011). Zhang et al. (2015) sequenced the transcriptome of rhizome and stem tissues and identified various differentially expressed genes between these two types of issues. Results of this analysis suggested that genes that control hormone production might have a role in rhizome development. Further analysis of the transcriptome led to the identification of a candidate gene (Olong01m10027813) for rhizome production. Sequence analysis between the rhizome-bearing *O. longistaminata* and other non-rhizome-bearing *Oryza* species revealed that one non-synonymous substitution is likely to be the mutation responsible for rhizome production (Zhang et al. 2015). Similarly, Hu et al. (2011) conducted a genome-wide differential analysis of various tissues among them rhizome tips and rhizome internodes. This analysis led to the identification of a total of 58 and 61 unique genes that were expressed specifically in the rhizomes tips (RT) and internodes (RI), respectively, and were co-localized with QTLs for rhizome traits. They also reported the involvement of genes regulating the production of various plant hormones particularly gibberellins, for the initiation and development of rhizomatous trait. These studies provide a good foundation for further exploratory work on the genetic and molecular basis of rhizomatousness in *O. longistaminata*. To uncover the molecular mechanism of self-incompatibility, Zhang et al. (2015) conducted transcriptomic analysis of stamen and pistil tissues, each obtained from the self-incompatible *O. longistaminata* and the self-compatibility hybrid line obtained from a cross between *O. longistaminata* and *O. sativa* ssp. *indica*. Analysis of these tissues revealed one gene (Olong01m10012815) with remarkably high expression in the pistils of the hybrid line compared to those of *O. longistaminata*. This gene is located in a region syntenic to one where the gene for self-incompatibility had been identified in perennial ryegrass (*Lolium perenne*) (Yang et al. 2009). This gene is annotated as EF-hand calcium-binding protein gene, suggesting that calcium-dependent signaling may be involved in self-incompatibility mechanisms.

15.3.3 Non-coding RNAs

MicroRNAs also called MiRANs are small, highly conserved non-coding RNA molecules involved in the regulation of gene expression (MacFarlane and Murphy 2010). MiRNAs are reported to play an important role in plant growth and development. Several hundreds of MiRNAs are available in rice miRBase.

A comparative analysis of two small RNA libraries obtained from aerial shoots and rhizomes using high-throughput RNA sequencing identified 380 known rice miRNAs in addition to 72 conserved miRNAs and 151 putative novel miRNAs. This shows that the majority of the identified rice miRNAs could be expressed in *O. longistaminata*. In addition, 144 potential targets were predicted for the differentially expressed miRNAs in the rhizomes.

15.3.4 Organellar Genome

Traditionally, plant molecular phylogenetic analysis involved amplifying, sequencing, and analyzing one or a few genes from many species. This approach lends itself well to broad taxon sampling. Currently, an alternative approach involving the use of chloroplast genome sequencing is gaining popularity. This approach provides much larger amounts of data per taxon, and usually a smaller number of species are analyzed. Although the two strategies are somewhat related, their results sometimes conflict for reasons that are currently debated (Albar et al. 2006; Asaf et al. 2017). The inheritance of mitochondrial and chloroplast genes differs from that of nuclear genes in showing vegetative segregation, uniparental inheritance, intracellular selection, and reduced recombination (Birky 2001). To date, significant progress has been made in the sequencing of organelle genomes, particularly those of economically important crop plants such as rice. Asaf et al. (2017) reported that a total of 300 mitochondrial (mt) and 342 complete chloroplast (cp) genomes have been submitted to GenBank Organelle Genome Resources. The organellar genome of *O. longistaminata* was recently reported by Wambugu et al. (2015) where it was used to reconstruct the phylogenetic relationships between various *Oryza* AA-genome species among which were two accessions of *O. longistaminata*.

The chloroplast genome possesses a typical quadripartite structure which is typical of angiosperms with its size ranging from 134,563 to 134,567 bp. The genome has a total of 124 unique genes comprising of 83 protein-coding genes, 33 transfer RNA (tRNA) and 8 ribosomal RNA (rRNA) genes. The structure and size of *O. longistaminata* chloroplast genome was recently confirmed by another study which compared the complete chloroplast genome of *O. minuta* with those of another eleven *Oryza* species (Asaf et al. 2017).

15.4 Genetic Resources

Oryza longistaminata genetic resources are conserved in various genebanks globally (Table 15.2).

Analysis of germplasm collection data for the various countries compared with the available herbarium records (Wambugu et al. 2013) points to the presence of collection gaps in the existing collections. Collection gaps in ex situ conservation facilities have previously been reported (Maxted and Kell 2009; Hay et al. 2013). These

Table 15.2 *O. longistaminata* germplasm conserved in various international genebanks

Genebank	Number of accessions
IRRI, the Philippines	285
Africa Rice Center	19
Australian Plant DNA Bank	2
Millennium Seed Bank Project	3
USDA	9
ILRI, Ethiopia	5
National Institute of Genetics	149
SPGRC	54

Adapted from Wambugu et al. (2013)

collection gaps in terms of few numbers of accessions could be attributed to a number of challenges, namely constrained national resources allocated to germplasm collection missions especially wild crop relatives and the high degree of sterility of the species, with only a few populations producing viable seeds. There is therefore need for concerted global efforts to conduct more planned and targeted germplasm collection efforts in order to capture the global diversity of *O. longistaminata* species. There is also need for a robust in situ conservation program for *O. longistaminata* gene pool. Advances in genomics are offering opportunities for identifying germplasm collection gaps. Analysis of genomes for example using SNP markers helps assess the range of alleles present in a collection. Germplasm acquisition should only be done after it has been established that such germplasm is going to add novel alleles to a collection. However, the lack of sequence data for conserved germplasm hinders the determination of collection gaps using molecular tools.

15.5 *O. Longistaminata* in Rice Improvement

15.5.1 Genetic Potential

Oryza longistaminata is directly used by local people as famine food and for grazing in countries like Ethiopia and Sudan (Vaughan and Sitch 1991) and as traditional food in West Africa. It also possesses useful alleles that are being used to improve agronomically important traits. As will be highlighted later, *O. longistaminata* possesses resistance to bacterial blight having contributed the *Xa21* which has been used to improve *O. sativa*. Recently, Neelam et al. (2016) screened a large set of 1176 accessions from various species including *O. longistaminata* for tolerance to Xoo pathotypes, namely PbXo-10 and PbXo-8. This study revealed broader resistance to bacterial blight than earlier known, suggesting the presence of bacterial blight resistance gene other than *Xa21*. Two of the *O. longistaminata* accessions IRGC92624 and IRGC92644 were found to have resistance against both the Xoo pathotypes (Table 15.3). These resistance genes can be transferred to elite cultivars of *O. sativa* for better management of bacterial blight.

In spite of its overall inferior phenotypic appearance, *O. longistaminata* has been reported to have drought avoidance mechanisms, good weed suppression ability via allelopathy, and high nitrogen-use efficiency (Zhanq et al. 2008; Yang et al. 2010). It also has known resistance to brown plant hopper (BPH), nematodes, yellow stem borer, rice tungro bacilliform virus, blast, and bacterial blight (BB) (Kobayasi et al. 1993; Kobayashi et al. 1994; Soriano et al. 1999; Brar and Khush 2002, 2003; Chen et al. 2009).

Table 15.3 Useful traits found in *O. longistaminata*

Trait	References
Resistance to bacterial blight	Ronald et al. (1992), Brar and Khush (2002), Neelam et al. (2016)
Blast resistance	Xu et al. (2015)
High yield-related traits	Ramos et al. (2016), Gichuhi et al. (2016)
Resistance to nematodes	Soriano et al. (1999), Brar and Khush (2002)
Weed competitiveness	Xu et al. (2014), Shen et al. (2016)
Yellow stem borer resistance	Brar and Khush (2006)
Drought avoidance	Brar and Khush (2002)

15.5.1.1 Potential for Yield Improvement

Using chromosome segment substitution lines (CSSLs) developed by crossing *O. longistaminata* and *O. sativa* cv *japonica*, 10 major putative QTLs controlling yield and various yield-related attributes were identified. A major putative QTL associated with increased grain number per panicle showed pleiotropic effects on other traits such as plant height, days to flowering, tiller number, number of branches per panicle, and grain length. *Oryza longistaminata* therefore seems to have potential for acting as a good source of new alleles that can be used to improve yield-related traits in cultivated rice varieties (Ramos et al. 2016). The potential of *O. longistaminata* in improving yield and yield-related attributes in cultivated rice was also recently reported by Gichuhi et al. (2016) who identified various QTLs for these traits.

Undoubtedly, *O. longistaminata* genetic resources have numerous traits of potential value in the improvement of cultivated rice. However, the lack of characterization data remains the greatest challenge in using this valuable germplasm for rice improvement. In order to enhance the utilization of *O. longistaminata* genetic resources in rice improvement, there is need for concerted global efforts to systematically characterize and evaluate *O. longistaminata* collections held in various national gene banks. This will help to identify accessions with novel traits thereby increasing their value to rice breeders.

15.5.2 Deployment of *Oryza longistaminata* Genetic Potential in Rice Improvement

Though *O. longistaminata* has great genetic potential as already highlighted, it remains underutilized in rice improvement due to hybrid sterility or hybrid breakdown (Nevame et al. 2014). One of the most remarkable cases of the use of genes from rice wild relatives to improve cultivated rice is the transfer of Xa*21* gene conferring resistance against bacterial blight resistance. This gene was successfully introgressed from *O. longistaminata* into *Oryza sativa* (Khush et al. 1990). Some of the improved varieties developed by transferring the *Xa21* gene have been released in the Philippines, India, China, and Thailand (Sanchez et al. 2013) (Table 15.4).

When grown on hilly lands, annual upland rice has been found to encourage soil erosion in Southeast Asia. This problem of soil erosion can be overcome through the use of perennial cultivars of upland rice (Zhang et al. 2014). Breeding for perennial rice therefore remains an important breeding objective especially in areas with major soil erosion problems.

Various efforts have been made in developing perennial rice with the long-term goal of ultimately breeding perennial rice to overcome the problem of soil erosion in various rain-fed lowland and rain-fed upland rice-based farming systems (Schmit 1996, Hu et al. 2003, Sacks

Table 15.4 Released rice varieties carrying Xa21 gene from *O. longistaminata*

Inbreds/hybrids	Year	BB resistance gene(s)	Institute/country
NSICRc 142 (Tubigan 7)	2006	Xa4 þ Xa21	PhilRice, the Philippines
NSICRc 154 (Tubigan 11)	2007	Xa4 þ Xa21	the Philippines
Improved Samba Mahsuri	2007	xa5þxa13þXa21	India
Improved Pusa Basmati 1	2007	xa5þxa13þXa21	India
Xieyou 218	2002	Xa21	China
Zhongyou 218	2002	Xa21	China
Guodao 1	2002	Xa4þxa5þxa13þXa21	China
Guodao 3	2004	Xa4þxa5þxa13þXa21	China
Neizyou	2004	Xa4þxa5þxa13þXa21	China
Ilyou 8006	2005	Xa4þxa5þxa13þXa21	China
Ilyou 218	2005	Xa21	China
ZhongbaiYou 1	2006	Xa21	China

Source Brar and Singh (2011)

et al. 2006, Zhang et al. 2014). In closely related efforts, Tao and Sripichitt (2000) reported the first successful hybridization between *O. sativa* and *O. longistaminata* through embryo rescue. These efforts have, for example, led to the development of high-yielding perennial rice varieties such as NSIC Rc 112, which has been successfully released in the Philippines (Brar and Khush 2006).

15.5.3 Genetic Mapping

Though the genetic potential of *O. longistaminata* is generally well known, the genetic mechanisms underlying various traits are poorly studied. With the continued advances in genetics and genomics, coupled with the increased understanding of the importance of wild species of rice relatives in rice improvement, interest in these studies is growing. To harness the genetic potential of wild rice species, various genetic resources such as recombinant inbred lines, backcross inbred lines, near isogenic lines, and CSSLs that are derived from wild rice relatives including *O. longistaminata* have been developed (Sanchez et al. 2013). These resources have also been instrumental in efforts aimed at studying the genetic mechanisms underlying various traits (Table 15.5). Despite the great potential of *O. longistaminata* in rice improvement as already highlighted, transferring genes from *O. longistaminata* into *O. sativa* is hampered by the presence of hybrid sterility (Zhao et al. 2012). In order to understand the nature of this hybrid sterility, Chen et al. (2009) studied pollen and spikelet fertility in a cross between *O. longistaminata* and *O. sativa*. These authors identified one main-effect QTL for pollen and spikelet fertility, *qpsf6,* which coincides with the gamete eliminator, *S1*. These findings point to the presence of an orthologous locus that is responsible for seed sterility between *O. sativa* and other AA-genome species. In the same study, a QTL for plant height *qph1*, which coincides with the semi-dwarf gene, *sd-1,* was identified.

Genetic mapping of the genes responsible for blast resistance in *O. longistaminata* was recently conducted by Xu et al. (2015) using an interspecific cross between *O. sativa spp. indica* and *O. longistaminata* (Table 15.5). This study identified a novel dominant blast-resistant gene, designated as Pi57 (t), which is located on chromosome 12. This gene conferred a broad-spectrum resistance to *Magnaporthe oryzae* isolates, and its resistance spectrum could be distinguished from that of five known blast R genes located on chromosome 12. Recently,

Table 15.5 Mapping populations developed between *O. longistaminata* and *O. sativa*

Trait	Population type	Parents	References
Yield traits	CSSLs	*O. longistaminata* × *Japonica* (Taichung 65)	Ramos et al. (2016)
Yield-related traits	F_1 (RIL)	*O. longistaminata* × *O. sativa* (Norin 18)	Gichuhi et al. (2016)
Hybrid sterility	BC_7F_2	*O. sativa* (RD23) × *O. longistaminata*	Chen et al. (2009)
Height	BC_7F_2	*O. sativa* (RD23) × *O. longistaminata*	Chen et al. (2009)
Blast resistance	BC_3F_7 introgression lines (ILs)	*O. sativa* (RD23) × *O. longistaminata*	Xu et al. (2015)

Gichuhi et al. (2016) identified a total of 36 QTLs located on various chromosomes for yield and various yield-related attributes. This study showed the potential of *O. longistaminata* in improving yield in *O. sativa* even under non-fertilized conditions.

15.6 Conclusion and Research Perspectives

Oryza longistaminata is one of the three African AA-genome rice species (*O. glaberrima*, *O. barthii*, and *O. longistaminata*). It is the most diverse *Oryza* species in Africa, capable of reproducing both asexually via rhizome and sexually by seeds where it behaves primarily as an outcrossing species. It hybridizes easily with other AA-genome species of the genus *Oryza* enabling its use as a source of novel genes being used for rice improvement (Lu et al. 2003). It is the species from which the first disease-resistant gene (*Xa21* gene) was cloned and introgressed into rice through a cross between *O. sativa* and *O. longistaminata* (Ikeda et al. 1990; Song et al. 1995). Despite this success in discovering genes in *O. longistaminata*, the exploitation of its diversity conserved in several genebanks including AfricaRice genebank remains limited.

This situation is, however, changing as advanced genomic tools are being used to characterize genetic resources at molecular level (Sakai et al. 2011). With the availability of SNP markers, genomic selection is being used in breeding to rapidly identify important genes and traits hidden in wild species. Currently, the collection of wild species including *O. longistaminata* conserved at AfricaRice is being characterized using SNP markers. This useful information will be used through the breeding strategy of the institute.

References

Albar L, Bangratz-Reyser M, Hébrard E, Ndjiondjop M-N, Jones M, Ghesquière A (2006) Mutations in the eIF(iso)4G translation initiation factor confer high resistance of rice to Rice yellow mottle virus. Plant J Cell Mol Biol 47:417–426. https://doi.org/10.1111/j.1365-313X.2006.02792.x

Asaf S, Waqas M, Khan AL, Khan MA, Kang S-M, Imran QM, Shahzad R, Bilal S, Yun B-W, Lee I-J (2017) The Complete chloroplast genome of wild rice (*Oryza minuta*) and its comparison to related species. Front Plant Sci. https://doi.org/10.3389/fpls.2017.00304

Banaticla-Hilario MCN (2012) An ecogeographic analysis of *Oryza* series Sativae in Asia and the Pacific. Wageningen Academic Publishers, Wageningen

Birky CW (2001) The inheritance of genes in mitochondria and chloroplasts: laws, mechanisms, and models. Annu Rev Genet 35:125–148. https://doi.org/10.1146/annurev.genet.35.102401.090231

Brar DS, Khush GS (1995) Fragile lives in fragile ecosystems. In: Proceedings of the international rice research conference, 13–17 Feb 1995, International Rice Research Institute, Los Baños, Laguna, Philippines. In: the International Rice Research Conference. International Rice Research Institute, pp 901–910

Brar DS, Khush GS (2002) transferring genes from wild species into rice. In: Kang MS (ed) Quantitative genetics, genomics and plant breeding. CABI, New York, pp 197–217

Brar DS, Khush GS (2003) Utilization of wild species in rice. In: Nanda, JS, Sharma SD (eds) Monograph on genus *Oryza*. Science Publishers, Enfield, pp 283–309

Brar D, Khush G (2006) Cytogenetic Manipulation and germplasm enhancement of rice (*Oryza sativa* L.). In: Singh R, Jauhar P (eds) Genetic resources, chromosome engineering, and crop improvement. CRC Press, Boca Raton, pp 115–158

Brar DS, Singh K (2011) In: Kole C (ed) Wild crop relatives: genomic and breeding resources, cereals. Springer, Berlin, pp 321–336

Callaway E (2014) Domestication: the birth of rice. Nature 514:S58–S59. https://doi.org/10.1038/514S58a

Chen ZW, Hu FY, Xu P, Li J, Deng XN, Zhou JW, Li F, Chen SN, Tao DY (2009) QTL analysis for hybrid sterility and plant height in interspecific populations derived from a wild rice relative, *Oryza longistaminata*. Breed Sci 59:441–445

Cissé F, Khouma MP (2016) Improvement of wild rice *Oryza longistaminata* through mutation induction. 6:82–89. https://doi.org/10.17265/2159-5828/2016.02.004

Duan S, Lu B, Li Z, Tong J, Kong J, Yao W, Li S, Zhu Y (2007) Phylogenetic analysis of AA-genome *Oryza* species (Poaceae) based on chloroplast, mitochondrial, and nuclear DNA sequences. Biochem Genet 45:113–129. https://doi.org/10.1007/s10528-006-9062-x

Fuller DQ, Qin L, Zheng Y, Zhao Z, Chen X, Hosoya LA, Sun G-P (2009) The domestication process and domestication rate in rice: spikelet bases from the Lower Yangtze. Science 323:1607–1610. https://doi.org/10.1126/science.1166605

Gichuhi E, Himi E, Takahashi H, Zhu S, Doi K, Tsugane K, Maekawa M (2016) Identification of QTLs for yield-related traits in RILs derived from the cross between pLIA-1 carrying *Oryza longistaminata* chromosome segments and Norin 18 in rice. Breed Sci 66:720–733. https://doi.org/10.1270/jsbbs.16083

Goicoechea JL, Ammiraju JSS, Marri PR, Chen M, Jackson S, Yu Y, Rounsley S, Wing RA (2010) The future of rice genomics: sequencing the collective *Oryza* genome. Rice 3:89–97. https://doi.org/10.1007/s12284-010-9052-9

Harlan JR, Stemler A (1976) The races of sorghum in Africa. In: Harlan JR, De Wet JM, Stemler AB (eds) Origins of African plant domestication, Reprint 2011 edition. Mouton, The Hague, Netherlands

Hay F, Hamilton NRS, Furman BJ, Reddy UK, Singh S (2013) Cereals. In: Normah MN, Chin HF, Reed BM (eds) Conservation of tropical plant species. Springer, New York, pp 293–315

He R, Salvato F, Park J-J, Kim M-J, Nelson W, Balbuena TS, Willer M, Crow JA, May GD, Soderlund CA, Thelen JJ, Gang DR (2014) A systems-wide comparison of red rice (*Oryza longistaminata*) tissues identifies rhizome specific genes and proteins that are targets for cultivated rice improvement. BMC Plant Biol 14:46. https://doi.org/10.1186/1471-2229-14-46

Heinrichs EA, Medrano FG, Rapusas HR (1985) Genetic evaluation for insect resistance in rice

Hu FY, Tao DY, Sacks E, Fu BY, Xu P, Li J, Yang Y, McNally K, Khush GS, Paterson AH, Li Z-K (2003) Convergent evolution of perenniality in rice and sorghum. Proc Natl Acad Sci USA 100:4050–4054. https://doi.org/10.1073/pnas.0630531100

Hu F, Wang D, Zhao X, Zhang T, Sun H, Zhu L, Zhang F, Li L, Li Q, Tao D, Fu B, Li Z (2011) Identification of rhizome-specific genes by genome-wide differential expression Analysis in *Oryza longistaminata*. BMC Plant Biol 11:18. https://doi.org/10.1186/1471-2229-11-18

Ikeda R, Khush GS, Tabien R (1990) A new resistance gene to bacterial blight derived from *O. longistaminata*. Jpn J Breed 40:280–281

Jacquemin J, Bhatia D, Singh K, Wing RA (2013) The International Oryza Map Alignment Project: development of a genus-wide comparative genomics platform to help solve the 9 billion-people question. Curr Opin Plant Biol 16:147–156. https://doi.org/10.1016/j.pbi.2013.02.014

Jones M, Dingkuhn M, Johnson DE, Fagade S (1996) Inter-specific hybridization: progress and prospects. In: Africa/asia joint research on inter-specific hybridization between the African and Asian rice species (*O. glaberrima* and *O. sativa*). West Africa Rice Development Association, Bouaké, Côte d'Ivoire, pp 44–49

Kanya JI, Hauser TP, Kinyamario JI, Amugune NO (2012) Hybridization potential between cultivated rice *Oryza sativa* and African wild rice *Oryza longistaminata*. Int J Agric Res 7:291–302. https://doi.org/10.3923/ijar.2012.291.302

Kellogg EA (2009) The evolutionary history of Ehrhartoideae, Oryzeae, and Oryza. Rice 2:1–14. https://doi.org/10.1007/s12284-009-9022-2

Khush GS (1997) Origin, dispersal, cultivation and variation of rice. Plant Mol Biol 35:25–34

Khush GS, Balacangco E, Ogawa T (1990) A new gene for resistance to bacterial blight from *Oryza longistaminata*. Rice Genet Newsl 7:121–122

Khush GS, Brar DS, Hardy B (2003) Advances in rice genetics. Int Rice Res, Inst

Kiambi DK, Newbury HJ, Ford-Lloyd BV, Dawson I (2005) Contrasting genetic diversity among *Oryza longistaminata* (A. Chev et Roehr) populations from different geographic origins using AFLP. Afr J Biotechnol 4:308–317. https://doi.org/10.5897/AJB2005.000-3060

Kiambi DK, Newbury HJ, Maxted N, Ford-Lloyd BV (2008) Molecular genetic variation in the African wild rice *Oryza longistaminata* A. Chev. et Roehr. and its association with environmental variables

Kobayashi N, Ikeda R, Vaughan DA, others (1994) Screening wild species of rice (*Oryza* spp.) for resistance to rice tungro disease. JARQ Jpn Agric Res Q 28:230–236

Kobayasi N, Ikeda R, DOMINGO IT, Vaughan DA (1993) Resistance to infection of rice tungro viruses and vector resistance in wild species of rice (*Oryza* spp.). Jpn J Breed 43:377–387. https://doi.org/10.1270/jsbbs1951.43.377

Londo JP, Chiang Y-C, Hung K-H, Chiang T-Y, Schaal BA (2006) Phylogeography of Asian wild rice, *Oryza rufipogon*, reveals multiple independent domestications of cultivated rice, *Oryza sativa*. Proc Natl Acad Sci 103:9578–9583. https://doi.org/10.1073/pnas.0603152103

Lu B, Song Z, Chen J (2003) Can transgenic rice cause ecological risks through transgene escape?*. Prog Nat Sci 13:17–24. https://doi.org/10.1080/10020070312331343070

MacFarlane L-A, Murphy PR (2010) MicroRNA: biogenesis, function and role in cancer. Curr Genomics 11:537–561. https://doi.org/10.2174/138920210793175895

Maxted N, Kell S (2009) Establishment of a global network for the in situ conservation of crop wild relatives: status and needs

Melaku G, Haileselassie T, Feyissa T, Kiboi S (2013) Genetic diversity of the African wild rice (*Oryza longistaminata* Chev. et Roehr) from Ethiopia as revealed by SSR markers. Genet Resour Crop Evol 60:1047–1056. https://doi.org/10.1007/s10722-012-9900-0

Neelam K, Lore JS, Kaur K, Pathania S, Kumar K, Sahi G, Mangat GS, Singh K (2016) Identification of resistance sources in wild species of rice against two recently evolved pathotypes of *Xanthomonas oryzae* pv oryzae. Plant Genet Resour 1–5. https://doi.org/10.1017/s1479262116000149

Nevame AYM, Emon RM, Nassirou TY, Mamadou G, Samoura AD (2014) The more knowledge about heterotic loci is a pre-requisite for enhancing grain yield potential in distant crosses. Int J Agron Agric Res 5:200–215

Phillips S (1995) Poaceae (Gramineae). In: Hedberg I, Edwards S (eds) Flora of Ethiopia and Eritrea. Uppsala, Sweden, pp 9–11

Portères R (1962) Berceaux Agricoles Primaires Sur Le Continent Africain. J Afr Hist 3:195–210. https://doi.org/10.1017/S0021853700003030

Portères R (1976) African cereals: eleusine, fonio, black fonio, teff, Brachiaria, Paspalum, Pennisetum, and African rice. In: Harlan JR, de Wet JMJ, Stemler ABL (eds) Origins of African plant domestication. Mouton, The Hague, Netherlands, pp 409–452

Ramos JM, Furuta T, Uehara K, Chihiro N, Angeles-Shim RB, Shim J, Brar DS, Ashikari M, Jena KK (2016) Development of chromosome segment substitution lines (CSSLs) of *Oryza longistaminata* A. Chev. & Röhr in the background of the elite japonica rice cultivar, Taichung 65 and their evaluation for yield traits. Euphytica 210:151–163. https://doi.org/10.1007/s10681-016-1685-3

Ronald PC, Albano B, Tabien R, Abenes L, Wu KS, McCouch S, Tanksley SD (1992) Genetic and physical analysis of the rice bacterial blight disease resistance locus, Xa21. Mol Gen Genet MGG 236:113–120

Sacks EJ, Dhanapala MP, Tao DY, Cruz MTS, Sallan R (2006) Breeding for perennial growth and fertility in an *Oryza sativa*/*O. longistaminata* population. Field Crops Res 95:39–48. https://doi.org/10.1016/j.fcr.2005.01.021

Sakai H, Ikawa H, Tanaka T, Numa H, Minami H, Fujisawa M, Shibata M, Kurita K, Kikuta A, Hamada M, Kanamori H, Namiki N, Wu J, Itoh T, Matsumoto T, Sasaki T (2011) Distinct evolutionary patterns of *Oryza glaberrima* deciphered by genome sequencing and comparative analysis. Plant J Cell Mol Biol 66:796–805. https://doi.org/10.1111/j.1365-313X.2011.04539.x

Sanchez PL, Wing RA, Brar DS (2013) The Wild Relative of Rice: Genomes and Genomics. In: Zhang Q, Wing RA (eds) Genetics and genomics of rice. Springer, New York, pp 9–25

Schmit V (1996) Improving sustainability of the uplands through development of a perennial upland rice. In: Piggin C, Courtois B, Schmit V(eds) Upland rice research in partnership. International Rice Research Institute, Manila, Philippines, pp 265–273

Shen S, Xu G, Clements DR, Jin G, Zhang F, Tao D, Xu P (2016) Competitive and allelopathic effects of wild rice accessions (*Oryza longistaminata*) at different growth stages. Pak J Biol Sci 19:82–88. https://doi.org/10.3923/pjbs.2016.82.88

Song WY, Wang GL, Chen LL, Kim HS, Pi LY, Holsten T, Gardner J, Wang B, Zhai WX, Zhu LH, Fauquet C, Ronald P (1995) A receptor kinase-like protein encoded by the rice disease resistance gene, Xa21. Science 270:1804–1806

Soriano IR, Schmit V, Brar DS, Prot J-C, Reversat G (1999) Resistance to rice root-knot nematode *Meloidogyne graminicola* identified in *Oryza longistaminata* and *O. glaberrima*. Nematology 1:395–398. https://doi.org/10.1163/156854199508397

Sweeney M, McCouch S (2007) The complex history of the domestication of rice. Ann Bot 100:951–957. https://doi.org/10.1093/aob/mcm128

Tao D, Sripichitt P (2000) Preliminary report on transfer traits of vegetative propagation from wild rice species to *Oryza sativa* via distant hybridization and embryo rescue. Witthayasan Kasetsart Sakha Witthayasat 34:1–11

Vaughan DA (1994) The wild relatives of rice: a genetic resources handbook. IRRI, International Rice Research Institute, Manila

Vaughan DA, Sitch LA (1991) Gene flow from the jungle to farmers. Bioscience 41:22–28. https://doi.org/10.2307/1311537

Vaughan DA, Morishima H, Kadowaki K (2003) Diversity in the *Oryza* genus. Curr Opin Plant Biol 6:139–146. https://doi.org/10.1016/S1369-5266(03)00009-8

Vaughan DA, Ge S, Kaga A, Tomooka N (2008) Phylogeny and Biogeography of the Genus Oryza. In: Hirano PDH-Y, Sano PDY, Hirai PDA, Sasaki DT (eds) Rice biology in the genomics era. Springer, Berlin, pp 219–234

Waiho K, Fazhan H, Shahreza MS, Moh JHZ, Noorbaiduri S, Wong LL, Sinnasamy S, Ikhwanuddin M (2017) Transcriptome analysis and differential gene expression on the testis of Orange Mud Crab, *Scylla olivacea*, during sexual maturation. PLoS ONE 12: e0171095. https://doi.org/10.1371/journal.pone.0171095

Wambugu PW, Furtado A, Waters DL, Nyamongo DO, Henry RJ (2013) Conservation and utilization of African *Oryza* genetic resources. Rice. https://doi.org/10.1186/1939-8433-6-29

Wambugu PW, Brozynska M, Furtado A, Waters DL, Henry RJ (2015) Relationships of wild and domesticated rices (*Oryza* AA genome species) based upon whole chloroplast genome sequences. Sci Rep. https://doi.org/10.1038/srep13957

Wang M, Yu Y, Haberer G, Marri PR, Fan C, Goicoechea JL, Zuccolo A, Song X, Kudrna D, Ammiraju JSS, Cossu RM, Maldonado C, Chen J, Lee S, Sisneros N, de Baynast K, Golser W, Wissotski M, Kim W, Sanchez P, Ndjiondjop M-N, Sanni K, Long M, Carney J, Panaud O, Wicker T, Machado CA, Chen M, Mayer KFX, Rounsley S, Wing RA (2014) The genome sequence of African rice (*Oryza glaberrima*) and evidence for independent domestication. Nat Genet 46:982–988. https://doi.org/10.1038/ng.3044

Xu G, Shen S, Zhang F, Zhang Y (2014) Relationships among weed suppression effect, allelopathy and agronomic characteristics of *Oryza longistaminata* and related descendants. Chin J Eco-Agric 22:1348–1356

Xu P, Dong L, Zhou J, Li J, Zhang Y, Hu F, Liu S, Wang Q, Deng W, Deng X, Tharreau D, Yang Q, Tao D (2015) Identification and mapping of a novel blast resistance gene Pi57(t) in *Oryza longistaminata*. Euphytica 205:95–102. https://doi.org/10.1007/s10681-015-1402-7

Yang B, Thorogood D, Armstead IP, Franklin FCH, Barth S (2009) Identification of genes expressed during the self-incompatibility response in perennial ryegrass (*Lolium perenne* L.). Plant Mol Biol 70:709–723. https://doi.org/10.1007/s11103-009-9501-2

Yang H, Hu L, Hurek T, Reinhold-Hurek B (2010) Global characterization of the root transcriptome of a wild species of rice, *Oryza longistaminata*, by deep sequencing. BMC Genom 11:705. https://doi.org/10.1186/1471-2164-11-705

Zhang S, Wang W, Zhang J, Ting Z, Huang W, Xu P, Tao D, Fu B, Hu F (2014) The progression of perennial rice breeding and genetics. Perenn Crops Food Secur 27

Zhang Y, Zhang S, Liu H, Fu B, Li L, Xie M, Song Y, Li X, Cai J, Wan W, Kui L, Huang H, Lyu J, Dong Y, Wang W, Huang L, Zhang J, Yang Q, Shan Q, Li Q, Huang W, Tao D, Wang M, Chen M, Yu Y, Wing RA, Wang W, Hu F (2015) Genome and comparative transcriptomics of African wild rice *Oryza longistaminata* provide insights into molecular mechanism of rhizomatousness and self-incompatibility. Mol Plant 8:1683–1686. https://doi.org/10.1016/j.molp.2015.08.006

Zhanq F, Li T, Shan Q, Guo Y, Xu P, Hu F, Tao D (2008) Weed-suppression ability of *Oryza longistaminata* and *Oryza sativa*. FAO-China Seminar on Rice Allelopathy, 2007/10/9-2007/10/10, pp 345-351, Haikou, PEOPLES R CHINA, 10/2008, pp 345–351

Zhao J, Li J, Xu P, Zhou J, Hu F, Deng X, Deng W, Tao D (2012) A new gene controlling hybrid sterility between *Oryza sativa* and *Oryza longistaminata*. Euphytica 187:339–344. https://doi.org/10.1007/s10681-012-0691-3

16 Oryza meridionalis N.Q.Ng

Ali Mohammad Moner and Robert J. Henry

Abstract

Oryza meridionalis is an AA genome species found in Northern Australia. Phylogenetic analysis places this as the most distant of the AA genome species from domesticated rice (*Oryza sativa*). This makes it a key genetic resource for rice improvement. A draft nuclear genome sequence is available, and also the chloroplast genome has been sequenced from many genotypes. The high amylose starch content in these taxa may be useful for developing new rice grain characteristics. Here we have reviewed the all the research advancements that are made till today on this species.

16.1 Economic and Academic Importance

Oryza meridionalis is an Australian wild species of rice in the AA genome group which is close relatives of domesticated rice. The academic interest in this species is associated with it being the most distant from domesticated rice of the species within the AA genome group making it an important resource for the improvement of rice and to study the evolution of rice. Apart from high starch content in the grain (Tikapunya et al. 2017), this species is also known to tolerate high temperature stress (Scafaro et al. 2009), both these traits are useful for varietal improvement of rice.

16.2 Botanical Description and Distribution

Oryza meridionalis was recognised as a species by Ng et al. in 1981. It is found across the Northern Australia from the Kimberley region in Western Australia to Queensland (Fig. 16.1). *Oryza meridionalis* (Fig. 16.2) has also been reported from New Guinea. This is one of the four *Oryza* species, *O. meridionalis*, *Oryza rufipogon*, *Oryza officinalis,* and *Oryza australiensis*, that are reported in Australia (Henry et al. 2010). A summary of the morphological description from Flora of Australia (Kodela 2009) is tabulated below (Table 16.1). It can be distinguished from other *Oryza* species found in Northern Australia on the basis of the closed panicles and small anthers. *Oryza meridionalis* was originally described as an annual (Ng et al. 1981). The presence of populations with similar appearance but apparent perennial habit led to some uncertainty about the identity of these

A. M. Moner · R. J. Henry (✉)
Queensland Alliance for Agriculture and Food Innovation, University of Queensland, Brisbane, QLD 4072, Australia
e-mail: robert.henry@uq.edu.au

T. K. Mondal and R. J. Henry (eds.), *The Wild Oryza Genomes*,
Compendium of Plant Genomes, https://doi.org/10.1007/978-3-319-71997-9_16

perennial populations. Therefore, while *O. meridionalis*-like plants were designated as Taxa B yet, *O. rufipogon*-like plants were designated as Taxa B (Brozynska et al. 2014). Subsequent analysis (Moner et al. unpublished) has suggested that the Taxa B are all part of one clade supporting the description of *O. meridionalis* as an annual or perennial as in the Flora of Australia (Kodela 2009). Julia et al. (2016) reported details of the morphology of some ex situ collections of *O. meridionalis*. Some Herbarium samples of *O. meridionalis* may be labelled as *O. rufipogon* especially if collected before *O. meridionalis* was described.

16.3 Cytological Study

Oryza meridionalis is a diploid with a chromosome number of $2n = 24$.

16.4 Physiological Studies

The grain physical traits (Kasem et al. 2010, 2012; Tikapunya et al. 2016) and starch properties (Kasem et al. 2014; Tikapunya et al. 2017) have been investigated. Starch gene sequences were reported by Kasem et al. (2011). *Oryza meridionalis* has high amylose content

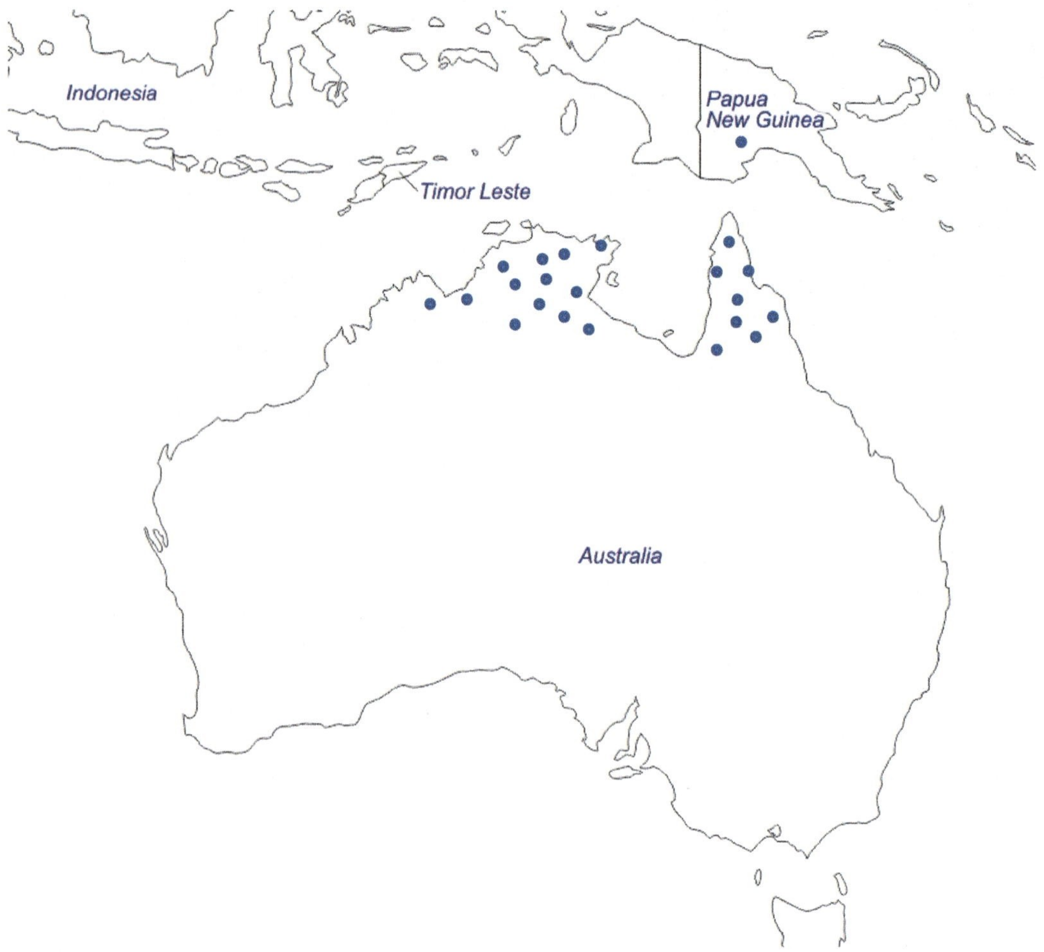

Fig. 16.1 Distribution of *O. meridionalis*

Fig. 16.2 Adult *O. meridionalis* plants growing in Northern Australia under natural condition

Table 16.1 Description of *O. meridionalis* (Kodela 2009)

Life cycle	Annual or perennial
Clums	0.3–2 m
Leaves	ligule 5–20(−30) mm, blade 6–47 cm long 4–14 mm wide
Panicles	9–30 cm long
Spikelets	6.5–10 mm long
Awn	(30−) 60–150 mm long
Anthers	1.3–2.5 (−3) mm long
Caryopsis	oblanceoloid or oboid-ellipsoidal laterally compressed (5−) 5.5–7.5 (−8.3) mm long

(around 35%) relative to domesticated rice (Tikapunya et al. 2017) but the genetic basis of this is not resolved. An understanding of the genetic basis of these starch properties could be useful in rice breeding.

16.5 Enumeration of Sequences

The genome has been sequenced using Illumina and PacBio sequencing techniques based upon 47.1 Gbp of shotgun Illumina sequence data and 15.0 Gbp of PacBio sequence data representing an estimated 127X and 41X coverage, respectively, of the estimated 370 Mbp genome (Brozynska et al. 2016).

16.6 Assembly

Brozynska et al. (2016) reported both hybrid (Illumina and PacBio) and PacBio only assemblies (Table 16.2). Hybrid assemblies covered 446 Mbp and PacBio alone, 355 Mbp of the genome.

16.7 Repetitive Sequences

Total repeats found in this species made up 36.5 and 46.4% of the Taxon A and Taxon B genomes, respectively. The most abundant group of transposable elements was found to be the *Gypsy*

Table 16.2 Hybrid and PacBio assembly statistics (Brozynska et al. 2016)

	Hybrid only	PacBio
Assembler	Sparse assembler + DBG2OLC	Celera assembler
Number of scaffolds	4718	3242
Total size of scaffolds	446,369,637	354,906,376
Longest scaffold	2,079,733	3,232,522
Mean scaffold size	94,610	109,135
N50 scaffold length	163,003	159,640
Number of contigs	4808	–
Total size of contigs	446,351,110	–
Longest contig	1,449,836	–
Median contig size	54,495	–
N50 contig length	159,759	–

family representing almost 40% of all repeats with *Copia* elements accounting for 9.3%. The classes and fractions of other repetitive elements were similar in both taxa; however, the numbers and lengths were significantly higher in the Taxon B genome (Brozynska et al. 2016).

16.8 Gene Annotation

Brozynska et al. (2016) identified 21,169 protein-encoding genes and 5,624 non-coding RNA genes (including 615 tRNA, 4,892 miRNA, 453 snoRNA, 87 sRNA and 129 rRNA). Improvements in that annotation should be possible with better assemblies and the availability of transcript sequences to confirm the predicted genes.

16.9 Organelle Genome

The complete chloroplast genome of *O. meridionalis* was reported by Nock et al. (2011). Wambugu et al. (2015) used the whole chloroplast genome sequence to relate *O. meridionalis* to other taxa. Some of the variation in the chloroplast genome within the species has been explored (Waters et al. 2012; Brozynska et al. 2014). Brozynska et al. (2014) reported a de novo assembled chloroplast genome of 134,557 bp. This showed 122–168 polymorphisms relative to the *O. sativa* reference and 41 polymorphisms within the populations. The mitochondrial genome has not been reported.

16.10 Impact on Plant Breeding Including Pre-breeding Work

Sanchez et al. (2013) produced hybrids between *O. sativa* and *O. meridionalis* that had heat and drought tolerance in extreme temperature conditions. Introgression from *O. meridionalis* into japonica cv. Taichung 65 leads to the identification of genes that control awn length on chromosomes 1, 4, 5. Awn length is controlled largely by a single dominant gene. However, other genes increase the expression and produce longer awns (Matsushita, et al. 2003). Arbelaez et al. (2015) reported introgression lines with *O. sativa* cv. Curinga as the recurrent parent.

16.11 Comparative Genomics

Comparison of the genome with that of domesticated rice by mapping of sequence reads suggests that *O. meridionalis* has more diversity in regions of the genome that lack variation in the domesticated rice gene pool (Krishan et al. 2014). The gene deserts in the domesticated rice gene pool (such as the large region on chromosome 5) were found to show significant diversity in this wild rice (Krishan et al. 2014). This suggests that useful genetic variation for rice breeders could be sourced from this species.

16.12 Future Prospects

Oryza meridionalis represents an important genetic resource for rice improvement providing a potential source of abiotic stress tolerance (Atwell et al. 2014) including heat tolerance (Scafaro et al. 2009, 2011, 2016). Photosynthesis traits may also be useful (Giuliani et al. 2013). Grain quality traits including starch properties may also add useful diversity to the rice gene pool. Further sequencing of this species will be of value (Henry 2014) especially to explore diversity within the species. This resource will be important in developing rice for production in new or altered environments (Henry et al. 2016).

References

Arbelaez JD, Moreno LT, Singh N, Tung C-W, Maron LG, Ospina Y, Martinez CP, Grenier C, Lorieux M, McCouch S (2015) Development and GBS-genotyping of introgression lines (ILs) using two wild species of rice, *O. meridionalis* and *O. rufipogon*, in a common recurrent parent, O. sativa cv. Curinga. Mol Breeding 35:81

Atwell BJ, Wang H, Scafaro AP (2014) Could abiotic stress tolerance in wild relatives of rice be used to improve *Oryza sativa*? Plant Sci 215–216:48–58

Brozynska M, Copetti D, Furtado A, Wing RA, Crayn D, Fox G, Ishikawa R, Henry RJ (2016) Sequencing of Australian wild rice genomes reveals ancestral relationships with domesticated rice. Plant Biotech J https://doi.org/10.1111/pbi.12674

Brozynska M, Omar ES, Furtado A, Crayn D, Simon B, Ishikawa R, Henry RJ (2014) Chloroplast genome of novel rice germplasm identified in Northern Australia. Tropical Plant Biol 7:111–120

Giuliani R, Koteyeva N, Voznesenskaya E, Evans MA, Cousins AB, Edwards GE (2013) Coordination of leaf photosynthesis, transpiration, and structural traits in rice and wild relatives (genus *Oryza*). Plant Physiol 162:1632–1651

Henry RJ (2014) Sequencing crop wild relatives to support the conservation and utilization of plant genetic resources. Plant Genetic Resour Charact Utilization 12:S9–S11. https://doi.org/10.1017/S1479262113000439

Henry RJ, Rice N, Waters DLE, Kasem S, Ishikawa R, Dillon SL, Crayn D, Wing R, Vaughan D (2010) Australian *Oryza*: utility and conservation. Rice 3:235–241

Henry RJ, Rangan P, Furtado A (2016) Functional cereals for production in new and variable climates. Curr Opin Plant Biol 30:11–18

Julia CC, Waters DL, Wood RH, Rose TJ (2016) Morphological characterisation of Australian *ex situ* wild rice accessions and potential for identifying novel sources of tolerance to phosphorus deficiency. Genetic Res Crop Evol 63:327–337

Kasem S, Waters DLE, Rice N, Shapter FM, Henry RJ (2010) Whole grain morphology of Australian rice species. Plant Genetic Resour Charact Utilization 8:74–81

Kasem S, Waters DLE, Rice N, Shapter FM, Henry RJ (2011) The endosperm morphology of rice and its wild relatives as observed by scanning electron microscopy. Rice 4:12–13

Kasem S, Waters DLE, Henry RJ (2012) Analysis of genetic diversity in starch genes in the wild relatives of rice. Tropi Plant Biol 5:286–308

Kasem S, Waters DLE, Ward RM, Rice N, Henry RJ (2014) Wild *Oryza* grain physico-chemical properties. Trop Plant Biol 7:13–18

Kodela PG (2009) Oryza. Flora of Australia, vol 44A. ABRS/CSIRO Australia, Melbourne, pp 361–368

Krishnan SG, Waters DL, Henry RJ (2014) Australian wild rice reveals pre-domestication origin of polymorphism deserts in rice genome. PLoS ONE 9:e98843

Matsushita S, Kurakazu T, Sobrizal DK, Yoshimura A (2003) Mapping of genes for awn in rice using *Oryza meridionalis* introgression lines. Rice Genet Newsl 20:17

Nock C, Waters DLE, Edwards MA, Bowen S, Rice N, Cordeiro GM, Henry RJ (2011) Chloroplast genome sequence from total DNA for plant identification. Plant Biotech J. 9:328–333

Sanchez PL, Wing RA, Brar DS. (2013) The wild relative of rice: genomes and genomics. In: Genetics and genomics of rice. Springer, Berlin, pp 9–25

Scafaro AP, Haynes PA, Atwell BJ. (2009) Physiological and molecular changes in *Oryza meridionalis* Ng., a heat-tolerant species of wild rice. J Exp Bot erp294

Scafaro AP, Von Caemmerer S, Evans JR, Atwell BJ (2011) Temperature response of mesophyll conductance in cultivated and wild *Oryza* species with contrasting mesophyll cell wall thickness. Plant Cell Env 34:1999–2008

Scafaro AP, Gallé A, Van Rie J, Carmo-Silva E, Salvucci ME, Atwell BJ (2016) Heat tolerance in a wild *Oryza* species is attributed to maintenance of Rubisco activation by a thermally stable Rubisco activase ortholog. New Phytolt 211:899–911

Tikapunya T, Fox G, Furtado A, Henry R (2016) Grain physical characteristic of the Australian wild rices. Plant Genetic Res 1–12

Tikapunya T, Zou W, Yu W, Powell P, Fox G, Furtado A, Henry RJ, Gilbert R (2017) Molecular structures and properties of starch of Australian wild rice. Carbohydrate Polymer 172:213–222

Wambugu PW, Brozynska M, Furtado A, Waters DL, Henry RJ (2015) Relationships of wild and domesticated rice species based on whole chloroplast genome sequences. Sci Rep 5:13957. https://doi.org/10.1038/srep13957

Waters DL, Nock CJ, Ishikawa R, Rice N, Henry RJ (2012) Chloroplast genome sequence confirms distinctness of Australian and Asian wild rice. Ecol Evol 2:211–217

Oryza meyeriana Baill

17

Kutubuddin Ali Molla, Subhasis Karmakar, Johiruddin Molla, T. P. Muhammed Azharudheen and Karabi Datta

Abstract

Wild relatives of cultivated rice (*Oryza sativa* and *Oryza glaberrima*) are treasure trove to the modern breeders as they contain valuable characteristics to improve cultivated rice specially to make them adaptable to climate change. *Oryza meyeriana* is one of the key species of Meyeriana complex of *Oryza* genus, mostly grown in South Asian countries, which is well known for its tolerance to many different biotic and abiotic stresses. The species was mostly explored for bacterial blight resistance. As the species contains GG genome, it is not readily crossable to the cultivated rice (AA). The chapter describes and discusses the probable reason for crossability barrier of *O. meyeriana* with *O. sativa,* way to overcome the barrier by asymmetric somatic hybridization (ASH), development of hybrid progenies and their utilization to map QTL and gene for bacterial blight resistance. All recent transcriptomic and proteomic studies with *O. meyeriana* to identify novel genomic resources for disease resistance are reviewed, and potential future prospects were mentioned.

17.1 Academic and Economic Importance

Rice is the most important cereal in terms of human consumption. *Oryza sativa* (Asian rice) and *Oryza glaberrima* (African rice) are the two *Oryza* species cultivated worldwide. However, there are around 24 other known species designated to *Oryza* genus. Those species are of tremendous potential to be used as valuable genetic resources for the improvement of cultivated rice. Due to the absence of distinct taxonomic key characters, sometimes it is very difficult to clearly set boundary lines among the members of a group of species; this group is called a species complex. Broadly, there are four different species complexes, viz. Sativa, Officinalis, Ridleyi and Meyeriana, designated under the genus *Oryza* (Tateoka 1962; Vaughan 1994). Names of a total of five species, viz. *Oryza granulata, Oryza meyeriana, Oryza indandamanica, Oryza abromeitiana* and *Oryza neocaledonica,* have

K. A. Molla (✉) · T. P.M. Azharudheen
ICAR-National Rice Research Institute, Cuttack 753006, Orissa, India
e-mail: kutubuddin.molla@icar.gov.in

S. Karmakar · J. Molla · K. Datta
Department of Botany, University of Calcutta, 35-Ballygunge Circular Road, Kolkata 700019, West Bengal, India

T. K. Mondal and R. J. Henry (eds.), *The Wild Oryza Genomes*,
Compendium of Plant Genomes, https://doi.org/10.1007/978-3-319-71997-9_17

been validly published for Meyeriana complex (Vaughan et al. 2008). We will discuss here about the species *O. meyeriana,* well known for its bacterial blight resistance properties.

O. meyeriana (Zoll. et Mor. ex Steud.) Baill is one of three native wild rice species in China, growing in the south-western Hainan Province and southern Yunnan Province. It belongs to the tertiary gene pool of cultivated rice (Bellon et al. 1998). It is a photoperiod-insensitive species which grows in grey sand, laterite soil and alluvium in full shade (Vaughan 1994).

17.2 Distribution

O. meyeriana is distributed in South Asian countries or countries of south-eastern Asia such as China, Indo-China, Indonesia, Malaysia, Philippines and Thailand (Fig. 17.1).

17.3 Importance

The species possesses many important traits, such as resistance to major diseases like rice blast, bacterial blight and sheath blight, tolerance to abiotic stresses like drought, shade, cold and barren soil, and it is well adapted to aerobic soil (Vaughan 1994). Specially, *O. meyeriana* has well known for its bacterial blight resistance and become subjected to study for the trait. It exhibited highest level of resistance against bacterial blight pathogen among 871 accessions of 13 different wild rice species (Zhang et al. 1994).

17.4 Botanical Descriptions

The plant (Fig. 17.2) is perennial with short rhizome and 60–70-cm-long-erect culms. Leaf sheaths are 6–8 cm long and striately veined.

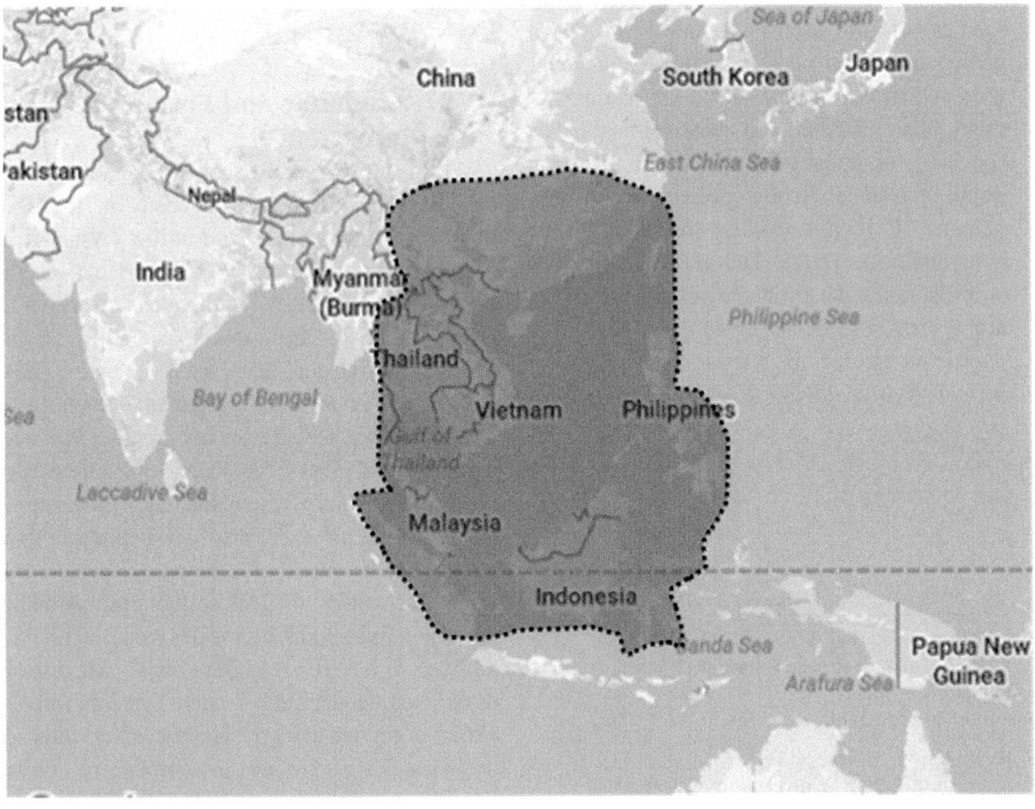

Fig. 17.1 Distribution of *O. meyeriana* species. The range of distribution is highlighted with dark grey colour and outlined with black dotted line

Leaf-blade base is broadly rounded with a brief petiole-like connection to sheath. Leaf-blades are linear, or lanceolate; 15–22 cm long; 16–20 mm wide.

Panicles are open, linear, 3–4 cm long, with few spikelets. Fertile spikelets are pedicelled. Pedicels are linear, angular, and bibracteate. Spikelets comprise two basal sterile florets and one fertile floret without rachilla extension. Spikelets are elliptic or ovate, laterally compressed and 5–5.5 mm long. Both glumes are absent or obscure. Basal sterile florets are similar, barren without significant palea. Fertile lemma is elliptic, laterally compressed, 5–5.5 mm long, coriaceous, keeled and 5-veined. Lemma surface is granulose. Lemma margins are interlocking with palea margins. Lemma apex is acute and muticous. Palea is elliptic, 5 mm long, coriaceous, 5-veined and 1-keeled. Palea surface is smooth. Palea apex is acute. Two membranous lodicules present. Anthers and stigma number 6 and 2, respectively. Fruit is a caryopsis with adherent pericarp. Disseminule comprises a floret (Clayton et al. 2010).

17.5 Genome Structure of *O. meyeriana* and Its Relationship with *O. sativa* Genome

O. meyeriana is a diploid wild rice species with GG genome (Aggarwal et al. 1997). Karyotype analysis using seed of *O. meyeriana* is difficult as it possesses early seed shattering and high seed sterility characters. Karyotype analysis using calli developed from in vitro culture of young panicle revealed $2n = 24$ chromosomes with metacentric (four pairs), submetacentric (seven pairs) and satellite-submetacentric (one pair) chromosomes (Wang et al. 1996). The average length of mitotic chromosome of *O. meyeriana* is 1.51–1.69 times longer than that of *O. sativa* (Xiong et al. 2006; Tan et al. 2006). *O. meyeriana* exhibited well-distinguished genetic distance with *O. granulata* and *O. indandamanica,* the other two species of the meyeriana complex (Agarwal et al. 1999). However, a biosystematic study by Gong et al. (2000) suggested that *O. meyeriana* and *O. granulata* should be combined into one species on the basis of very limited

Fig. 17.2 A pot grown tuft of *O. meyeriana* bail plants. "The photograph is kindly provided by Prof. Hong FENG, Molecular genetics, College of Life Sciences Sichuan University, China and Prof. Zai-Quan CHENG, Biotechnology and Genetic Resources Institute, Yunnan Academy of Agricultural Sciences, Yunnan, China"

reproductive isolation and very high genomic affinity between the two species. In order to study the cytological features of *O. meyeriana* (GG) and the genomic relationship between A and G genomes, Xiong et al. (2006) applied comparative GISH to distinguish between *O. sativa* and *O. meyeriana* chromosomes in their interspecific hybrid and analysed the spatial distribution and organization of the two genomes in interphase nuclei. A clear discrimination was observed between the chromosomes of *O. sativa* and *O. meyeriana* in the F_1 hybrid, and co-hybridization was barely detected. The finding of the study is in accordance with the consideration that *O. meyeriana* is the most distinct species from those with an AA genome (Ge et al. 1999). However, they have predicted a genome size of more than 727 Mb as *O. meyeriana* contains a large amount of heterochromatin (Xiong et al. 2006). In an another study using comparative genomic hybridization (CGH) and fluorescence in situ hybridization (FISH), it has been observed that among 1123-Mb probe signals obtained when *O. sativa* genomic DNA hybridized with *O. meyeriana*, 591 Mb came from *O. sativa* genomic DNA which are not repetitive (Lan et al. 2007). When labelled human $C_0t - 1$ DNA probe hybridized with *O. sativa* and *O. meyeriana* genomes, the coverage percentages were $\sim$47 and $\sim$44, respectively, indicating that repetitive sequences in *Oryza* genus were conserved as the functional genes during the evolution process and reduplication of repetitive sequence might be one of the reason of genome enlargement in *O. meyeriana* (Lan et al. 2007).

17.6 Impact on Plant Breeding Including Pre-breeding Work

As *O. meyeriana* is having many important characteristics including tolerance to many biotic and abiotic stresses, it is considered a precious genetic resource for rice improvement breeding programme. Generation of interspecific hybrids through hybridization combined with embryo rescue technique is a prerequisite for introgressing alien genes from the wild species to the cultivated one. However, a major bottleneck to transfer desired gene from wild rice genome to the cultivated one is the interspecific crossability barrier. Several attempts have been made to generate interspecific hybrids between *O. sativa* and *O. meyeriana,* but F_1 hybrids were completely sterile when hybrid embryos were cultured in vitro (He 1998; Tan et al. 2006; Fu et al. 2009). The reasons behind this severe crossability barrier between *O. sativa* and *O. meyeriana* were inviability of hybrid embryo, embryo development stagnation and degeneration (Fu et al. 2009). However, successful hybrid plants obtained when hybrid embryo was used as explants to develop calli and subsequent calli-mediated in vitro regeneration (Huang et al. 2001). Lin et al. (2008) reported regeneration of *O. meyeriana* plant from protoplast culture. Asymmetric somatic hybridization using protoplast fusion is another technique to avoid the problems related to "gene conflict" which may result in regeneration of hybrids with very low fertility, abnormal and slow hybrid growth, and recalcitrant calli or microcalli development (Eeckhaut et al. 2006; Shankar et al. 2013). In asymmetric fusion, only partial genome is transferred to the hybrid as the donor protoplast is fragmented before fusion. Use of asymmetric somatic hybridization yielded interspecific hybrids for *O. sativa* and *O. meyeriana* (Yan et al. 2004a, b; Zhu et al. 2004). In asymmetric hybridization, usually *O. meyeriana* protoplast (donor) is treated with X-ray for fragmentation, whereas *O. sativa* protoplast with iodoacetamide to inactivate the cytoplast. After many years of breeding, a near isogenic line (Y73) have been generated through asymmetric somatic hybridization between *O. meyeriana* and *O. sativa* L. ssp. japonica cv. Dalixiang, which showed resistance to bacterial blight pathogen *Xanthomonas oryzae* pv. *oryzae* (Wang et al. 2013). Similarly, a stable somatic hybrid line SH76 was also developed and it had been shown to exhibit wide-spectrum resistance against *Xoo* strains (Chu et al. 2008).

17.7 Genome Sequence

In order to use the potential of this wild rice species, more genomic information is required. Unfortunately, only few batches of mRNAs or full-length cDNAs are available in public databases for *O. meyeriana,* yet no genome sequence is available.

17.8 Transcriptomics

Transcriptomics are the study of whole set of RNA transcripts produced by a particular genome in a specific condition using high-throughput technology. Identification of differentially expressed genes in distinct populations in response to different treatments can be carried out by comparative transcriptome. Transcriptome study usually generates huge data, and survey of those data eases up the identification of putative novel genes/transcripts associated with disease and stress tolerance.

Recently, Illumina sequencing technology was used to analyse the first global transcriptome of *O. meyeriana* using root, stems and leaves (He et al. 2015). In the study, about 13% of the *O. meyeriana* contigs were found to be uniquely expressed when compared with genotypes of *O. sativa,* namely Nipponbare and 93–11. Genes for resistance to bacterial blight, blast, fusarium head blight, cyst nematode were mined from the *O. meyeriana* transcriptomic database. Moreover, they identified 52 unique contigs as disease resistance protein which was not present in *O. sativa*. Furthermore, the enrichment of the *O. meyeriana* transcriptome enabled them to identify 13 genes annotated to be involved in plant–pathogen interaction. Candidate-gene-based SSR (cgSSR) markers are highly useful for their greater linkage with important loci, suitability in functional diversity analysis and high transferability to related species (Molla et al. 2015). He et al. (2015) identified a total of 60,498 SSRs in 23.27% of *O. meyeriana* transcripts with highest percentage of trinucleotide repeats.

Unlike the study of transcriptome in native condition by He et al. (2015), early and late transcriptional responses of *O. meyeriana* inoculated with the bacterial blight pathogen *Xanthomonas oryzae* pv. *oryzae* (*Xoo*) have been investigated recently (Cheng et al. 2016). As an early response, a total of about 9561 unitranscripts were found to be differentially expressed (4263 down-regulated and 5298 up-regulated) when compared with the non-inoculated control, while as a late response the number of differentially expressed unitranscripts reduced to 6507 (3079 down-regulated and 3428 up-regulated) (Cheng et al. 2016). They also found 19 up-regulated DEGs in plant–pathogen interaction pathway at the early stage in response to *Xoo* inoculation. *Non-expressor of PR1* (*NPR1*), a key regulatory gene in defence signalling pathway and which provide enhanced tolerance in rice against *Xoo* (Chern et al. 2001) and sheath blight pathogen *Rhizoctonia solani* (Molla et al. 2016), was found to be up-regulated at the early stage in *O. meyeriana* (Cheng et al. 2016). Phenylalanine ammonia-lyase (PAL) mediates an important step in the shikimic acid pathway for the synthesis of phenolic compounds, including lignin, which play a role in the plant defence mechanism (Molla et al. 2013). All 14 transcripts of PAL were found to be up-regulated as late response in the study by Cheng et al. (2016). In *O. meyeriana*, two *Xa* genes (*xa13* and *Xa21*) exhibited differential expression after *Xoo* infection (Cheng et al. 2016), whereas expression of Xa1 and Xa26 was recorded in *O. meyeriana* transcriptome in native condition (He et al. 2015). In an another study, a near isogenic lines (Y73) developed through asymmetric somatic hybridization between *O. meyeriana* and *O. sativa* were subjected to transcriptome analysis after *Xoo* infection using Affymetrix Rice GeneChip microarrays, and only 0.22% of genes exhibited differential regulation when compared with control (Wang et al. 2013). The study also reported more than fivefold upregulation of NAC domain-containing protein, an elicitor-responsive protein involved in ubiquitination, and a glycosyl transferase gene, two transcription factors (TFs) and two unknown genes were up-regulated.

17.9 Proteomics Study

Study of proteomes, i.e. the whole set of proteins of a particular species or cell type, is known as proteomics study. Secreted proteins play significant role in the rice–Xoo interaction (Chatterjee et al. 2003). In order to analyse the secretome (secreted proteins) of *O. meyeriana* suspension culture in response to *Xoo* infection, a proteomic study has identified 34 proteins using 2D-DIGE coupled with MALDI-TOF (Chen et al. 2016). Those proteins have been predicted to be involved in different pathways like signal transduction, defence, cell wall modification and ROS Homeostasis. GDS-like lipase, a PP2C and a LysM receptor-like kinase were found to be up-regulated in *O. meyeriana* suspension culture in response to *Xoo* inoculation. *O. meyeriana* exhibited ~3 times higher peroxidase activity than susceptible rice variety Nipponbare, indicating a possible role of peroxidases in *O. meyeriana* in early response to *Xoo* (Chen et al. 2016). A total of 64 protein spots were identified by 2D/MS from a stable somatic hybrid line (SH76) of *O. meyeriana* and *O. sativa* after *Xoo* infection (Chu et al. 2008). Western blot analysis confirmed that intact Rubisco large subunit (RcbL) and RcbL big fragments degraded following *Xoo* attack (Chu et al. 2008). A complex regulation was found to be in existence in the somatic hybrid line SH76 which possibly increases resistance to *Xoo* (Chu et al. 2008). Similarly, Protein gel blot analysis and immuno-gold electron microscopy of *O. meyeriana* infected with *Xoo* indicated a possible role of Rubisco activase protein in protecting thylakoid membrane from *Xoo* damage (Yang et al. 2011).

Seed storage proteins of *O. meyeriana* were analysed, and its comparison with the protein profile of two cultivated and two wild rice varieties was done in a recent proteomic study (Jiang et al. 2014). Although in most cereal seeds, prolamines are the major storage proteins, and rice grains contain glutelin as the major storage protein. Glutelin accounts for 60–80% of total rice seed protein and, prolamine accounts for 20–30% (Kawakatsu et al. 2008). Jiang et al. (2014) found significantly higher content of total seed protein including glutelin in *O. meyeriana* than that of other two cultivated and two wild rice varieties. Hence, *O. meyeriana* could be used as potential genetic resources to improve seed protein content and quality of cultivated rice.

17.10 QTL and Gene Identification from *O. meyeriana*

Although *O. meyeriana* has long been known as one of the most important genetic resources for different abiotic and biotic stress tolerance specially bacterial blight, utilization of this material in rice breeding programme is delayed because of the inability to generate fertile hybrid by conventional crossing method between cultivated rice (AA) and *O. meyeriana* (GG). However, continuous effort has been made to overcome the sexual incompatibility through asymmetric somatic hybridization, and as a result, hybrid progenies like Y73 (Wang et al. 2013), SH76 (Chu et al. 2008) and ASH1 (Han et al. 2014) were generated. Generation of mapping population is a prerequisite to carry out any QTL/gene mapping study. Asymmetric hybridization paves the way to identify important trait-related QTL contributed by *O. meyeriana* (Table 17.1). ASH1 exhibited high level of bacterial blight resistance inherited from *O. meyeriana*. Han et al. (2014) mapped quantitative trait loci (QTLs) for bacterial blight resistance in ASH1 using bulked segregant analysis (BSA) method. They have used an F2 population of 101 plants developed from self-pollinations of the F_1 plants derived from ASH1 and IR24 (susceptible) and 17 polymorphic SSR/STS markers to identify 3 QTLS, viz. *qBBR1*, *qBBR3* and *qBBR5* with 21.5, 12.3 and 39.2% variance, respectively. The identified QTLs were not found similar with the previously identified QTLs for bacterial blight (Han et al. 2014). Similarly, a RIL population was developed from a cross of Y73 and IR24, and a genetic linkage map was constructed with 155 SSR and 56 STS markers (Yang et al. 2012). Utilizing the RIL population, Chen et al. (2012) mapped 5 QTLs for bacterial blight resistance on chromosome number 1, 3, 5, 10 and 11. Two QTLs, *qBBR5* and *qBBR1,* have been found to explain 37–26% phenotypic variance respectively, whereas other three were minor explaining 6–14%. In addition, a total of 21 QTLs for eight different agronomic traits have also been identified in the same study. One QTL for plant height, *qPH1* and one QTL for heading date, *qHD11* were co-localized with *qBBR1* and *qBBR11,* respectively. Allele from Y73 delayed HD and decreased plant height at qPH1 and qPH2.

Till date, only a single gene has been identified from *O. meyeriana.* First clue for existence of novel bacterial blight resistance gene came up in a study conducted by Huang et al. (2008) when they could not find any known gene responsible for the resistance in two germplasms, SH5 and SH76, derived from somatic hybridization of *O. meyeriana* and cultivated rice. A new resistance gene named as *xa32(t)* has been identified from somatic hybrid generation derived from *O. meyeriana* and mapped on long arm of chromosome 12 at a position of 1.7 cM from the marker RM20A (Ruan et al. 2008).

17.11 Structural and Functional Genomics Resources Developed

Functional and structural genomics study in *O. meyeriana* is scanty and need to be explored. Serine/threonine protein kinase gene and nucleotide binding sites and leucine-rich repeats (NBS-LRR) containing genes were sequenced from *O. meyeriana* (Liu et al. 2003). A suppression subtractive hybridization study of *O. meyeriana* inoculated with *Xoo* identified induction of genes responsible for cell wall lignifications and secondary metabolism indicating important role of lignifications to restrict the bacterial invasion (Cheng et al. 2010). Also one leucine-rich repeat containing gene ME281 was found to exhibit peak expression at 48–72 h post-inoculation in the study suggesting its role in the late response to *Xoo* infection.

17.12 Physiological and Anatomical Studies

Physiological and anatomical studies are done in *O. meyeriana* to investigate the basis of its resistance towards bacterial blight pathogen *Xanthomonas oryzae* pv. *oryzae.* Inhibition of chloroplast deformation, steady rate of photosynthetic activity and induced thickening of secondary wall of xylem were found to be the probable basis of high level of resistance of *O. meyeriana* to *Xoo* (Yang et al. 2012). However, large numbers of *Xoo* bacterial cells were found in the xylem vessel of *O. meyeriana,* suggesting that the pathogen can enter and reproduce in the vessel but the wall thickening restricts them from further invasion into adjacent cell (Yang et al. 2012). Similar result of induction of secondary wall thickening in the wild rice species in response *Xoo* infection was obtained by Cheng et al. (2014). In addition, nitric oxide was found to play important role in resistance as evidenced by a post-inoculation elevation in nitric oxide accumulation in xylem cell wall of *O. meyeriana* (Cheng et al. 2014).

17.13 Future Prospects

In spite of having many important characters such as blast, bacterial blight and sheath blight resistance, drought, shade, cold tolerance and being well known for adaptation to aerobic soil, the potential of *O. meyeriana* is only explored for bacterial blight resistance. In a changing scenario of world climate, other characters such as drought and cold tolerance, sheath blight tolerance for which no known resistant rice germplasms are identified, the species could be utilized further in rice improvement programme. *O. meyeriana* could be proved as an excellent

Table 17.1 QTLs and gene identified from *O. meyeriana*

Name of QTLs/gene	Chromosome	Nearest/flanking markers	Mapping population (Resistant × susceptible)	LOD value/genotypic variance	Correlated traits	References
qBBR1	1	RM128–R01D144	101 F_2 population derived from a cross of ASH1 (somatic hybrid progeny of *O. meyeriana* × *O. sativa* cv. Dalixiang) and *O. sativa* cv. IR24	10.1	BB—bacterial blight	Han et al. (2014)
qBBR3	3	R03D158–RM85		6.2		
qBBR5	5	RM7081–RM3616		17.8		
qBBR1	1	R01D124–RM1361	112 F_7 RIL derived from a cross of Y73 (somatic hybrid progeny of *O. meyeriana* × *O. sativa*) and *O. sativa* pv IR24	13.60	BB—bacterial blight PH—plant height HD—heading date PNPP—panicle number per plant SPP—spikelets per panicle GPP—grains per panicle SSR—seed setting rate TGW-1000-grain weight PYD—plant yield	Chen et al. (2012)
qBBR3	3	R03D143–R03D159		8.12		
qBBR5	5	RM7081–RM233B		16.90		
qBBR10	10	R10D45–RM239		3.63		
qBBR11	11	RM3428–RM287		4.91		
qPH1	1	RM128–RM5310		3.11		
qPH2	2	RM6465–R02D144		2.85		
qPH6	6	RM5814–RM439		4.87		
qPH11	11	R11D80–RM254		6.25		
qHD6a	6	RM3805–RM510		3.50		
qHD6b	6	RM510–RM50		2.63		
qHD7	7	R07D44–R07D68		3.36		
qHD8a	8	RM3572–R08D50		3.44		
qHD8b	8	R08D54–RM72		2.46		
qHD11	11	RM3428–RM5349		4.20		
qPNPP3	3	RM6349–RM6301		4.18		
qPNPP7	7	RM5344–R07D25		2.75		
qPNPP8	8	R08D17–RM3572		3.63		
qSPP6	6	RM7434–RM162		2.93		
qGPP2	2	R02D23–R02D55		1.89a		
qGPP6	6	RM585–RM539		1.93a		
qSSR1	1	RM5310–R01D182		1.72a		
qSSR8	8	RM152–RM1376		1.73a		
qTGW3a	3	RM16–R03D146		2.98		
qTGW3b	3	RM168–RM514		2.89		
qTGW8	8	RM284–RM230		4.65		
qTGW9	9	R09D75–R09D92		3.02		
qTGW11	11	RM7557–R11D24		2.25		
qTGW12	12	RM7619–R12D51		7.19		
qPYD6	6	RM50–R06D51		2.16		
Xa32 (t)	12	RM20A	Somatic hybrid progeny of *O. meyeriana* × *O. sativa* cv. Dalixiang			Ruan et al. (2008)

resource for improving different important characters of cultivated rice once the concern related to the crossability barrier is addressed.

References

Aggarwal RK, Brar DS, Khush GS (1997) Two new genomes in the *Oryza* complex identified on the basis of molecular divergence analysis using total genomic DNA hybridization. Mol Gen Genet 254(1):1–12

Aggarwal RK, Brar DS, Nandi S, Huang N, Khush GS (1999) Phylogenetic relationships among *Oryza* species revealed by AFLP markers. Theor Appl Genet 98 (8):1320–1328

Bellon MR, Brar DS, Lu BR, Pham JL, Dowling NGGSM (1998) Rice genetic resources. Sustain Rice Glob Food Syst 16:251–283

Chatterjee S, Sankaranarayanan R, Sonti RV (2003) PhyA, a secreted protein of *Xanthomonas oryzae* pv. *oryzae*, is required for optimum virulence and growth on phytic acid as a sole phosphate source. Mol Plant-Microbe Interact 16(11):973–982

Chen LN, Yang Y, Yan CQ, Wang XM, Yu CL, Zhou J, Zhang WL, Cheng Y, Cheng XY, Chen JP (2012) Identification of quantitative trait loci for bacterial blight resistance derived from *Oryza meyeriana* and agronomic traits in recombinant inbred lines of *Oryza sativa*. J Phytopathol 160(9):461–468

Chen X, Dong Y, Yu C, Fang X, Deng Z, Yan C, Chen J (2016) Analysis of the proteins secreted from the *Oryza meyeriana* Suspension-cultured cells induced by *Xanthomonas oryzae* pv. *oryzae*. PLoS One 11(5): e0154793

Cheng Z, Tang K, Yan H, Fu J, Ying F, Huang X (2010) Analysis of differentially expressed genes in *Oryza meyeriana* in response to infection by *Xanthomonas oryzae* pv *Oryzae*. Plant Breed 129(4):393–399

Cheng X, Wang X, Yang Y, Zhou J, Yu C, Han X, Cheng Y, Yan C, Chen J (2014) Nitric oxide involved in defense responses of *Oryza meyeriana* to rice bacterial blight. Acta Agric Zhejiangensis 26(1):0–6

Cheng XJ, He B, Chen L, Xiao SQ, Fu J, Chen Y, Yu TQ, Cheng ZQ, Feng H (2016) Transcriptome analysis confers a complex disease resistance network in wild rice *Oryza meyeriana* against *Xanthomonas oryzae* pv. *oryzae*. Sci Rep 6:38215

Chern MS, Fitzgerald HA, Yadav RC, Canlas PE, Dong X, Ronald PC (2001) Evidence for a disease-resistance pathway in rice similar to the NPR1-mediated signaling pathway in Arabidopsis. Plant J 27(2):101–113

Chu LY, Yan SP, Wang CC, Hu HT, Sun WN, Yan CQ, Chen JP, Yang L (2008) Pathogenesis-related proteins in somatic hybrid rice induced by bacterial blight. Phytochemistry 69(10):1989–1996

Clayton WD, Vorontsova M, Harman KT, Williamson H (2010) GrassBase—the online world grass flora. *Oryza*. RBG, Kew

Eeckhaut T, Van Laere K, De Riek J, Van Huylenbroeck J (2006) Overcoming interspecific barriers in ornamental plant breeding. Floriculture, ornamental and plant biotechnology: advances and topical issues, 1st edn. Global Science Books, London, pp 540–551

Fu XL, Lu YG, Liu XD, Li JQ (2009) Crossability barriers in the interspecific hybridization between *Oryza sativa* and *O. meyeriana*. J Integr Plant Biol 51 (1):21–28

Ge S, Sang T, Lu BR, Hong DY (1999) Phylogeny of rice genomes with emphasis on origins of allotetraploid species. Proc Natl Acad Sci 96(25):14400–14405

Gong Y, Borromeo T, Lu BR (2000) A biosystematic study of the *Oryza meyeriana* complex (Poaceae). Plant Syst Evol 224(3):139–151

Han X, Yang Y, Wang X, Zhou J, Zhang W, Yu C, Cheng C, Cheng Y, Yan C, Chen J (2014) Quantitative trait loci mapping for bacterial blight resistance in rice using bulked segregant analysis. Int J Mol Sci 15(7):11847–11861

He GC (1998) Cellular and molecular approaches in exploitation of useful genes in wild rice. Prog Biotechnol 18(2):41–45

He B, Tao X, Gu Y, Wei C, Cheng X, Xiao S, Cheng Z, Zhang Y (2015) Transcriptomic analysis and the expression of disease-resistant genes in *Oryza meyeriana* under native condition. PLoS ONE 10(12): e0144518

Huang YL, Shu LH, Zhu LL (2001) Characteriztion of the intersectional hybrid between *Oryza sativa* and *O. meyeriana* indigenous to China. J Wuhan Univ Nat Sci Ed 46:739–744

Huang JN, Wang CC, Hu HT, Ma BJ, Yan CQ, Yang L (2008) Primary identification of a new resistance gene to bacterial blight from *Oryza meyeriana*. Chin J Rice Sci 22(1):33–37

Jiang C, Cheng Z, Zhang C, Yu T, Zhong Q, Shen JQ, Huang X (2014) Proteomic analysis of seed storage proteins in wild rice species of the *Oryza* genus. Proteome Sci 12(1):51

Kawakatsu T, Yamamoto MP, Hirose S, Yano M, Takaiwa F (2008) Characterization of a new rice glutelin gene GluD-1 expressed in the starchy endosperm. J Exp Bot 59(15):4233–4245

Lan WZ, He GC, Wang CY, Wu SJ, Rui Q (2007) Comparative analysis of genomes in *O. sativa*, *O. officinalis*, and *O. meyeriana*, with $C_0t - 1$ DNA and genomic DNA of cultivated rice. Agric Sci China 6 (9):1027–1034

Lin ZL, Bai XG, Lei LG, Wei Wei LI, Yin FY, Huang XQ et al (2008) Establishment of embryogenic cell suspension cultures and plants regeneration from protoplast culture of wild rice *Oryza meyeriana*. 44 (6):1181–1184

Liu JM, Cheng ZQ, Yang MZ, Wu CJ, Wang LX, Sun YD, Huang XQ (2003) Cloning and sequence analysis of disease resistance gene analogues from three wild rice species in Yunnan. Agric Sci China 2 (3):265–272

Molla KA, Karmakar S, Chanda PK, Ghosh S, Sarkar SN, Datta SK, Datta K (2013) Rice oxalate oxidase gene driven by green tissue-specific promoter increases tolerance to sheath blight pathogen (*Rhizoctonia solani*) in transgenic rice. Mol Plant Pathol 14 (9):910–922

Molla KA, Debnath AB, Ganie SA, Mondal TK (2015) Identification and analysis of novel salt responsive candidate gene based SSRs (cgSSRs) from rice (*Oryza sativa* L.). BMC Plant Biol 15(1):122

Molla KA, Karmakar S, Chanda PK, Sarkar SN, Datta SK, Datta K (2016) Tissue-specific expression of Arabidopsis NPR1 gene in rice for sheath blight resistance without compromising phenotypic cost. Plant Sci 250:105–114

Ruan HH, Yan CQ, An DR, Liu RH, Chen JP (2008) Identifying and mapping new gene xa32 (t) for resistance to bacterial blight (*Xanthomonas oryzae* pv. *oryzae*, Xoo) from *Oryza meyeriana* L. Acta Agric Bor Sin 17:170–174

Shankar LP, Tom E, Dieter D (2013) Asymmetric somatic plant hybridization: status and applications. Am J Plant Sci 4:1–10

Tan GX, Xiong ZY, Jin HJ, Li G, Zhu LL, Shu LH, He GC (2006) Characterization of interspecific hybrids between *Oryza sativa* L. and three wild rice species of China by genomic in situ hybridization. J Integr Plant Biol 48(9):1077

Tateoka T (1962) Taxonomic studies of Oryza II Several species complexes. Bot Mag Tokyo 75:455–461

Vaughan DA (1994) The wild relatives of rice: a genetic resources handbook. International Rice Research Institute, Manila, p 50

Vaughan DA, Ge S, Kaga A, Tomooka N (2008) Phylogeny and biogeography of the genus Oryza. In: Hirano H-Y et al (eds) Rice biology in the genomics era. Springer, Berlin, pp 219–234

Wang XL, Shu LH, Yuan WJ, Liao LJ (1996) Panicle culture and karyotype analysis from callus cells of a diploid wild rice *Oryza meyeriana*. International Rice Research Notes, Philippines

Wang XM, Zhou J, Yang Y, Yu FB, Chen J, Yu CL, Wang F, Cheng Y, Yan CQ, Chen JP (2013) Transcriptome analysis of a progeny of somatic hybrids of cultivated rice (*Oryza sativa* L.) and wild rice (*Oryza meyeriana* L.) with high resistance to bacterial blight. J Phytopathol 161(5):324–334

www.knowledgebank.irri.org

Xiong ZY, Tan GX, He GY, He GC, Song YC (2006) Cytogenetic comparisons between A and G genomes in *Oryza* using genomic in situ hybridization. Cell Res 16(3):260–266

Yan CQ, Qian KX, Xue GP, Wu ZC, Chen YL, Yan QS, Zhang XQ, Wu P (2004a) Production of bacterial blight resistant lines from somatic hybridization between *Oryza sativa* L. and *Oryza meyeriana* L. J Zhejiang Univ Sci 5(10):1199–1205

Yan CQ, Qian KX, Yan QS, Zhang XQ, Xue GP, Huangfu WG, Wu YF, Zhao YZ, Xue ZY, Huang J, Xu GZ, Wu P (2004b) Use of asymmetric somatic hybridization for transfer of the bacterial blight resistance trait from *Oryza meyeriana* L. to *O. sativa* L. ssp. japonica. Plant Cell Rep 22(8):569–575

Yang Y, Chen LN, Yan CQ, Cheng Y, Cheng XY, Chen JP (2012a) Construction of a genetic linkage map of a bacterial blight resistance rice line derived from *Oryza Meyeriana* L. J Zhejiang Agric Coll 24 (5):846–852

Yang Y, Xie L, Yan CQ, Wang XM, Yu CL, Cheng XY, Cheng Z, Chen JP (2012b) Xylem secondary cell-wall thickening involved in defense responses of *Oryza meyeriana* to *Xanthomonas oryzae* pv. *oryzae*. Acta Phytopathol Sin 42:505–514

Yang Y, Yu CL, Wang XM, Yan CQ, Cheng Y, Chen JP (2011) Inoculation with *Xanthomonas oryzae* pv. *oryzae* induces thylakoid membrane association of Rubisco activase in *Oryza* meyeriana. J Plant Physiol 168(14):1701–1704

Zhang Q, Wang CL, Shi AN, Bai JF, Ling SC, Li DY, Chen CB, Pang HH (1994) Evaluation of resistance to bacterial blight (*Xanthomonas oryzae* pv. *oryzae*) in wild rice species. Sci Agric Sinica 27(5):1–9

Zhu Y, Chen B, Yu S, Zhang D, Zhang X, Yan Q (2004) Transfer of bacterial blight resistance from *Oryza meyeriana* to *O. sativa* L. by asymmetric somatic hybridization. Chin Sci Bull 49(14):1481–1484

Oryza minuta J. Presl. ex C. B. Persl

18

Walid Hassan Elgamal
and Mostafa Mamdouh Elshenawy

Abstract

Oryza minuta is an allotetraploid with 48 chromosomes of genome type (BBCC) belonging to the *Oryza officinalis* complex species. It has a genome size of 1124 Mb, the second largest among the *Oryza* species. *O. minuta* is found in southeast Asia in semi-shaded or occasionally shaded locations, along streams and in other wet areas where soil is clay or loamy. It is not economically important but has an academic importance due to its biotic and abiotic stresses tolerance characters such as resistance to brown planthopper, white-backed planthopper, green leafhopper, yellow stem borer, bacterial blight, sheath blight, and blast. This provides a gene pool that could be used for improving rice cultivars. In addition, many genes related to yield and other agronomic traits have been identified in this species. In the present chapter, we have reviewed the work related to breeding and biotechnology of this species.

W. H. Elgamal (✉) · M. M. Elshenawy
Rice Research and Training Center, Agricultural Research Center, Field Crops Research Institute, Gizah, Egypt
e-mail: elgamal.rrtc@gmail.com

W. H. Elgamal
Institute of Food Crops, Yunnan Academy of Agricultural Sciences, Kunming, China

18.1 Academic and Economic Importance

Oryza minuta is a wild species that is not cultivated, so it has no direct economic importance. However, it has academic importance due to the presence of tolerant alleles for various pests and diseases and ability to survive under extreme environmental conditions. Thus, it could serve as a source of genes that could be used in improving cultivated rice varieties (Jena 2010). It exhibits the potential to resist blast, bacterial blight (BB), white-backed plant hoper (WBPH) and brown planthopper (BPH) infections. Furthermore, various resistance genes have been transferred successfully from *O. minuta* to cultivated rice (Rahman et al. 2009; Amante et al. 1992).

18.2 Origin, Distribution, and Taxonomy

All information suggests that Philippines is the main center of origin of *O. minuta*, but the species also extends to Thailand and the Fly River Delta, Papua New Guinea, (Fig. 18.1). It grows in semi-shaded or occasionally shaded forest margins and sago swamps, along streams and in other wet areas where the soil is clay or loamy (Ackerman 2004; Linh et al. 2008). *O. minuta* is classified as part of the *O. officinalis* complex with an allotetraploid (BBCC) genome

T. K. Mondal and R. J. Henry (eds.), *The Wild Oryza Genomes*,
Compendium of Plant Genomes, https://doi.org/10.1007/978-3-319-71997-9_18

(2n = 4x = 48 chromosomes, Ge et al. 1999). A phylogenetic analysis based upon *Adh* and *matk* has determined the origins of allotetraploid genomes of *Oryza* and suggested that these genomes originated at different times. The BC genome originated most recently, while the HJ and HK genomes were the most ancient (Khush et al. 2001). The allotetraploid BBCC genome is located between the BB and CC genomes which are considered maternal parents in the phylogenetic analysis (Ge et al. 1999). The *officinalis* complex, including *O. minuta*, is closely related to the *sativa* complexes including *O. sativa*. *O. minuta* is closer to *O. malampuzhaensis* and *O. punctata* than to other members of the *O. officinalis* complex. The same phylogenetic conclusions were suggested by both SSR analysis and AFLP analysis (Aggarwal et al. 1999; Tomatro et al. 2005; Paweena et al. 2015). A taxomonic description is given below (Table 18.1).

18.3 Botanical Descriptions

A brief morphological description of various parts of the plants is given here. Roots have strong secondary fibrous roots, culm: The outside tillers are scrambling while the interior tillers are ascending with (50–100 cm) height, (Fig. 18.2) panicle: Open ovate panicles (8–17 cm) long without branches on 1/3 basal part, each panicle has from 6 to 20 fertile spikletes. *O. minuta* has a loopy leaf blade of 14–28 cm long, 6–13 mm width, and rough on both sides. The leaf sheaths are smooth, glabrous on the surface. Auricles are absent. Ligule has a ciliate membrane; 3–5 mm long and obtuse erose. Flower lodicules; two membranous, six anthers with 2 mm long, and two dark brown or black stigmas. Fertile spikelets (seeds) comprise two basal sterile florets and one fertile floret. Spikelets are ovate, lightly compressed about 3.8–4 mm long and 1.75 mm width, easy falling. Awns length are ranged between 3 mm and 12 mm; glumes are absent or obscure (Clayton et al. 2006). Pollen morphology of *O. minuta* is described as granulose pollen with size axis; 33–45 μm, pore diameter; 3.75–4 μm, annulus diameter is 9.1 μm, and exine with a thickness of 1.1 μm (Chaturvedi et al. 1998).

18.4 In Vitro Culture

Tissue culture plays an important role for all wild species, because this technique is used for overcoming the difficulties of interespecific

Fig. 18.1 Geographical distribution of *O. minuta* showed by dark spot points, the wild specie *O. minuta* distributing in Philippines, Papua New Guinea, and Thailand

Table 18.1 Taxonomic classification of *O. minuta*

Kingdom	Plantae (plants)
Subkingdom	Tracheobionta (vascular plants)
Super division	Spermatophyte (seed plants)
Division	Magnoliophyta (flowering plants)
Class	Liliopsida (monocotyledons)
Subclass	Commelinidae
Order	Cyperales
Family	Poaceae (Grass family)
Genus	*Oryza L.* (rice)
Species	*Oryza minuta*

Fig. 18.2 An adult plants of *O. minuta* [*photograph courtesy* Dr. C. Gireesh, Scientist, IIRR, India]

hybridization by embryo rescue or cell culture after transformations. *O. minuta* has a very low ability of callus induction from anthers (Tang et al. 1998). The optimum time for embryo rescue after crossing between *O. minuta* and *O. sativa* ranged between 9 and 14 days after pollination. The regenerated plantlets are completely sterile, so using colchicine for chromosome pairing was attempted (Suputtitada et al. 1994; Rodrangboon et al. 2002). After crossing between *O. minuta* and *O. sativa* cultivars, the seed set ranged between 9 and 20% and the F_1 plants had a similar normal domesticated plant type except the panicles and seeds which were similar to the wild-type species (Pongtongkam et al. 1993).

18.5 Chloroplast and Mitochondria

The study of mitochondria and chloroplast genetics is useful in determining maternal genetic effects, in phylogenetic studies and to identify non-nuclear genes for various traits. The complete mitochondrial genome of *O. minuta* has been sequenced. The size of the mtDNA genome is 515,022 bp, containing 60 protein-coding genes, 31 tRNA genes, and two rRNA genes. The *O. minuta* mitochondrial genome organization and the gene content at the nucleotide level are highly similar (89%) to that of *O. rufipogon*. Comparison with other related species revealed that most of the genes with known function are conserved among the Poaceae members. Similarly, the *O. minuta* mt genome shared tween four protein-coding genes, fifteen tRNA genes, and one rRNA gene with domesticated rice. Phylogenetic analysis of evolutionary relationships revealed that *O. minuta* is more closely related to *O. rufipogon* than to any other related species (Asaf et al. 2017). Within the genomes of B, BC, C and CD, RFLP patterns were similar to each other and showed a closer affinity except for *O. minuta* (BBCC). Within the BC genome species, the patterns of *O. punctata* and *O. minuta* were largely different from each other and separated into two different subclusters. Thus, the mitochondrial genomes of the two BC species (*O. punctata* and *O. minuta*) apparently evolved independently (Abe et al. 1999).

The *O. minuta* chloroplast genome also has been sequenced recently. The size of mtDNA genome is 135,094 bp, containing 91 protein-coding genes, 40 tRNA genes, and 8 rRNA genes (Asaf et al. 2017). The *O. minuta* cp genome has been used in many studies to identify the relationship between *Oryza* species and other species in the Poaceae. Using microsatellite markers in genetic analysis showed that *O. minuta* is more closely related to *O. officinalis* CC genome, *O. latifolia* CCDD genome, and *O. sativa,* respectively, and has along a genetic distance from *O. ridleyi* which is closer to another species such as maize, barley, and wheat (Ishii and Mc Couch 2000). Both the chloroplast and mitochondria give us more information about relationships between *Oryza* species and maternal resources and indicate that species with the CC genome are the maternal parents for BBCC genome species such as *O. minuta* and also for CCDD genome species like *O. latifolia* (Nishikawa et al. 2005).

18.6 Cytogenetic and Karyotype

O. minuta is wild relative of rice in the genus *Oryza* and belongs to the *O. officinalis* complex of species in the Poaceae family. It is an allotetraploid with 48 chromosomes and the genome type is BBCC. *O. minuta* has a genome size of 1124 Mb. This is the second largest genome in the *Oryza* species. Cytogenetic and karyotype studies of the *O. minuta* genome constitution by Genome In Situ Hybridization (GISH) indicated that *O. minuta* was an allotetraploid with clearly distinct BBCC genomes (Fukuii et al. 1997; Li et al. 2001)

Karyotypic differences were found among the tetraploid rice species (Nandi 1936, 1938). Karyotype studies showed that differences in chromosome size between the B and C genomes were obvious in *O. minuta*, as the chromosomes of the C-genome are larger than the chromosomes of the B-genome. Thus, the ratio of the total chromosome lengths of the BB to CC genomes is about 1:1.5 in *O. minuta* (Li et al. 2001). Subsequently, a two-step in situ hybridization procedure including GISH and FISH was used to distinguish the A, B, and C genomes in a spontaneous hybrid between *O. sativa* and *O. minuta* to investigate chromosome pairing at meiotic metaphase 1 in the pollen mother cells of the hybrid. As expected in most intergenomic hybrids, meiotic chromosome pairing in the *O. sativa* × *O. minuta* spontaneous hybrid was irregular, with mostly univalents, a low frequency of bivalents and occasional trivalents. However, A, B, and C genomes could not be distinguished from each other. In addition, the occurrence of a low frequency of bivalents in the spontaneous hybrid indicates that there was a possibility for gene transfer from the *O. minuta* genome to the *O. sativa* genome through genetic recombination (Mariam et al. 1996; Yi et al. 2008). The cytological analysis of the interespecific hybrid between *O. sativa* AA genome and *O. minuta* BBCC genome showed irregular meiosis with predominant occurrence of univalents and unequal distribution of chromosomes at anaphase one of meiosis (Eung-Gi et al. 2005).

18.7 Isozyme

Isozyme markers have been important in detecting the genes for resistance to biotic stresses in *O. minuta* and integration lines with resistance genes to blast(*Magnaporthe grisea*), bacterial blight (*Xanthomonas oryzae pv.oryza*), and brown planthopper (*Nilaparvata lugens*) in addition to supporting taxonomic studies and classification in diversity studies (Chee et al. 2006). The mitogen-activated protein kinas (MAPK) signaling cascade is critical for regulating plant defense systems against various kinds of pathogens and environmental stresses. One component of this cascade, the MAP kinase kinases (MAPKK), has not previously been shown to be induced in plants following biotic attacks, such as those by insects and fungi. A gene coding for a blast and brown planthopper-responsive putative MAPK kinase, OmMKK1 (*O. minuta* MAPKK 1), was identified in a library of *O. minuta* expressed sequence tags (ESTs). Two copies of OmMKK1 are present in the *O. minuta* genome. They encode a predicted protein with molecular mass 39 kDa and pI of 6.2 transcript patterns following imbibition of plant hormones such as methyl jasmonic acid (MeJA), ethephone, salicylic acid (SA), and abscisic acid (ABA), as well as exposure to methyl viologen (MV), revealed that the expression of OmMKK1 is related to defense response signaling pathways. A comparative analysis of OmMKK1 and its *O. sativa* ortholog OsMKK1 showed that both were induced by stress-related hormones and biotic stresses, but that the kinetics of their responses differed despite their high amino acid sequence identity (96%) (You et al. 2007). Starch gel electrophoresis was used to compare the isozyme phenotypes of *O. sativa* IR31917 (AA genome) and *O. minuta* accessions Om101089 and Om101141 (BBCC genome) for ten enzyme systems. Polymorphism between *O. minuta* accessions was detected for shikimate dehydrogenase and glutamate oxaloacetate transaminase [aspartate aminotransferase]. The quaternary structure of the *O. minuta* isozymes was comparable to that of *O. sativa*. Interspecific variation, along with chromosome data, was employed to monitor the relative genetic contribution of the two parents in the IR31917/Om101141 F_1 hybrids and recurrent (IR31917) backcross progenies. The isozyme content of the F_1 hybrids reflected their triploid nature (ABC genome composition), while that of the backcross progenies paralleled the duplication of the A genome and the gradual loss of *O. minuta* chromosomes during the backcrossing process. Evidence was provided for a degree of homoeology between the A, B, and C genomes, and for introgression from *O. minuta* into *O. sativa* (Romero et al. 1993).

18.8 Ability of Crossing *O. minuta* with *O. sativa*

Crossing ability between *O. sativa* and other *Oryza* species is difficult but it is very important, because the cultivated species *O. sativa* has a very narrow genetic variation and we need to widen the genetic base with new resources. The production of viable embryos from hybrids is higher when cultivated rice was used as the female and the wild species as male parent. Emasculation with artificial spikelet-cut and continuous pollination for 3 days resulted in higher rates of embryo development. During the process of embryo rescue, using young embryo after pollinating at 10–12 days and sterilizing with 0.1% $HgCl_2$ for 10 min followed by culture inoculation with ovary-slice to 1/4 MS medium resulted in a higher germination rate of young embryos, and hybrid plants were obtained. There are many ways to cross *O. sativa* and other wild species like *O. minuta*, the canonical one is manual crossing, followed by embryo rescue using tissue culture after pollination at 9–14 days to get F_1 hybrid plants, then self-crossing or back crossing to get introgression lines (Rahman et al. 2009; Guo and Lin 2010; Guo et al. 2013). In addition, transgenic methods for gene transfer from *O. minuta* to *O. sativa* overcame the barriers to chromosomes pairing. The "spike-stalk injection method (SIM)" was useful for genome transfer to improve grain traits (Jena 2010; Yan et al. 2016).

18.9 Resistance of Stresses

O. minuta is considered to be a good source for biotic stresses tolerance. It is a very important source of resistance to some of the major pests, such as brown planthopper, yellow stem borer, green leafhopper, and white-backed planthopper (Panigrahi and Rajmani 2008; Asaf et al. 2016; Neelam et al. 2016). Similarly, for diseases, *O. minuta* has resistance genes for bacterial blight, fungal blast, and sheath blight. Table 18.2 shows

Table 18.2 *O. minuta* is resistant to the following pest and diseases

Name of the pest and diseases	References
White-backed planthopper	Asaf et al. (2016)
Green planthopper	Neelam et al. (2016)
Yellow stem borer	Panigrahi and Rajamani (2008)
Bacterial blight	Guo Si-bin et al. (2010), Jeung et al. (2011a, b)
Sheath blight	Jena (2010)

common pests and diseases which *O. minuta* has resistance against (Rahman et al. 2009; Jena 2010; Sanchez et al. 2014).

18.10 Impact of *O. minuta* in Breeding

Wild species have good gene pools for biotic stresses. Breeders have used them by indirect methods for improving cultivars of *O. sativa*. Wild progenitors provide potential gene sources for complex traits such as yield, multiple resistances to biotic and abiotic stresses and thus are expected to contribute to sustainable food supplies (Rahman et al. 2009; Jena 2010; Ram et al. 2014; Wei et al. 2015). More useful genes for pests and diseases resistance in addition to some panicle traits, spikletes number grain quality, have been used to improve *Oryza sativa* grain yield by crossing followed by embryo rescue using immature embryo culture to get new introgression lines that have the target genes. Genetic engineering has a role in improving *Oryza* cultivars by gene transformation using the spike-stick injection method (Jena 2010; Yan et al. 2016). New CMS lines, restorers, and maintainers have good traits derived from *O. minuta* and use for hybrid rice studies to improve rice yield and grain quality (Xing et al. 2004; Zhao et al. 2005; Wang et al. 2006).

18.11 Mapping of Important Genes/QTLs

Initially, mapping was done to identify some useful genes in *O. minuta* to help breeders improve cultivars of *O. sativa* using advanced backcrossing and segregation populations. Pest and disease resistance are very important objectives for breeders, and hence, *O. minuta* was used as a donor parent (Rahman et al. 2007). In a brief account of some examples of classical QTLs; for number of days to heading and some grain traits, including 1000-seed weight, a total of 19 QTLs were identified using single point analysis. Four of these QTLs were detected on chromosomes 7, 10, and 11 in the F_2 population, and 15 QTLs were detected on chromosomes 3, 5, 7, 8, 10, and 11 in the F_3 population. Four QTLs associated with number of days to heading, 1000-seed weight, seed length, and L/W ratio of seed were shared in both populations; the remaining 11 QTLs were shared only in the F_3 population. Among the QTLs identified, 9 have not been detected in previous QTL studies between *Oryza* cultivars, indicating potentially novel alleles from *O. minuta* (Eung et al. 2005). BC_5F_3 families derived from a cross between *O. minuta* and *O. sativa* were characterized to map QTLs for awn length and heading date which yielded 4 QTLs, 2 QTLs each for days to heading and awn length on chromosomes 6 and 9, respectively. Among the QTLs identified, the QTLs for heading date (dth9) and awn length (awn6) had not been detected in previous QTL studies, indicating the existence of potentially novel alleles in *O. minuta*.

For bacterial blight resistance, *O. minuta* has a major dominant resistant gene located on the subterminal region of chromosome 4, where at least seven BB resistance genes, *Xa1*, *Xa2*, *Xa3*, *Xa4*, *Xa22*, *Xa26*, and *Xa31* were reported previously (Jeung et al. 2011a, b). Five QTLs on chromosomes 6, 7, 9, and 11 and seven epistatic QTLs were identified against two blast isolates (KI307 and KI209). Two QTLs (*qKI307-2* and *qKI209-3*) shared a similar position on chromosome 11. *O. minuta* introgression contributed the resistance alleles for all of these QTLs (Rahman et al. 2011). Advanced backcrossing substitution lines were constructed based on *O. minuta* and cultivated rice IR24, thirteen favorable QTLs of *O. minuta* are detected out in two seasons, six of which, viz. *qPLH-2*, *qTGW-1.1*, *qTGW-9*, *qTGW-12*, *qSPP-1*, and *qYGP-12* may be play important role in the super-high-yield rice breeding (Wei et al. 2015).

18.12 Molecular Mapping of Genes and QTLs

Fine mapping of a *O. minuta* x *O. sativa* cross was used to identify new useful genes from *O. minuta* to obtain introgression lines having new useful genes against blast, brown planthopper, bacterial blight resistance, and high number of spikelets per panicle, as presented in Table 18.3.

18.13 Transposable Elements in *O. Minuta*

There is a strong positive relationship between transposable elements and genome size as tetraploids genomes have more transposable elements than diploids genomes. Repetitive elements are known to play a significant role in genome size variation. Transposable elements are divided into different classes. The most important classes of transposable elements such as long terminal repeats Retrotransposons, long interspersed nuclear elements, helitrons, and DNA transposable elements have a major role in shaping the genome and its size variation. A paradigmatic case was that of two Ty3-gypsy families; Atlantys and RIRE2 (Piegu et al. 2006).

The *O. minuta* [BBCC; 1124 Mbp] genome represents the only polyploid *Oryza* genome for which a comparison with both the original genome counterparts BB (*O. punctata*) and CC (*O. officinalis*) is possible. All LTR-RT families identified in *O. minuta* are found in at least one of the diploid counterparts, and there were only a few cases where an element was significantly more represented in the polyploid genome than in the two diploid counterparts. The most convincing example was that of the Ty1- copia element RIRE1 (0.16, 0.1, and 0.6% of total reads in *O. punctata*, *O. officinalis,* and *O. minuta, respectively*). The most abundant lineages contained representatives of all the different species, and there was not a single species-specific lineage with the partial exception of RIRE1 in *O. australiensis* [EE; 965 Mbp]. This lineage, also including a few sequences from *O. minuta*,

Table 18.3 Important traits along with details of information from *O. minuta* X *O. sativa*

Trait	Locus	Chr no.	Locus mapping	References
Blast resistance	Pi9	6	Between RG 46 and R2123	Amante-Bordeos et al. (1992), Jena (2010)
Blast resistance	qK1307-1 qK1307-2 qK1209-1 qK1209-2 qK1209-3	9 11 6 7 11	S9026 S11077.4 S6065 S7048 S110774	Rahman et al. (2011)
Spikelets per panicle	SPP 7	7	Between RM 445 and RM 21615	Linh et al. (2008)
Spikelets per panicle	QSPP7	7	Between RM 9452 and RM 21605	Balkunde et al. (2013)
Bacterial blight resistance	Xa35(t)	11	(RM 144) between RM 6293 and RM 7654	Gue Sibin et al. (2010)

was separated from the others, and it was possible to see how the branch lengths are short suggesting that amplification of this element took place after speciation.

18.14 Conclusion

O. minuta has a wide distributing on southeast Asia, growing is related with semi-shaded or shaded conditions. It has an academic importance due to the genetic background for biotic and abiotic stresses tolerance providing a good gene pool could be used for improving rice cultivars for some traits like resistance to BPH, WBPH, GLH, YSB, BB, SHB, and blast. By specific methods, hybridization between *O. minuta* and *O. sativa* has been achieved using transduction, transformation, and manual crossing followed by embryo rescue and doubling chromosomes to overcome the F_1 complete sterility to transfer useful genes. Many researches have used wild rice species including *O. minuta* for mining new genes related to rice plant architecture, agronomic, and yield components traits and are still working to broaden the genetic base of rice.

O. minuta has a large nuclear genome, large mt genome, and large cp genome providing an abundance of genes. *O. minuta* originated and developed from two different genomes BB and CC. BB came from *O. punctata* (an African species) has adaptation and tolerance to drought and heat, while CC genome came from *O. officinalis* (tropical Asian species) and has ability to tolerate heat. *O. minuta* needs more research for drought and heat tolerance to identify new QTLs for new target genes. MicroRNA, long non-coding RNA, and proteomics studies are need to maximize the benefits of *O. minuta* species.

References

Abe T, Edanami T, Adachi E, Sasahara T (1999) Phylogenetic relationships in the genus *Oryza* based on mitochondrial RFLPs. Genes and Genetic Syst 74 (1):23–27

Ackerman J (2004) Granes. National Geographic, http://magma.Nationalgeographic.com/ngm/0404/feature2/index.html

Aggarwal RK, Brar DS, Nandi S, Huang N, Khush GS (1999) Phylogenetic relationships among *oryza* species revealed by AFLP markers. Theor Appl Genet 98:1320–1328

Amante-Bordeos A, Sitch LA, Nelson R, Damacio RD, Oliva NP, Aswidinnoor H, Leung H (1992) Transfer of bacterial blight and blast resistance from the tetraploid wild rice *Oryza minuta* to cultivated rice, *Oryza sativa*. Theor Appl Genet 84:345–354

Asaf S, Khan AL, Khan AR, Waqas M, Kang S-M, Khan MA et al (2016) Mitochondrial genome analysis of wild Rice (*Oryza minuta*) and tts comparison with other related species. PLoS ONE 11(4):e0152937

Asaf S, Waqas M, Khan AL, Khan MA, Kang S-M et al (2017) The complete chloroplast genome of wild rice (*Oryza minuta*) and its comparison to related Species. Front Plant Sci 8:304

Balkunde S, Le H, Lee H, Kim D, Kang J, Ahn S (2013) Fine mapping of a QTL for the number of spikelets per panicle by using near-isogenic lines derived from an Interspecific cross between *Oryza sativa* and *Oryza minuta*. Plant Breeding 132(1):70–76

Chaturvedi M, Datta K, Nair PKK (1998) Pollen morphology of *Oryza* (Poaceae). Grana 37(2):79–86. https://doi.org/10.1080/00173139809362647

Chee FT, Mariam AL, Brar DS, Khush GS (2006) Analysis of introgression from a tetraploid wild species (*Oryza minuta*) into cultivated rice (*O. sativa*). Borneo Science 19:7–15

Clayton WD, Vorontsova MS, Harman KT, Williamson H (2006) Grass base the on line world grass flora. http://www.kew.org/data/grasses-db.html

Eung GJ, Darshan SB, Kyong HK, Heung GH, Kshirod KJ, Ho YK, Sang NA, Gihwan Y, Min HN (2005) Cytological characterization of interespecific hybrids in rice (*Oryza sativa* L.). Korean J Breeding 37(1): 52–56

Fukui K, Shishido R, Kinoshita T (1997) Identification of the rice D-genome chromosomes by genomic in situ hybridization. Theor Appl Genet 95:1239–1245

Ge S, Sang T, Lu BR, Hong DY (1999) Phylogeny of rice genomes with emphasis on origins of allotetraploid species. Proc Natl Acad Sci USA 96:14400–14405

Guo SB, Lin XH (2010) Discussions on wide-crossing between *Oryza minuta* and *Oryza sativa*. Guangxi Agri Sci 41(2):95–99

Guo S, Zhang D, Lin X (2010) Identification and mapping of a novel bacterial blight resistance gene Xa35(t) originated from *Oryza minuta*. Scientia Agricultura Sinica 43(13):2611–2618

Guo S, Wei Y, Li X, Liu K, Huang F, Chen C, Gao G (2013) Development and identification of introgression lines from cross of *Oryza sativa* and *Oryza minuta*. Rice Sci 20(2):95–102

Ishii T, Mc Couch SR (2000) Microsatellites and microsynteny in the chloroplast genome of *oryza* and eight other gamine species. Theor Appl Gent 100:1266–1275

Jena KK (2010) The species of the genus *Oryza* and transfer of useful genes from wild species into cultivated rice *O. sativa*. Breed Sci 60(5):518–523

Jeung J, Roh T, Kang K, Jeong J, Kim M, Kim Y (2011a) Genetic analysis on the bacterial blight resistance of Suweon497, a rice breeding line developed through wide hybridization. Korean J Breed Sci 43(1):81–91

Jeung J, Ron T, Kang K, Shin Y, Kim Y (2011b) Genetic analysis on the bacterial blight resistance gene from a wild relative, *Oryza minuta*. Korean J Crop Science 56:124–133

Khush GS, Brar S, Hardy B (2001) Rice genetics IV. International Rice Research Institute, Science Publishers Inc., Enfield

Li CB, Zhang DM, Ge S, Lu BR, Hong DY (2001) Identification of genome constitution of *Oryza malampuzhaensis*, *O. minuta*, and *O. punctata* by multicolor genomic in situ hybridization. Theor Appl Genet 103:204–211

Linh LH, Hang NT, Jin FX, Kang KH, Lee YT, Kwon SJ, Ahn SN (2008) Introgression of a quantitative trait locus for spikelets per panicle from *Oryza minuta* to the *O. sativa* cultivar Hwaseongbyeo. Plant Breed 127 (3):262–267

Mariam AL, Zakri AH, Mahani MC, Normah MN (1996) Interspecific hybridization of cultivated rice, *Oryza sativa* L. with the wild rice, *O. minuta* Presl. Theor Appl Genet 93(5/6):664–671

Nandi HK (1936) The chromosome morphology, secondary association and origin of cultivated rice. J Genet 33:315–336

Nandi HK (1938) Interspecific hybridization in *Oryza* I Cytogenetical evidence of the hybrid origin of *O minuta* Presl. Trans Bose Res Inst 2:99 (abstract in Pl Breed Abstr 1939, 13: 35)

Neelam K, Kumar K, Dhaliwal HS, Singh K (2016) Introgression and exploitation of QTL for yield and yield components from related wild species in rice cultivars. In: Rajpal VR et al (eds) Molecular breeding for sustainable crop improvement, sustainable development and biodiversity 11. Springer International Publishing, Switzerland

Nishikawa T, Vaughan DA, Kadowaki K (2005) Phylogenetic analysis of *Oryza* species, based on simple sequence repeats and their flanking nucleotide sequences from the mitochondrial and chloroplast genomes. Theor Appl Genet 110(4):696–705

Panigrahi D, Rajamani S (2008) Genetic evaluation of the wild *Oryza* species for resistance against the yellow stem borer, *Scirpophaga incertulas* wlk. J Plant Port Environ 5(1):26–29

Paweena T, Mounthon N, Stephan W (2015) A taxonomic revision of the tribe Oryzeae (Poaceae) in Thailand. Sci Asia 41:363–376

Piegu B et al (2006) Doubling genome size without polyploidization: dynamics of retrotransposition-driven genomic expansions in *Oryza australiensis*, a wild relative of rice. Genome Res 16:1262–1269

Pongtongkam P, Klakhaeng K, Ratisoontorn P, Suputtitada S, Peyachoknagul S, Sangduen N, Maneepongse C, Uraivong H (1993) Immature embryo culture in hybrid rice. J Natural Sciences 27(1):15–19

Rahman ML, SangHo C, MinSun C, YongLi Q, WenZhu J, RiHua P, Khanam S, YoungI C, JiUng J, Jena KK, HeeJong K (2007) Identification of QTLs for some agronomic traits in rice using an introgression line from *Oryza minuta*. Mol Cell 24:16–26

Rahman ML, Jiang W, Chu SH, Qiao Y, Ham TH, Woo MO et al (2009) High-resolution mapping of two rice brown planthopper resistance genes, Bph20(t) and Bph21(t), originating from *Oryza minuta*. Theor Appl Genet 119(7):1237–1246

Rahman L, Khanam S, JaeHwan R, HeeJong K (2011) Mapping of QTLs involved in resistence to rice blast (*Magnaporthe grisea*) using *Oryza minuta* introgression lines. Czech J Geneti Plant Breed 47:85–94

Ram T, Prasad MS, Laha GS, Lakshmi VJ, Ram D, Dinesh C, Yamini D, Akanksha S, Sohanvir S (2014) Widening the genetic base of rice varieties by introgressing novel genes/traits from wild species. In: International conference on enhanced genepool utilization—capturing wild relative and landrace diversity for crop improvement. Cambridge, UK, 16–20 June 2014

Rodrangboon P, Pongtongkam P, Suputtitada S, Adachi T (2002) Abnormal embryo development and efficient embryo rescue in interspecific hybrids, *Oryza sativa x O. minuta* and *O. sativa x O. officinalis*. Breed Sci 52 (2):123–129

Romero GO, Amante-Bordeos AD, Dalmacio RD, Elloran R, Sitch LA (1993) Comparative studies of isozymes in *Oryza sativa*, *O. minuta*, and their interspecific derivatives: evidence for homoeology and recombination. Theor Appl Genet 87:609–615

Sanchez PL, Wing RA, Brar DS (2014) The wild relative of rice: genomes and genomics. In Zhang Q, Wing RA (eds) Genetics and genomics of rice, plant genetics and genomics: 9 crops and models 5. https://doi.org/10.1007/978-1-4614-7903-1_2. © Springer

Suputtitada S, Adachi T, Pongtongkam P, Ratisoontorn P, Peyachoknagul S, Apisitwanich S, Klakhaeng K, Rodrangboon P, Lertvichai R (1994) Rice improvement through tissue culture. In: Proceedings of the international colloquium on impact of plant biotechnology on agriculture, Rogla, Slovenia, 5th–7th December, pp 73–84

Tang K, Sun X, He Y, Zhang Z (1998) Anther culture response of wild *Oryza* species. Plant Breed 117: 443–446

Tomotaro N, Duncan A, Vaughan KK (2005) Phylogenetic analysis of *Oryza* species, based on simple sequence repeats and their flanking nucleotide sequences from the mitochondrial and chloroplast genomes. Theor Appl Genet 110:696–705

Wang SW, Zhao BR, Xing QH, Yang HH, Jin DM, Liu X, Wang B (2006) Molecular verification of DNA flow from wild rice (*O. minuta*) to cultivated rice. Scientia Agricul Sinica 39(11):2170–2177

Wei Y, Li X, Huang K, Chen Y, Gao G, Deng G, Guo S (2015) Mapping of yield-related quantitative trait locus of *Oryza minuta*. J South Agricul 46:958–963

Xing Q, Zhao B, Xu K, Yang H, Liu X, Wang S, Jin D, Yuan LP, Wang B (2004) Test of agronomic characteristics and amplified fragment length polymorphism analysis of new rice germplasm developed from transformation of genomic DNA of distant relatives. Inter Soc Plant Mol Biol 22:155–164

Yan P, Yuanyi H, Bigang M, Haitao X, Ye S, Yinlin P, Xiabing S, Yaokui L, Xuemei N, Yumei X et al (2016) Genetic analysis for rice grain quality traits in the YVB stable variant line using RAD-seq. Mol Genet Genom 291:297–307

Yi CD, Cheng X, Wang BB, Liang GH, Gong ZY, Tang SZ, Gu MH (2008) Molecular cytogenetic analysis of spontaneous interspecific hybrid between *Oryza sativa* and *Oryza minuta*. Rice Sci 15:283–288

You MK, Oh SI, Ok SH, Cho SK, Shin HY, Jeung JU, Shin JS (2007) Identification of putative MAPK Kinases in *Oryza minuta* and *O sativa* responsive to biotic stresses. Mol Cells 23(1):108–114

Zhao BR, Xing QH, Xia HA, Yang HH, Jin DM, Liu X, Wang SW, Wang B, Yuan LP (2005) DNA polymorphism among Yewei B, V20B, and *Oryza minuta* J. S. Presl. ex C. B Presl J Integ Plant Biol 47: 1485–1492

Oryza neocaledonica Morat

19

Kutubuddin Ali Molla, Subhasis Karmakar, Johiruddin Molla, T. P. Muhammed Azharudheen and Karabi Datta

Abstract

Rice is the primary dietary source of energy for nearly half of the world's population. Rice genetic improvement is the need of the hour to feed the world's growing population. Wild rice species are treated as a useful source of genetic variability for different important traits. Besides two cultivated rice species (*Oryza sativa* and *Oryza glaberrima*), there are a total of 24 wild species belong to four species complex, viz. Sativa, Officinalis, Ridleyi and Meyeriana. This chapter deals with the most unexplored wild rice species *Oryza neocaledonica,* an endemic and endangered species with very few known populations. The species belongs to Meyeriana complex and is only confined to an island, Neo Caledonia of southwest Pacific Ocean. Scientific study subjected to *O. neocaledonica* is rarely available. This member of tertiary gene pool of rice could be proved as a valuable genetic resource for drought tolerance as it grows in much lesser water than that is required by cultivated rice.

19.1 Academic and Economic Importance

Oryza neocaledonica is the most recent wild species reported in the genus *Oryza*. The species was first described by Philippe Morat in 1994 from an overseas territory of France called Neo Caledonia, an island situated at 1210 km east of Australia in the southwest Pacific Ocean (Morat et al. 1994). It grows mostly on humid area and shady grounds of dry forests, and much less water is needed for its growth than that is needed by cultivated rice. Therefore, the species could be potentially utilized for drought tolerance trait. The species has been reported only from four different locations in the northern half of New Caledonia, forming four sub-populations (Fig. 19.1). The exact population size in these four different sub-populations is unknown. The Pouembout sub-population, in the fenced reserve of Tiea, is small with a few scattered individuals. The two sub-populations in Poum, in the banks of Nehoue River, consist of many individuals distributed all along the stream. The fourth sub-population present in the Ougne forest is known only from the herbarium

K. A. Molla (✉) · T. P. M. Azharudheen
ICAR-National Rice Research Institute, Cuttack 753006, India
e-mail: kutubuddin.molla@icar.gov.in

S. Karmakar · J. Molla · K. Datta
Department of Botany, University of Calcutta, 35-Ballygunge Circular Road, Kolkata 700019, India

T. K. Mondal and R. J. Henry (eds.), *The Wild Oryza Genomes*,
Compendium of Plant Genomes, https://doi.org/10.1007/978-3-319-71997-9_19

Fig. 19.1 Distribution of *O. neocaledonica* species. The range of distribution is highlighted with dark grey and outlined with circle

collection. With all these four locations taken together, the species occupies an area of only 6 km^2 and its extent of occurrence is 785 km^2 (Hequet 2010).

19.2 Conservation Status

O. neocaledonica is endemic to the tropical dry forests of the Neo Caledonia. The dry forests have reduced dramatically in the past owing to wild fire, grazing and the conversion for agricultural purposes. Though the species, which is usually found along river banks or valley in these dry forests, has good regeneration capacity after wild fire, grazing, especially by the introduced Rusa Deer led to reduction in the population size and is now considered as an endangered species. A 32 ha reserve land has been created to protect the Tiéa sub-population in Pouembout (Hequet 2010).

19.3 Botanical Description

O. neocaledonica is usually found along river banks or valley within the tropical dry forests of the Neo Caledonia at an altitude of about 10–20 m. *O. neocaledonica* is a perennial grass which grows as scattered tufts. It has short rhizomes; culms are 60–80 cm tall with glabrous culm nodes. Leaves are narrowly lanceolate, dark green. Oral hairs are scanty on leaf sheath. Leaf-sheath auricles are absent. Leaf-blades are 17–21 cm long and 7–11 mm wide. Leaf blade surface is smooth, or scaberulous. The inflorescence is a panicle comprising 7–40 fertile spikelets. Spikelets are elliptical or oblong, laterally compressed, 7–9 $\times$ 1.5–1.8 mm. Open panicle is lanceolate or ovate, 30–55 cm long and 3–8 cm wide. Primary panicle branches are simple. Spikelets are solitary. Fertile spikelets are pedicelled. Pedicels are linear, sinuous, bibracteate. Spikelets comprise of 2 basal sterile florets and 1 fertile floret without rachilla extension. Glumes are either absent or obscure.

Basal sterile florets are similar, barren without significant palea. Lemma of upper and lower sterile floret is linear, 2–2.3 mm long and single-veined. Fertile lemma is oblong, laterally compressed, 7–8 mm long, 3.5 mm wide, coriaceous, red (pink) in colour, keeled, and 5-veined. Lemma surface is granulose, hispidulous (spinulose) with hooked hairs. Lemma apex is acute and muticous. The palea is elliptic in nature;

7 mm long; coriaceous; 3-veined; 1-keeled. Palea apex is acute. Lodicules are 2 in number, 0.8 mm long, membranous. Anthers are 6 in number, 2.8–3 mm long and yellow. Stigmas are 2 in number. Fruit is a Caryopsis with adherent pericarp, dark brown. Embryo is 0.2 length of caryopsis. Hilum is linear. Disseminule comprises of a floret (Clayton et al. 2010).

19.4 Genome Affiliation

Cytogenetic and genome analysis revealed that *O. neocaledonica* is diploid, with GG genome complement (2n = 2x = 24, GG). The species is included in the *Oryza meyeriana* complex, because of the morphological similarities it shares with other members of the complex, namely *O. meyeriana* and *Oryza granulata*. Being the most eastern wild rice species in the Pacific, it is found very far from other populations of *O. meyeriana* complex. The nearest known location to New Caledonia at which *O. meyeriana* complex grow is on the Maluku islands, Indonesia. This is also the only species of *O. meyeriana* complex, which is reported east of the Wallace Line. Based on the phylogenetic analysis of the sequences of 20 chloroplast fragments, Tang et al (2010) reported a close affinity of *O. neocaledonica* with *O. granulata*. They also confirmed the report of Vaughan and Morishima (2003) that *O. neocaledonica* is sister to *O. granulata*. Zodinpuii et al. (2013) studied the phylogenetic relationship in wild and cultivated rice using the chloroplast *matK* gene, which revealed a close evolutionary relationship among the GG genome species representing *O. meyeriana* (Malaysia), *O. granulata* (China) and *O. neocaledonica* (Neo Caledonia) with the three forming a single clad. The main distinguishing criteria for these species are spikelet size and shape.

Though inclusion of the species in the *O. meyeriana* complex was unambiguous, the independent species status of *O. neocaledonica* was disputed. Vaughan and Morishima (2003) considered the species as a diminutive version of *O. granulata,* and the two can be distinguished primarily on the basis of microscopic epidermal characters. Later, in 2008, they had suggested that this taxon might be considered as a subspecies or variant of *O. granulata* or *O. meyeriana* (Vaughan et al. 2008). Clayton et al. (2010) however have recognized *O. neocaledonica* as a valid *Oryza* species.

19.5 Genomic Resources

Sequences and other genomic resources of *O. neocaledonica* are rare in the public database. The study of Tang et al. (2010) resulted in the sequencing of partial fragment of twenty chloroplast genes like *ndhA, ndhC, atpB, psbH, atpI, atpF, trnL, trnV, trnD, trnG, trnQ* and some others. Based on the sequence, they constituted a phylogenetic tree revealing a close relationship with *O. granulata.* No other genomic resources of this species are available in public database. Sequence-based alternative genomic resources like non-coding RNA, and different molecular markers could be developed utilizing the easiness of next-generation sequencing technology.

19.6 Future Prospects

O. neocaledonica still remains one of the most unexplored and least studied wild relative of rice. The usefulness of the species in the genetic enhancement of cultivated rice was not given any attention in the past owing mainly to the endemic distribution of the species in a very remote part of the globe. In a juncture where, rice genetic improvement is stuck with yield barriers and emerging biotic and abiotic stresses, this species is a potential candidate for future research. Recent days witnessed a precipitous drop in sequencing cost, and considering the fact the genome of this unplumbed wild species is needed to be explored which will in turn facilitate to generate more and more valuable genomic resources.

References

Clayton WD, Vorontsova M, Harman KT, Williamson H (2010) GrassBase—the online world grass Flora. Oryza. RBG, Kew, U.K

Hequet V (2010) *Oryza neocaledonica*. The IUCN red list of threatened species 2010: e.T177819A7464375. http://dx.doi.org/10.2305/IUCN.UK.2010-4.RLTS.T177819A7464375.en

Morat P, Deroin T, Coudere H (1994) Presence in New Caledonia of an endemic species in the genus *Oryza* L. (Gramineae). Bull Mus Natl Hist Nat, B, Adansonia 16:3–10

Tang L, Zou XH, Achoundong G, Potgieter C, Second G, Zhang DY, Ge S (2010) Phylogeny and biogeography of the rice tribe (Oryzeae): evidence from combined analysis of 20 chloroplast fragments. Mol Phylogenet Evol 54(1):266–277

Vaughan DA, Ge S, Kaga A, Tomooka N (2008) Phylogeny and biogeography of the genus Oryza. Rice Biology in the Genomics Era. Springer, Berlin Heidelberg., pp 219–234

Vaughan DA, Morishima H (2003) Biosystematics of the genus Oryza. In: Smith CW (ed) Rice: origin, history, technology, and production. Wiley, NJ, USA, pp 27–65

Zodinpuii D, Ghatak S, Mukherjee S, Kumar NS (2013) Genetic relatedness of genus Oryza from Eastern Himalayan region as revealed by chloroplast matK gene. Asian J Conserv Biol 2:144–151

Oryza nivara Sharma et Shastry 20

Guttikonda Haritha, Surapaneni Malathi, Balakrishnan Divya, B. P. M. Swamy, S. K. Mangrauthia and Neelamraju Sarla

Abstract

Oryza nivara is the closest annual wild progenitor of *O. sativa* subspecies *indica* distributed in South and Southeast Asia. The accessions of *O. nivara* are well known to contribute genes for resistance to biotic stresses; grassy stunt virus, bacterial leaf blight, sheath blight, neck blast, brown planthopper and tolerance to abiotic stresses; drought, salinity and heat for rice improvement. In addition, loci for yield, seedling vigour, quality and biofortification were also reported from *O. nivara*. It has accessions with high leaf photosynthetic efficiency. DRRDhan 40 is the first *O. nivara*-derived variety released in 2014 in India for commercial cultivation in three states Maharashtra, Tamil Nadu and West Bengal. Chromosome segment substitution lines have been developed using *O. nivara* to map genes and quantitative trait loci. Genome sequencing and annotation of *O. nivara* in Oryza map alignment project (OMAP) revealed a genome size of 448 Mb and 36,313 coding and 1648 noncoding genes. This will help further in gene editing and silencing technology based on future breeding programmes. The development of CSSLs from *O. nivara* will help in more precise utilization for basic and applied studies.

G. Haritha · S. Malathi · B. Divya ·
S. K. Mangrauthia · N. Sarla (✉)
ICAR-Indian Institute of Rice Research (ICAR),
Rajendranagar, Hyderabad 500030, India
e-mail: sarla_neelamraju@yahoo.com

G. Haritha
e-mail: harithaguttikonda@gmail.com

S. Malathi
e-mail: malathisurapaneni@gmail.com

B. Divya
e-mail: divyab0005@gmail.com

S. K. Mangrauthia
e-mail: skmdrr@gmail.com

B. P. M. Swamy
Plant Breeding, Genetics and Biotechnology Division, International Rice Research Institute, DAPO Box 7777, Metro Manila, Philippines
e-mail: swamy.m@irri.org

20.1 Introduction

The genus *Oryza* has 22 wild and 2 cultivated species represented by 10 distinct genome groups (AA, BB, CC, BBCC, CCDD, EE, FF, GG, HHJJ and KKLL) (Ammiraju et al. 2010; Jacquemin et al. 2013; Sanchez et al. 2013). The Asian cultivated species *Oryza sativa* L. is grown globally, and African cultivated species *O. glaberrima* is grown in West African regions only. There are four wild species of *Oryza* distributed in India, and those are *O. nivara* Sharma et Shastry, *O. rufipogon* Griff, *O. officinalis* Wall.

T. K. Mondal and R. J. Henry (eds.), *The Wild Oryza Genomes*,
Compendium of Plant Genomes, https://doi.org/10.1007/978-3-319-71997-9_20

Ex Watt and *O. granulata* Nees et Arn. Ex Watt. *O. nivara* and its perennial ancestor *O. rufipogon* are widely distributed over eastern Indian regions (Chouhan et al. 2014).

20.2 Economic and Academic Importance

Wild relatives of *Oryza* are an excellent reservoir of genetic diversity or allelic variability for several useful traits such as resistance to biotic stress (disease and pest), abiotic stress (drought, salinity, environmental factors, etc.), quality and productivity-related traits (Tanksley and McCouch 1997), which can provide significant economic benefits to crop when incorporated into elite varieties. Only 10–20% genetic diversity from wild species was retained in *O. sativa* subspecies *indica* and *japonica,* respectively (Zhu et al. 2007). The nucleotide diversity $\theta_{\text{sil}} = 0.0077$ and outcrossing rate of *O. nivara* are lesser (5–25%) than its ancestral species *O. rufipogon* ($\theta_{\text{sil}} = 0.0095$, 10–30%) as the panicles of *O. nivara* are compact and partially overlapped by flag leaves to prevent cross-pollination (Oka 1988; Grillo et al. 2009). The accessions of *O. nivara* are a source of abundant genetic diversity (Joshi et al. 2000; Sarla et al. 2003; Juneja et al. 2006). This is the first rice species identified for resistance to grassy stunt virus (Ling et al. 1970; Khush and Ling 1974). This virus threatened the rice crop during the 1970s and damaged around 116,000 ha of cultivated rice land in Asia (Khush 1977). Out of 9540 *O. sativa* and 100 wild accessions screened, only one accession from Uttar Pradesh, India, showed resistance to grassy stunt virus (Chang et al. 1975; Khush 1971). This disease no longer exists as resistance has been transferred into elite breeding lines through conventional and marker-assisted breeding programmes (Khush et al. 1977b; Brar and Khush 1997).

The major destructive monophagous insect pest of rice is brown planthopper (BPH), which causes up to 60% yield loss throughout Asia under epidemic conditions. Seven accessions of *O. nivara* were identified as a source of BPH (*Nilaparvata lugens Stål*) resistance. These are Acc no IRGC81859, CR100313A from India; IRGC104646 from Thailand; IRGC93198 from Nepal; and IRGC92945, IRGC92960 and IRGC93092 from Cambodia (Sarao et al. 2016). Further, the gene *bph2* for BPH resistance was identified from 5 *O. nivara* accessions WRAC 02, WRAC 04, WRAC 07, WRAC 21 and WRAC 25 from Sri Lanka. The lines harbouring *bph2* gene showed substantial levels of antixenosis and antibiosis effects on BPH (Madurangi et al. 2011, 2013). Six BC_2F_6 introgression lines RPBio4918-212S, RPBio4918-215S, RPBio4918-221S, RPBio4918-224S, RPBio4918-228S and RPBio4918-230S from *O. sativa* (Swarna)/ *O. nivara* (IRGC81848) were found to be highly resistant to BPH (Lakshmi et al. 2010). Of these, introgression line 228S was also resistant to multiple pests. Another major insect pest of rice is green rice leafhopper (GRH) caused by *Nephotettix cincticeps* Uhler, distributed throughout the temperate regions of East Asia (Ghauri 1971). The gene *Grh6* for GRH resistance has been identified from *O. nivara* accession IRGC105715 (Fujita et al. 2004). Pyramided lines for different combinations of GRH genes were developed that showed higher level of resistance than those containing each of the gene alone (Fujita et al. 2010). Likewise, a serious threat to rice yield in Southeast Asia is bacterial leaf blight caused by *Xanthomonas oryzae* pv *oryzae.* Two bacterial leaf blight resistance genes *Xa30* (*t*)/ *Xa38* and *Xa33* have been identified from two *O. nivara* accessions, i.e. IRGC81825 and IRGC105710, respectively (Cheema et al. 2008a; Kumar et al. 2012), and it also showed resistance to bacterial leaf streak, caused by *Xanthomonas oryzae* pv. (International Rice Research Institute 1968). Similarly, three *O. nivara* accessions, IRGC104705, IRGC100898 and IRGC104443, showed moderate resistance to sheath blight, caused by a soilborne necrotrophic fungus *Rhizoctonia solani Kühn* (Prasad and Eizenga 2008), and some accessions showed moderate resistance to rice leaf folder, caused by *Cnaphalocrocis medinalis* (Guenee) (Khan et al. 1989). One backcross introgression line RPBio4918-10-3 (BC_2F_6) derived from the cross between *O. sativa* (Swarna) and *O. nivara* (IRGC81848) showed moderate resistance to sheath blight (Dey et al. 2016).

Reproductive barriers prevent gene flow from one species to another species. These are frequently observed in progenies derived from interspecific crosses. Hybrid breakdown locus *hbd (1t)* was identified on short arm of chromosome 2 of *O. nivara* (IRGC105444) which induces interspecific hybrid breakdown in Koshihikari genetic background. Some of its progeny plants showed extremely weak phenotype such as short plant height, few tillers and weak root growth and died before heading (Miura et al. 2008). The same locus was reported by Matsubara et al. (2007) in an intraspecific cross between Koshihikari (*japonica*) and Nonabokra (*indica*). Nonabokra induces hybrid breakdown in Koshihikari genetic background. Furthermore, a novel loss-of-function allele of *S27* (S27-niv^s) was identified in *O. nivara* (IRGC105444), and it leads to F_1 pollen sterility in hybrids derived from a cross between *O. sativa* (Taichung 65) and *O. nivara* (IRGC105444). This is due to nucleotide mutations in coding and promoter regions of two tandem copies of mitochondrial ribosomal protein L27 genes *mtRPL27* (mtRPL27a and mtRPL27b) located at S27-niv^s, which makes it inactive (Win et al. 2010). In wild rice, the frequency of complete restoration *Rf* for gametophytic Honglian cytoplasmic male sterility (HL-CMS) was higher (60%) than sporophytic Wild-abortive cytoplasmic male sterility (WA-CMS) which is as low as 41.9%. Interestingly, the wild progenitor species *O. rufipogon* and *O. nivara* have also been mined for fertility restoration *Rf* genes of HL-CMS and WA-CMS. The accessions of *O. rufipogon* and *O. nivara* could restore the fertility of both HL-CMS and WA-CMS. However, some accessions such as W46 (106321) from *O. rufipogon* and W20 (103836), W37 (105704) from *O. nivara*, W13 (101855) from *O. glaberrima* and W35 (105561) from *O. glumaepatula* restore the fertility of only HL-CMS, whereas W39 (105887) from *O. rufipogon* and W38 (105736) from *O. nivara* were complete restorers for HL-CMS, but partial restorers for WA-CMS indicating that the complete restorer accessions aggregated mainly in the two species of *O. nivara* and *O. rufipogon* and the *Rf* genes in *O. sativa* have originated from *O. rufipogon*/*O. nivara* complexes (Li et al. 2005).

Genetic variability for resistance to abiotic stress tolerance is limited in cultivated genotypes but abundant in wild species. Drought is one of the major devastating abiotic stresses which limits rice production throughout the world (Lanceras et al. 2004). *O. nivara* (Acc. 16150) was reported as a potential source of genes for drought avoidance (Khush 1997; Eizenga and Rutger 2003). These genes for drought avoidance were transferred to rice cultivars by backcross breeding. Six BC_1F_2 lines from a cross between rice (cv. RD23) and *O. nivara* (acc. 16150) and seven from a cross between CN1 and *O. nivara* (acc. 16150) showed high grain yield per plant under drought (Thanh et al. 2006). The transcription factor genes of WRKY family are involved in plant responses to various biotic and abiotic stresses (Banerjee and Roychoudhury 2015). A total of 97 *O. nivara* WRKY (*OnWRKY*) genes were identified and were highly expressed in roots. These are induced in response to the application of salicylic acid, abscisic acid and drought (Xu et al. 2016). Two advanced backcross introgression lines eg. 75S and 166S derived from Swarna x *O. nivara* IRGC81848 and 3K, 3-1K and 231K derived from Swarna x *O. nivara* IRGC81832 were found to be drought-tolerant under transplanted and direct seeded condition (Rai et al. 2010). Similarly, salinity is the second major abiotic stress, which affects 20% of global irrigated area and leads to considerable yield loss. BC_2F_6 introgressed lines 14S, 75S and 166S derived from a cross between Swarna and *O. nivara* (IRGC81848) and one IL 3-1K and from a cross between Swarna and *O. nivara* (IRGC81832) showed significantly higher yield compared to control under salinity treatment of 150 mM NaCl (Ganeshan et al. 2016). Flooding is a widespread environmental stress particularly in lowlands of South, East and Southeast Asia. Tolerance to flooding is an important agronomical trait. Three accessions of *O. nivara* IRGC80507, IRGC101508 and ONIVA01G39120 (cultivar IRGC: 100897) have

SUB1A-1 allele and were highly tolerant to submergence (Niroula et al. 2012; dos Santos et al. 2017). Likewise, heavy metal stress is also an environmental stress in present days. The *O. nivara* accessions from Central Rice Research Institute (CRRI), Cuttack, were screened under different chromium levels (10–100 μM Cr^{+6}), and two accessions IC-283169 and IC-336684 showed the ability to withstand 100 μM Cr^{+6} (Nayak et al. 2014).

Apart from the contribution to biotic and abiotic stresses, *O. nivara* accessions also provide favourable alleles for improving yield and yield-related traits, heterosis and seedling vigour in elite varietal background. For example, the *O. nivara* accession IRGC81832 collected from Bihar was used in crossing programme with Swarna, an elite lowland rice variety. In this population, two major yield-enhancing QTLs *qyldp2.1* and *qyldp11.1* were mapped from BC_2F_2 mapping population (Kaladhar et al. 2008; Swamy et al. 2014). Further, two major yield QTLs *qyldp2.1* and *qyldp9.1* were mapped from the cross between Swarna and another accession from Uttar Pradesh *O. nivara* (IRGC81848) (Swamy et al. 2011a, b, 2014). Recently, Ma et al. (2016) developed a set of 131 *O. nivara* Acc. W2014 introgression lines in the background of 93-11 and mapped QTLs for 13 yield and yield-related traits using whole-genome resequencing approach. Out of 65 QTLs detected, around 36.9% were derived from *O. nivara* and showed positive allelic effects for yield and yield component traits. Malathi et al. (2017) detected two major effect QTLs in BC_2F_8 in both years for days to maturity (DM) and plant height (PH) with 13 and 20% phenotypic variance (PV). Only 3 out of 15 QTLs were also reported previously in BC_2F_2 of the same cross. Further, a set of 74 CSSLs was identified and 22 of these CSSLs showed significantly higher values than Swarna for 5 yield traits. Two parental lines Pusa44 and PR114 were improved by introgression from *O. nivara* (IRGC100593 and CR100142A) for the development of high yielding hybrids. These backcross introgression lines were crossed with WA-CMS line PMS17A to develop hybrids. Two hybrids (ILH921 and ILH951) having introgressions from *O. nivara* showed significant heterosis over parental introgression line, recurrent parents and check hybrids for grain yield-related traits. Thus, introgressions from *O. nivara* not only improve the parental lines but also its derived hybrids (Gaikwad et al. 2014). Early seedling vigour is another important trait for uniform and rapid establishment of crop which may influence yield in direct seeded condition (Anandan et al. 2016; Borjas et al. 2016). A recent study on backcross inbred lines derived from a cross between M-202 and *O. nivara* (IRGC100195) showed that *O. nivara* introgressions helped improve seedling vigour by increasing both coleoptile and shoot length which led to increase in panicles per plant and seed weight (Eizenga et al. 2016).

In addition to yield and heterosis, grain quality is an important quantitative trait that can benefit from introgressions of wild species and needs to be improved. The QTLs *qmp1.2* for milling percentage, *qkw6.1* for kernel width and *qklac12.1* for kernel length after cooking were mapped from the BC_2F_2 populations of Swarna and *O. nivara* (IRGC81832) and Swarna and *O. nivara* (IRGC81848), and it was observed that increasing effect was from *O. nivara* (Swamy et al. 2012). Likewise, the progeny derived from IR64 and *O. nivara* showed higher total seed protein content and arginine than parents. *O. nivara* (Acc. 104444) showed significantly higher values than IR64 for 8 out of 17 amino acids. Thus, seed protein content can be improved using *O. nivara* (Mahmoud et al. 2008).

20.3 Brief Botanical Descriptions Including Distribution

O. nivara is an annual wild species inhabiting temporary swamps, ponds, lake shores, riverbanks, in the vicinity of rice fields with shallow water, and is seasonally adapted to dry habitats (Chang 1976; Vaughan 1989; Li et al. 2006). It is predominantly self-fertilized, late flowering and photoperiod insensitive (Oka 1988; Sharma et al. 2000; Grillo et al. 2009; Kim et al. 2016). In general, its plant height ranges from 82 to 140 cm with a mean of 114 cm, flag leaf length varies from 19 to 52.4 cm, and leaf width found to be from 1.2 to 1.7 cm. Tillers are mostly semierect, and tiller number per plant ranges from 8 to 27 and productive tillers from 6 to 27. Panicle length varies from 17.8 to 28 cm with an average of 22.7 cm. It has more spikelets than its perennial species *O. rufipogon* (Sang and Ge 2007). But, it sheds the seeds naturally soon after seed maturation and also has stronger seed dormancy (Li et al. 2006) (Fig. 20.1). Morphological traits of 20 *O. nivara* accessions collected from different states of India are given in Table 20.1. The gene bank at IRRI, the Philippines, and *Oryza*base at the National Institute of Genetics, Japan, maintain nearly 1260 accessions of *O. nivara*, which is distributed throughout tropical and subtropical regions of Asia (Sanchez et al. 2013).

20.4 Geographical Origin, Distribution and Diversity of *Oryza nivara*

O. nivara ($2n = 24$ AA) is a distinct species and direct wild progenitor of *O. sativa* (Sharma and Shastry 1965; Lu et al. 2001). It is geographically distributed in South and Southeast Asia (Fig. 20.2). Taxonomic studies in genus *Oryza* were conducted by Shastry and Sharma and coined the nomenclature *Oryza nivara* in 1965 (Fig. 20.3). The origin of *O. nivara* is still unclear (Liu et al. 2015). Phylogenetic analysis shows that it evolved from the perennial ancestor resembling *O. rufipogon* (Zhu and Ge 2005; Grillo et al. 2009). However, the two differ distinctly in their morphology, life-history traits, mating system, flowering time, climate preferences and growth habitats (Oka 1988; Vaughan 1989; Grillo et al. 2009). These two species are phenotypically different but there is no significant genetic differentiation or no reproductive barriers between them (Morishima 2001). The life-history traits exhibit continuous variation in nature and segregate in an F_2 population (Morishima 2001). Consequently, these two wild species were often treated as two different ecotypes/siblings (Oka 1988; Morishima et al. 1992; Lu et al. 2000; Cheng et al. 2003; Sarao et al. 2016) or subspecies (Vaughan and Morishima 2003) within a single species of *O. rufipogon* (Zhu et al. 2007). A recent study on ten nuclear and two chloroplast sequences from 26 wild populations across the geographic ranges also showed that *O. nivara* might have independently originated multiple times from different *O. rufipogon* populations. Climatic factors played a significant role in adaptation, persistence and expansion of *O. nivara* (Liu et al. 2015).

It was believed that AA genome-wide relatives are direct ancestors of *O. sativa* in Asia. However, there are eight species of *Oryza* categorized into AA genome group ($2n = 24$ AA) under *O. sativa* complex, which includes both perennial and annual wild species, also known as primary gene pool. These are highly compatible with *O. sativa* and easily crossable (Oka 1988; Brar and Singh 2011). Previous studies demonstrated that *O. sativa* diverged from the intermediate type of *O. rufipogon* (Sano et al. 1980; Oka 1988). Some showed that it is an intermediate between perennial and annual wild species (Doebley and Stec 1991; Yamanaka et al. 2004). There are three hypotheses proposed to trace the immediate ancestor of *O. sativa*. The monophyletic hypothesis proposed that wild rice has a potentiality to evolve into *indica* as well as *japonica* types (Oka and Morishima 1982) and that *indica* versus *japonica* differentiation was not found in *O. rufipogon* (Chang 1976; Lu et al. 2002). Contradicting this, the diphyletic hypothesis was proposed based on the distribution of retrotransposon *p-SINE1-r2*, at the Waxy locus, which leads to *indica–japonica* differentiation. The annual form *O. nivara* has this element, and perennial forms of *O. rufipogon* lack this element indicating that *indica* subspecies evolved from annual *O. nivara* and *japonica* subspecies evolved from perennial form of *O. rufipogon* (Yamanaka et al. 2003). In contrast, polyphyletic hypothesis was postulated by Zhu and Ge (2005) based on intron sequences of four nuclear genes, but it did not resolve the phylogeny debate. Finally, the evidence obtained from the study of Li et al. (2006) showed that

Fig. 20.1 **a** Seeds of *O. nivara* showing dark hull and long awns. **b** Dehulled seeds of three *O. nivara* accessions

O. nivara is an immediate ancestor for the origin of *indica* rice. The phenotypic similarities in life history, mating system and reproductive allocation between *O. nivara* and *O. sativa* support this hypothesis (Chang 1976; Oka 1988; Sharma et al. 2000).

Table 20.1 Details of 20 *O. nivara* accessions

S. no	IRGC accession	Origin	PH	NT	NP	LL	LW	LA	PL	Ligule length	Ligule colour	Awn type	Awn colour	Panicle type	Apiculus colour	Stigma colour
1	IRGC80548	MP, India	109	15	7	38.1	1.3	37.9	23.2	0.5	Brown	Long	Pink	Semi compact	Red	Purple
2	IRGC80611	Orissa, India	105	9	9	30.4	1.4	31	5	0.6	Brown	Long	White	Compact	Red	Purple
3	IRGC80713	MP, India	97	11	10	25	1.5	27.4	5	0.8	Brown	Long	White	Semi compact	White	White
4	IRGC81832	Bihar, India	117	9	9	46.7	1.4	48.8	24	0.5	Brown	Long	White	Semi compact	White	Purple
5	IRGC81848	UP, India	120	11	10	43	1.4	46	24.1	0.3	Brown	Long	White	Semi compact	White	Purple
6	IRGC81852	UP, India	110	14	11	41.7	1.3	39.4	22	0.8	Brown	Long	White	Semi compact	White	White
7	IRGC100903	AP, India	116	9	6	49.7	1.3	48.3	22.9	0.3	Brown	Long	White	Open	White	Purple
8	IRGC105318	Kerala, India	130	11	11	54.2	1.5	60.5	22.1	0.7	Brown	Small	White	Semi compact	White	White
9	IRGC105320	Kerala, India	140	12	11	36.2	1.3	35.1	22.1	0.8	Brown	Long	Pink	Semi compact	Red	Purple
10	IRGC105333	Kerala, India	91	11	6	34.6	1.3	34.5	20.4	0.4	Brown	Long	Pink	Semi compact	White	White
11	IRGC105623	Orissa, India	130	10	9	25.5	1.3	25.4	25.1	1.2	Brown	Small	White	Semi compact	White	White
12	IRGC105708	TN, India	114	27	27	33.8	1.3	32	20.4	1.2	Brown	Long	White	Semi compact	White	White
13	IRGC105710	TN, India	82	12	11	31.5	1.4	33	8	1	Brown	Small	White	Compact	Red	Purple
14	IRGC106044	Assam, India	100	8	7	37.9	1.5	42.5	23.6	0.5	Brown	Long	White	Compact	White	White
15	IRGC106052	Assam, India	114	16	11	19	1.3	18.8	24.1	0.7	Brown	Long	White	Semi compact	White	White
16	IRGC106069	Assam, India	105	14	11	33.5	1.4	35	17.8	0.6	Brown	Long	Pink	Semi compact	Red	Purple
17	IRGC106094	WB, India	139	20	14	38	1.5	41.7	28	0.7	Brown	Long	Pink	Semi compact	Red	Purple
18	IRGC106099	WB, India	128	11	8	33.2	1.7	41.4	23.8	0.5	Brown	Small	Pink	Semi compact	White	Purple
19	IRGC106187	WB, India	111	12	9	37.7	1.2	32.9	5	0.7	Brown	Long	Pink	Semi compact	Red	Purple
20	IRGC106495	Kerala, India	120	14	9	32.6	1.6	39.8	24.6	0.8	Brown	Long	Pink	Semi compact	Red	Purple

DFF Days to 50% flowering, *DM* Days to maturity, *PH* Plant height, *NT* Number of tillers, *NPT* Number of productive tillers, *PW* Panicle weight, *YLDP* Yield per plant, *BY* Bulk yield and *BM* Biomass. LW in mm, all other length in cm

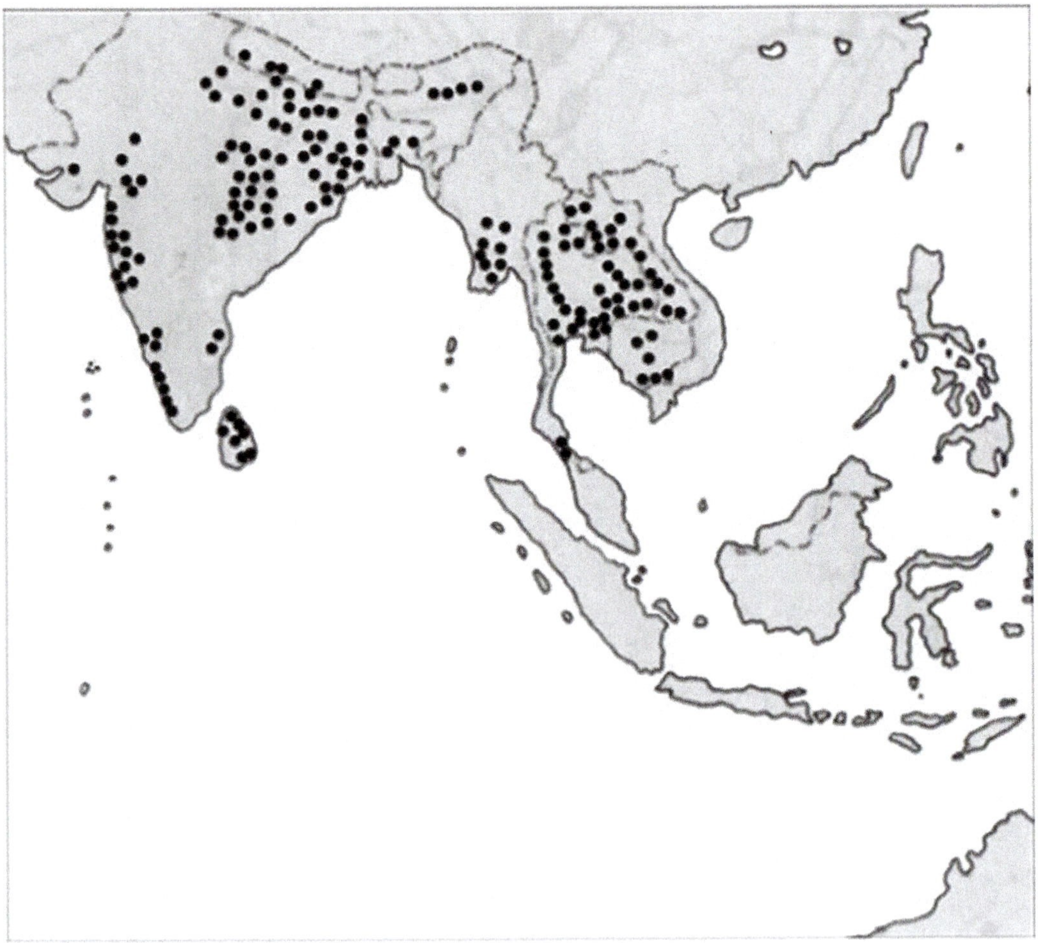

Fig. 20.2 Geographical distribution of *O. nivara*. Adopted from Vaughan et al. (2008)

Fig. 20.3 **Shastry SVS** and **Sharma SD** conducted taxonomic studies in genus *Oryza* and coined the nomenclature ***Oryza nivara*** in 1965

20.5 Classical Genetics on Gene Mapping

O. nivara is considered as immediate progenitor of cultivated *indica* rice, and numerous accessions are available, but classical genetics studies are lacking in wild species, since the value of each wild accession cannot be determined as selfed progenies. Wild progenitors provide an experimental system for characterizing the genetic basis of adaptation in any crop species. The genetics studies are mainly focused on evolution and domestication of *O. sativa* in relation to both of its wild progenitors *O. nivara* and *O. rufipogon.*

O. nivara and *O. rufipogon* are the closest wild relatives of *O. sativa,* but evidences on immediate progenitor of the cultivated rice remain contentious. It was reported that *O. sativa* was domesticated from *O. nivara* based on their phenotypic similarity (Khush 1997; Sharma et al. 2000; Chang 2003; Grillo et al. 2009). Oka (1988) reported *O. rufipogon* as the ancestor of Asian cultivated rice, as a direct wild progenitor. There are two schools of thought on the evolution of *O. sativa,* i.e. monophyletic evolution (Oka 1988; Wang et al. 1992; Joshi et al. 2000; Zhu and Ge 2005) and diphyletic evolution (Oka and Morishima 1982; Second 1982; Glaszmann 1987). Phylogenetic analysis has indicated that the annual species *O. nivara* was derived from a perennial ancestor resembling *O. rufipogon* (Zhu and Ge 2005).

Dolores et al. (1979) crossed *O. nivara* (Acc. 101508), which is resistant to grassy stunt virus, made direct as well as reciprocal crosses with six cultivars, viz. Mudgo, TKM-6, Taichung Native 1 (TNl), Dee-geo-woo-gen (Dgwg), IRS and IR8 and identified that *O. nivara* has same genome combination as of *O. sativa* from cytological studies. They also observed partial sterility in some of the crosses. Niroula et al. (2009) crossed *O. nivara* with *O. sativa* (IR64, Kalanamak and Manshara) and found same genome composition between species and chromosome pairing through meiotic analyses from F_1 hybrids. Recent studies on rice chloroplast genes (Takahashi et al. 2008; Fuller 2011), hybrid sterility gene (Du et al. 2011) and major genes of the photoperiod pathway (Huang et al. 2012) confirmed the independent origins of *indica* and *japonica* subspecies from wild progenitors *O. nivara* and *O. rufipogon,* respectively.

The rice chloroplast genetics gives evidence for shared characters between *indica* and *O. nivara* and fundamental difference between *indica* and *japonica.* Similarly, tri-allelic system of hybrid sterility gene provided a prospect to study the evolutionary changes underlying reproductive isolation in plants through sequence diversity and evolutionary history in *O. sativa* and its wild relatives *O. rufipogon* and *O. nivara* indicating their independent evolution. Miura et al. (2008) identified hybrid breakdown *hbd1* (*t*) locus from *O. nivara* chromosome, in a similar locus as previously reported by Matsubara et al. (2007) in *O. sativa* reflecting the domestication steps of *O. nivara* to *O. sativa* L. Spp *indica.* Based on QTL analysis, Hd1 was identified as one of the candidate genes for flowering time variation from a segregating mapping population between *O. rufipogon* and *O. nivara.* About 80% of QTL alleles of *O. nivara* were in the same direction of phenotypic evolution, showing direct selection and fixation of the traits in *O. nivara* during evolution (Grillo et al. 2009). Huang et al. (2012) studied the patterns of natural and artificial selections in domestication process and found that four major genes of the photoperiod pathway caused the subdivisions, among cultivated and wild rice.

Although a variable number of fertility restorer genes (*Rf*) for WA and HL-CMS have been identified in various cultivars, information on *Rf* in *Oryza* species with the AA genome is sparse. Therefore, the distribution and heredity of *Rf* for WA and HL-CMS in wild rice species of *Oryza* with the AA genome were investigated (Li et al. 2005). The fertility restorer genes exist widely in *Oryza* species with the AA genome and found that *Rf* in *O. sativa* originated from the *O. rufipogon*/*O. nivara* complex. The origin and evolution of *Rf* are tightly linked to those of CMS in wild rice, and fertility of a given CMS type is controlled by several *Rf* alleles in various wild restorer accessions. The fertility restorer accessions, especially those with complete restoring ability, aggregated mainly in two species of *O. rufipogon* and *O. nivara.* The wild rice accessions with *Rf* for HL-CMS were distributed in Asia, Oceania, Latin American and Africa, but were centred mainly in Asia, while the wild restorer accessions for WA-CMS were limited only to Asia and Africa.

Grillo et al. (2009) studied the F_2 mapping population derived from a cross between *O. rufipogon* (IRGC80506) and *O. nivara* (IRGC80470) identified QTLs for phenotypic differentiation in life history, mating system and flowering time between both the species. The study helped to identify the genetic architecture for the adaptive origin of *O. nivara* and the population genetic theories of adaptation. Choi and Purugganan (2017) proposed multiple origin from wild progenitor subpopulations of *O. nivara* or *O. rufipogon,* but single domestication led to domesticated Asian rice; however, domestication alleles were transferred between rice subpopulations through introgression.

Hurwitz et al. (2010) made a comparative analysis of structural variation between rice and three of its closest relatives *O. nivara, O. rufipogon* and *O. glaberrima* and found that BAC end sequences of *O. nivara* have 16% sequence similarity with *O. sativa* genome. *O. sativa* showed the most contracted and inverted regions when compared to *O. nivara.* These regions are to be studied for detailed mechanisms of speciation and domestication in rice, and artificial and natural selection in these genomes. Zhang et al. (2014) conducted comparative genomic analyses among closely related species by de novo assembling of AA genome sequences of *O. nivara, O. glaberrima, O. barthii, O. glumaepatula* and *O. meridionalis* and revealed a close similarity of *O. nivara* with *O. sativa* genome.

The major challenges in using any wild species for classical genetics studies are lack of data on species' basic biology (ploidy levels, pedigree, progenitors, taxonomy, etc.). The phenotype of donor is not always the predictor for progenies and impact of allelic combinations in different cultivated crop backgrounds. So, interspecific crosses from wild species are rarely screened for genetic studies.

20.6 Cytological Details of Genome Including Karyotype Data

Karyotyping helps to explore the chromosomal aberrations, speciation, evolutionary aspects and transfer of genes from wild to cultivars of rice. The report on cytological studies of *O. nivara*, *O. sativa* (IR64, Kalanamak and Manshara) and its hybrids (*O. sativa* cv. IR64/*O. nivara*, Kalanamak/*O. nivara* and Manshara/*O. nivara*) at different stages of meiosis was first given by Niroula et al. (2009). The outcrossing rate of *O. nivara* with cultivars varied from 35 to 44.75%, and it depends on the genetic make-up of female parents and geographic race of wild species.

The chromosomal pairing was found to be normal between *O. nivara* and *O. sativa*. At metaphase I, the number of univalents (0.06) and quadrivalents (0.01) was very low in *O. nivara*, whereas it was slightly high in hybrids. All hybrids had more than 11.8 bivalents and 22 chiasmata/PMC, but these were quite high in *O. nivara* (bivalents 11.94, chiasmata 23/PMC). However, few abnormalities such as unequal segregation, bridges, laggards and late disjunction were high in IR64/*O. nivara*. Only 3.08% of PMCs had bridges and fragments in Kalanamak/*O. nivara* hybrid, absent in parents and remaining hybrids in anaphase and telophase I. The chromosomal rearrangements and genome expansions may lead to the new phenotypes which are beneficial for breeding programme (Niroula et al. 2009).

20.7 Physiological Studies

Rice is a model crop to study structural and physiological properties of leaf, which will change over the course of plant developmental stages or in response to several environmental stresses. Leaves play a major role in long-term adaptation to environment (Tian et al. 2016). The size, shape, morphology and anatomy of leaf vary greatly in different environments (Evans et al. 2004). The leaf anatomical traits (structural) such as size/density, thickness and surface area of stomata and mesophyll were associated with the functional (physiological) traits such as net photosynthetic rate, stomatal and mesophyll conductance, transpiration efficiency, carboxylation efficiency and water use efficiency (Giuliani et al. 2013; Baker et al. 2017). Most of the plant dry matter is derived from the carbon originated from photosynthesis (Poorter and Remkes 1990). But, grain yield was stagnant in past 10 years, compared with the Green Revolution period in the 1980s (FAOSTAT 2010). Further improvement of grain yield could be achieved by the identification of lines with higher photosynthetic efficiency (Makino 2011). Since the diversity available for trait improvement became very less in cultivars, the worth of wild species was recognized for improving photosynthesis efficiency that can lead to increase in grain yield. For example, the *O. nivara* accession CR100097 from India had highest net photosynthetic rate (Pn) of 24.2 μmol/(m^2 s), transpiration efficiency (Pn/E) 2.24 μmol/mmol, carboxylation efficiency (Pn/Ci) 0.127 μmol/(m^2 s) and intrinsic water use efficiency (Pn/gs) 109.0 μmol/mol. Similarly, two *O. nivara* accessions CR100008 and IR104650 showed highest leaf thickness (0.13 mm), and one *O. nivara* accession IR104650 showed highest specific leaf mass of 3.7 mg cm^{-2} (Kiran et al. 2013). These can be used as potential donors for improving leaf photosynthetic efficiency in cultivated varieties, thereby increasing its yield and biomass. Three of these accessions are being used to develop chromosome segment substitution lines (CSSLs) in the background of two popular cultivars Swarna and MTU010 at IIRR (Sarla N, ICAR-National Profesor Project, Indian Institute of Rice Research (IIRR)).

O. nivara was reported to have maximum number (28) of stomata in both adaxial and abaxial sides of leaf and was positively correlated with stomatal conductance and carbon balance (Kondamudi et al. 2016). The size and density of stomata contribute to the leaf stomatal conductance to gas diffusion (g_s), whereas leaf mesophyll architecture contributes to CO_2 diffusion from the intercellular air space to the chloroplast stroma (Giuliani et al. 2013). Highest fraction of mesophyll cell wall covered by chloroplast (94%) and cell volume occupied by chloroplasts (64.4%) was recorded in *O. nivara* accession PI590405. This can provide more pathways acting in parallel for CO_2 diffusion to and from chloroplasts which reduce the resistance to CO_2 movement into mesophyll cells and increase the mesophyll conductance. This accession also showed lowest intrinsic transpiration efficiency. Generally, leaf net photosynthetic rate is positively correlated with mesophyll conductance (gm) and negatively correlated with intrinsic water use efficiency. This could be beneficial for preserving water, when rice is cultivated in warm climates (Giuliani et al. 2013). Thus, *O. nivara* accessions need to be collected from different climatic areas to get good combination of genes for physiological adaptability.

20.8 Molecular Mapping of Genes and QTLs

Wild progenitor species of Asian cultivated rice are important genetic resources for rice improvement. The IRRI gene bank holds more than 1200 accessions of *O. nivara,* but they are less exploited in breeding programmes, especially for yield improvement due to their poor phenotypic appearances. *O. nivara* is an AA genome species and easily crossable with the cultivated rice germplasm. Even though some sterility barriers exist for use of *O. nivara*, advanced backcross QTL analysis (AB-QTL) has been extensively used to develop wild species-derived mapping populations, which helps to restore spikelet fertility and get rid of undesirable traits through successive backcrosses to cultivated rice (Tanksley and Nelson 1996; Swamy et al. 2011b). This process helps in simultaneous QTL identification and introgression to develop backcross inbred lines (BILs). In addition to providing

Table 20.2 QTLs/genes mapped for disease resistance and spikelet fertility from *O. nivara*

Trait	Donor	Gene/QTL	Marker	References
BLB	IRGC81825	*Xa38*	Oso4g53050-1	Bhasin et al. (2012)
BLB	IRGC105710	*Xa33*	RMWR7.1, RMWR7.6	Kumar et al. (2012)
BPH	Acc. no. 102, 165	–	RM16655, RM3317	Wu et al. (2009)
GLH	IRGC105715	*Grh6*	RM8213	Fujita et al. (2004)
GS	IRGC101508	*GS*	–	Khush and Ling (1974)
DT	*O. nivara*	*OsDREB1F1*	–	Singh et al. (2015)
SB	IRGC100898, IRGC104705	*qShb1*, *qShb3*, *qShb6*, *qShb7*		Eizenga et al. (2013)
BL	IRGC100898, IRGC104705	*qBlast8*, *qBlast12*		Eizenga et al. (2013)
PF	IRGC102167	*qHS6a*	RM190-RM510	Yang et al. (2016)
SF	IRGC102167	*qHS6a*	RM190-RM510	Yang et al. (2016)
PS	IRGC105444	*S36*	M1S36, RM3483	Win et al. (2012)
PS	*O. nivara*	*S27-niv*	–	Win et al. (2011)
WUE	IRGC105715	*BKL*	RM30	Hamaoka et al. (2017)
Awn	W0054	*QTLs on chr4 and 8*	–	Uehara et al. (2016)

important alleles which were left behind during domestication, the crosses with wild species also generate genome-wide de novo genetic variations resulting in lot of transgressive segregants for various traits of economic importance in rice (Dong et al. 2006).

Several genes and major effect QTLs have been identified for pest and disease resistance, yield and yield component traits, grain quality and nutritional traits from different accessions of *O. nivara* (Tables 20.2, 20.3 and 20.4). It is interesting to note that *O. nivara* is the only known source of resistance for grassy stunt virus (GSV) among thousands of accessions screened (Khush and Ling 1974). Two genes *Xa33* and *Xa38* for bacterial leaf blight (BLB) resistance have been mapped on chromosomes 7 and 4, respectively, and for green leaf hopper (GLH) resistance on chromosome 4. Several major loci for sheath blight resistance have been mapped on chromosomes 1, 3, 6 and 7, and blast resistance QTLs were identified on chromosomes 8 and 12. A drought-responsive promoter *OsDRRBF1* was found only in *O. nivara* accessions among the 136 Indian wild rice accessions, which were resequenced for allele mining and association analysis (Singh et al. 2015). Major loci for pollen and spikelet fertility have also been identified from *O. nivara* (Melonek et al. 2016). The *O. nivara* accessions can also be a source of cytoplasmic male sterility (CMS) and heterotic hotspots, which are useful in hybrid rice breeding programmes (Hoan et al. 1998; Gaikwad et al. 2014). Phylogentic studies have clearly shown that many of the abiotic and biotic stress tolerance genes in the land races or cultivated germplasm are derived from *O. nivara* (Menguer et al. 2017). The outcrossing nature of *O. nivara* has made it a good resource to study the flow of transgenes and their environmental escape as well (Chen et al. 2004).

QTL analysis has been carried out using *O. nivara* to understand rice domestication, which revealed significant gene losses and severe genetic bottlenecks during the domestication of cultivated rice from its wild progenitors (Li et al.

Table 20.3 QTLs for yield, yield components and grain quality traits from *O. nivara*

Trait	QTL	References
Days to flowering	*qDH8, qDH3, qDH4, qDH6*	Eizenga et al. (2016)
	DF1, DF6, DF9	
Days to maturity	*qdm2.1*	Malathi et al. (2017)
Plant height	*qph1.1*	Malathi et al. (2017)
	qPh1, qPh12	Eizenga et al. (2016)
	Ph4, Ph7	Li et al. (2006)
	Ph1.1, Ph2.1, Ph4.1, Ph6.1	Swamy et al. (2014)
Number of tillers	*nt3.1, nt5.1, nt11.1, nt12.1*	Swamy et al. (2014)
	qPT1, qPT9	Eizenga et al. (2014)
Number panicles	*qPNNB7*	Eizenga et al. (2016)
	npt1.1, npt6.1, npt9.1, npt11.1, npt12.1,	Swamy et al. (2014)
Panicle length	*pl1*	Gaikwad et al. (2014)
Number of spikelets	*sn1, sn4, sn6, sn7, sn12*	Li et al. (2006)
	nsp1.1, nsp2.1, nsp3.1, nsp9.1	Swamy et al. (2014)
	spp1, spp4, spp5	Gaikwad et al. (2014)
Number of filled spikelets	*nfg1.1, nfg2.1, nfg12.1*	Swamy et al. (2014)
	gpp5, gpp7, gpp10	
Number of primary branches	*pbr1, pbr4, pbr6, pbr7*	Li et al. (2006)
	npb2.1	Swamy et al. (2014)
Number of secondary branches	*sbr1, sbr4, sbr6, sbr7.1, sbr7.2, sbr9*	Li et al. (2006)
	nsb1.1, nsb2.1	Swamy et al. (2014)
Grain weight (g)	*gw1.1, gw2.1, gw4.1*	Swamy et al. (2014)
	gw10, gw11	Gaikwad et al. (2014)
Hundred seed weight (g)	*qHSDWT10*	Eizenga et al. (2016)
Culm length	*qCL5*	Eizenga et al. (2016)
	qSL5	Eizenga et al. (2016)
	qCa9	Eizenga et al. (2016)
Seed shattering	*sh3, sh4, sh8*	Li et al. (2006)
Seed maturation	*sm9*	Li et al. (2006)
Seed dormancy	*sd1, sd6*	Li et al. (2006)
Spikelet fertility (%)	*sf1, sf4, sf5, sf7*	
Tiller angle	*ta3, ta4, ta7, ta9, ta10, ta11*	Li et al. (2006)
Vegetative biomass (g)	*bm1.1, bm2.1, bm5.1, bm9.1, bm11.1,*	Swamy et al. (2014)
Plot yield (Kg)	*yldp2.1, yldp3.1, yldp8.1, yldp9.1, yldp11.1, yldp12.1*	Swamy et al. (2014)
	yld2, yld8	Gaikwad et al. (2014)
Stem diameter (mm)	*sd6.1, sd7.1, sd7.2, sd8.1, sd9.1*	Swamy et al. (2011b)
Rachis diameter (mm)	*rd2.1, rd2.2, rd6.1, rd9.1, rd11.1*	Swamy et al. (2011b)

Table 20.4 QTLs for grain quality and grain micronutrient traits from *O. nivara*

Trait	QTL	References
Kernel width (mm)	*qKLRW1, qKLRW5*	Eizenga et al. (2016)
	kw1.1, kw1.2, kw1.3, kw1.4, kw3.1, kw4.1, kw5.1, kw6.1, kw8.1	Swamy et al. (2012)
Grain weight (g)	*gw1.1, gw2.1, gw4.1*	Swamy et al. (2012)
	gw10, gw11	Gaikwad et al. (2014)
Kernel width (mm)	*qKL1*	Eizenga et al. (2016)
Amylose content	*qAAC6*	Eizenga et al. (2016)
	ac2.1, ac5.1	Swamy et al. (2012)
Milling potential (%)	*mp1.2*	Swamy et al. (2012)
Length and width ratio	*lwr12.1*	Swamy et al. (2012)
	qKLWR-1, qKLWR-5	Eizenga et al. (2016)
Water uptake	*wup2.1*	Swamy et al. (2012)
Kernel length after cooking (mm)	*klac5.1, klac12.1*	Swamy et al. (2012)
Elongation ratio	*er3.1*	Swamy et al. (2012)
Alkali spreading value	*asv11.1*	Swamy et al. (2012)
Grain Zn (ppm)	$qZn_{3.1}$, $qZn_{5.1}$, $qZn_{12.1}$	Swamy et al. (2011a, b)
Grain Fe (ppm)	$qFe_{1.2}$, $qFe_{3.2}$, $qFe_{4.1}$, $qFe_{11.1}$, $qFe_{11.2}$	Swamy et al. (2011a, b)

2006; Kovach and McCouch 2008). In most of the QTL mapping studies using *O. nivara* as one of the parents, it has been reported that more than 50% of the positive alleles were derived from *O. nivara* (Swamy et al. 2014). These results proved that even phenotypically poor wild progenitors have favourable alleles for yield, grain quality and nutritional improvement of rice (Swamy et al. 2014; Ma et al. 2016). QTLs for these traits were found to be distributed on almost all the 12 chromosomes, and several of them had phenotypic variance (PV) of more than 10% (Table 20.3). In several QTL hotspot regions, colocation of QTLs for different traits was observed. Some of the important QTL hot spot regions were on chromosomes 1, 2, 3, 4, 5, 8, 9, 11 and 12. Meta-analysis of QTLs derived from *O. nivara* and other wild relatives also showed significant trait colocations indicating the conservation of these alleles among different accessions and also with other AA genome species such as *O. rufipogon* and *O. grandiglumis*, but the direction of their allelic effects varied in different accessions and in different recipient genetic backgrounds (Swamy et al. 2011a, b). Some of these consistent and major QTLs can be explored in marker-assisted breeding (MAB) through a well-designed QTL pyramiding programme with the main goal of breaking yield barriers and to diversify the genetic base of cultivated rice varieties.

20.9 Structural and Functional Genomics Resources Developed for *O. nivara*

There are very scanty structural and functional genomics resources available for *O. nivara*. However, considering the potential of wild rice as a source of novel alleles for rice improvement, significant efforts are being made to develop genomics resources for wild species of rice. The OMAP was one of the major rice research initiatives of the last decade to decode the wild rice genomes, which also included *O. nivara* accession IRGC100897 (Wing et al. 2005). In this project, BAC end sequences were produced, sequenced and assembled to develop a physical map and the sequences were functionally annotated (Ammiraju et al. 2006). The *O. nivara* chloroplast genome has also been sequenced and is being extensively used in comparative genomics studies (Masood et al. 2004). Comparative genomics of *O. sativa* and *O. nivara* showed that almost 5% (1000 genes) of the genes were lost in *O. sativa* during evolution (Sakai and Itoh 2010), while chromosome 3 showed high degree of synteny and colinearity, but *O. nivara* chromosome 3 was 21% smaller than that of *japonica* (IRGSP 2005).

With the availability of cheaper and faster sequencing technologies and advanced computing facilities, resequencing of several thousand accessions from rice gene bank has become possible (Deschamps et al. 2012). Xu et al. (2012) resequenced 50 rice accessions including five accessions of *O. nivara* and identified millions of single nucleotide polymorphisms (SNPs). Several genetic linkage maps have also been constructed using simple sequence repeats (SSRs) and SNPs in different populations such as F_2, backcross populations and BILs, which were developed using different accessions of *O. nivara* (Li et al. 2006; Swamy et al. 2012; Ma et al. 2016). A genome-wide and across different rice species transposable elements data base (RITE-db) has been established (Copetti et al. 2015). Most of the genomics resources developed for cultivated rice *O. sativa* can also be easily applied to *O. nivara* since both the species share high level of synteny and colinearity. All these recent advances in rice genomics have opened up new avenues for characterizing the wild rice accessions at the molecular level and for their better exploitation in rice improvement.

Several molecular tools have been used to understand the phylogeny and evolutionary pattern of *O. nivara* and cultivated rice. The SSR-based diversity analysis among *O. nivara* accessions revealed wider intra- and interpopulation variations (Sarla et al. 2003; Juneja et al. 2006; Kaladhar et al. 2008; Swamy et al. 2014; Malathi et al. 2016). Extensive gene losses from wild rice to cultivated rice during domestication have also been reported (Kovach and McCouch 2008). Diversity analyses and phylogenetic studies using genes such as *OsDREB1F1*, dehydrins, *Aldh1*, *Aldh2*, *WRKY* and *RTPOSON* transcription factors, *Sub1* gene and *OsFRDL4* showed species-specific gene conservation and wider diversity among different *O. nivara* accessions (Ammiraju et al 2007; Singh et al. 2015; Yokosho et al. 2016; dos Santos et al. 2017). Diversity using *Xa5*, *Xa21* and *Xa25* gene-specific markers also showed more diversity among *O. nivara* compared to cultivated rice for resistance to BLB (Bimolata et al. 2015). Molecular analysis also clearly showed that wild species-derived crosses generate lot of *de novo* variations and transgressive segregants, and one example is the high protein content in hybrids derived between *O. nivara* and *O. sativa*, where progenies had 18% more protein than both the parents (Mahmoud et al. 2008), for yield and yield-related traits also transgressive segregants have been reported (Li et al. 2006; Swamy et al. 2014; Ma et al. 2016).

There are fewer attempts to characterize *O. nivara* accessions using genomics, metabolomics and proteomics approaches. In an expression analysis using *O. nivara*, *indica* and *japonica* accessions revealed that 4489 genes differentially were expressed between *O. nivara* and *japonica*, while 1767 genes differentially expressed with *indica* (Peng et al. 2009). *Wsi18* promoter from wild rice genotype, *O. nivara*, shows enhanced expression under soil water stress in contrast to elite cultivar, IR20. Xu et al. (2016) identified 97 WRKY transcription factors specific to *O. nivara*. Whole-genome sequencing is being carried out for identifying polymorphic SNPs, QTLs and genes of economic importance (Silva et al. 2012; Duitama et al. 2015; Ma et al. 2016). All these studies have revealed new useful information on *O. nivara* accessions and their potential for use in rice improvement.

20.10 International Genomics Initiatives of *O. nivara*

To better understand the wild species of *Oryza*, a comparative genomics programme OMAP was initiated at AGI to study evolution, development, genome organization, polyploidy, domestication, gene regulatory networks and crop improvement (Wing et al. 2005; Sanchez et al. 2013). A total of 12 (11 wild and 1 cultivated rice species) BAC (bacterial artificial chromosome) libraries were constructed (Wing et al. 2005; Ammiraju et al. 2006) which included *O. nivara* (Ac. no W0106), the closest wild progenitor species of *O. sativa* with a genome size of 448 Mb ($2n = 12$). Its genome has 42.8% of GC content and constitutes the total organellar DNA of 4.1%. The BAC clones were fingerprinted using modified Snapshot method described by Luo et al. (2003). These clones were end sequenced with an average high-quality sequence read length of 656 bp. The overlapping sequences of clones were assembled using FPC (clone-based fingerprint contig) software described by Soderlund et al. (2000), and these contigs were aligned to the IRGSP reference genome sequence using BAC end sequences (BESs) to develop physical maps (Wing et al. 2005). The coverage of genome after subtracting the organellar and non-insert containing clones was 19 fold (Wing et al. 2005; Ammiraju et al. 2006). By the year 2005, the consortia of AGI completed the fingerprinting and end sequencing of 3 OMAP BAC libraries: *O. brachyantha* (OB-Ba); *O. nivara* (OR_BBa); and *O. rufipogon* (OR_CBa) (Wing et al. 2005).

Further investigations done under OMAP produced huge data which were difficult to handle by consortium alone; therefore, it was proposed to establish the international *Oryza* map alignment project (I-OMAP) with the collaboration of Arizona Genomics Computational Laboratory, Cold Spring Harbor Laboratory, Purdue University and National Center of Gene Research (Wing et al. 2005). I-OMAP was initiated in 2007 to generate reference sequence and transcriptome data for all eight AA genome species and generate map and phenotype-advanced ABC, CSSL and RIL populations for the AA genome species for functional and breeding studies (Wing et al. 2005). To date, 16 *Oryza* species BAC-based physical maps have been constructed in I-OMAP. This constitutes 15 wild species and 1 cultivated species.

20.11 Strategies and Tools Used for Sequencing

Total genomic DNA was extracted using cetyltrimethylammonium bromide (CTAB) method, and it was partially digested with the restriction enzyme HindIII. The DNA fragments were then ligated into pIndigoBAC536 SwaI (pAGIBAC1) and transformed into Escherichia coli. The libraries were sequenced on Illumina HiSeq 2000 and were arrayed in 384-well microtiter plate (MTP). The Illumina reads of different library sizes were assembled with Allpaths-LG/Snapshot, scaffolds were constructed with SSPACE, and gaps were closed with GapFiller (Ammiraju et al. 2006). Similarly, the repeat sequences, transposable elements, LTR-retrotransposons, LINEs, SINEs, DNA elements and unclassified sequences were analysed and annotated using similarity based database RepeatMasker (http://www.repeatmasker.org/). Protein-coding gene annotation was performed with evidence-based MAKER-P genome annotation pipeline. Noncoding RNA genes were predicted with infernal and tRNA genes with tRNA scan (http://plants.ensembl.org/Oryza_nivara/Info/Annotation).

20.11.1 Genome Sequencing of *O. nivara*

The complete genome sequence and assembly of *O. nivara* accession IRGC100897 was performed by AGI. The Illumina reads of different BAC library sizes were assembled. The overall estimated coverage from whole-genome shotgun (WGS) method was 102×; total genome sequence length was 337,950,324 bp. Protein-coding gene annotation was performed with evidence-based MAKER-P genome annotation pipeline. It contains 36,313 coding genes and 1648 noncoding genes including 1647 small noncoding RNAs. The total number of gene transcripts reported was 50,032. The SNP variation of 40 cultivated and 10 wild accessions revealed that *indica* was closely related to *O. nivara* (Xu et al. 2012).

20.12 Assembly

20.12.1 FPC/BES Contig Assembly

The complete genome sequence and assembly of *O. nivara* accession IRGC100897 was carried out by AGI. The Illumina reads of different BAC library sizes were assembled with Allpaths-LG, scaffolds were constructed with SSPACE, and the gaps were closed with GapFiller. A 19× coverage BAC library was used to generate a fingerprint contig (FPC) map (http://www.ebi.ac.uk/ena/data/view/GCA_000576065.1). The overall estimated coverage from whole-genome shotgun (WGS) method was 102×; total genome sequence length was 337,950,324 bp.

20.12.2 Repetitive Sequences

Most of the repetitive DNA sequences in rice are from nuclear DNA which are primarily transposable elements (TEs) that constitute 25.7% of total genome content (Li et al. 2017). The copy number and the total coverage of TE in genome are given in Table 20.5 (Li et al. 2017). The ancestral LTR-transposon family RWG (RIRE2, Wallabi and Gran3) has a high implication on lineage-specific regulation which can dramatically change the genome size and organization (Ammiraju et al. 2007). The copy number of RWG family is 3240 in comparison with the reference sequence of international rice genome sequencing project. In addition to this, class II transposable elements CACTA (Rim2/Hipa-TD) have been used as marker system for genetic diversity studies. These were highly polymorphic in Asian taxa *O. nivara* (93%) and *O. rufipogon* (95%) (Kwon et al. 2006). The comparative analysis of five rice genomes *O. nivara*, *O. glaberrima*, *O. barthii*, *O. glumaepatula* and *O. meridionalis* with *O. sativa* suggested that LTR retrotransposons are most abundant and occupy roughly 18.0, 18.7, 19.3, 18.9, 19.6 and 30.6% of genomes, respectively (Zhang et al. 2014).

20.13 Gene Annotation

Protein-coding gene annotation was performed with evidence-based MAKER-P genome annotation pipeline, and number of contigs found was 16,484 bp. It contains 36,313 coding genes and 1648 noncoding genes including 1647 small noncoding and 1 misc noncoding genes. The total number of gene transcripts reported was 50032. In addition, some noncoding RNA genes were predicted with infernal and tRNA genes with tRNAscan. The repeated sequences like transposable elements with *Oryza*-specific de novo repeat libraries were annotated using RepeatMasker software (http://plants.ensembl.org/Oryza_nivara/Info/Annotation/) (Ammiraju et al. 2006). The SNP variation of 40 cultivated and 10 wild accessions revealed that *indica* was closely related to *O. nivara* (Xu et al. 2012).

20.14 Genome Duplication

Gene duplications play a vital role in acquiring new genes which synthesize novel proteins in organisms and contribute tremendously to evolutionary process (Magadum et al. 2013). The duplicated genes are fixed through the natural selection, but its expression depends on nucleotide substitution, *cis*-regulation and post-translational regulation (Shiu et al. 2006; Hittinger and Carroll 2007). The plant miRNAs play important role in post-transcriptional gene silencing which are usually derived from inverted duplication of their target genes (Sunkar and Jagadeeswaran 2008). The members of distinct miRNA families have been found to be located in duplicated genomic sequences of rice (Jiang et al. 2006). The gene families are formed by the duplication of single gene with similar biochemical function. The largest gene families involved in plant development and stress responses are WRKY transcription factors. Overall, 97 OnWRKY genes were identified, OnWRKY 1 to OnWRKY 125, and their length ranged from 149 to 1389 amino acids. In total, 17 *OnWRKY* genes were considered as tandem

Table 20.5 Copy number of the transposable elements in genome, genes and total coverage

Transposable elements	Copy number of TE in the genome	Copy number of TE in the genes	Coverage of TE in Mbp
Class I (Retroposons)			
SINEs	5166	1238	1.2
LINEs	5283	1308	3.7
LTR elements			
Copia	7337	1236	8.8
Gypsy	30,685	3567	39.3
Class II (DNA transposons)			
TcMar-Stowaway	49,820	12,783	9.5
PIF-Harbinger	48,286	9440	10.4
MULE-MuDR	36,946	8840	11.4
CMC-EnSpm	22,671	4171	9.1
hAT	9143	1975	2.2
RC/Helitron	8020	1713	4.4
Total interspersed	244,092	51,021	86.8

Adopted from Li et al. (2017)

duplication genes in three clusters on chromosomes 1, 5 and 11 (Xu et al. 2016).

20.15 Synteny with Allied and Model Genomes

A comparative physical map of *O. nivara* chromosome 3 revealed a high degree of sequence identity and synteny with *japonica*, but it was 21% smaller than *japonica* which diverged ~10,000 years ago. It shares large blocks of synteny with maize and sorghum genome. *Oryza* was separated from maize and sorghum approximately 50 MYA, whereas from wheat and barley approximately 40 MYA. Importantly, the short arm was highly collinear to the short arm of maize chromosome 1 (1S) and the inverted long arm of maize chromosome 9 (9L), while long arm of *japonica* is highly conserved with maize 1L and the inverted maize 5S (Minx et al. 2005). The similarity searches revealed the proteome of *Arabidopsis thaliana* showed putative homologs for 67% of the chromosome 3 proteins of *O. sativa* (Minx et al. 2005). The Sh1 gene for seed shattering has undergone parallel selection during the process of domestication in rice, maize and sorghum. For example, the Sh1 gene in sorghum was conserved and showed colinearity in several cereals (Lin et al. 2012).

20.16 Comparison of Gene Families

The gene families are formed by the duplication of single gene with similar biochemical function. The largest gene families which are involved in plant development and stress responses are WRKY transcription factors. The expression of these WRKY families was highest in roots and lowest in panicles. In total, 17 *OnWRKY* genes were considered as tandem duplication genes in three clusters on chromosomes 1, 5 and 11. The phylogenetic analyses showed that ancient subgroup IIc WRKY genes were the ancestors of all WRKY genes (Xu et al. 2016).

Micro-RNAs (miRNA) are small noncoding RNAs play a major role in post-transcriptional level gene expressions. Approximately 21 miRNA families are conserved between the dicots and monocots of higher plants (Sunkar and Jagadeeswaran 2008). 682 miRNAs were found in 155 diverse plant species, and in these, 15 miRNA families were conserved in 11 plant species using all publicly available nucleotide databases.

The whole-genome sequencing of *O. nivara, O. glaberrima, O. barthii, O. glumaepatula, O. meridionalis* compared with *O. sativa* identified 39,293 orthologous gene families comprising 211,718 genes (Zhang et al. 2014). The expansions/contractions in gene families led to diversification (Purugganan et al. 1995). Excluding lineage-specific families, 17563 gene families were identified which contain 130,636 genes. A total of 276, 271, 276, 263 and 251 miRNA genes were identified which belong to 203, 125, 124, 133, 125 and 123 miRNA families in *O. nivara, O. glaberrima, O. barthii, O. glumaepatula, O. meridionalis* compared with *O. sativa,* respectively. These gene families were involved in biological processes such as biotic or abiotic stress responses, or the control of reproductive isolation through pollen recognition. The expansion of agronomically relevant gene families in six AA genome species was associated with disease resistance and flower development (Zhang et al. 2014). MITE (miniature inverted-repeat transposable element) sequences are also one of the major components of repetitive DNA families in the rice genome. The genetic variations in wild species were higher than *O. sativa* and *O. glaberrima.*

20.17 Impact on Germplasm Characterization and Gene Discovery

The narrow genetic base of existing germplasm is causing a serious threat to food security due to their vulnerability to new biotic and abiotic stresses. *O. nivara* is a superior source to enhance the diversity in the existing genepool due to its allelic diversity and its compatibility with *O. sativa* cultivars in hybridization programmes.

The impact of wild species cannot be assessed in breeding programmes based on their phenotype, and the genetic distance or allelic diversity is a rational approach to estimate their value. Prebreeding is a key step in bringing the valuable traits of common wild rice CWR to cultivars in a more precise and usable form (Stander 1993; Valkoun 2001; Haussmann et al. 2004; Sharma et al. 2013) and characterization. Genetic diversity through genotyping has been reported in rice germplasm sets including *O. nivara* (Sarla et al. 2003; Juneja et al. 2006; Zhu et al. 2007). Knowledge of genetic diversity and genetic relationships between germplasm accessions is the basic foundation for crop improvement programmes (Thomson et al. 2007). Park et al. (2003) studied genetic variations of AA genome *Oryza* species using MITE-AFLP (miniature inverted-repeat transposable element) technique to identify genetic variations, species relationships and detected high polymorphism within and between species. In comparison between geographical lineages of the AA genome species, African taxa *O. glaberrima* and *Oryza barthii* showed lower variation than the Asian taxa *O. sativa*, *O. rufipogon* and *O. nivara*, and Australian taxon *O. meridionalis*. They found that clustering pattern of the Asian taxa supports diphyletic evolution showing the independent clustering of *japonica* and *indica* accessions with a closer affinity to different wild accessions of *O. rufipogon* and *O. nivara* (Park et al. 2003).

Zhu et al. (2007) conducted multilocus analysis of nucleotide variation of *Oryza sativa* and its wild relatives and found varying degrees of reduction of genetic diversity in crops relative to their wild progenitors occurred during the process of domestication. Higher nucleotide diversity is observed in *O. rufipogon* than in *O. nivara* because the former has wider distribution and is largely outcrossing species, whereas *O. nivara* is primarily inbreeding (Vaughan 1989). Higher LD (linkage disequilibrium) in the annual *O. nivara* than in the perennial *O. rufipogon* is also consistent with the higher outcrossing rate in *O. rufipogon* than in *O. nivara* (Morishima et al. 1984; Barbier 1989). Over all loci, it was found that 66.7% of *O. rufipogon* individuals are heterozygous, whereas 23.1% of *O. nivara* individuals are heterozygous due to their varying outcrossing rates (Morishima et al. 1984; Barbier 1989). Genetic variation in self-pollinating species like *O. nivara* is comparatively lower due to less rate of recombination (Charlesworth 2003) and self-fertilization. It also showed that LD mapping by genome scans may not be feasible in wild rice due to the high density of markers needed.

Hilario et al. (2013) found that genetic variation patterns within and between species may change along geographic gradients and at different spatial scales. This was revealed by microsatellite data at 29 loci obtained from 119 accessions from *O. sativa*, *O. nivara*, *O. rufipogon* and *O. meridionalis*. Genetic similarities between *O. nivara* and *O. rufipogon* across their distribution are evident in the clustering and ordination results and in the large proportion of shared alleles between these taxa. Kovach et al. (2009) examined the occurrence of the *badh2.1* allele for aroma in 280 accessions of wild rice (*O. rufipogon*/*O. nivara*) and found that it was absent from all wild genotypes except for a single accession that was heterozygous for the allele and suggested that the *badh2.1* allele was selected as a de novo mutation in *O. sativa* after domestication from its wild progenitor.

The diverse chromosomal segments harbouring unique alleles present in wild accessions of *O. nivara* can be harnessed for identifying and transferring valuable genes from wild to cultivated rice to meet the future food grain requirements. Distant genotypes identified in the study can be used in hybridization programmes for further improvement of rice germplasm. These diverse materials can be utilized for interspecific breeding programmes making use of the identified polymorphic markers. Breeding activities can now be more focused between distant genotypes or from untapped resources to enhance the diversity and identification of new genes and QTLs for further crop improvement.

20.18 Gene Discovery

The annual wild species, *O. nivara,* adapted to seasonally dry habitats is the closest progenitor of *O. sativa* (Sharma and Shastry 1965). It is also a potential source of favourable alleles for agriculturally important traits. One *O. nivara* accession IRGC101508 collected from Uttar Pradesh, India, was identified as the only accession resistant to grassy stunt virus after screening 5000 accessions and 100 breeding lines (Khush et al. 1977). *O. nivara* also contributed resistance to bacterial leaf blight, blast, brown planthopper and drought tolerance (Brar and Khush 1997; Khush 2000; Thanh et al. 2006; Ali et al. 2010). Dey et al. (2016) reported an introgression line 10^{-3} derived from a cross between Swarna and *O. nivara* cross was moderately resistant to sheath blight among 1013 germplasm comprising of cultivars, mutants, wild rice accessions, introgression lines, landraces, based on three years of testing under field conditions. A gene *Xa30* for bacterial blight resistance was identified from *O. nivara* IRGC81825 on chromosome 4L and transferred to PR114 (Cheema et al. 2008b), and *Xa33* on chromosome 7 was identified from another *O. nivara* accession IRGC105710 (Kumar et al. 2012). QTLs for sheath blight (*qShB3*, *qShB6*) and blast (*qBLAST8-1*, *qBLAST12*) were identified using two *O. nivara* accessions IRGC100898 and IRGC104705 as donors using AB-QTL strategy (Eizenga et al. 2013). QTLs were reported for seedling vigour, yield and quality traits using advanced backcross population derived from a cross between M-202 and *O. nivara* (acc. IRGC100195) (Eizenga et al. 2016). One introgression line (IL-3) derived from a cross between PR114 and *O. nivara* (acc. 105410) was identified as a potential donor for neck blast resistance breeding programmes in India (Devi et al. 2015). More recently, a set of 131 ILs derived from 93-11 and *O. nivara* W2014 were developed, and 65 QTLs were identified for 13 agronomic traits using whole-genome resequencing (Ma et al. 2016).

Some *O. nivara* accessions were identified as a source of resistance to grassy stunt virus, sheath blight, blast, stem borer and whorl maggot and also a source of drought avoidance, CMS, hybrid breakdown and pollen sterility locus (Hoan et al. 1998; Khush 2000; Brar and Khush 2003; Thanh et al. 2006; Ali et al. 2010). Because of their outcrossing nature, *O. nivara* accessions are used extensively to study the flow of transgenes (Chen et al. 2004). Furuta et al. (2016) evaluated three sets of CSSLs that harbour genomic fragments from *O. nivara*, *O. rufipogon* and *O. glaberrima* in a common *O. sativa* genetic background (cv. Koshihikari). Their phenotypic analyses of these libraries revealed the existence of three genes, Regulator of Awn Elongation 1 (*RAE1*), *RAE2* and *RAE3*, which are involved in the loss of long awns in cultivated rice. They demonstrated that *O. sativa* is awnless due to two dysfunctional genes *RAE1* and *RAE2*, whereas *O. glaberrima* (Acc. IRGC104038) achieved an awnless phenotype through mutation(s) in *RAE3*. Evaluation of *O. nivara* CSSLs and *O. rufipogon* CSSLs showed that *RAE1* and *RAE2* can independently induce the formation of long awns.

CSSLs in *indica* and *japonica* subgroups using accessions of *O. nivara* and *O. rufipogon* are in the development stage (Huang et al. 2012; Shakiba and Eizenga 2014). In India, CSSL libraries are being developed in ICAR-National Professor Project at IIRR, Hyderabad, India, using *O. rufipogon* as well as *O. nivara* as donors and popular Indian rice varieties as recipient parents. Introgression of *O. nivara* alleles into elite cultivars was also reported by several researchers. Li et al. (2006) mapped QTLs for domestication traits in F_2 population derived from a cross between CL16 and *O. nivara* (Acc. IRGC80470). At IIRR, India, two accessions of *O. nivara* IRGC81832 and IRGC81848 which were genetically distinct from 22 other accessions were extensively used for QTL mapping of yield and quality traits in BC_2F_2 and BC_2F_3 (Sarla et al. 2003). Kaladhar et al. (2008) identified 17 major effect QTLs for different yield traits in BC_2F_2 population derived from a cross between Swarna and *O. nivara* (Acc. IRGC81832) including *qyldp8.1* with LOD score of 8.76 which increased yield by 5.8 g per plant and grain number by 426 grains per plant. Significant yield-enhancing QTLs *qyldp2.1*, *qyldp3.1*, *qyldp8.1*, *qyldp9.1*, *qyldp11.1* were reported from another BC_2F_2 population derived from a cross between Swarna and *O. nivara* (Acc. IRGC81848) (Swamy et al. 2011b and 2014). QTLs for grain quality traits such as milling percentage *qmp1.2*, kernel width *qkw3.1*, *qkw6.1*, kernel length after cooking *qklac12.1* were also identified in these two populations (Swamy et al. 2012). Also, QTLs for stem diameter *qSD7.2*, *qSD8.1*, *qSD9.1*, rachis diameter *qRD9.1* and number of secondary branches *qNSB1.1* were identified as good targets for use in marker-assisted selection (MAS) (Swamy et al. 2011b). Meta-analysis of 76 yield QTLs reported in 11 studies involving interspecific crosses resulted in 23 independent meta-QTLs on ten chromosomes (Swamy and Sarla 2011). These can be helpful in MAS and candidate gene identification.

20.19 Organelle Genome

The recent study of Kim et al. (2015) showed the complete chloroplast genome *O. nivara* (Acc. IRGC88812) is 134,483 bp including the ribosomal DNAs of 45S (5823 bp) and 5S (322 bp). However, the chloroplast genome of other accession of *O. nivara* (Acc. IRGC100897) was 134516 bp, and 45S and 5S ribosomal DNA sequences were of 7904 and 322 bps, respectively. There are 100 unique genes identified in chloroplast genome of *O. nivara*. It has high number of protein-coding genes in length of 68598 bp with GC content of 39.9%. It also has transfer RNA (tRNA) genes of 2865 bp length with GC content of 53% and noncoding intergenic region of 53841 bp with 36% of GC content. An overall A + T content was 61% (Masood et al. 2004). The chloroplast genome analysis of nine species revealed maternal genome of *indica* may have been derived from *O. nivara* (Brozynska et al. 2016; Tong et al. 2016; Wambugu et al. 2015).

The *O. nivara* chloroplast genome encodes genes of similar function like *O. sativa*. But, it showed 57 insertions, 61 deletions and 159 base substitutions. Most of these insertions/deletions and few substitutions were in coding regions (Masood et al. 2004). Garaycochea et al. (2015) identified nine indels specific to *O. nivara* and three of these indels were common between *O. sativa indica* and *O. nivara*. The distribution of indels grouped *O. sativa indica* with *O. nivara*.

20.20 Impact on Plant Breeding Including Prebreeding Work

O. nivara has contributed many high-value QTLs and genes for rice improvement. One classic example is the introgression of GSV-resistant gene into several popular rice varieties grown in different countries of Asia (Khush 1977). For the first time, an *O. nivara*-derived IL has been released as a high yielding rice variety DRR Dhan 40 (RPBio4918-248S) in India (Table 20.6). The IL had a yield advantage of 25% over the national check variety Jaya, and it is released for commercial cultivation in Maharashtra, Tamil Nadu and West Bengal (Fig. 20.4). *O. nivara*-derived novel BLB-resistant genes *Xa33* and *Xa38* have been introgressed in the genetic backgrounds of RPHR1005 and PB1121 (Ellur et al. 2016), while green rice leafhopper (GLH) resistance gene *Grh6 was* transferred into Taichung 65 from *O. nivara* IRGC105715 (Fujita et al. 2004). Several ILs and CSSLs have been developed for various traits using *O. nivara* accessions (Table 20.7).

20.21 Comparative Genomics

Comparative genomics is a powerful tool to identify gene duplications in closely related species or across the entire genome (Wang et al. 2006). Previous reports showed that novel genetic variations are created by retroposition, exon shuffling, tandem gene duplication and transposon-mediated gene duplications in numerous organisms (Long et al. 2003; Fan et al. 2007). An orthologous region (1.5 Mb) of AA genome of *O. glaberrima* and BB genome of *O. punctata* was sequenced and compared with *O. sativa* subspp *japonica* reference sequence and identified a 60-kb segment located in the middle of the subtelomeric region of chromosome 3 which was unique to the species *O. sativa*. This analysis also showed that part of the unique 60-kb sequence was also found in its putative progenitor species *O. nivara* and *O. rufipogon* (Fan et al. 2008). In addition, the structural and functional variations (expansions, contractions and inversions) among closely related species *O. rufipogon*, *O. nivara* and *O. glaberrima* in relation to *O. sativa* were also determined. The repeats and transposable elements were 15% more in *O. sativa* than *O. nivara*. Among the TE, the expanded regions constitute intact LTR elements which correlate with the AA genome size diversification (Ammiraju et al. 2006; Hurwitz et al. 2010). Xu et al. (2012) resequenced the 40 cultivated and 10 wild rice accessions with >15× raw data coverage to identify important rice genes for serving as molecular markers for breeding. A total of ~15 million candidate single nucleotide polymorphisms (SNPs) were reported in all 50 accessions. After excluding the SNPs with missing data 6,496,456, high-quality SNPs were obtained. In this 103,321 synonymous and 132,819 nonsynonymous, SNPs were identified. Population structure and phylogenetic analysis showed *indica* was very close to *O. nivara* and *japonica* was close to *O. rufipogon*. The *de novo* assembly of *O. nivara*, *O. glaberrima*, *O. barthii*, *O. glumaepatula* and *O. meridionalis* with

Table 20.6 Details of *O. nivara*-derived high yielding rice variety DRRDhan 40

Traits	Performance of DRRDhan 40
Parentage	Swarna x *O. nivara* (Acc. IRGC81848)
Yield potential	5.5 t/ha
Days to maturity	138
Yield improvement	25% over national check Jaya
Grain quality	Short bold, Head rice 55.7%
Biotic stress tolerance	Moderately resistant to blast, brown spot, tungro, sheath blight and stem borer
States released	Maharashtra, Tamil Nadu and West Bengal

Fig. 20.4 Field view of DRRDhan 40 (IET21542 RPBio4918-248)

Table 20.7 Evaluation of *O. nivara* introgression lines for various traits

Traits screened	Introgression lines	References
Salinity	166S, 3-1S, 14S, K463, K467, K478	Ganeshan et al. (2016)
Yield and yield components	26 CSSLs	Furuta et al. (2016)
	220S, 10-2S	Malathi et al. (2017)
	166S, 14S	Balakrishnan et al. (2016)
Heat tolerance	S116-2, S175-2, S3-1, K50, K13-7	Prasanth et al. (2012)
Fe and Zn	Swarna BILs	Swamy et al. (2012)
Drought tolerant	166S, 248S, 75S,	Rai et al. (2010)
BPH	212S, 215S, 221S, 224S, 228S, 230S	Lakshmi et al. (2010)
BLB (Xa33)	RPHR1005 ILs	Kumar et al. (2012)
BLB (Xa38)	PB1121 NILs	Ellur et al. (2016)
Awn	WBSL-10, WBSL-18	Uehara et al. (2016)
GLH	Taichung 65 BILs	Fujita et al. (2004)
Sheath blight	RPBio4918-10-3	Dey et al. (2016)

O. sativa showed the genomic structure was well conserved in all species with less segmental duplication variations (~1.5%) and significantly enriched with genes involved in specific biological functions, especially cell death and response to stress (Zhang et al. 2014). A comparative physical map of *O. nivara* chromosome 3 revealed a high degree of sequence identity and synteny with *japonica*, though it was 21% smaller than *japonica* which diverged ~10,000 years ago. It shares large blocks of synteny with maize and sorghum genome also.

20.22 Micro-RNAs and Transcriptome

Micro-RNAs (miRNAs) are small noncoding RNAs that have emerged as important players in post-transcriptional gene silencing (Jones-Rhoades et al. 2006). They play a vital role during plant development and adaptation to environmental stress (Navarro et al. 2006; Sunkar et al. 2007; Agarwal et al. 2015; Mangrauthia et al. 2013; Mangrauthia et al. 2017). The miRNAs direct post-transcriptional gene silencing by triggering the cleavage or translational repression of the target transcripts (Brodersen et al. 2008).

Nuclear genome sequencing of *O. nivara* was performed using a whole-genome shotgun sequencing (WGS) analysis with the next-generation sequencing platform from Illumina (Zhang et al. 2014). The study showed a large number of positively selected genes, especially those involved in flower development, reproduction and resistance-related processes. They predicted 41,490 protein-coding genes for *O. nivara*. In total, 276 miRNA genes belonging to 203 miRNA families were identified in the genome of *O. nivara*. Nucleotide substitutions in miRNA genes and target sites across the six rice species were compared which revealed that most were conserved without mutations.

The whole-genome sequences of *O. nivara*, *O. glaberrima*, *O. barthii*, *O. glumaepatula* and *O. meridionalis* were compared with *O. sativa* to identify 39,293 orthologous gene families comprising 211718 genes (Zhang et al. 2014). A total of 276, 271, 276, 263 and 251 miRNA genes were identified which belong to 203, 125, 124, 133, 125 and 123 miRNA families in *O. nivara*, *O. glaberrima*, *O. barthii*, *O. glumaepatula* and *O. meridionalis* compared with *O. sativa*, respectively. Using a combination of bioinformatic tools and expression analyses, Baldrich et al. (2016) performed genome-wide analysis to discover polycistronic miRNAs in the genome of *Oryza* species such as *O. rufipogon*, *O. nivara*, *O. glumaepatula*, *O. barthii* and African rice *O. glaberrima*. Most of the polycistronic miRNAs exhibited a pattern of conservation in the lengths, sequence and chromosomal location in AA genome species such as *O. sativa* spp. *indica* and *japonica*, *O. rufipogon* but to a lesser extent in *O. nivara*. The variation in the organization of miRNA genes suggests a versatile mechanism for the control of gene expression in different species.

Gene expression investigation of two *Oryza* species, *O. rufipogon* and *O. nivara*, was conducted using digital gene expression technology (DGE) and paired-end RNA sequencing method (Guo et al. 2015). 21,415 expressed genes were identified across three reproduction-related tissues-flag leaves at the heading stage, panicles at the heading stage and panicles at the flowering stage. Of 21,415 expressed genes, approximately 8% (1717) differed significantly in expression levels between the two species. About 62% (1064) of the differentially expressed genes exhibited a signature of directional selection in at least one species. The study suggested that *cis*-regulatory changes play an important role in the ecological divergence of the two species. Gene expression analysis revealed ecological divergences might be associated with substantial alteration of the expression levels of a large number of genes across the genome.

20.23 *O. nivara*, a Source for Biofortification

A significant increase was observed in seed protein content in an interspecific hybrid developed by crossing the *indica* group cultivar IR64 and the wild species *O. nivara*. The hybrid showed 12.4% protein content, which was 28 and 18.2% higher than those of the parents *O. nivara* and IR64, respectively. The elevated protein content of the hybrid was due to significant increase in prolamins and glutelins. Amino acid analysis of seed proteins revealed that the hybrid had net gain of 19.5% in lysine and 19.4% in threonine over the *O. nivara* parent on a seed dry weight basis. Thus, using *O. nivara*, it was suggested that interspecific hybrid could serve as the initial breeding material for the selection of new rice genotypes that combine the high yield potential and superior cooking quality with high seed protein content (Mahmoud et al. 2008). Identification and introgression of major effect QTLs for grain Fe and Zn concentrations can assist in faster and more precise development of micronutrient dense rice varieties through marker-assisted QTL pyramiding. QTLs were mapped for Fe and Zn concentration in two BC_2F_3 mapping populations derived from the crosses of Swarna with two different accessions of *O. nivara* (IRGC81832 and IRGC81848). In all 17 and 13 QTLs were identified for grain Fe and Zn concentrations, respectively. These QTLs were located on chromosomes 1, 2, 3, 4, 8, 9 and 12. They contributed to 4 to 36% of the total phenotypic variance, and their additive effect varied from 0.2 to 6.6 ppm. Most of the trait-increasing alleles were derived from *O. nivara*. Five QTLs for Fe and three QTLs for Zn each explained more than 15% phenotypic variance. The location of *O. nivara*-derived QTLs such as $Fe_{2.1}$, $Fe_{3.1}$, $Fe_{8.1}$, $Fe_{8.2}$ and $Zn_{12.1}$ was consistently identified in both the populations. Epistatic interaction was observed only between RM106 and RM6 on chromosome 2 and RM270 and RM7 on chromosome 3 for Fe concentration in population 1. Nineteen candidate genes for metal homoeostasis were found to colocate with 13 QTLs for Fe and Zn in both the populations. The 14 QTLs identified in the present study have been reported in earlier studies in the same chromosomal locations. Some of the major effect and consistent QTLs are worthy of use in MAB to improve grain Fe and Zn in rice. Two lines 4918-24K and 4918-233K with 20–22 ppm zinc in polished grains have been obtained. Useful introgression lines have been developed for several agronomic traits from these two populations (Swamy et al. 2011a, b, 2014).

Expression profiles of metal homoeostasis-related genes were characterized in leaf and root tissues of *O. nivara* (Chandel et al. 2010). Twenty-one rice genes belonging to five gene families (*OsNRAMPs*, *OsFROs*, *OsZIPs*, *OsFERs* and *OsYSLs*) and four rice gene homologues (*OsNAAT1*, *OsNAS2*, *OsNAC*, *OsVIT1*) were analysed at maximum tillering and mid grain fill stages by semi-quantitative RT-PCR analysis. High-level expression of 6 genes (*OsFRO2*, *OsZIP8*, *OsYSL4*, *OsNAAT1*, *OsNRAMP5* and *OsVIT1*) was recorded in *O. nivara* from root transcription analysis.

20.24 Future Prospects

Exploration to collect more *O. nivara* accessions specially from rice-growing areas with adverse conditions is required considering the potential of this species in crop improvement and gene discovery. A systematic characterization and utilization of *O. nivara* will help in rice improvement with the help of recent advances in rice genomics. Prebreeding programmes involving development of ILs and CSSLs in multiple elite genetic backgrounds will bridge the gap of wild species utilization in breeding programmes while helping the widening of the genetic base in existing germplasm resources. It will also facilitate marker-assisted breeding of high-value QTLs and genes identified from *O. nivara* into popular cultivar backgrounds. Generation of more structural and functional genomics resources specifically for *O. nivara* has great value. Large-scale and systematic evaluation of *O. nivara*-derived lines is essential in multi location trials as a major part of multi-institute research programmes to evaluate their stability and adaptability across ecosystems.

Acknowledgements Authors gratefully acknowledge Department of Biotechnology, Government of India, project DBT No. BT/AB/FG-2 (PHII) IA/2009 and ICAR-National Professor Project to NS for financial support.

References

Agarwal S, Mangrauthia SK, Sarla N (2015) Expression profiling of iron deficiency responsive microRNAs and gene targets in rice seedlings of Madhukar x Swarna recombinant inbred lines with contrasting levels of iron in seeds. Plant Soil 396:137–150. https://doi.org/10.1007/s11104-015-2561-y

Ali ML, Sanchez PL, Yu SB, Lorieux M, Eizenga GC (2010) Chromosome segment substitution lines: a powerful tool for the introgression of valuable genes from *Oryza* wild species into cultivated rice (*O. sativa*). Rice 3:218–234

Ammiraju JSS, Luo M, Goicoechea JL, Wang W, Kudrna D, Mueller C, Talag J, Kim H, Sisneros NB, Blackmon B, Fang E, Tomkins JB, Brar D, MacKill D, McCouch S, Kurata N, Lambert G, Galbraith DW, Arumuganathan K, Rao K, Walling JG, Gill N, Yu Y, SanMiguel PS, Soderlund C, Jackson S, Wing RA (2006) The *Oryza* bacterial artificial chromosome library resource: construction and analysis of 12 deep-coverage large-insert BAC libraries that represent the 10 genome types of the genus *Oryza*. Genome Res 16:140–147

Ammiraju JSS, Zuccolo A, Yu Y, Song X, Piegu B, Chevalier F, Walling JG, Ma J, Talag J, Brar DS, SanMiguel PJ, Jiang N, Jackson SA, Panaud O, Wing RA (2007) Evolutionary dynamics of an ancient retrotransposon family provides insights into evolution of genome size in the genus *Oryza*. Plant J 52:342–351

Ammiraju JS, Fan C, Yu Y, Song X, Cranston KA, Pontaroli AC, Lu F, Sanyal A, Jiang N, Rambo T, Currie J (2010) Spatio-temporal patterns of genome evolution in allotetraploid species of the genus *Oryza*. Plant J 63:430–442

Anandan A, Anumalla M, Pradhan SK, Ali J (2016) Population structure, diversity and trait association analysis in rice (*Oryza sativa* L.) Germplasm for early seedling vigor (ESV) using trait linked SSR Markers. PLoS ONE 11:e0152406. https://doi.org/10.1371/journal.pone.0152406

Baker RL, Yarkhunova Y, Vidal K, Ewers BE, Weinig C (2017) Polyploidy and the relationship between leaf structure and function: implications for correlated evolution of anatomy, morphology, and physiology in *Brassica*. BMC Plant Biol 17:3

Balakrishnan D, Subrahmanyam D, Badri J, Raju AK, Rao YV, Beerelli K, Mesapogu S, Surapaneni M, Ponnuswamy R, Padmavathi G, Babu VR, Sarla N (2016) Genotype x environment interactions of yield traits in backcross introgression lines derived from *Oryza sativa* cv. Swarna/*Oryza nivara*. Front Plant Sci 7:1530. https://doi.org/10.3389/fpls.2016.01530

Baldrich P, Hsing YIC, Segundo BS (2016) Genome-wide analysis of polycistronic MicroRNAs in cultivated and wild rice. Genome Biol Evol 8:1104–1114

Banerjee A, Roychoudhury A (2015) WRKY proteins: signaling and regulation of expression during abiotic stress responses. Sci World J 2015:807560

Barbier P (1989) Genetic variation and ecotypic differentiation in the wild rice *Oryza rufipogon*. II. Influence of the mating system and life-history traits on the genetic structure of populations. Jpn J Genet 64:273–285

Bhasin H, Bhatia D, Raghuvanshi S, Lore JS, Sahi GK, Kaur B, Vikal Y, Singh K (2012) New PCR-based sequence-tagged site marker for bacterial blight resistance gene *Xa38* of rice. Mol Breed 30:607–611

Bimolata W, Kumar A, Sundaram RM, Laha GS, Qureshi IA, Ghazi IA (2015) Nucleotide diversity analysis of three major bacterial blight resistance genes in rice. PLoS ONE 10:e0120186

Borjas AH, De Leon TB, Subudhi PK (2016) Genetic analysis of germinating ability and seedling vigor under cold stress in US weedy rice. Euphytica 208:251–264

Brar DS, Khush GS (1997) Alien introgression in rice. Plant Mol Biol 35:35–47

Brar DS, Khush GS (2003) Utilization of wild species of genus *Oryza* in rice improvement. In: Nanda JS, Sharma SD (eds) Monograph on Genus *Oryza*. Science Publishers, Enfield, New Hampshire, pp 283–309

Brar DS, Singh K (2011) *Oryza*. In: Kole C (ed) Wild crop relatives: genomic and breeding resources, cereals. Springer, Berlin, Heidelberg, pp 321–365

Brodersen P, Sakvarelidze-Achard L, Bruun-Rasmussen M, Dunoyer P, Yamamoto YY, Sieburth L, Voinnet O (2008) Widespread translational inhibition by plant miRNAs and siRNAs. Science 320:1185–1190

Brozynska M, Copetti D, Furtado A, Wing RA, Crayn D, Fox G, Ishikawa R, Henry RJ (2016) Sequencing of Australian wild rice genomes reveals ancestral relationships with domesticated rice. Plant Biotech J 15:1–10

Chandel G, Banerjee S, Verulkar SB (2010) Expression profiling of metal homeostasis related candidate genes in rice (*Oryza spp.*) using semi quantitative RT-PCR analysis. Rice Genet Newl 25:44–47

Chang TT (1976) The origin, evolution, cultivation, dissemination, and diversification of the Asian and African rices. Euphytica 25:425–441

Chang TT (2003) Origin, domestication, and diversification. In: Smith CW, Dilday RH (eds) Rice: origin, history, technology and production. Wiley, Hoboken, pp 3–25

Chang TT, Ou SH, Pathak MD, Ling KC, Kauffman HE (1975) The search for disease and insect resistance in rice germplasm. In: Frankel OH, Hawkes JG (eds) Crop genetic resources for today and tomorrow. Cambridge University Press, Cambridge, pp 183–200

Charlesworth D (2003) Effects of inbreeding on the genetic diversity of populations. Philos Trans R Soc Lond 358:1051–1070

Cheema KK, Navtej SB, Mangat GS, Das A, Vikal Y, Brar DS, Khush GS, Singh K (2008a) Development of high yielding IR64 x *Oryza rufipogon* (Griff.) introgression lines and identification of introgressed alien chromosome segments using SSR markers. Euphytica 160:401–409

Cheema KK, Grewal NK, Vikal Y, Sharma R, Lore JS, Das A, Bhatia D, Mahajan R, Gupta V, Bharaj TS, Singh K (2008b) A novel bacterial blight resistance gene from *Oryza nivara* mapped to 38 kb region on chromosome 4L and transferred to *Oryza sativa* L. Genet Res 90:397–407

Chen LJ, Lee DS, Song ZP, Suh HS, Lu BR (2004) Gene flow from cultivated rice (*Oryzasativa*) to its weedy and wild relatives. Ann Bot 93:67–73

Cheng C, Motohashi R, Tsuchimoto S, Fukuta Y, Ohtsubo H, Ohstubo E (2003) Polyphyletic origin of cultivated rice: based on the interspersion pattern of SINEs. Mol Biol Evol 20:67–75

Choi JY, Purugganan M (2017) Multiple origin but single domestication led to domesticated Asian rice. bioRxiv, p 127688

Chouhan SK, Singh AK, Aparajita S, Ram M, Singh PK, Singh NK (2014) Characterization and evaluation of *Oryza nivara* and *Oryza rufipogon*. The Bioscan 9:853–858

Copetti D, Zhang J, El Baidouri M, Gao D, Wang J, Cossu RM, Angelova A, Maldonado LCE, Roffler S, Barghini E, Ohyanagi H, Fan C, Zuccolo A, Chen Costa, de Oliveira A, Han B, Henry R, Hsing YI, Kurata N, Wang W, Jackson SA, Panaud O, Wing RA (2015) RiTE database: a resource database for genus-wide rice genomics and evolutionary biology. BMC Genom 16:538

Deschamps S, Llaca V, May GD (2012) Genotyping-by-sequencing in plants. Biology 1:460–483

Devi SR, Singh K, Umakanth B, Vishalakshi B, Renuka P, Sudhakr KV, Prasad MS, Viraktamath BC, Babu VR, Madhav MS (2015) Development and identification of novel rice blast resistant sources and their characterization using molecular markers. Rice Sci 22:300–308

Dey S, Badri J, Prakasam V, Bhadana VP, Eswari KB, Laha GS, Priyanka C, Rajkumar A, Ram T (2016) Identification and agro-morphological characterization of rice genotypes resistant to sheath blight. Australas Plant Pathol 45:145–153

Doebley J, Stec A (1991) Genetic analysis of the morphological differences between maize and teosinte. Genetics 129:285–295

Dolores RC, Chang TT, Ramirez DA (1979) The cytogenetics of F_1 hybrids from *Oryzanivara* Sharma et Shastry x *O. sativa* L. Cytologia 44:527–540

Dong ZY, Wang YM, Zhang ZJ, Shen Y, Lin XY, Ou XF, Han FP, Liu B (2006) Extent and pattern of DNA methylation alteration in rice lines derived from introgressive hybridization of rice and Zizania latifolia Griseb. Theor Appl Genet 113:196–205

dos Santos RS, da Rosa Farias D, Pegoraro C, Rombaldi CV, Fukao T, Wing RA, de Oliveira AC (2017) Evolutionary analysis of the SUB1 locus across the *Oryza* genomes. Rice 10:4

Du H, Ouyang Y, Zhang C, Zhang Q (2011) Complex evolution of *S5*, a major reproductive barrier regulator, in the cultivated rice *Oryza sativa* and its wild relatives. New Phytol 191:275–287

Duitama J, Silva A, Sanabria Y, Cruz DF, Quintero C, Ballen C, Lorieux M, Scheffler B, Farmer A, Torres E, Oard J (2015) Whole genome sequencing of elite rice cultivars as a comprehensive information resource for marker assisted selection. PLoS ONE 10:e0124617

Eizenga GC, Rutger JN (2003) Genetics, cytogenetics, mutation, and beyond. In: Smith CW, Dilday RH (eds) Rice: origin, history, technology, and production. Wiley, Hoboken, pp 153–175

Eizenga GC, Prasad B, Jackson AK, Jia MH (2013) Identification of rice sheath blight and blast quantitative trait loci in two different *O. sativa*/ *O. nivara* advanced backcross populations. Mol Breed 31:889–907

Eizenga GC, Ali ML, Bryant RJ, Yeater KM, McClung AM, McCouch SR (2014) Registration of the rice diversity panel 1 for genome-wide association studies. J Plant Regist 8:109–116

Eizenga GC, Neves PCF, Bryant RJ, Agrama HA, Mackill DJ (2016) Evaluation of a M-202 x *Oryza nivara* advanced backcross mapping population for seedling vigor, yield components and quality. Euphytica 208:157–171

Ellur RK, Khanna A, Bhowmick PK, Vinod KK, Nagarajan M, Mondal KK, Singh NK, Singh K, Prabhu KV, Singh AK (2016) Marker-aided incorporation of xa38, a novel bacterial blight resistance gene, in PB1121 and comparison of its resistance spectrum with xa13 + xa21. Sci Rep 6:2918

Evans JR, Terashima I, Hanba Y, Loreto F (2004) CO_2 capture by the leaf. In: Smith WK, Vogelmann TC, Critchley C (eds) Photosynthetic adaptation: chloroplast to the landscape, vol 178. Ecol Stud. Springer, New York, pp 107–132

Fan C, Emerson JJ, Long M (2007) The origin of new gene. In: Pagel M, Pomiankowski A (eds) Evolutionary genomics and proteomics. Sinauer Associates Inc., Sunderland, Massachusetts, USA, pp 27–44

Fan C, Zhang Y, Yu Y, Rounsley S, Long M, Wing RA (2008) The subtelomere of *Oryza sativa* chromosome 3 short Arm as a hot bed of new gene origination in rice. Mol Plant 1:839–850

FAOSTAT (2010) FAO statistical databases. Food and Agriculture Organization of the United Nations, Rome, Italy http://www.fao.org

Fujita D, Doi K, Yoshimura A, Yasui H (2004) Introgression of a resistance gene for green rice leafhopper from *Oryza nivara* into cultivated rice, *Oryza sativa* L. Rice Genet Newsl 21:64–66

Fujita D, Yoshimura A, Yasui H (2010) Development of near-isogenic lines and pyramided lines carrying resistance genes to green rice leafhopper (Nephotettix cincticeps Uhler) with the Taichung 65 genetic background in rice (*Oryza sativa* L.). Breed Sci 60:18–27

Fuller DQ (2011) Pathways to Asian civilizations: tracing the origins and spread of rice and rice cultures. Rice 4:78–92

Furuta T, Uehara K, Angeles-Shim RB, Shim J, Nagai K, Ashikari M, Takashi T (2016) Development of chromosome segment substitution lines harbouring *Oryza nivara* genomic segments in Koshihikari and evaluation of yield-related traits. Breed Sci 66:845–850

Gaikwad KB, Singh N, Bhatia D, Kaur R, Bains NS, Bharaj TS, Singh K (2014) Yield-Enhancing heterotic QTL transferred from wild species to cultivated rice *Oryza sativa* L. PLoS ONE 9:e96939. https://doi.org/10.1371/journal.pone.0096939

Ganeshan P, Jain A, Parmar B, Rao AR, Sreenu K, Mishra P, Mesapogu S, Subrahmanyam D, Ram T, Sarla N, Rai V (2016) Identification of salt tolerant rice lines among interspecific BILs developed by crossing *Oryza sativa* × *O. rufipogon* and *O. sativa* *O. nivara*. Aus J Crop Sci 10:220–228

Garaycochea S, Speranza P, Alvarez-valin F (2015) A strategy to recover a high-quality, complete plastid sequence from low-coverage whole-genome sequencing. Appl Plant Sci 3:1500022

Ghauri MSK (1971) Revision of the genus nephotettix matsumura (homoptera: cicadelloidea: euscelidae) based on the type material. Bull Ent Res 60:481–512

Giuliani R, Koteyeva N, Voznesenskaya E, Evans MA, Cousins AB, Edwards GE (2013) Coordination of leaf photosynthesis, transpiration and structural traits in rice and wild relatives (genus *Oryza*). Plant Physiol 162:1632–1651

Glaszmann (1987) Isozyme classification of asian rice varieties. Theor Appl Genet 74:21–30

Grillo MA, Li C, Fowlkes AM, Briggeman TM, Zhou A, Schemske DW, Sang T (2009) Genetic architecture for the adaptive origin of annual wild rice, *Oryza nivara*. Evolution 63:870–883

Guo J, Liu R, Huang L, Zheng XM, Liu PL, Du YS, Cai Z, Zhou L, Wei XH, Zhang FM, Ge S (2015) Widespread and adaptive alterations in genome-wide gene expression associated with ecological divergence of two *oryza* species. Mol Biol Evol 33(1):62–78

Hamaoka N, Yasui H, Yamagata Y, Inoue Y, Furuya N, Araki T, Ueno O, Yoshimura A (2017) A hairy-leaf gene, BLANKET LEAF, of wild *Oryza nivara* increases photosynthetic water use efficiency in rice. Rice 10:20

Haussmann BIG, Parzies HK, Presterl T, Susic Z, Miedaner T (2004) Plant genetic resources in crop improvement. Plant Genet Resour 2:3–21

Hilario MNB, van den Berg RG, Ruaraidh N, Hamilton S, McNally KL (2013) Local differentiation amidst extensive allele sharing in *Oryza nivara* and *O. rufipogon*. Ecol Evol 3(9):3047–3062

Hittinger CD, Carroll SB (2007) Gene duplication and the adaptive evolution of a classic genetic switch. Nature 449:677–681

Hoan NT, Sarma NP, Siddiq EA (1998) Wide hybridization for diversification of CMS in rice. Int Rice Res Not 23:5–6

Huang CL, Hung CY, Chiang YC, Hwang CC, Hsu TW, Huang CC, Hung KH, Tsai KC, Wang KH, Osada N, Schaal BA (2012) Footprints of natural and artificial selection for photoperiod pathway genes in *Oryza*. Plant J 70:769–782

Hurwitz BL, Kudrna D, Yu Y, Sebastian A, Zuccolo A, Jackson SA, Ware D, Wing RA, Stein L (2010) Rice structural variation: a comparative analysis of structural variation between rice and three of its closest relatives in the genus *Oryza*. Plant J. 63:990–1003

International rice genome project- IRGSP (2005) The map-based sequence of the rice genome. Nature 436:793–800

International Rice Research institute (IRRI) (1968) Annual report for 1968. Los Baños, Philippines, p 405

Jacquemin J, Bhatia D, Singh K, Wing RA (2013) The international *Oryza* map alignment project: development of a genus-wide comparative genomics platform to help solve the 9 billion-people question. Curr Opin Plant Biol 16:147–156

Jiang D, Yin C, Yu A, Zhou X, Liang W, Yuan Z, Xu Y, Yu Q, Wen T, Zhang D (2006) Duplication and expression analysis of multicopy miRNA gene family members in Arabidopsis and rice. Cell Res 16:507–518

Jones-Rhoades MW, Bartel DP, Bartel B (2006) MicroRNAS and their regulatory roles in plants. Annu Rev Plant Biol 57:19–53

Joshi SP, Gupta VS, Aggarwal RK, Ranjekar PK, Brar DS (2000) Genetic diversity and phylogenetic relationship as revealed by inter simple sequence repeat (ISSR) polymorphism in the genus *Oryza*. Theor Appl Genet 100:1311–1320

Juneja S, Das A, Joshi SV, Sharma S, Vikal Y, Patra BC, Bharaj TS, Sidhu JS, Singh K (2006) *Oryza nivara* (Sharma et Shastry) the progenitor of *O. sativa* (L.) subspecies *indica* harbours rich genetic diversity as measured by SSR markers. Curr Sci 91(8):1079–1085

Kaladhar K, Swamy BPM, Babu AP, Reddy CS, Sarla N (2008) Mapping quantitative trait loci for yield traits in BC_2F_2 population derived from Swarna x *O. nivara* cross. Rice Genet Newsl 24:10

Khan ZR, Reuda BP, Caballero P (1989) Behavioral and physiological responses of rice leaf folder Cnaphalocrocis to selected wild rices. Entomol Exp Appl 52:7–13

Khush GS (1971) Rice breeding for disease and insect resistance at IRRI. Oryza 8:111–119

Khush GS (1977) Breeding for resistance in rice. Ann N Y Acad Sci 287:296–308

Khush GS (1997) Origin, dispersal, cultivation and variation of rice. Plant Mol Biol 35:25–34

Khush GS (2000) Taxonomy and origin of rice. In: Singh RK, Singh US, Khush GS (eds) Aromatic Rices. Oxford and IBH Publishing Co. Pvt. Ltd., New Delhi, India, pp 5–13

Khush GS, Ling KC (1974) Inheritance of resistance to grassy stunt virus and its vector in rice. J Hered 65:134–136

Khush GS, Ling KC, Aquino RC, Aguiero VM (1977) Breeding for resistance to grassy stunt in rice. In: International seminar of society for advancement breeding research in Asia and Oceania. Canberra, Australia 1(4b):3–9, 12–13 Feb 1997

Kim K, Lee SC, Lee J, Yu Y, Yang K, Choi BS, Koh HJ, Waminal NE, Choi HI, Kim N, Jang W (2015) Complete chloroplast and ribosomal sequences for 30 accessions elucidate evolution of *Oryza* AA genome species. Sci Rep 5:15655

Kim H, Jung J, Singh N, Greenberg A, Doyle JJ, Tyagi W, Chung JW, Kimball J, Hamilton SR, McCouch SR (2016b) Population dynamics among six major groups of the *Oryzarufipogon* species complex, wild relative of cultivated Asian rice. Rice 9:56

Kiran TV, Rao YV, Subrahmanyam D, Rani NS, Bhadana VP, Rao PR, Voleti SR (2013) Variation in leaf photosynthetic characteristics in wild rice species. Photosynthetica 51:350–358

Kondamudi R, Swamy KN, Rao YV, Kiran TV, Suman K, Rao DS, Rao PR, Subrahmanyam D, Sarla N, Kumari BR, Voleti SR (2016) Gas exchange, carbon balance and stomatal traits in wild and cultivated rice (*Oryza sativa* L.) genotypes. Acta Physiol Plant 38:160

Kovach MJ, McCouch SR (2008) Leveraging natural diversity: back through the bottleneck. Curr Opin Plant Biol 11:193–200

Kovach MJ, Calingacion MN, Fitzgerald MA, McCouch SR (2009) The origin and evolution of fragrance in rice (*Oryza sativa* L.). PNAS 106: 14444–14449

Kumar PN, Sujatha K, Laha GS, Rao KS, Mishra B, Viraktamath BC, Hari Y, Reddy CS, Balachandran SM, Ram T, Madhav MS, Rani NS, Neeraja CN, Reddy GA, Shaik H, Sundaram RM (2012) Identification and fine-mapping of *Xa33*, a novel gene for resistance to *Xanthomonas Oryzae* pv. *oryzae*. Phytopathol 102:222–228

Kwon SJ, Lee JK, Hong SW, Park YJ, McNally KL, Kim NS (2006) Genetic diversity and phylogenetic relationship in AA Oryza species as revealed by rim2/hipa cacta transposon display. Genes Genet Syst 81:93–101

Lakshmi VJ, Swamy BPM, Kaladhar K, Sarla N (2010) BPH resistance in introgression lines of Swarna/*Oryza nivara* and KMR3/*O. rufipogon*. DRR News Lett 8:4

Lanceras JC, Pantuwan G, Jongdee B, Toojinda T (2004) Quantitative trait loci associated with drought tolerance at reproductive stage in rice. Plant Physiol 135:384–399

Li SQ, Yang GL, Li SB, Li YS, Chen ZY, Zhu YG (2005) Distribution of fertility restorer genes for wild-abortive and Honglian CMS lines of rice in the AA genome species of genus *Oryza*. Ann Bot 96:461–466

Li C, Zhou A, Sang T (2006) Genetic analysis of rice domestication syndrome with the wild annual species, *Oryza nivara*. New Phytol 170:185–194

Li X, Guo K, Zhu X, Chen P, Li Y, Xie G, Wang L, Wang Y, Persson S, Peng L (2017) Domestication of rice has reduced the occurrence of transposable elements within gene coding regions. BMC Geno 18:55

Lin Z, Li X, Shannon LM, Yeh CT, Wang ML, Bai G, Peng Z, Li J, Trick HN, Clemente TE, Doebley J (2012) Parallel domestication of the Shattering1 genes in cereals. Nat Gen 44:720–724

Ling KC, Aguicro VM, Lee SH (1970) A mass screening method for testing resistance to grassy stunt disease of rice. Plant Dis Rep 54:565–569

Liu R, Zheng XM, Zhou L, Zhou HF, Ge S (2015) Population genetic structure of *Oryza rufipogon* and *O. nivara*: Implications for the origin of *O. nivara*. Mol Ecol 24:5211–5228

Long M, Betran E, Thornton K, Wang W (2003) The origin of new genes: glimpses from the young and old. Nat Rev Genet 4:865–875

Lu BR, Naredo MBE, Juliano AB, Jackson MT (2000) Preliminary studies on taxonomy and biosystematics of the AA genome *Oryza* species (Poaceae). In: Jacobs SWL, Everett J (eds) Grasses: systematics and evolution. CSIRO, Melbourne Australia, pp 51–58

Lu BR, Song G, Sang T, Chen JK, Hong DY (2001) The current taxonomy and perplexity of the genus *Oryza* (Poaceae). Acta Phytotax Sin 39:373–388

Lu BR, Zheng KL, Qian HR, Zhuang JY (2002) Genetic differentiation of wild relatives of rice as referred by the RFLP analysis. Theor Appl Genet 106:101–106

Luo MC, Thomas C, You FM, Hsiao J, Ouyang S, Buell CR, Malandro M, McGuire PE, Anderson OD, Dvorak J (2003) High-throughput fingerprinting of bacterial artificial chromosomes using the snapshot labeling kit and sizing of restriction fragments by capillary electrophoresis. Genomics 82:378–389

Ma X, Fu Y, Zhao X, Jiang L, Zhu Z, Gu P, Xu W, Su Z, Sun C, Tan L (2016) Genomic structure analysis of a set of *Oryza nivara* introgression lines and identification of yield-associated QTLs using whole-genome resequencing. Sci Rep 6:27425. https://doi.org/10.1038/srep27425

Madurangi SAP, Samarasinghe WLG, Senanayake SGJN, Hemachandra PV, Ratnesekara D (2011) Resistance of *Oryza nivara* and *Oryza eichingeri* derived lines to brown planthopper, *Nilaparvata lugens* (*Stal*). J Natl Sci Found Sri Lanka 39:175–181

Madurangi SAP, Ratnasekera D, Senanayake SGJN, Samarasinghe WLG, Hemachandra PV (2013) Antixenosis and antibiosis effects of *Oryza nivara* accessions harbouring bph2 gene on brown planthopper [*Nilaparvata lugens* (Stal)]. J Natl Sci Found Sri Lanka 41:147–154

Magadum S, Banerjee U, Murugan P, Gangapur D, Ravikesavan R (2013) Gene duplication as a major force in evolution. J Genet 92:155–161

Mahmoud AA, Sukumar S, Krishnan HB (2008) Interspecific rice hybrid of *Oryza sativa* x *Oryza nivara* reveals a significant increase in seed protein content. J Agric Food Chem 56:476–482

Makino A (2011) Photosynthesis, grain yield, and nitrogen utilization in rice and wheat. Plant Physiol 155:125–129

Malathi S, Divya B, Sukumar M, Krishnam Raju A, Venkateswara Rao YV, Tripura Venkata VGN, Sarla N (2016) Genetic characterization and population structure of Indian rice cultivars and wild genotypes using core set markers. 3 Biotech 6:95

Malathi S, Divya B, Sukumar M, Krishnam Raju A, Venkateswara Rao Y, Tripura Venkata VGN, Neelamraju Sarla (2017) Identification of major effect QTLs for agronomic traits and CSSLs in rice from Swarna/*O. nivara* derived backcross inbred lines.

Front Plant Sci 8:1027. https://doi.org/10.3389/fpls.2017.01027

Mangrauthia SK, Agarwal S, Sailaja B, Madhav MS, Voleti SR (2013) MicroRNAs and their role in salt stress response in plants. In: Ahmad P, Azooz MM, Prasad MNV (eds) Salt stress in plants. Springer, New York, NY, pp 15–46

Mangrauthia SK, Sailaja B, Agarwal S, Prasanth VV, Voleti SR, Sarla N, Subrahmanyam D (2017) Genome-wide changes in microRNA expression during short and prolonged heat stress and recovery in contrasting rice cultivars. J Exp Bot 68(9):2399–2412. https://doi.org/10.1093/jxb/erx111

Masood MS, Nishikawa T, Fukuoka S, Njenga PK, Tsudzuki T, Kadowak K (2004) The complete nucleotide sequence of wild rice (*Oryza nivara*) chloroplast genome: first genome wide comparative sequence analysis of wild and cultivated rice. Gene 340:133–139

Matsubara K, Ito S, Nonoue Y, Ando T, Yano M (2007) A novel gene for hybrid breakdown found in a cross between *japonica* and *indica* cultivars in rice. Rice Genet Newsl 23:11–13

Melonek J, Stone JD, Small I (2016) Evolutionary plasticity of restorer-of-fertility-like proteins in rice. Sci Rep 6:35152

Menguer PK, Sperotto RA, Ricachenevsky FK (2017) A walk on the wild side: *Oryza* species as source for rice abiotic stress tolerance. Genet Mol Biol 40:238–252

Minx P, Cordum H, Wilson R (2005) Sequence, annotation, and analysis of synteny between rice chromosome 3 and diverged grass species. Genome Res 15:1284–1291

Miura K, Yamamoto E, Morinaka Y, Takashi T, Kitano H, Matsuoka M, Ashikari M (2008) The hybrid breakdown 1(t) locus induces interspecific hybrid breakdown between rice *Oryza sativa* cv. Koshihikari and its wild relative *O. nivara*. Breed Sci 58:99–105

Morishima H (2001) Evolution and domestication of rice. In: Khush GS, Brar DS, Hardy B (eds) Rice genetics IV. Proceedings of the fourth international rice genetics symposium. IRRI, Los Banos, The Philippines, pp 63–77

Morishima H, Sano Y, Oka HI (1992) Evolutionary studies in cultivated rice and its wild relatives. Oxf Surv Evol Biol 8:135–184

Morishima H, Shimamoto Y, Sano Y, Sato YI (1984) Observations on wild and cultivated rices in Thailand for ecological-genetic study. Report of study-tour in 1983. Rep Nat Inst Genet, Japan

Navarro L, Dunoyer P, Jay F, Arnold B, Dharmasiri N, Estelle M, Voinnet O, Jones JD (2006) A plant miRNA contributes to antibacterial resistance by repressing auxin signaling. Science 312(80):436–439

Nayak J, Mathan J, Mohanty M, Pradhan C (2014) An in vitro hydroponic study on physiological and biochemical responses of indian wild rice to varying doses of hexavalent chromium. Int Res J Env Sci 3:20–28

Niroula RK, Subedi LP, Upadhyay MP (2009) Cytogenetic analyses of intragenomic rice hybrids derived from *Oryza sativa* L. and *O. nivara* Sharma et Shastry. Bot Res Int 4:277–283

Niroula RK, Pucciariello C, Ho VT, Novi G, Fukao T, Perata P (2012) SUB1A-dependent and independent mechanisms are involved in the flooding tolerance of wild rice species. Plant J 72:282–293

Oka HI (1988) Origin of cultivated rice. Japan Scientific Societies Press, Tokyo, Elsevier, Amsterdam Oxford, New York, Tokyo

Oka HI, Morishima H (1982) Phylogenetic differentiation of cultivated rice, XXIII. Potentiality of wild progenitors to evolve the *indica* and *japonica* types of rice cultivars. Euphytica 31:41–50

Park KC, Kim NH, Cho YS, Kang KH, Lee JK, Kim NS (2003) Genetic variations of AA genome Oryza species measured by MITE-AFLP. Theor Appl Genet 107:203–209

Peng ZY, Zhang H, Liu T, Dzikiewicz KM, Li S, Wang X, Hu G, Zhu Z, Wei X, Zhu QH, Sun Z (2009) Characterization of the genome expression trends in the heading-stage panicle of six rice lineages. Genomics 93:169–178

Poorter H, Remkes C (1990) Leaf area ratio and net assimilation rate of 24 wild species differing in relative growth rate. Oecologia 83:553–559

Prasad B, Eizenga GC (2008) Rice sheath blight disease resistance identified in *Oryza* spp. accessions. Plant Dis 92:1503–1509

Prasanth VV, Chakravarthi DV, Kiran TV, Rao YV, Panigrahy M, Mangrauthia SK, Viraktamath BC, Subrahmanyam D, Voleti SR, Sarla N (2012) Evaluation of rice germplasm and introgression lines for heat tolerance. Ann Biol Res 3:5060–5068

Purugganan MD, Rounsley SD, Schmidt RJ, Yanofsky MF (1995) Molecular evolution of flower development: diversification of the plants MADS-box regulatory gene family. Genetics 140:345–356

Rai V, Sreenu K, Pushpalatha B, Babu Brajendra, Sandhya G, Sarla N (2010) Swarna/*Oryza nivara* and KMR3/*O. rufipogon* introgression lines tolerant to drought and salinity. DRR News Lett 8:4

Sakai H and Itoh T (2010) Massive gene losses in Asian cultivated rice unveiled by comparative genome analysis. BMC Genomics 11:121

Sanchez PL, Wing RA, Brar DS (2013) The wild relative of rice: genomes and genomics. In: Zhang Q, Wing RA (eds) Genetics and genomics of rice, vol 5. Plant genetics and genomics: crops and models. Springer, New York, pp 9–25. https://doi.org/10.1007/978-1-4614-7903-1_2

Sang T, Ge S (2007) The puzzle of rice domestication. J Int Plant Bio 49:760–768

Sano Y, Morishima H, Oka HI (1980) Intermediate perennial-annual populations of *Oryza perennis* found in Thailand and their evolutionary significance. J Plant Res 93:291–305

Sarao PS, Sahi GK, Neelam K, Mangat GS, Patra BC, Singh K (2016) Donors for resistance to Brown

planthopper Nilaparvata lugens (Stål) from wild rice species. Rice Sci 23:219–224
Sarla N, Bobba S, Siddiq EA (2003) ISSR and SSR markers based on AG and GA repeats delineate geographically diverse *Oryza nivara* accessions and reveal rare alleles. Curr Sci 84:683–690
Second G (1982) Origin of the genetic diversity of cultivated rice (*Oryza* ssp.). Study of the polymorphism scored at 40 isozyme loci. Jpn J Genet 57:25–57
Shakiba E, Eizenga GC (2014) Unraveling the secrets of rice wildspecies. In: Yan WG, Bao JS (eds) Rice: germplasm, genetics and improvement. Intech, Croatia, pp 1–58
Sharma SD, Shastry SVS (1965) Taxonomic studies in genus *Oryza* L: III. *O. rufipogon* griff.sensu stricto and *O. nivara* Sharma et Shastry nom. nov. Indian J Genet Plant Breed 25:157–167
Sharma SD, Tripathy S, Biswal J (2000) Origin of *O. sativa* and its ecotypes. In: Nanda JS (ed) Rice breeding and genetics: research priorities and challenges. Science Publications, Enfield, NH, USA, pp 349–369
Sharma S, Upadhyaya HD, Varshney RK, Gowda CLL (2013) Pre-breeding for diversification of primary gene pool and genetic enhancement of grain legumes. Front Plant Sci 4:309
Shiu SH, Byrnes JK, Pan R, Zhang P, Li WH (2006) Role of positive selection in the retention of duplicate genes in mammalian genomes. Proc Natl Acad Sci USA 103:2232–2236
Silva J, ScheZer B, Sanabria Y, De Guzman CD, Galam D, Farmer A, Woodward J, May G, Oard J (2012) Identification of candidate genes in rice for resistance to sheath blight disease by whole genome sequencing. Theor Appl Genet 124:63–74
Singh BP, Jayaswal PK, Singh B, Singh PK, Kumar V, Mishra S, Singh N, Panda K, Singh NK (2015) Natural allelic diversity in *OsDREB1F* gene in the Indian wild rice germplasm led to ascertain its association with drought tolerance. Plant Cell Rep 34:993–1004
Soderlund C, Humphray S, Dunham A, French L (2000) Contigs built with fingerprints, markers, and FPC V4.7. Genome Res 10:1772–1787
Stander JR (1993) Pre-breeding from the perspective of the private plant breeder. J Sugar Beet Res 30:197–208
Sunkar R, Jagadeeswaran G (2008) *In silico* identification of conserved microRNAs in large number of diverse plant species. BMC Plant Biol 8:37
Sunkar R, Chinnusamy V, Zhu J, Zhu JK (2007) Small RNAs as big players in plant abiotic stress responses and nutrient deprivation. Trends Plant Sci 12:301–309
Swamy BPM, Sarla N (2011) Meta-analysis of yield QTLs derived from inter-specific crosses of rice reveals consensus regions and candidate genes. Plant Mol Biol Rep 29:663–680
Swamy BPM, Kaladhar K, Anuradha K, Batchu AK, Longvah T, Viraktamath BC and Sarla N (2011a) Enhancing iron and zinc concentration in rice grains using wild species. In: ADNAT convention and international symposium on genomics and biodiversity, CCMB, Hyderabad, 23–25 Feb 2011, p 71
Swamy BPM, Kaladhar K, Ramesha MS, Viraktamath BC, Sarla N (2011b) Molecular mapping of QTLs for yield and yield-related traits in *Oryza sativa* cv Swarna x *O. nivara* (IRGC81848) backcross population. Rice Sci 18:178–186
Swamy BPM, Kaladhar K, Shobharani N, Prasad GSV, Viraktamath BC, Reddy GA, Sarla N (2012) QTL analysis for grain quality traits in 2 BC_2F_2 populations derived from crosses between *Oryza sativa* cv Swarna and 2 accessions of *O. nivara*. J Heredity 103:442–452
Swamy BPM, Kaladhar K, Reddy GA, Viraktamath BC, Sarla N (2014) Mapping and introgression of QTL for yield and related traits in two backcross populations derived from *Oryza sativa* cv. Swarna and two accessions of *O. nivara*. J Genet 93:643–654
Takahashi H, Sato YI, Nakamura I (2008) Evolutionary analysis of two plastid DNA sequences in cultivated and wild species of *Oryza*. Breed Sci 58:225–233
Tanksley SD, McCouch SR (1997) Seed banks and molecular maps: unlocking genetic potential from the wild. Science 277:1063–1066
Tanksley SD, Nelson JC (1996) Advanced backcross QTL analysis: a method for the simultaneous discovery and transfer of valuable QTLs from unadapted germplasm into elite breeding lines. Theor Appl Genet 92:191–203
Thomson MJ, Septiningsih EM, Suwardjo F, Santoso TJ, Silitonga TS, McCouch SR (2007) Genetic diversity analysis of traditional and improved Indonesian rice (*Oryza sativa* L.) germplasm using microsatellite markers. Theor Appl Genet 114:559–568
Thanh PT, Sripichitt P, Chanprame S, Peyachoknagul S (2006) Transfer of drought resistant character from wild rice (*Oryza meridionalis* and *Oryza nivara*) to cultivated rice (*Oryza sativa* L.) by backcrossing and immature embryo culture. Kasetsart J (Nat Sci) 40:582–594
Tian M, Yu G, He N, Hou J (2016) Leaf morphological and anatomical traits from tropical to temperate coniferous forests: Mechanisms and influencing factors. Sci Rep 6:19703
Tong W, Kim TS, Park YJ (2016) Rice chloroplast genome variation architecture and phylogenetic dissection in diverse *Oryza* species assessed by whole-genome resequencing. Rice 9:57
Uehara K, Furuta T, Komeda N, Ashikari M (2016) Identification of responsible genes involved in Awn development and discussion about rice domestication process. In: Poster 0726, Plant and animal genome XXIV conference 8–13 Jan 2016, San Diego, USA
Vaughan DA (1989) The genus *Oryza* L.: current status of taxonomy. IRRI, Los Baños, The Philippines
Valkoun JJ (2001) Wheat pre-breeding using wild progenitors. Euphytica 119:17–23
Vaughan DA, Morishima H (2003) Biosystematics of the genus Oryza. In: Smith CW, Dilday RH (eds) Rice:

origin, history, technology and production. Wiley, New York, pp 27–65

Vaughan DA, Lu BR, Tomooka N (2008) The evolving story of rice evolution. Plant Sci 174:394–408

Wambugu PW, Brozynska M, Furtado A, Waters DL, Henry RJ (2015) Relationships of wild and domesticated rices (*Oryza* AA genome species) based upon whole chloroplast genome sequences. Sci Rep 5:13957

Wang ZY, Second G, Tanksley SD (1992) Polymorphism and phylogenetic relationships among species in the genus Oryza as determined by analysis of nuclear RFLPs. Theor Appl Genet 83:565–581

Wang W, Zheng H, Fan C, Li J, Shi J, Cai Z, Zhang G, Liu D, Zhang J, Vang S, Lu Z, Wong GKS, Long M, Wang J (2006) High rate of chimeric gene origination by retroposition in plant genomes. Plant Cell 18:1791–1802

Win KT, Yamagata Y, Miyazaki Y, Doi K, Yasui H, Yoshimura A (2010) Independent evolution of a new allele of F1 pollen sterility gene S27 encoding mitochondrial ribosomal protein L27 in *Oryza nivara*. Theor Appl Genet 122:385–394

Win KT, Yamagata Y, Miyazaki Y, Doi K, Yasui H, Yoshimura A (2011) Independent evolution of a new allele of F1 pollen sterility gene S27 encoding mitochondrial ribosomal protein L27 in *Oryza nivara*. Theor Appl Genet 122:385–394

Win KT, Kubo T, Miyazaki Y, Doi K, Yasui H, Yamagata Y, Yoshimura A (2012) Molecular mapping of two loci conferring F_1 pollen sterility in inter and intra specific crosses of rice. Genes Gen Genom 6:16–21

Wing RA, Ammiraju JSS, Luo M, Kim D, Goicoechea JL, Wang W, Nelson W, Rao K, Brar D, Mackill DJ, Han B, Soderlund C, Stein L, SanMiguel P, Jackson S (2005) The *Oryza* map alignment project: the golden path to unlocking the genetic potential of wild rice species. Plant Mol Biol 59:53–62

Wu MT, Li CP, Chen JR, Huang SH, Ku HM (2009) Mapping of brown plant hopper resistance gene introgressed from *Oryza nivara* into cultivated rice, *O. sativa*. In: International symposium. Rice research in the era of global warming, pp 56–65

Xu X, Liu X, Ge S, Jensen JD, Hu F, Li X, Dong Y, Gutenkunst RN, Fang L, Huang L, Li J, He W, Zhang G, Zheng X, Zhang F, Li Y, Yu C, Kristiansen K, Zhang X, Wang J, Wright M, McCouch S, Nielsen R, Wang J, Wang W (2012) Resequencing 50 accessions of cultivated and wild rice yields markers for identifying agronomically important genes. Nat Biotech 30:105–111

Xu H, Watanabe KA, Zhang L, Shen QJ (2016) WRKY transcription factor genes in wild rice *Oryza nivara*. DNA Res 23:311–323

Yamanaka S, Nakamura I, Nakai H, Sato YI (2003) Dual origin of the cultivated rice based on molecular markers of newly collected annual and perennial strains of wild rice species, *Oryza nivara* and *O. rufipogon*. Genet Res Crop Evol 50:529–538

Yamanaka S, Nakamura I, Nakai H, Sato YI (2004) Annual-perennial and *indica-japonica* differentiations related to retrotransposon polymorphism at waxy locus in wild relatives of rice. Online Available: http://www.carleton.ca/~bgordon/Rice/papers/yama20.htm

Yang Y, Zhou J, Li J, Xu P, Zhang Y, Tao D (2016) Mapping QTLs for hybrid sterility in three AA genome wild species of Oryza. Breed Sci 66:367–371

Yokosho K, Yamaji N, Fujii-Kashino M, Ma JF (2016) Retrotransposon-mediated aluminum tolerance through enhanced expression of the citrate transporter *OsFRDL4*. Plant Physiol 172:2327–2336

Zhang QJ, Zhu T, Xia EH, Shi C, Liu YL, Zhang Y, Liu Y, Jiang WK, Zhao YJ, Mao SY, Zhang LP (2014) Rapid diversification of five *Oryza* AA genomes associated with rice adaptation. Proc Natl Acad Sci 111:4954–4962. https://doi.org/10.1073/pnas.1418307111

Zhu Q, Ge S (2005) Phylogenetic relationships among A-genome species of the genus *Oryza* revealed by intron sequences of four nuclear genes. New Phytol 167:249–265

Zhu Q, Zheng X, Luo J, Gaut BS, Ge S (2007) Multilocus analysis of nucleotide variation of *Oryza sativa* and its wild relatives: severe bottleneck during domestication of rice. Mol Biol Evol 24:875–888

Oryza officinalis Complex

21

Soham Ray, Joshitha Vijayan, Mridul Chakraborti,
Sutapa Sarkar, Lotan Kumar Bose and Onkar Nath Singh

Abstract

Oryza officinalis complex is the largest species complex within the genus *Oryza*, consisting of both diploid and allotetraploid species. It contains at least ten reported perennial species (*O. officinalis, O. rhizomatis, O. eichengeri, O. minuta, O. malampuzhaensis, O. punctata, O. latifolia, O. alta, O. grandiglumis, O. schweinfurthiana* and *O. australiensis*) distributed throughout the tropics. Most of the allotetraploid species in this group have originated recently, and their progenitors are also naturally available in many cases. This provides a unique opportunity to study evolution of polyploidy within the genus *Oryza*. Classical cytogenetic techniques like karyotyping, meiotic pairing have provided valuable information in this regard. Most of the diploid and allotetraploid species contain 'C' genome except diploid *O. punctata* (BB) and *O. australiensis* (EE). Till date, whole genome sequence is available only for *O. punctata* and rest are under progress. Considerable success has been achieved in terms of developing genetic stock through wide hybridization and mapping agronomically important genes from the species of *O. officinalis* complex. Successful introgressions of brown plant hopper resistance and bacterial leaf blight resistance to the cultivated varieties have been achieved from these genetic stocks. Molecular marker resources to facilitate such alien gene introgression events are becoming more and more available. In order to address the problem of world food security, global importance has been laid on utilization of wild resource in rice. With focused objectives and rapidly emerging scientific and technological know-how, we can expect much more efficient utilization of the *O. officinalis* complex in future rice improvement.

21.1 Introduction

Wild relatives of cultivated rice belonging to the genus *Oryza* are grown in different ecologies around the world. These wild relatives are considered as virtually untapped reservoir of agronomically important genes. By analysing a vast herbaria collection and detailed field studies carried out in Asia and Africa, Tateoka (1963)

S. Ray (✉) · M. Chakraborti · S. Sarkar · L. K. Bose
O. N. Singh
Crop Improvement Division, National Rice Research Institute, Cuttack 753006, Odisha, India
e-mail: sohamray27@gmail.com

J. Vijayan
Integrated Rural Development and Management, Ramakrishna Mission Vivekananda University, Kolkata, India

T. K. Mondal and R. J. Henry (eds.), *The Wild Oryza Genomes*,
Compendium of Plant Genomes, https://doi.org/10.1007/978-3-319-71997-9_21

demonstrated the basic groups within the genus which he designated as species complexes. It may be noted that the term species complex is used for a group of species where distinct taxonomic keys are lacking and the categorization to species or subspecies level is rather arbitrary (Vaughan et al. 2005). There are four major complexes in the genus *Oryza,* viz. *O. sativa* complex (contains AA genome), *O. officinalis* complex, *O. ridleyi* complex (HHJJ genome) and *O. granulata* complex (GG genome) and a prominent out-group consisting of a lone species *O. brachyantha* (FF genome). Taken together, they share an evolutionary history of approximately 15 million years (Jacquemin et al. 2013). Among these, *O. officinalis* complex (also known as *O. latifolia* complex) is the largest group consisting of at least ten reported species (Tateoka 1964) with both diploid (comprising BB, CC, DD and EE) and allotetraploid (comprising BBCC and CCDD) genomes (Vaughan et al. 2003) which are widely distributed throughout the tropics. Hence, the *Officinalis* complex is represented by six different genome groups, BB, CC, DD, BC, CD and EE (Table 21.1). Besides, all are perennial in nature (Maclean 2002). The allotetraploid species of the complex are recent in origin and the progenitor species are also naturally available (Zou et al. 2015). Hence, this complex provides an ideal platform to study evolution of polyploids in the genus *Oryza*. The present chapter deals upon various aspects related to the genomes of the member species under *O. officinalis* complex based on the information available till date. A graphical overview of the content of the chapter has been represented in Fig. 21.1.

21.2 Distribution and Evolution of *O. Officinalis* Complex

Lu (1999) divided the genus *Oryza* into three different sections, namely *Padia*, *Brachyantha* and *Oryza*. *Padia* section consists of *O. ridleyi* complex and *O. granulata* complex (Sharma et al. 2003) is known to be distributed in Asia, Australia, New Caledonia and Papua New Guinea. *Oryza brachyantha,* which is restricted to Africa, is the only species of *Brachyantha* section. *O. officinalis* complex is the largest among four species complexes in the genus *Oryza*. Palaeontological studies suggest that the tribe *Oryzae* differentiated from other related grasses about 36–40 million years ago (Mya) by the end of Eocene age (Tang et al. 2010). Section *Oryza* with *Officinalis* and *Sativa* complexes seems to be most recent in origin in the genus (Sharma et al. 2003). The timescale for divergence of *O. officinalis* complex from rest of the species under genus '*Oryza*' was predicted to be 7.9 Mya (Wang et al. 2009). The ten species in this group are widely distributed throughout the tropics (Fig. 21.2). The specific features which distinguish the different species of *O. officinalis* complex from rest of the '*Oryza*' species are truncated ligules, straight rachilla, small-sized spikelets and linearly arranged tubercles on the surface of lemma and palea. The specific morphological features of all species within this complex have been represented in Table 21.2 along with their photographs in Fig. 21.3.

O. officinalis, *O. rhizomatis* and *O. eichingeri* are the three diploid species with CC genome. The genome of *O. officinalis* was designated as CC ($2n = 2x = 24$) by Morinaga and Kuriyama (1959) based on the meiotic pairing behaviour of chromosomes in interspecific crosses. Moreover, CC genome is present in all the six allotetraploids of the complex (*O. minuta*, *O. malampuzhaensis*, *O. punctata/O. schweinfurthiana*, *O. latifolia*, *O. alta* and *O. grandiglumis*) and therefore is regarded as the 'pivotal genome' of the *Officinalis* complex (Vaughan 2003). Information regarding the origin and spread of the genome is scanty (Buso et al. 2001; Sui et al. 2014). Wang et al. (2009) have suggested that CC genome might have evolved from BB genome about 4.8 Mya. This was followed by a series of speciation events of CC genome diploids (ca. 1.8–0.9 Mya). The wider geographical distribution of CC genome across the tropics of three continents (Asia, Africa and America) might have resulted in huge genomic variation leading to wider adaptation of this complex. High sterility reported among

Table 21.1 Brief information about the species reported under *O. officinalis* complex

Reported species	Genome type	Chr. no. (2*n*)	DNA cont. (pg/2C)	Distribution	Natural habitat	Potential use
O. officinalis	CC	24	1.45	Asia to New Guinea and rarely in Australia	Seasonally dry area, open sunlight	Pest resistance, sheath rot resistance
O. rhizomatis	CC	24	NA	Sri Lanka (dry zone)	Seasonally dry area, open sunlight	Drought tolerance, brown plant hopper resistance
O. eichingeri	CC	24	1.47	Sri Lanka (wet or intermediate zone), East Africa (Uganda and Ivory Coast)	Stream sides, forest floor; semi-shady	Pest resistance
O. minuta	BBCC	48	2.33	The Philippines and Papua New Guinea	Stream sides; semi-shady	Disease resistance
O. malampuzhaensis	BBCC	48	NA	Southern India	Seasonally dry forest pools; shady	Pest resistance
O. punctata	BB BBCC	24 48	1.1 N.A.	Africa	Seasonally dry area; open sunlight (diploid) Forest floor; semi-shady (tetraploid)	Resistance to bacterial blight and brown plant hoppers
O. latifolia	CCDD	48	2.32	Latin America	Seasonally dry; open sunlight	Salt tolerance
O. alta	CCDD	48	NA	Belize, Brazil, Colombia, Guyana, Paraguay and Venezuela	Seasonally inundated; open sunlight	Useful for forage and biomass production
O. grandiglumis	CCDD	48	1.99	Latin America	Seasonally inundated; open sunlight	Useful for forage and biomass production
O. australiensis	EE	24	1.96	Tropical Australia	Seasonally dry; open sunlight	Pest resistance and drought tolerance

Adapted from Vaughan et al. (2003) and the Database of National Agricultural and Food Research Organization, Japan (https://www.gene.affrc.go.jp/databases-plant_images_en.php)

interspecific hybrids of CC genome diploids confirms their classification into three different species (Ogawa 2003). Among these, *O. eichingeri* is presumed to be the first species to evolve followed by *O. rhizomatis* and *O. officinalis* (Shcherban et al. 2000a, b). *O. eichingeri* is distributed largely in Africa, besides some parts of Asia (Wang et al. 2009). *O. officinalis* is most

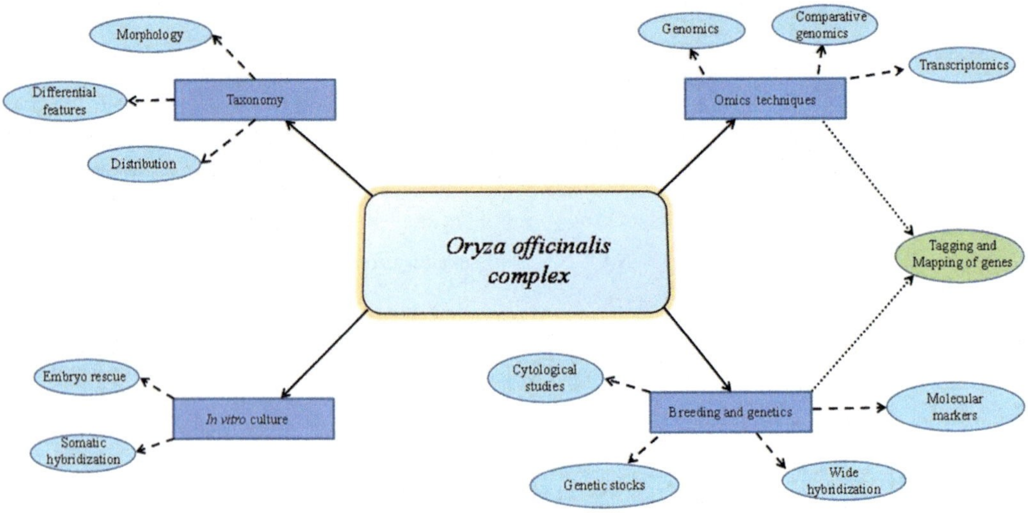

Fig. 21.1 Utilization of *O. officinalis complex* in rice improvement. The bold arrows represent the major areas of research while the dashed arrows depict the subareas. Dotted arrows represents merger of two research areas

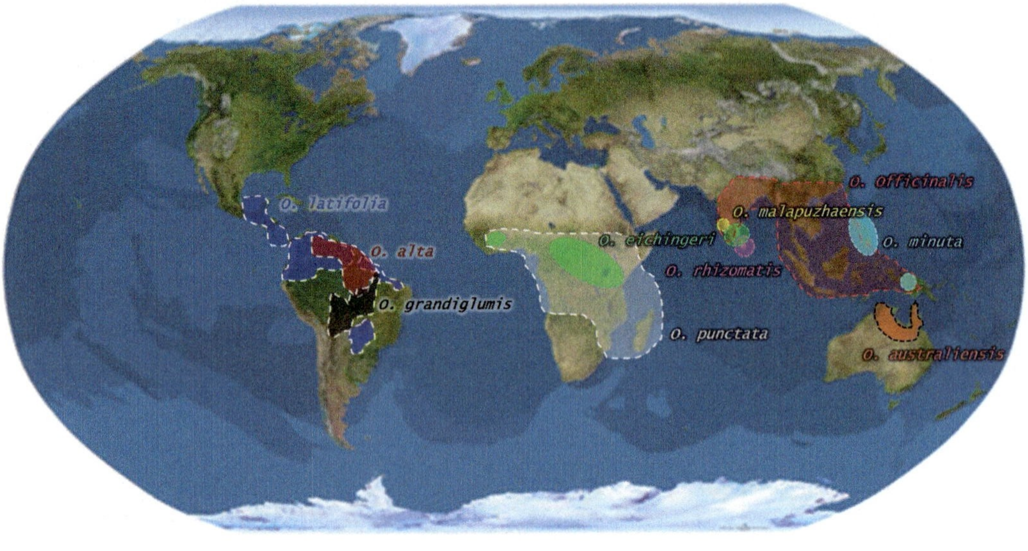

Fig. 21.2 Global distribution of the species belonging to *O. officinalis* complex

widely distributed species in this complex and found in South China, South and Southeast Asia, and Papua New Guinea. *O. rhizomatis* is closely related to *O. officinalis* and is endemic to Sri Lanka.

O. punctata is found in Africa and comprises both diploid (BB, $2n = 2x = 24$) (Katayama 1967) and allotetraploid (BBCC, $2n = 4x = 48$) genomes. Despite of extensive exploration efforts undertaken by researchers worldwide, diploid DD genome progenitors have not been discovered till date and any of the CC genome diploid progenitors do not exist in America (Jena and Kochert 1991; Li et al. 2001). *O. australiensis* with EE genome (Li et al. 1963) is genetically most diverse species in the complex which is found only in Australia. DNA sequence studies based on sequence of *Adh2* gene suggested that

Table 21.2 Specific morphological features of the species of *O. officinalis* complex (Adapted from Vaughan 1994)

O. officinalis	*O. latifolia*	*O. eichingeri*	*O. minuta*
Lowest panicle node almost always has a whorl of branches, generally non-rhizomatous	Broad leaf (up to 5 cm) with features almost similar to two other species *O. alta* and *O. grandiglumis*. Both dwarf and tall forms exist. Spikelet generally less than 7.0 mm	Lowest panicle node almost always has only one branch, not rhizomatous. Perennial	Stoloniferous herbs showing ascending growth habit. Panicle branches lack whorl at base, spikelets small (length 4.7 mm, width <2.0 mm)
O. malampuzhaensis	*O. punctata*	*O. rhizomatis*	*O. alta*
Erect plant, spikelets generally larger and more elongated than *O. officinalis*. Some workers consider it as an ecotype of *O. officinalis*	Ligule about 3.0 mm, truncated, glabrous and split. Panicles branched and spreading type. Diploid races are weakly perennial whereas tetraploid races are strongly perennial	Perennial, erect and rhizomatous grass, 1–3 m tall; panicles open without whorled basal panicle branches and diploid	Only tall (up to 4.0 m) form exists when compared to *O. latifolia*. Other features are almost similar
O. grandiglumis		*O. australiensis*	
Tall plants (up to 4 m) with broad leaves. Ligules are pubescent. The length of empty glumes similar to palea and lemma		Plants taller than 2.0 m in height. Leaves are grey-green coloured, strap shaped. Length of the spikelets varies from 6.5 to 9.0 mm. Pear shaped with wispy awn (<5 cm). Panicles have scabrous axes	

the species might have originated approximately 8.5 Mya which is much earlier than evolution of allotetraploids in this complex (Piegu et al. 2006 and Wang et al. 2009).

Among allotetraploids, BBCC genome species (*O. minuta, O. malampuzhaensis* and 2X + 2X *O. punctata/O. schweinfurthiana*) evolved between 0.2 and 0.6 Mya and the CCDD species (*O. latifolia, O. alta* and *O. grandiglumis*) about 0.9–1.0 Mya in stepwise manner (Vaughan 1989, 1994; Zou et al. 2015). The three BBCC genome allopolyploids have evolved through three independent events of allopolyploidization. The three CCDD genome species are assumed to have evolved from a single hybridization event (Ge et al. 1999; Buso et al. 2001; Bao and Ge 2004). Interestingly, these allopolyploids have originated during the era of drastic climate changes which includes several ice ages. Under such circumstances, allopolyploidization might have been a preferred strategy for creation and retention of sufficient heterogeneity in the genome. This in turn might have helped the plants to produce more permutations and combinations of metabolites imparting better adaptation in the process (Wendel 2000). *O. minuta* originated in Asia from the hybridization of diploid *O. punctata* (BB, maternal parent) and *O. officinalis* (CC, paternal parent) followed by allopolyploidization. The BBCC genome of *O. minuta* was first designated by Morinaga (1943) through chromosome pairing behaviour studies. However, what remains most interesting is that its diploid BB genome progenitor *O. punctata* is found only in Africa. The plausible origin may be explained either by assuming that a presently extinct diploid species with BB genome was available in Asia or the BBCC allotetraploid may have originated in Africa and later dispersed to Asia. The tetraploid *O. punctata* is assumed to have evolved through hybridization of *O. eichingeri* (CC) as maternal parent and diploid *O. punctata* (BB) as paternal parent followed by chromosome doubling. Many taxonomists now designate the tetraploid *O. punctata* as a separate species *O. schweinfurthiana* based on their growth habit, morphological features and molecular evidences.

Fig. 21.3 Photographs of the species belonging to *O. officinalis* complex. Photograph in insets depict the panicle of corresponding species. Photograph courtesy: Dr. B. C. Patra, ICAR-NRRI, Cuttack (*O. latifolia* and *O. grandiglumis*), Dr. A. Das, ICAR-IIPR, Kanpur (*O. rhizomatis*) and Dr. P. Kaushal, ICAR-NIBSM, Raipur (*O. malamphuzaensis*)

There is a controversy about recognizing *O. malampuzhaensis* as a separate valid species which is exclusively found in the Malampuzha area of India. However, Joseph et al. (2008) provided molecular and morphological evidences to distinguish this as a separate species within the *O. officinalis* complex. *O. punctata* (BB, as maternal parent) and *O. officinalis* (CC, as paternal parent) are the probable diploid progenitors of *O. malampuzhaensis* (Zou et al. 2015).

All the CCDD genome species are distributed in Latin America (Vaughan 1989, 1994) and show very little differences in morphology and genome content and considered as conspecific (Tateoka 1962; Nayar 1973; Jena and Kochert 1991; Aggarwal et al. 1996; Fukui et al. 1997; Buso et al. 2001; Ying and Song 2003). CCDD genome of *O. latifolia* was first described by Morinaga in 1943.

The evolution of several species with BBCC and CCDD genomes has been schematically depicted in Fig. 21.4 based on the hypothesis proposed by several research groups.

The occurrence of both diploid and tetraploid species in *O. officinalis* complex makes it an ideal system for study of polyploidy-driven speciation and evolution.

21.3 Cytological Features of *O. officinalis* Complex

21.3.1 Karyotype Development

Like all other crops, initial studies on genome analysis of *O. officinalis* complex were based on karyotype analysis. Karyotypic studies involve length measurements of chromosomes and their arms. Initial studies in karyotyping were mostly

<table>
<tr><th>♀ / ♂</th><th>O. punctata (BB)</th><th>O. officinalis (CC)</th><th>Unknown/Extinct (DD)</th></tr>
<tr><td>O. punctata (BB)</td><td></td><td>O. minuta (BBCC)
O. malampuzhaensis (BBCC)</td><td></td></tr>
<tr><td>O. officinalis (CC)</td><td></td><td></td><td rowspan="3">O. grandiglumis,
O. alta & O. latifolia
(CCDD)</td></tr>
<tr><td>O. eichingeri (CC)</td><td>Tetraploid O. punctata
(O. schweinfurthiana)
(BBCC)</td><td></td></tr>
<tr><td>O. rhizomatis (CC)</td><td></td><td></td></tr>
</table>

Fig. 21.4 Schematic representation of polyploid evolution in *O. officinalis* complex

limited to counting the chromosome number which was later extended as identification key for individual chromosomes leading to development of karyotypes. The details of chromosome number and genome types of different species of *O. officinalis* complex have been presented in Table 21.1. Wild species of rice show more symmetric karyotypes than cultivated species (Shastry 1963). Hu (1961) compared karyotypes of several *Oryza* species and found that the karyotypes were almost identical among those species except *O. australiensis* and *O. officinalis*. The morphology of pachytene bivalents differs significantly among the species of *O. officinalis* complex. Among the species, *O. australiensis* shows highest degree of heterochromatinization with small darkly stained segments alternating with lighter ones. The centromere of this species is also highly distinct (Shastry and Rao 1961). The karyotype of *O. officinalis* is comparatively uniform and is much less heterochromatic when compared with *O. austraiensis* (Shastry 1963). Kurata and Omura (1982) constructed karyotype of *O. punctata* and *O. officinalis*. With the advent of molecular tools, genomic *in situ* hybridization (GISH) and fluorescent *in situ* hybridization (FISH) are increasingly being used for identification of individual chromosomes. Tang et al. (2007) used a set of chromosomal gene-enriched region-derived arm-specific markers in pachytene chromosomes of *O. punctata* (BB genome) and *O. officinalis* (CC genome) which enabled identification of different chromosomes of these two species. Such studies also helped to reconstruct the karyotypes.

21.3.2 Meiotic Pairing Behaviour in F_1 Hybrids

21.3.2.1 Meiotic Pairing Among Species of *O. officinalis* Complex

Interspecific hybridization among members of *O. officinalis* complex helped in genome analysis of this group. In the hybrids of *O. officinalis* × *O. australiensis* and *O. minuta* × *O. australiensis,* the chromosomes of *O. australiensis* were found to be larger as compared to *O. minuta* and *O. officinalis* (Morinaga et al. 1960, 1962). Studies by other researchers (Kihara and Nezu 1960; Li et al. 1963) also reported similar results. Interspecific hybridization followed by meiotic chromosome pairing behaviour was studied by several workers to understand the genomic affinities among different species under this complex. The relevant details have been listed in Table 21.3.

21.3.2.2 Meiotic Pairing of *O. officinalis* Complex with Cultivated Rice

Besides study of crossability and chromosome pairing behaviour among the members of *O. officinalis* complex, several studies were undertaken in hybrids of the species *O. officinalis* with the two cultivated AA genome species, namely *O. sativa* and *O. glaberrima* (Table 21.4). The hybrids were mostly sterile with rare bivalents. These studies have helped in understanding the genomic relationships to a great extent. Besides providing basic information on the relationship of the genomes of respective species of *O. officinalis* complex with AA genome of *O. sativa* and *O. glaberrima*, these studies also helped in development of valuable genetic stocks especially monosomic alien addition lines (MAALs) where one extra chromosome from the wild species is present besides full complement of the cultivated species. MAALs are useful cytogenetic stocks in mapping genes introgressed from wild species.

21.4 Genomic Resources of *O. officinalis* Complex

21.4.1 Background History and Status of Genome Initiatives

Genetic variations available within the wild *Oryza* species are considered as an untapped resource for the improvement of cultivated rice.

Table 21.3 Crossability behaviour among members of *O. officinalis* complex

S. No.	Crosses studied	Specific features	References
1	*O. officinalis* (CC) × *O. australiensis* (EE)	Somatic cell chromosome number of F_1—24. Number of univalents—16 to 24 (Mean—21.84 with 01 to 04 loosely paired chromosomes). Pollen and seed completely sterile. In case of amphiploid CCEE plants, range and mean of bivalents per cell were 23.6	Katayama (1982)
2	Hybrid CE plant × *O. latifolia* (CCDD)	Hexaploids—73/92, tetraploid—9/92 and heteroploids—20/92. All configurations ranging from univalents to tetravalents with the highest frequency of bivalents were observed. Pollen and seeds were completely sterile	Katayama (1982)
3	Amphidiploid CCEE × *O. latifolia* (CCDD)	Number of bivalents ranged from 17.6 to 20.9. Ring- or rod-shaped tetravalents recorded in the many of the F_1 hybrids suggested the presence of reciprocal translocation	Katayama (1982)
4	*Oryza officinalis* (4x, CCCC) × *O. minuta (BBCC)* and *O. grandiglumis* (CCDD)	Chromosome pairing differed depending upon the presence of B or D genomes. More homology among C and B genomes. D genome has a suppressing effect	Hu (1970)
5	*O. eichengeri* × *O. punctata*	Pollen fertility 73% and good seed set	Hu and Chang (1965)
6	*O. punctata* × *O. eichengiri*	Hybrids completely sterile. Only two out of 12 chromosomes displayed non-perfect bivalent	Ranganadhacharyulu and Raj (1974)
7	*O. minuta* × *O. punctata*	Sterile with 4.1% pollen fertility	Hu and Chang (1965)

Table 21.4 Crossability behaviour between cultivated species of *Oryza* and members of *O. officinalis* complex

S. No.	Crosses studied	Specific features	References
1	*O. sativa* × *O. punctata* (2n = 4 × = 48)	Mostly univalent with occasional bivalent association. F_1 plants were completely sterile	Nwokeocha et al. (2007)
2	*O. sativa* × *O. punctata* (2n = 2x = 24)	F_1 was highly sterile. Amphidipolids were fertile.	Wang et al. (2013)
3	Amphidiploid of *O. sativa* and *O. punctata* (AABB) developed by Wang et al. (2013) × Tetraploid *Oryza sativa* (AAAA)	Seed set ranged up to 71.31%. Grain morphology similar to cultivated rice	Zhang et al. (2013)
4	*O. glaberrima* × *O. punctata* (20 = 4x = 48)	Mostly univalent with occasional bivalent and trivalent associations. F_1 plants were completely sterile with malformed microspores	Nwokeocha et al. (2007)
5	*O. sativa* × *O. eichingeri*	–	Yan et al. (1997)
		Mostly univalent with occasional bivalent association. F1 plants were completely sterile	
6	*O. sativa* + *O. eichingeri and O. sativa* + *O. officinalis*	Produced by cell fusion. Both somatic hybrids produced viable pollen. The second one also produced viable progeny	Hayashi et al. (1988)
7	*O.sativa* × *O. officinalis*	Hybrids obtained by using male-sterile lines without need of any embryo culture	Xuexing (1993)
8	*O. sativa* × *O. officinalis*	Average of 0.77 bivalents found. High level of pollen sterility	Yan et al. (1995)
9	A series of crosses of *O. sativa* with *O. alta, O. latifolia, O. minuta and O. eichingeri, O. australiensis* and *O. officinalis*	When crosses were attempted between 2x and 4x species, 2x ♀ × 4x ♂ matings were more successful than the reciprocals. Genome B in O.minuta and O. eichingeri was found to be partially homologous with A	Nezu et al. (1960)
10	A series of crosses of *Oryza sativa* with *O. punctata, O. malampuzhaensis, O. minuta* and *O. latifolia*	Hybrids were obtained from crosses of *O. malampuzhaensis, O. latifolia* and *O. punctata* using embryo rescue and were completely sterile showing very high frequency of univalents	Kaushal (1998)
11	*O. sativa* × *O. officinalis*	Monosomic alien addition lines (MAALs) containing the complete genome of *Oryza sativa* and 12 individual chromosomes of *Oryza officinalis*	Tan et al. (2005)
12	*O. sativa* × *O. officinalis*	Monosomic alien addition line (MAAL) containing chromosome 4 of *Oryza officinalis*	Jin et al. (2006)
13	*O. sativa* × *O. officinalis*	Monosomic alien addition lines (MAALs) containing the complete genome of *Oryza sativa* and 12 individual chromosomes of *Oryza officinalis*	Jena and Khush (1986)
14	*O. sativa* × *O. latifolia*	Monosomic alien addition lines (MAALs) containing the complete genome of *Oryza sativa* and 12 individual chromosomes of *Oryza latifolia*	Multani et al. (2003)

(continued)

Table 21.4 (continued)

S. No.	Crosses studied	Specific features	References
15	*O. sativa* × *O. punctata*	Monosomic alien addition lines (MAALs) containing the complete genome of *Oryza sativa* and 11 individual chromosomes of *Oryza latifolia*	Yasui and Iwata (1991)

But, in order to utilize the available variation in rice improvement programme, it is of utmost importance to mine the useful variations from the genomes of the wild relatives. Presently, within the *O. officinalis* complex, the genome sequence is available only for diploid *O. punctata* (BB) (pre-publication data available at Gramene database (http://www.Gramene.org), at EMBL-ENA by assembly accession number GCA_000573905.1 and at NCBI-Genbank by accession number AVCL00000000.1) while genome sequencing for the rest is in progress. *O. punctata* contains genes for bacterial blight and brown plant hopper resistance. Its genome size is 423 Mb (as estimated by flow cytometry) out of which about 394 Mb has been sequenced. According to the available whole genome sequence, chromosome 1 of the species is the largest (~46 Mb) while the chromosome 10 is smallest (~26 Mb). The whole genome of *O. punctata* contains 31,762 protein-coding and 5168 non-coding genes. Its reference proteome is available at Uniprot database (Apweiler et al. 2004) with the proteome ID UP000026962. *O. officinalis* is most common species in this complex which is prevalent almost everywhere in Asia and Papua New Guinea. The whole genome sequence is not available till now for this species, but the sequence of chromosome 3 short arm (~26 Mb) is available in the public domain (NCBI accession number CM002825). Apart from this, the sequences of bacterial artificial chromosome (BAC) clones and fingerprinted contigs (FPC) of *O. officinalis* genome are also available at marker database of Gramene (http://www.gramene.org). Among the tetraploid species, sequences of short arm of chromosome 3 are available publicly for *O. minuta* only at National Centre for Biotechnology Information (NCBI) under the accession numbers CM002693 (BB3 short arm of about 20 Mb) and CM002694 (BB3 short arm of about 25 Mb). Whole genome sequencing of the other species in this complex is also in progress under OMAP. Once achieved, it will yield a comprehensive physical framework to undertake basic and strategic research on discovering and introgressing new genes and QTLs form wild *Oryza* species to cultivated rice by genome guided breeding approach.

21.4.2 Comparative Genome Mapping with *O. Sativa*

In order to understand evolutionary relationship between the species, discover alien genes and alleles and to understand their function, comparative genome mapping is a highly useful tool. The earliest effort of constructing comparative physical map between cultivated rice and the members of *O. officinalis* complex was made during 1994 when restriction fragment length polymorphism (RFLP)-based comparative mapping was carried out between *O. sativa* (AA) and *O. officinalis* (CC) at International Rice Research Institute (IRRI) (Jena et al. 1994). This RFLP-based mapping was carried out using 139 genomic- and cDNA-based probes which were previously used for the mapping of rice genome. The results suggested that 9 out of 12 *O. officinalis* chromosomes were highly similar to that of *O. sativa*. Among the rest of the chromosomes, chromosome 1 contains maximum variation depicting major rearrangement of gene order. Chromosomes 3 and 11 depicted small inversions whereas chromosome 11 hosted majority of the translocated markers.

An effort for developing a comparative physical map of AA and BB genome was made by aligning 63,942 BAC end sequences and 34,224 fingerprinted contigs of *O. punctata* against *O. sativa* genome sequence (Kim et al. 2007). Out of these, 40% of BSEs and 87% of FPCs could be aligned which covered 98% of *O. sativa* genome. The results suggested macro-colinearity between these two *Oryza* species which diverged from each other about 2 million years ago (Mya), besides indicating the locations of expansions, contractions, inversions and transpositions which probably shaped the evolution of these two species. Expectedly, the sequence divergences in the intergenic and repeat regions were found to be higher than in the genic regions.

With the advent of OMAP, comparative genomics of *Oryza* genus has got tremendous impetus. One of the major goals of OMAP is to develop a comparative framework of genomes representing the genus '*Oryza*' so that it may emerge as a genome-level comparative experimental system (Kim et al. 2008; Goicoechea et al. 2010). Collaborative efforts under I-OMAP have already completed construction of BAC-based physical maps for 17 *Oryza* species, out of 23 reported, which represents all the 10 genome types including the genome coming under *O. officinalis* complex (Ammiraju et al. 2006, 2010; Kim et al. 2008; Jacquemin et al. 2013). The data obtained from this large-scale comparative genomic study suggested that a single reference genome might be inadequate to comprehensively capture the total genomic and allelic diversity present in the genus '*Oryza*' (Goicoechea et al. 2010).

21.4.3 Transcriptome Studies in *O. Officinalis* Complex

Transcriptome studies involving the species of *officinalis* complex are scanty. An interesting study based on transcriptome analysis was carried out involving artificially created allotriploid F_1 hybrid (ABC, $2n = 3x = 36$) of *O. sativa* (AA, $2n = 2x = 24$) and *O. punctata* (BBCC, $2n = 4x = 48$) (Wu et al. 2016). Transcriptome profiling of the parents and the hybrid demonstrated a 'transcriptome shock' in the triploid hybrid which was manifested by immediate and aberrant changes in gene expression patterns. A clear-cut homeologous gene expression bias was observed between in silico predicted hybrid and the real hybrid. For example, about 16% of the homeologous genes were expressed non-additively in the hybrid. Non-random maternal gene silencing was found to be one of the principle contributors of this homeologous gene expression bias in the hybrid. In fact, such 'transcriptome shock' (Hegarty et al. 2006; Buggs et al. 2012) and homeologous gene expression bias (Wendel 2000; Doyle et al. 2008; Jackson and Chen 2010) are a common phenomenon both in natural or artificial hybrids due to the combining of distant genomes.

Among the wild *Oryza* species, *O. officinalis* exhibit high degree of resistance against several diseases (Bao et al. 2015). Some of the transcriptome studies have been carried out involving *O. officinalis* where the problem of not having a reference genome has been circumvented using de novo sequence assembly method. Bao et al. (2015) carried out one such study to understand the composition of leaf transcriptome of two-week-old *O. officinalis* seedling sunder normal (without stress) condition using RNA-Seq. De novo assembly of approximately 23 million reads yielded 687,123 unigenes which were functionally annotated using different databases. Further, they have identified 476 unique disease resistance-related unigenes which might be important for improving the cultivated rice. In a similar study, transcriptome obtained from the leaf, stem and root tissues of *O. officinalis* was sequenced and differentially expressed gene enriching the pathways conditioning disease resistance was examined (He et al. 2015). They found that most of the genes related to disease resistance pathway were expressed ubiquitously in all three tissues under study. Further, they could identify two unique type of bacterial blight resistance genes *Xa1* (two copies) and *Xa26* (three copies). Interestingly, tissue-specific expression bias was observed in case of these genes. Highest expressions of both *Xa1* genes were found in stem, whereas one of the three *Xa26* genes showed maximum expression in leaf. These studies enriched the transcriptome database of *O. officinalis* which provides an opportunity to understand and utilize novel disease resistance genes of *O. officinalis* for the genetic improvement of rice.

21.4.4 Genomic Resources for *O. officinalis* Complex

Discovery of genome-wide markers in *O. officinalis* complex was largely impeded due to the unavailability of whole genome sequence in the public domain. The initial efforts were made towards the cross-transferability of well-annotated AA genome-specific DNA markers *O. officinalis* complex. For an example, 14 *O. sativa*-specific restriction fragment length polymorphism (RFLP) markers were cross-transferred in BBCC allotetraploid species, namely *O. latifolia, O. alta and O. grandiglumis* (Jena and Kochert 1991). These markers were able to discriminate different species in the phylogenetic tree according to their geographical origin. Besides, with these markers, it was possible to make CC and DD genome-specific banding (Jena and Kochert 1991). Another similar study carried out using RFLP markers with 34 wild accessions of almost all the species of *O. officinalis* complex (except *O. malampuzhaensis*) from Asia, Africa, America and Australia comprising BB, CC, EE, BBCC and CCDD genomes (Federici et al. 2002). The study revealed important findings related to divergence and differentiation of the species under *O. officinalis* complex and also showed that retrotransposon-like sequence pOe.49 was useful for undermining phylogenetic relationship within *Oryza* genus. Apart from this, PCR-RFLP of *Adh1* and *Adh2* genes using *ScaII* and *EcoNI* restriction enzymes, respectively, also throw light in the species diversification of *O. officinalis* complex (Bao et al. 2005). But probably the most robust and reliable method to study the diversification and differentiation of the species under '*Oryza*' genus came through amplified fragment length polymorphism (AFLP) markers. In a study by Aggarwal et al. (1999), AFLP-based DNA fingerprinting was carried out for 77 different accessions including reported 23 species of *Oryza*, five related genera and three out-group taxa. The study yielded several significant findings regarding the phylogeny of the genus '*Oryza*', including the species that come under *O. officinalis* complex. Besides, there are successful examples of DNA fingerprinting studies using multilocus dominant markers like random amplified polymorphic DNA (RAPD) and inter-simple sequence repeat (ISSR) markers revealing species genetic diversity in *O. officinalis* complex (Joshi et al. 2000; Bautista et al. 2006). Among the locus-specific co-dominant markers, sequence-tagged microsatellite (STMS) marker database is very rich in cultivated rice. These markers are well distributed throughout the rice genome and also densely saturate the genome. But, since the microsatellite has genome specificity, STMS markers show genome biasness, and hence, cross-transferability of STMS markers across genomes has always been a questionable issue (Ray et al. 2016). Efforts have been made to cross-transfer *O. sativa* (AA)-specific STMS markers (commonly known as SSR markers) in species of *O. officinalis* complex, but the results were conflicting (Ray et al. 2016). Further, Jin et al. (2006) reported low cross-transferability of *O. sativa*-derived STMS markers to *O. officinalis*. However, Li et al. (2008) found that the cross-transferability between these two species was high. This anomaly in observation is possibly due to the use of different sets of STMS markers by the two groups of workers (Ray et al. 2016). Recently, during performing a genome survey of *O. latifolia* (CCDD) using 293 *O. sativa*-derived STMS markers, only about 13% cross-transferable markers were recorded which were polymorphic (Angeles-Shim et al. 2014). Like the other markers, *O. sativa*-derived STMS markers have also been utilized to assess genetic diversity within and between the species of *O. officinalis* complex with limited success (Guo and Ge 2005; Bao et al. 2005; Bautista et al. 2006). Wang et al.

(2014) have developed tetraploid BBCC genome-specific STMS markers from genomic library of *O. minuta*. They could identify 13,991 SSR motifs from where they could design 14,508 STMS markers of *O. minuta*. Further, they found that among the microsatellite repeats, hexanucleotide repeats were the most abundant (31.39%), followed by trinucleotide (27.67%) and dinucleotide (19.04%) repeats. Some of these STMS markers showed good cross-transferability between species, which was demonstrated by PCR assay with 600 selected markers and 23 accessions belonging to different *Oryza* species. Hence, these markers can be useful to study phylogeny as well as for the purpose of tagging, mapping and introgression of genes from *O. officinalis* complex to cultivated rice.

21.4.5 Mapping of Genes/QTLs and Their Utilization in the Improvement of Cultivated Rice

Most of the breeding projects till date in rice have utilized intraspecific hybridization between and within ecotypes of *Oryza sativa* (e.g. indica-indica, japonica-japonica, indica-japonica, indica-tropical japonica). Utilization of wild species remained limited although in several cases, genetic variability for target agronomic traits was lacking in the primary gene pool.

Barring a few exception, wild species of rice have been utilized as valuable source of tolerant genes for various biotic and abiotic stresses. The *O. officinalis* complex is also not an exception. Indeed this complex is recognized as a storehouse of major genes for resistance to brown plant hopper, a devastating pest of rice. Beside, genes for tolerance to white backed plant hopper (WBPH), bacterial leaf blight (BLB), sheath rot and leaf/neck blast have been identified and transferred. Alien introgressed lines developed using *O. officinalis* as donor have been released as BPH-resistant varieties (MTL98, MTL103 and MTL110) in Vietnam (Brar and Singh 2011).

However, the transfer is practically very difficult as the wild species are associated with several weedy traits like grain shattering, poor grain characteristics, low grain yield and undesirable plant type. Besides, there are several pre- and post-fertilization barriers. With advance techniques of plant tissue culture especially embryo rescue and protoplast fusion, wild species are increasingly being used. Cytogenetic techniques along with availability of cross-transferrable markers derived from genome sequencing projects of wild species have created further opportunities in wide hybridization of rice.

Strategy for transferring genes from wild to cultivated species depends on various factors such as nature of trait (quantitative/qualitative or monogenic/oligogenic/polygenic), genetic relatedness of cultivated species to the wild species and their incompatibility barriers. In crosses involving *O. eichingeri* and *O. officinalis* as pollen parents and *O. sativa* as female, pollen germination was found to be normal and it could penetrate the stigma. However, further growth was inhibited in the style or ovary wall, particularly in *O. eichingeri* crosses (Sitch and Romero 1990). In few cases embryos developed, but finally aborted. However, several protocols are available to overcome such barriers and have been utilized for successful wide hybridization (Brar and Khush 2002). With *O. officinalis* complex, major focus was to develop the MAALs and advance introgressed lines containing segments of alien chromosomes. Unlike AA genome species, chromosome segment substitution line development is not possible in *O. sativa* background with *O. officinalis* complex. Therefore, advance introgression lines (AILs) have been produced for gene transfer from species with CC, BBCC, CCDD and EE genomes to cultivated rice.

21.4.5.1 CC Genome Introgression Lines

Among the three CC genome species of *O. officinalis* complex, successful gene transfer has been achieved only from two species, namely *O.*

officinalis and *O. eichingeri*. Many of the introgression lines from *O. officinalis* have shown resistance to BPH. Beside this, genes for resistance to WBPH, BLB and sheath rot have been identified and transferred. Four varieties with BPH resistance gene derived from *O. officinalis* have been released for commercial cultivation in Vietnam. Screening of *O. eichingeri* accessions helped in identification of BPH resistance genes and their subsequent transfer to rice.

21.4.5.2 BBCC Genome Introgression Lines

Interspecific hybrids have been produced between *O. sativa* and tetraploid wild species *O. minuta*, besides *O. punctata* and *O. malampuzhaensis*. However, developments of AILs have only been reported from *O. minuta* by several workers. The traits which were transferred are resistance to BPH, BLB and Blast.

21.4.5.3 CCDD Genome Introgression Lines

Among the three CCDD genome species of *O. officinalis* complex, *O. alta* is yet to be utilized in rice breeding. However, reports are available about development of AILs involving *O. sativa* × *O. latifolia* and *O. sativa* × *O. grandiglumis*. The major problems with development of CCDD-based AILs are phenotypic instability and abnormal segregation behaviour where abnormal plants with many wild traits suddenly appear in disomic lines (Brar and Singh 2011). AILs with resistance to BPH, WBPH and BLB have been developed by transfer of genes from *O. latifolia*. QTLs for yield contributing traits have been mapped from backcross progeny lines of *O. sativa* × *O. grandiglumis*.

21.4.5.4 EE Genome Introgression Lines

Advance introgression lines with genes for resistance to BPH and leaf blast have been developed, and important genes like *Bph10*, *Bph18* and *Pi40(t)* have been tagged in *O. australiensis* (the only species with EE genome). *Bph18* is a major gene for BPH resistance and is now being utilized for marker-assisted selection in rice (Brar and Singh 2011). A comprehensive list of the reports along with their references has been presented in Table 21.5.

21.4.6 Future Prospects

From the above discussions, this is aptly clear that the species of *O. officinalis* complex carry a large numbers of useful genes for genetic improvement of cultivated rice. In several cases, success was achieved in transfer of those genes. However, there still remains lot of scope for precise interspecific gene transfer. Lack of availability of whole genome sequences information for most of these species is a major bottleneck. Presently, within the *O. officinalis* complex, the whole genome sequence is available only for diploid *O. punctata* while genome sequencing for the rest is in progress. Once the genome sequences of all other species are available, a significant progress in utilization of these wild genetic resources is expected. However, useful genomic resources may still be developed from the available whole genome sequence of *O. punctata* and from other DNA sequence data available for rest of the species. Cross-transferrable markers with respect to AA genome may be immediately identified in order to utilize these wild species in rice improvement through mapping of genes or QTLs of interest and their utilization through marker-assisted introgression. The whole genome sequencing initiative undertaken by OMAP and OGEP promises development of high-throughput molecular marker-assisted genetic analysis, gene discovery and breeding programs involving the wild species of *O. officinalis* complex. In our view, the road map for utilization of *O. officinalis* complex for rice improvement might be as follows:

1. Utilize the sequences available in BAC libraries, and the whole genome sequence of *O. punctata* to isolate novel genes/QTLs for desired traits is necessary. The cytogenetic stocks like MAALs or introgression lines will be specifically useful for tagging or mapping of genes as well as functional genomics.

Table 21.5 Progress in development of AILs of *O. sativa* with members of *O. officinalis* complex

Genome	Cross combination	Specific features	References
CC	*O. sativa* × *O. officinalis.*	Introgression line with BPH resistance. Gene was tagged in chromosome 11 by RAPD marker	Jena et al. (2003)
		BPH biotype-4 resistance gene [Bph13(t)] derived from *O. officinalis* was tagged on chromosome 3 by RAPD marker which was subsequently converted to a reproducible sequence-tagged sites (STS) marker	Renganayaki et al. (2002)
		BPH-10 was tagged and mapped on chromosome 12. From linked RFLP, one STS marker was developed	Lang and Buu (2003)
		High-resolution genetic map of *Bph15* was developed and it was possible to assign this to a genomic segment of approximately 47 kb	Yang et al. (2004)
		Two brown plant hopper resistance genes were tagged on chromosomes 3 and 4	Huang et al. (2001)
		BPH resistance was transferred from *O. officinalis* to *Oryza sativa* by conventional backcrossing and phenotypic selection	Zhong et al. (1997)
		Eleven different traits from *O. officinalis* were found to have been transferred in BC_2 progenies. Among the traits, BPH resistance and WBPH resistance were economically important	Jena and Khush (1990)
		Sheath rot (ShR) resistance in the breeding lines of *Oryza sativa* derived from *O. officinalis*	Lakshmanan and Veluswamy (1991)
		Bph11 gene tagged on chromosome 3 by RFLP marker	Hirabayashi et al. (1998)
		Wbph7(t) and *Wbph8(t)* genes tagged using RILs derived from introgression lines	Tan et al. (2004b)
		Bacterial leaf blight (BLB) resistance gene Xa29 has been transferred	Tan et al. (2004a)
		Advanced-backcross progenies from crosses of an elite BLB susceptible breeding line of New Plant Type (NPT) rice with two accessions of *O. officinalis* were produced. Several BLB-resistant progenies identified	Sanchez et al. (2000)
CC	*O. sativa* × *O. eichingeri*	Molecular screening revealed positive results for *Bph13* in *O. eichingeri*	Madurangi et al. (2011)
		Transfer of BPH resistance through conventional breeding from two accessions	Yan et al. (1997)
BBCC	*O. sativa* × *O. minuta*	High-resolution mapping of *Bph20(t)* in chromosome 4 and *Bph21(t)* in chromosome 12 from *O. minuta*	Rahman et al. (2009)
		Bacterial leaf blight (BLB) resistance gene Xa29 has been transferred. The gene was further cloned	Gu et al. (2004, 2005)
		Introgression lines with resistance to bacterial blight and blast were identified. The introgressed gene for resistance to blast was later designated as Pi9(t)	Amante-Bordeos et al. (1992) and Liu et al. (2002)
		BPH resistance genes with wide spectrum of resistance has been transferred	Brar et al. (1996)
		Advanced-backcross progenies from crosses of an elite BLB susceptible breeding line of New Plant Type	Brar and Singh (2011)

(continued)

Table 21.5 (continued)

Genome	Cross combination	Specific features	References
		(NPT) rice with one accessions of *O. minuta* were produced. Several BLB-resistant progenies identified	
BB/BBCC	*O. sativa* × *O. punctata*	Dominant resistance gene for BLB from wild species was detected in F_1	Kaushal (1998)
BBCC	*O. sativa* × *O. malampuzhaensis*	Dominant resistance gene for BLB from wild species was detected in F_1	Kaushal (1998)
CCDD	*O. sativa* × *O. latifolia*	Introgression lines for resistance to BPH, WBPH and BLB resistance. Four plant progenies with resistance to BPH, WBPH and specific race of BLB	Multani et al. (2003)
		Dominant resistance gene for BLB from wild species was detected in F_1	Kaushal (1998)
		BPH resistance gene tagged in chromosome 4	Yang et al. (2002)
CCDD	*O. sativa* × *O. grandiglumis*	Quantitative trait loci (QTL) for yield component straits mapped in an advanced-backcross population	Yoon et al. (2006)
EE	*O. sativa* × *O. australiensis*	*Bph10* gene tagged on chromosome 12	Ishii et al. (1994)
		Bph18 gene tagged on chromosome 12	Jena et al. (2006)
		Major leaf and neck blast resistance gene, *Pi40(t)* tagged by CAPS marker on chromosome 6	Jeung et al. (2007)

2. Isolation and cloning of already reported, tagged or mapped genes from this complex is highly desirable. For example, this complex contains several genes for BPH resistance. Detailed characterization of these genes, identification of appropriate combinations for pyramiding, as well as, understanding the molecular mechanisms are necessary. Similar work can be carried out for other reported biotic stress resistance genes too. Completion of *Oryza* Map Alignment (OMAP) and *Oryza* Genome Evolution (OGEP) Projects is highly anticipated in this regard, which will expedite the whole process.
3. Evaluation of large number of accessions of all the species within this complex is required for identification and mining of new alleles of the target genes having higher efficacy. Specific accessions may then be used for transcriptome analysis to isolate useful genes.
4. Finally, detailed understanding of evolutionary pathways in this complex is necessary in order to discover the mechanism of polyploid evolution within this complex. With different ploidy levels among species, this species complex may serve as valuable resource for study of ploidy-regulated gene expression. This might also provide hints why polyploidy is absent in *O. sativa* complex.

References

Aggarwal RK, Brar DS, Huang N, Khush GS (1996) Differentiation within CCDD genome species in the genus *Oryza* as revealed by total genomic hybridization and RFLP analysis. Rice Genet Newslett 13:54–57

Aggarwal RK, Brar DS, Nandi S, Huang N, Khush GS (1999) Phylogenetic relationships among *Oryza* species revealed by AFLP markers. Theor Appl Genet 98:1320–1328

Amante-Bordeos A, Sitch LA, Nelson R, Dalmacio RD, Oliva NP, Aswidinnoor H, Leung H (1992) Transfer of bacterial blight and blast resistance from the tetraploid wild rice *Oryza minuta* to cultivated rice, *Oryza sativa*. Theor Appl Genet 84:345–354

Ammiraju JS, Luo M, Goicoechea JL, Wang W, Kudrna D, Mueller C, Fang E (2006) The *Oryza* bacterial artificial chromosome library resource: construction and analysis of 12 deep-coverage large-insert BAC libraries that represent the 10 genome types of the genus Oryza. Genome Res 16:140–147

Angeles-Shim RB, Vinarao RB, Marathi, Jena KK (2014) Molecular analysis of Oryza latifolia Desv.(CCDD genome)-derived introgression lines and identification of value-added traits for rice (*O. sativa L.*) improvement. J Hered 105(5):676–89. https://doi.org/10.1093/jhered/esu032

Apweiler R, Bairoch A, Wu CH, Barker C, Boeckmann B, Ferro S (2004) UniProt: the universal protein knowledgebase. Nucl Acid Res 32(suppl_1):D115–D119

Bao Y, Ge S (2004) Origin and phylogeny of *Oryza* species with the CD genome based on multiple-gene sequence data. Plant Sys Evol 249:55–66

Bao Y, Lu BR, Ge S (2005) Identification of genomic constitutions of *Oryza* species with the B and C genomes by the PCR-RFLP method. GenetResour Crop Evol 52:69–76

Bao Y, Xu S, Jing X, Meng L, Qin Z (2015) De novo assembly and characterization of *Oryza officinalis* leaf transcriptome by using RNA-Seq. Bio Med Res Int 2015:982065

Bautista NS, Vaughan D, Jayasuriya AH, Liyanage AS, Kaga A, Tomooka N (2006) Genetic diversity in AA and CC genome *Oryza* species in southern South Asia. Genet Resour Crop Evol 53:631–640

Brar DS, Khush GS (2002) Transferring genes from wild species into rice. In: Quantitative Genetics, Genomics and Plant Breeding, CABI, New York, pp 197–217

Brar DS, Singh K (2011) Oryza. In: Kole C (ed) Wild crop relatives: genomic and breeding resources: cereals, vol 1. Springer Science Business Media, Berlin, pp 321–366

Brar DS, Dalmacio R, Elloran R, Aggarwal R, Angeles R, Khush GS (1996) Gene transfer and molecular characterization of introgression from wild *Oryza* species into rice. In Third international rice genetics symposium, IRRI, Manila (Philippines), 16–20 Oct 1995

Buggs RJ, Chamala S, Wu W, Tate JA, Schnable PS, Soltis DE, Barbazuk WB (2012) Rapid, repeated, and clustered loss of duplicate genes in allopolyploid plant populations of independent origin. Curr Biol 22:248–252

Buso GSC, Rangel PHN, Ferreira ME (2001) Analysis of random and specific sequences of nuclear and cytoplasmic DNA in diploid and tetraploid American wild rice species (*Oryza* spp). Genome 44:476–494

Doyle JJ, Flagel LE, Paterson AH, Rapp RA, Soltis DE, Soltis PS, Wendel JF (2008) Evolutionary genetics of genome merger and doubling in plants. Ann Rev Genet 42:443–461

Federici MT, Shcherban AB, Capdevielle F, Francis M, Vaughan D (2002) Analysis of genetic diversity in the *Oryza officinalis* complex. Elect J Biotech 5:16–17

Fukui K, Shishido R, Kinoshita T (1997) Identification of the rice D-genome chromosomes by genomic in situ hybridization. Theor Appl Genet 95:1239–1245

Ge S, Sang T, Lu BR, Hong DY (1999) Phylogeny of rice genomes with emphasis on origins of allotetraploid species. Proc Nat Acad Sci 96:14400–14405

Goicoechea JL, Ammiraju JSS, Marri PR, Chen M, Jackson S, Yu Y, Wing RA (2010) The future of rice genomics: sequencing the collective *Oryza* genome. Rice 3:89–97

Gu K, Tian D, Yang FWUL, Wu L, Sreekala C, Wang D, Yin Z (2004) High-resolution genetic mapping of Xa27 (t), a new bacterial blight resistance gene in rice, *Oryza sativa* L. Theor Appl Genet 108:800–807

Gu K, Yang B, Tian D, Wu L, Wang D, Sreekala C, Yin Z (2005) R gene expression induced by a type-III effector triggers disease resistance in rice. Nature 435:1122–1125

Guo YL, Ge S (2005) Molecular phylogeny of *Oryzeae* (Poaceae) based on DNA sequences from chloroplast, mitochondrial, and nuclear genomes. AmerJ Bot 92:1548–1558

Hayashi Y, Kyozuka J, Shimamoto K (1988) Hybrids of rice (*Oryza sativa L*) and wild *Oryza* species obtained by cell fusion. Mol Gen Genet MGG 214:6–10

He B, Gu Y, Tao X, Cheng X, Wei C, Fu J, Zhang Y (2015) De novo transcriptome sequencing of *Oryza officinalis* wall ex watt to identify disease-resistance genes. Int J Mol Sci 16:29482–29495

Hegarty MJ, Barker GL, Wilson ID, Abbott RJ, Edwards KJ, Hiscock SJ (2006) Transcriptome shock after interspecific hybridization in Senecio is ameliorated by genome duplication. Curr Biol 16:1652–1659

Hirabayashi H, Kaji R, Okamoto M, Ogawa T, Brar DS, Angeles ER, Khush GS (1998) Mapping QTLs for brown planthopper (BPH) resistance introgressed from Oryza officinalis in rice. Advances in rice genetics, pp 268–270

Hu CH (1961) Comparative karyological studies of wild and cultivated species of *Oryza*. Taiwan Provincial College of Agriculture, Taiwan

Hu CH (1970) Cytogenetic studies of Oryza officinalis complex. 4 F1 hybrids of *O. minuta* and *O. grandiglumis* with 4x *O. officinalis*. Indian J Genet. Plant Breed 30(2):410–417

Hu CH, Chang CC (1965) A note of F1 hybrids between *Oryza punctata* and its related species. Jap J Breeding 15:281–283

Huang Z, He G, Shu L, Li X, Zhang Q (2001) Identification and mapping of two brown planthopper resistance genes in rice. Theor Appl Genet 102: 929–934

Ishii T, Brar DS, Multani DS, Khush GS (1994) Molecular tagging of genes for brown planthopper resistance and earliness introgressed from *Oryza australiensis* into cultivated rice, *O. sativa*. Genome 37:217–221

Jackson S, Chen ZJ (2010) Genomic and expression plasticity of polyploidy. Curr Opin Plant Biol 13: 153–159

Jacquemin J, Bhatia D, Singh K, Wing RA (2013) The International *Oryza* map alignment project: development of a genus-wide comparative genomics platform to help solve the 9 billion-people question. Curr Opin Plant Biol 16:147–156

Jena KK, Khush GS (1986) Production of monosomic alien addition lines of *Oryza sativa* having a single chromosome of *O. officinalis* In: Rice genetics proceedings of the international rice genetics symposium, 27–31 May 1985, IRRI, Manila, Philippines, pp 199–208

Jena KK, Khush GS (1990) Introgression of genes from *Oryza officinalis* Well ex Watt to cultivated rice, *O. sativa* L. Theor Appl Genet 80:737–745

Jena KK, Kochert G (1991) Restriction fragment length polymorphism analysis of CCDD genome species of the genus *Oryza* L. Plant Mol Biol 16:831–839

Jena KK, Khush GS, Kochert G (1994) Comparative RFLP mapping of a wild rice, *Oryza officinalis*, and cultivated rice, *O. sativa*. Genome 37:382–389

Jena KK, Pasalu IC, Rao YK, Varalaxmi Y, Krishnaiah K, Khush GS, Kochert G (2003) Molecular tagging of a gene for resistance to brown planthopper in rice (*Oryza sativa* L). Euphytica 129:81–88

Jena KK, Jeung JU, Lee JH, Choi HC, Brar DS (2006) High-resolution mapping of a new brown planthopper (BPH) resistance gene, Bph18 (t), and marker-assisted selection for BPH resistance in rice (*Oryza sativa* L). Theor Appl Genet 112:288–297

Jeung JU, Kim BR, Cho YC, Han SS, Moon HP, Lee YT, Jena KK (2007) A novel gene, Pi40 (t), linked to the DNA markers derived from NBS-LRR motifs confers broad spectrum of blast resistance in rice. Theor Appl Genet 115:1163–1177

Jin H, Tan G, Brar DS, Tang M, Li G, Zhu L, He G (2006) Molecular and cytogenetic characterization of an *Oryza officinalis–O sativa* chromosome 4 addition line and its progenies. Pl Mol Biol 62:769–777

Joshi SP, Gupta VS, Aggarwal RK, Ranjekar PK, Brar DS (2000) Genetic diversity and phylogenetic relationship as revealed by inter simple sequence repeat (ISSR) polymorphism in the genus *Oryza*. Theor Appl Genet 100:1311–1320

Joseph L, Kuriachan P, Thomas G (2008) Is *Oryza malampuzhaensis* Krish. et Chand. (Poaceae) a valid species? Evidence from morphological and molecular analyses. Genet Resour Crop Ev 270:75–94

Katayama T (1967) Cytogenetical studies on *Oryza*. Proc Jap Acad Sci 43:327–331

Katayama T (1982) Cytogenetical studies on the genus *Oryza*. Jap J Genet 57:613–621

Kaushal P (1998) Crossability of wild species of Oryza with *O. sativa* cvs PR 106 and Pusa Basmati 1 for transfer of bacterial leaf blight resistance through interspecific hybridization. J Agr Sci 130:423–430

Kihara H, Nezu M (1960) Genome analysis in the genus Oryza. Annual Report (1959). Natl Inst Genet Jpn 10:37–39

Kim H, San Miguel P, Nelson W, Collura K, Wissotski M, Walling JG, Wing RA (2007) Comparative physical mapping between *Oryza sativa* (AA genome type) and *O. punctata* (BB genome type). Genetics 176:379–390

Kim H, Hurwitz B, Yu Y, Collura K, Gill N, SanMiguel P, Braidotti M (2008) Construction, alignment and analysis of twelve framework physical maps that represent the ten genome types of the genus Oryza. Genome Biol 9:R45

Kurata N, Omura T (1982) Karyotype analysis in rice: III karyological comparisons among four *Oryza* Species. Jap J Breed 32:253–258

Lakshmanan P, Velusamy R (1991) Resistance to sheath rot (ShR) of breeding lines derived from *Oryza officinalis*. International Rice Research Newsletter 16:8–9

Lang NT, Buu BC (2003) Genetic and physical maps of gene Bph-10 controling brown plant hopper resistance in rice (*Oryza sativa* L). Omonrice 11:35–41

Li HW, Chen CC, Weng TS, Wuu KD (1963) Cytogenetical studies of *Oryza sativa* L and its related species 4 Interspecific crosses involving *O australiensis* with *O sativa* and *O minuta*. Bot Bull Acad Sinica 4:65–74

Li CB, Zhang DM, Ge S, Lu BR, Hong DY (2001) Differentiation and inter-genomic relationships among C, E and D genomes in the *Oryza officinalis* complex (Poaceae) as revealed by multicolor genomic in situ hybridization. Theor Appl Genet 103:197–203

Li G, Hu W, Qin R, Jin H, Tan G, Zhu L, He G (2008) Simple sequence repeat analyses of interspecific hybrids and MAALs of *Oryza officinalis* and *Oryza sativa*. Genetica 134(2):169–180

Liu G, Lu G, Zeng L, Wang GL (2002) Two broad-spectrum blast resistance genes, Pi9 (t) and Pi2 (t), are physically linked on rice chromosome 6. Mol Genet Genomics 267:472–480

Lu BR (1999) Taxonomy of the genus *Oryza* (Poaceae): historical perspective and current status. Int Rice Res Notes 243:4–8

Maclean JL (2002) Rice almanac: source book for the most important economic activity on earth. IRRI, Manila

Madurangi SAP, Samarasinghe WLG, Senanayake SGJN, Hemachandra PV, Ratnasekera D (2011) Resistance of *Oryza nivara* and *Oryza eichingeri* derived lines to brown planthopper, Nilaparvata lugens (Stal). J Natl Sci Found Sri 39:175–181

Morinaga T (1943) Cytogenetical studies on *Oryza sativa* L VI The cytogenetics of a hybrid of *O. minuta* Presl and *O. latifolia* Desv. Jap J Bot 12:347–357

Morinaga T, Kuriyama H (1959) Genomic constitution of*Oryza officinalis*. Jpn J Breed 9:259

Morinaga T, Kuriyama H (1960) Interspecific hybrids and genetic constitution of various species in the genus Oryza. Agr Hort 35:935–938

Morinaga T, Kuriyama Murty VVS (1962) On the interspecific hybrids of *Oryza officinalis* and *O australiensis*. Jpn J Breed 12:193 (abst in Japanese)

Morinaga T, Kuriyama H, Ono S (1962) On the interspecific hybrids of *Oryza officinalis* and *O australiensis*. Jpn J Genet 35:277–278

Multani DS, Khush GS, Delos Reyes BG, Brar DS (2003) Alien genes introgression and development of monosomic alien addition lines from *Oryza latifolia* Desv to rice, *Oryza sativa* L. TheorAppl Genet 107:395–405

Nayar NM (1973) Origin and cytogenetics of rice. Adv Genet 17:153–292

Nezu M, Katayama TC, Kihara H (1960) Genetic study of the genus Oryza I Crossability and chromosomal affinity among 17 species. Seiken Jiho Rep Kihara Inst biol Res 11:1–11

Nwokeocha CC, Faluyi JO, Aladejana FO (2007) Cytogenetic studies in the ABC genome of the genus *Oryza*. Linn Res J Bot 2:41–47

Ogawa T (2003) Genome research in genus *Oryza*. In: Nanda JS, Sharma SD (eds) Monograph on genus *Oryza*. Science Publishers, Enfield, New Hampshire, pp 171–212

Piegu B, Guyot R, Picault N, Roulin A, Saniyal A, Kim H, Panaud O (2006) Doubling genome size without polyploidization: dynamics of retro transposition-driven genomic expansions in *Oryza australiensis*, a wild relative of rice. Genome Res 16:1262–1269

Rahman ML, Jiang W, Chu SH, Qiao Y, Ham TH, Woo MO, Brar DS (2009) High-resolution mapping of two rice brown planthopper resistance genes, Bph20 (t) and Bph21 (t), originating from *Oryza minuta*. Theor Appl Genet 119:1237–1246

Ranganadhacharyulu N, Raj AY (1974) Pachytene analysis in an interspecific hybrid *Oryza punctata* Kotschy ex Steud × *O eichingeri* A Peter. Cytologia 39:233–243

Ray S, Bose LK, Ray J, Ngangkham U, Katara JL, Samantaray S, Wing RA (2016) Development and validation of cross-transferable and polymorphic DNA markers for detecting alien genome introgression in *Oryza sativa*. Mol Genet Genom 291:1783–1794

Renganayaki K, Fritz AK, Sadasivam S, Pammi S, Harrington SE, McCouch SR, Reddy AS (2002) Mapping and progress toward map-based cloning of brown planthopper biotype-4 resistance gene introgressed from into cultivated rice. Crop Sci 42:2112–2117

Sanchez AC, Brar DS, Huang N, Li Z, Khush GS (2000) Sequence tagged site marker-assisted selection for three bacterial blight resistance genes in rice. Crop Sci 40:792–797

Sharma SD, Dhua SR, Agrawal PK (2003) Species of genus *Oryza* and their interrelationships. Monograph on the genus *Oryza*. Science Publishers, Enfield, New Hampshire, pp 73–111

Shastry SVS (1963) Application of pachytene analysis to evolutionary problems in the genus *Oryza*. Memoirs of The Indian Botanical Society (Memoir-4), pp 82–89

Shastry SVS, Rao PM (1961) Pachytene analysis in Oryza. In: Proceedings of the Indian Academy of sciences-section B vol 54, pp 100–112

Shcherban AB, Vaughan DA, Tomooka N (2000a) Isolation of a new retrotransposon-like DNA sequence and its use in analysis of diversity within the *Oryza officinalis*. Genetica 108:145–154

Shcherban AB, Vaughan DA, Tomooka N, Kaga A (2000b) Diversity in the integrase coding domain of a gypsy-like retrotransposon among wild relatives of rice in the *Oryza officinalis* complex. Genetica 110:43–53

Sitch LA, Romero GO (1990) Attempts to overcome prefertilization incompatibility in interspecific and intergeneric crosses involving *Oryza sativa* L. Genome 33:321–327

Sui Y, Li B, Shi J, Chen M (2014) Genomic, regulatory and epigenetic mechanisms underlying duplicated gene evolution in the natural allotetraploid *Oryza minuta*. BMC Genom 15:11

Tan GX, Ren X, Weng QM, Shi ZY, Zhu LL, He GC (2004a) Mapping of a new resistance gene to bacterial blight in rice line introgressed from *Oryza officinalis*. Acta Genetica Sinica 31:724–729

Tan GX, Weng QM, Ren X, Huang Z, Zhu LL, He GC (2004b) Two white backed planthopper resistance genes in rice share the same loci with those for brown planthopper resistance. Heredity 92:212–217

Tan G, Jin H, Li G, He R, Zhu L, He G (2005) Production and characterization of a complete set of individual chromosome additions from *Oryza officinalis* to *Oryza sativa* using RFLP and GISH analyses. Theor Appl Genet 111:1585–1595

Tang X, Bao W, Zhang W, Cheng Z (2007) Identification of chromosomes from multiple rice genomes using a universal molecular cytogenetic marker system. J Integr Plant Biol 49:953–960

Tang L, Zou XH, Achoundong G, Potgieter C, Second G, Zhang DY, Ge S (2010) Phylogeny and biogeography of the rice tribe (*Oryzeae*): evidence from combined analysis of 20 chloroplast fragments. Mol Phylogenet Evol 54:266–277

Tateoka T (1962) Taxonomic studies of *Oryza* I *O. latifolia* complex II several species complexes. Bot Mag Tokyo 75:418–427

Tateoka T (1963) Taxonomic studies of *Oryza* III Key to the species and their enumeration. Bot Mag Tokyo 76:165–173

Tateoka T (1964) Taxonomic studies of the genus *Oryza*. In: IRRI (ed) Rice genetics and cytogenetics. Elsevier, Amsterdam, pp 15–21

Vaughan DA (1989) The genus *Oryza* L: current status of taxonomy. IRRI, Los Baños, The Philippines, pp 1–21

Vaughan DA (1994) The wild relatives of rice: a genetic resources handbook. IRRI, Manila

Vaughan DA (2003) Genepools of the genus *Oryza*. In: Sharma SD, Nanda JS (eds) Monograph on genus

Oryza. Science Publishers, Enfield, New Hampshire, pp 113–138
Vaughan DA, Morishima H, Kadowaki K (2003) Diversity in the *Oryza* genus. Curr Opin Pl Biol 6:139–146
Vaughan DA, Kadowaki KI, Kaga A, Tomooka N (2005) On the phylogeny and biogeography of the genus *Oryza*. Breed Sci 55:113–122
Wang B, Ding Z, Liu W, Pan J, Li C, Ge S, Zhang D (2009) Polyploid evolution in *Oryza officinalis* complex of the genus *Oryza*. BMC Evol Biol 9:250
Wang A, Zhang X, Yang C, Song Z, Du C, Chen D, Cai D (2013) Development and characterization of synthetic amphiploid (AABB) between *Oryza sativa* and *Oryza punctata*. Euphytica 189:1–8
Wang C, Liu X, Peng S, Xu Q, Yuan X, Feng Y, Hanyong Y, Yiping W, Wei X (2014) Development of novel microsatellite markers for the BBCC Oryza genome (Poaceae) using high-throughput sequencing technology. PloS one, 9(3), e91826
Wendel JF (2000) Genome evolution in polyploids. In: Doyle JJ, Gaut BS (eds) Plant molecular evolution. Springer, Netherlands, pp 225–249
Wu Y, Sun Y, Wang X, Lin X, Sun S, Shen K, Liu B (2016) Transcriptome shock in an interspecific F1 triploid hybrid of *Oryza* revealed by RNA sequencing. J Int Plant Biol 58:150–164
Xuexing YWLKG (1993) Obtaining the Hybrids of *Oryza sativa* × *O. officinalis* without embryo culture by using male-sterile lines. Acta Genetica Sinica 26:160–161
Yan H, Hu H, Fu Q, Yu H, Tang S, Xiong Z, Min S (1995) Morphological and cytogenetical studies of the hybrids between *Oryza sativa* and *Oryza officinalis*. Zhongguo Shuidao Kexue 10:138–142
Yan H, Xiong Z, Min S, Hu H, Zhang Z, Tian S, Fu Q (1997) The production and cytogenetical studies of *Oryza sativa–Oryza eichingeri* amphiploid. Acta Genetica Sinica 24:32–34
Yang H, Ren X, Weng Q, Zhu L, He G (2002) Molecular mapping and genetic analysis of a rice brown planthopper (*Nilaparvata lugens* Stål) resistance gene. Hereditas 136:39–43
Yang H, You A, Yang Z, Zhang F, He R, Zhu L, He G (2004) High-resolution genetic mapping at the Bph15 locus for brown planthopper resistance in rice (*Oryza sativa* L). Theor and Appl Genet 110:182–191
Yasui H, Iwata N (1991) Production of monosomic alien addition lines of *Oryza sativa* having a single *O. punctata* chromosome In: Rice genetics II: proceedings of the Second International Rice Genetics Symposium, 14–18 May 1990, IRRI, pp 147–155
Ying B, Song GE (2003) Identification of *Oryza* species with the CD genome using RFLP analysis on nuclear ribosomal ITS sequences. Acta Bot Sin 45:762–765
Yoon DB, Kang KH, Kim HJ, Ju HG, Kwon SJ, Suh JP, Ahn SN (2006) Mapping quantitative trait loci for yield components and morphological traits in an advanced backcross population between *Oryza grandiglumis* and the *O sativa* japonica cultivar Hwaseongbyeo. Theor Appl Genet 112:1052–1062
Zhang X, Wang W, Jin J, Liao S, Liu X, Song Z, Cai D (2013) Cytomorphological characterization of the backcross progeny of synthetic amphiploid rice (AABB) and tetraploid *Oryza sativa* (AAAA). J Agr Sci 5:30–37
Zhong DB, Luo LJ, Guo LB, Ying CS (1997) Studies on resistance to brown planthopper (BPH) of hybrid between *Oryza sativa* and *Oryza officinalis*. Southwest China J Agric Sci 10:5–9
Zou XH, Du YS, Tang L, Xu XW, Doyle JJ, Sang T, Ge S (2015) Multiple origins of BBCC allopolyploid species in the rice genus (*Oryza*). Sci Rep Uk 5:14876

22 Oryza perennis

Blanca E. Barrera-Figueroa and Julián M. Peña-Castro

Abstract

Oryza perennis is a common wild rice species complex composed by highly diverse members, with perennial, intermediate, and annual life cycles. *O. perennis* species are broadly distributed in four major geographical regions of the world. It is assumed that the Asian form of *O. perennis* is the ancestor of cultivated rice *Oryza sativa* and is important as a taxa model for studying the evolution of cultivated rice. From a biotechnology perspective, it may be a rich resource for breeding of new cultivated rice varieties. In this chapter, some aspects of *O. perennis* complex are reviewed.

22.1 Academic and Economic Importance

Oryza perennis is known for apporting a high amount of genetic elements to cultivated rice in the process of domestication. *Oryza perennis* species are important as models to investigate the origins of cultivated rice and phylogenetic relationships with wild rice species (Morishima 1969), the population structure and dynamics of domestication as a continuous process (Oka and Chang 1960; Oka and Morishima 1971). Furthermore, *O. perennis* has a broad range of life cycles and has adapted natural breeding systems to cope with environmental cues (Oka and Morishima 1965). All this knowledge is relevant for the selection and gene transfer of elements that determine agronomic traits into new varieties of cultivated rice, such as resistance to diseases and production of high-quality seeds and male sterile lines.

22.2 Description and Geographical Distribution

Oryza perennis as a species complex occurs in swampy places and is distributed in the humid tropics of four continents: Asia, Africa, the Americas, and Oceania (Fig. 22.1). Naming the wild rice species in the *O. perennis* complex has been historically a matter of controversy. Initially, *O. perennis* was regarded as a single species, but subsequent registers were contradictory to the original description. Therefore, the name *O. perennis* for a single species was removed (Tateoka 1964). Some authors continued to give the generic name of *O. perennis* to the wild taxa of common rice, rather than to a species (Morishima 1984). Later, it was remarked the distinction of geographical origins for each

B. E. Barrera-Figueroa (✉) · J. M. Peña-Castro
Laboratorio de Biotecnología Vegetal, Instituto de Biotecnología, Universidad Del Papaloapan, Tuxtepec, Oaxaca, Mexico
e-mail: bbarrera@unpa.edu.mx

T. K. Mondal and R. J. Henry (eds.), *The Wild Oryza Genomes*,
Compendium of Plant Genomes, https://doi.org/10.1007/978-3-319-71997-9_22

Fig. 22.1 Geographical distribution of *O. perennis* complex species. Black dots indicate the place of availability. Adapted from Morishima (1984)

O. perennis strain, and names of *O. rufipogon* and *O. longistaminata* were adopted based in results of taxonomic studies (Second 1985).

Currently, it is accepted that *O. perennis* species are part of the *Oryza sativa* complex with diploid AA genome ($2n = 24$) and include the cultivated *O. sativa* and *Oryza glaberrima*, and the wild *Oryza nivara*, *Oryza rufipogon*, *Oryza breviligulata*, *Oryza barthii*, *Oryza longistaminata*, *Oryza meridionalis*, and *Oryza glumaepatula* (Sanchez et al. 2014). This chapter refers to *O. perennis* as the wild species complex.

22.3 Evolution and Relationship with Cultivated Rice

The evolution of common wild rice has been intensively investigated. A phenetic analysis performed by Morishima (1969) based on morphological characteristics and F_1 sterility suggested that genotypes of *O. perennis* from different geographical origin evolved independently from one another, and that the Asian group (*O. perennis* Moench) has generated several advanced forms at a high evolutionary rate, which was in support of the idea of *O. perennis* being the original source of cultivated rice (Morishima 1969).

Other studies have investigated the relationship of *O. perennis* with cultivated rice. Oka and Chang (1960) analyzed three natural populations of wild rice consisting of *O. perennis* or intermediate *perennis-spontanea* types. The study revealed that most of the plants in those populations were highly heterozygous for genes of cultivated rice, thus implying the capability of *O. perennis* to bear and accumulate genes of cultivated rice and suggesting an active role of *O. perennis* species in the creation of new forms of cultivated rice in the natural habitat.

Different biochemical and molecular approaches have been applied to the phylogenetic analysis of *O. perennis* and cultivated rice. Second (1985), developed an isozyme analysis to study the evolutionary relationships in the *O. sativa* group, which includes the *O. perennis* complex. Based on the results, it was proposed a new classification of *O. perennis* complex and the elimination of *O. perennis* as a single species (Second 1985). *O. perennis* is renamed as *O. rufipogon* for Asian, Oceanian, and American forms, and as

O. longistaminata for the African form. Since then, *O. perennis* was referred as a complex species, rather than as a single species (Barnes and Pental 1986; Ghesquiere 1986).

Barnes and Pental (1986) also investigated the phylogenetic relationships of cultivated rice and *O. perennis* complex species. Analyzing isoelectric focusing patterns of ribulose biphosphate carboxylase/oxygenase and repetitive DNA sequences in rice from different geographical origin, it was confirmed the close relationship between *O. perennis* species. Moreover, *O. perennis* genotypes from South America differ greatly from the others. As a result, cultivated rices are more related to *O. perennis* than different *O. perennis* are to one another. Previously, it was proposed that cultivated rice species (*O. sativa* and *O. glaberrima*) had arisen independently from Asia and Africa at the time of continental drift (Chang 1976). However, the observations of Barnes and Pental (1986) do not support that and indicate that cultivated rice evolved just recently from *O. perennis* species.

22.4 Adaptive Strategies

Oryza perennis species have been useful as a model for the study of adaptive mechanisms in rice since the species are genetically diverse, colonize different habitats, and show a continuum of life cycles, from annuals to perennials. Sano and Morishima (1982) studied 54 genotypes of *O. perennis* with different life cycles and analyzed partition of photosynthate and stress tolerance parameters as means to evaluate adaptive strategies. It was found that annual strains preferently direct the energy to the production of seeds, while perennials accumulate biomass in the vegetative organs. It was also revealed that among *O. perennis* strains, tolerance to drought and submergence is positively correlated and that annual types are generally more tolerant than perennials to both drought and submergence (Sano and Morishima 1982).

In addition, other characteristics have been studied in *O. perennis* to dilucidate the mode of adaptation. Some of such characteristics are pollination system and type of propagation, vegetative or by seeds. Oka and Morishima (1965) studied the pollination system and type of propagation in *O. perennis* and reported that perennial strains depending on vegetative propagation are mainly allogamous. In contrast, most annual are autogamous. In that way, in perennials a high rate of cross-pollination ensures genetic diversity, while in annual, a high reproductive rate is favored by selfing. Regarding type of propagation, their results suggest that *O. perennis* species as a whole have the potential to differentiate into vegetative and seed-propagating types, thus enabling plants to adapt to diverse environmental conditions.

22.5 Biotechnology Potential of *Oryza perennis* Strains

Cytoplasmic male sterility (CMS) is a useful tool to develop commercial rice hybrids. Most of the CMS commercial hybrids are based on a Wild Abortive (WA) source of CMS, meaning that all male have sterile pollen. This trait renders a cytoplasmic uniformity that could lead to susceptibility to diseases and insect pests. In order to diversify the sources for introgressing CMS into cultivated rice, Dalmacio et al. (1995) used 46 *O. perennis* accessions as male parent to introduce cytoplasmic male sterility into *O. rufipogon* as a female parent, generating a new line IR66707A, which is different to WA and does not have any major rearrangement of mitocondrial DNA. However, there is not a good restorer of CMS for this line, and thus, its use in commercial hybrid rice breeding programs is limited. Further research to find good restorers of CMS will be determinant for the use of *O. perennis* in the development of new hybrid rice for commercial use. In addition, it has been identified resistance to zigzag leafhopper *Recilia dorsalis* in *O. perennis*, which is a desirable agronomic trait that can be exploited for the improvement of insect resistance in cultivated rice (He et al. 2013).

Besides CMS, the diversity of natural adaptation mechanisms, present in *O. perennis* species, such as different breeding systems, life cycles, and diversity of tolerance to abiotic and biotic stress, offers unique opportunities for improving rice. The use of molecular approaches such as transcriptome analysis and proteomics will broad the knowledge on the fundamentals of tolerance to adverse factors and other adaptive strategies, which in turn will facilitate the identification and transfer of desirable traits to new cultivated rice varieties.

References

Barnes SR, Pental D (1986) Repeated DNA sequences and ribulose biphosphate carboxylase/oxygenase as tools for the study of rice evolution. In: Rice genetics. Proceedings of the international rice genetics symposium, International Rice Institute. Manila, Philippines, pp 41–51

Chang TT (1976) The origin, evolution, cultivation, dissemination, and diversification of Asian and African rices. Euphytica 25:425–441

Dalmacio R, Brar DS, Ishii T, Sitch LA, Virmani SS, Khush GS (1995) Identification and transfer or a new cytoplasmic male sterility source from *Oryza perennis* into indica rice (*O. sativa*). Euphytica 82:221–225

Ghesquiere A (1986) Evolution of *Oryza longistaminata*. In: Rice genetics. Proceedings of the international rice genetics symposium, International Rice Institute. Manila, Philippines, pp 15–25

He G, Du B, Chen R (2013) Insect resistance. In: Zhang Q, Wing RA (eds) Genetics and genomics of rice. Springer, Heidelberg, pp 177–192

Morishima H (1969) Phenetic similarity and phylogenetic relationships among strains of *Oryza perennis*, estimated by methods of numerical taxonomy. Evolution 23:429–443

Morishima H (1984) Wild plants and domestication. In: Tsunoda S, Takahashi N (eds) Biology of rice. Japan Scientific Society Press/Elsevier, Tokyo/Amsterdam, pp 3–30

Oka H-I, Chang W-T (1960) Hybrid swarms between wild and cultivated rice species, *Oryza perennis* and *O. sativa*. Evolution 15:418–430

Oka H-I, Morishima H (1965) Variation in the breeding systems of a wild rice, *Oryza perennis*. Evolution 21:249–258

Oka H-I, Morishima H (1971) The dynamics of plant domestication: cultivation experiments with *Oryza perennis* and its hybrid with *O. sativa*. Evolution 25(356):364

Sanchez PL, Wing RA, Brar DS (2014) The wild relative of rice: Genomes and genomics. In: Zhang Q, Wing RA (eds) Genetics and genomics of rice, plant genetics and genomics: crops and models. Springer, New York, pp 9–25

Sano Y, Morishima H (1982) Variation in resource allocation and adaptive strategy of a wild rice, *Oryza perennis*, Moench. Bot Gaz 143:518–522

Second G (1985) Evolutionary relationships in the Sativa group of *Oryza* based on isozyme data. Genet Sel Evol 17:89–114

Tateoka T (1964) Taxonomic studies of the genus Oryza. In: Chang TT, Jennings PR (eds) Rice genetics and cytogenetics. Elsevier, Amsterdam, pp 15–21

Oryza rhizomatis Vaughan

23

S. Somaratne, S. R. Weerakoon and
K. G. D. I. Siriwardana

Abstract

The members of the genus *Oryza* are broadly classified into cultivated and wild rice. *Oryza sativa* complex includes the species with AA genome meanwhile *Oryza officinalis* complex with CC genome. The taxonomy of *O. officinalis* complex in Sri Lanka is complicated due to the problematic morphological variation, inadequacy of herbarium and living specimens. *Oryza rhizomatis* belongs to the same genome group (CC) as both *O. officinalis* and *Oryza eichingeri*. Recently, a new species of *Oryza* was described and named as *O. rhizomatis* justifying that this taxon is morphologically and geographically or ecologically isolated from all other *Oryza* species. The distribution of *O. rhizomatis* is restricted to the dry zone low country of Sri Lanka and appears to be intermediate between *O. officinalis* and *O. eichingeri*. *Oryza eichingeri* deserved to consider as a distinct species restricted to Africa. Once the disjunctive distribution of the *O. eichingeri* is resolved, Sri Lankan representative *O. eichingeri* is deserved to be renamed as *O. rhizomatis*. *Oryza rhizomatis* and other related wild rice species in Sri Lanka require thorough molecular/morphological/ecological studies with a larger samples from diverse ecological regions of the country. As *O. rhizomatis* is endemic to Sri Lanka, conservation measurers are essential to be adopted in future. In the present chapter, we have compiled all the relevant information related to this species. Further, research gap that needs to be filled to harness beneficial genomic resources from this species is highlighted.

23.1 Introduction

Rice is the most widely consumed staple food for a large part of the world's human population, especially tropical and subtropical areas in Asia. In general, rice categorized into cultivated rice and wild rice of the genus *Oryza*. The genus *Oryza* L. is placed under the subfamily Oryzoidea of the tribe Oryzeae of the grass family (Poaceae). This genus consists of two cultivated species *Oryza sativa* and *Oryza glaberrima* and 20 wild species (Morishima 1984; Brar and Khush 2003). *Oryza sativa* includes two subspecies: *indica* cultivated in Southeast Asian countries and *japonica* cultivated in Japan. *Oryza glaberrima* is restricted to African continent and

S. Somaratne · S. R. Weerakoon (✉) ·
K. G. D. I.Siriwardana
Department of Botany, Faculty of Natural Sciences,
The Open University of Sri Lanka, 21, Nawala,
Sri Lanka
e-mail: shyamaweerakoon@gmail.com

T. K. Mondal and R. J. Henry (eds.), *The Wild Oryza Genomes*,
Compendium of Plant Genomes, https://doi.org/10.1007/978-3-319-71997-9_23

is cultivated in savanna zones of West Africa and southern edge of the Sahara.

The morphological variation of the genus *Oryza* across different geographical and ecological ranges is highly complicated and it makes difficult to delimit the species within the genus. This variation resulted in varying opinion on the number of species included in the genus. Further, a number of species complexes were recognized and maintained by introducing a system of grouping the species (Tateoka 1962a, b). *Oryza sativa* complex includes the species with AA genome meanwhile *O. officinalis* complex with CC genome. Further, *Oryza* species consists of BB, CC, EE genomes and allotetraploid species with the BBCC and CCDD genomes are also included in the *O. officinalis* complex. In Sri Lanka, two species, *O. eichingeri* and O. *rhizomatis* are considered as the members of the *O. officinalis* complex.

23.2 Taxonomy, Genetic Diversity, and Species Relationship

The taxonomy of *O. officinalis* complex in Sri Lanka has confused the systematists due to the problematic morphological variation, inadequacy of herbarium specimens as well as the living plant materials. An effort has been made to resolve the problem of the morphological variation of the species within the complex and Biswal and Sharma (1987) considering the similarity between *O. collina* and *O. eichingeri* retract the name *O. collina*, considering these taxa to be identical. Thus, the authors have agreed with the previous works of both Bor (1960) and Tateoka (1962a, b) that *O. eichingeri* is the sole representative of the *O. officinalis* complex in Sri Lanka.

Based on one sample material studied in India, coming originally from Sri Lanka via USA, a form of the *O. officinalis* complex from Sri Lanka was named *O. collina* (Trimen 1889; Sharma and Shastry 1965). A successive study resulted in a retraction of the *O. collina* and agreed with Tateoka (1963) that this form of Sri Lanka encompasses the range of *O. eichingeri* Peter, an African species of the same complex (Biswal and Sharma 1987). *Oryza eichingeri* grows in the shade of forests in Uganda, whereas *O. officinalis* from Sri Lanka grows in both shaded and open habitats (Vaughan et al. 2003). However, due to inadequate field notes, taxonomists were unable to give much weight to the habitat of a taxon. Further, the majority of environmental parameters of plant are complex and change over time.

New material of wild *Oryza* species has been collected in recent years in cooperative germplasm collection missions by the International Rice Research Institute, Philippines and Central Agricultural Research Institute, Peradeniya, Sri Lanka. These novel collections led to identify diagnostic morphological and habitat differences in *O. eichingeri* and revealed that it is a larger taxon which occurs in the drier habitats in Sri Lanka (Vaughan 1989). This bulky rhizomatis taxon has formerly been named as *Oryza latifolia* and *O. officinalis*. *Oryza officinalis* is a rhizomatis diploid (CC genome) from Asia and Australia with smaller spikelets inserted away from the base of primary branches and shorter palea tip. *Oryza latifolia* is a 1–2 m tall non-rhizomatis tetraploid (CCDD) from South and Central America with broader leaves and whorled panicle branches. *Oryza officinalis* is also genetically different from this Sri Lankan taxon with which it can form sterile hybrids (Vaughan 1989). However, Sri Lankan taxon belongs to the same genome group as both *O. officinalis* and *O. eichingeri*, which is CC (Jean and Khush 1985). As a result, Vaughan in 1990 described a new species; *O. rhizomatis* justifying the necessity for the reason this taxon is morphologically and geographically or ecologically isolated from all other *Oryza* species.

The section *Oryza* within the genus *Oryza* consists of two species complexes, the *O. officinalis* complex and the *O. sativa* complex that includes all the AA genome species (Tateoka 1965; Vaughan and Morishima 2002). The three species *O. officinalis*, *O. eichingeri*, and *O. rhizomatis* are diploid CC genome.

There are two diploid CC genome species in Sri Lanka, *O. eichingeri* and *O. rhizomatis*

(Biswal and Sharma 1987; Vaughan 1990). *Oryza rhizomatis* appears to be intermediate between *O. officinalis* and *O. eichingeri*. Analysis of the nuclear and chloroplast genome of *O. rhizomatis* with RFLP and SSR markers reveals that *O. rhizomatis* differs from *O. eichingeri* and *O. officinalis* (Dally and Second 1990; Provan et al. 1997; Wang et al. 1992). *Oryza rhizomatis* shows a low activity of aryl acylamidase, an enzyme that hydrolyzes the herbicide propanil, in contrast to *O. eichingeri* and *O. officinalis* (Jun and Matsunaka 1990). Meanwhile, Harriman (1994) is of the opinion that *O. eichingeri* should be considered as a distinct species restricted to Africa and need further studies. If the disjunctive distribution of the *O. eichingeri* is resolved, Sri Lankan representative *O. eichningeri* is deserved to be renamed as *O. rhizomatis*.

23.3 Distribution, Habitat and Climatic Zones

The distribution of *O. rhizomatis* is restricted to the dry zone low country of Sri Lanka. It has been collected mainly from many areas of the island such as Ruhuna Wildlife Sanctuary, Hambantota District; Wilpattu National Park, Anuradhapura and other districts such as Puttalam, Trincomalee, Vavunia, and Moneragala. The distribution of *O. rhizomatis* overlaps *O. eichingeri* and populations can be found within about 20 km of each other in the north central dry zone in Sri Lanka (Vaughan et al. 2003) (Fig. 23.1). *Oryza rhizomatis* is often grown in primary and secondary forests and open, tall scrub with grassy openings and occurs in swampy or periodically flooded areas, usually open to full sun or partial shade. Plants are visible during the period of late December to May.

The dry zone of Sri Lanka where *O. rhizomatis* grows is characterized by rainfall of 850–1900 mm. The months of May to September are dry season with mean annual temperature is 30 °C. Within the dry zone, *O. rhizomatis* is found in areas considered arid (Vaughan et al. 2003).

23.4 Botanical Characterization

Morphologically, there are discrete differences between the key characters of *O. rhizomatis* and *O. eichingeri* (Table 23.1). *Oryza rhizomatis* is erect and rhizomatous grass, reaching 1–3 m tall while *O. eichingeri* is a short (usually <1 m) grass, with hard and slender culms. Both species are perennial in nature.

A typical *O. rhizomatis* has open panicles without whorled basal panicle branches; spikelets inserted near the base of lowest panicle branches; mature spikelets ca. 6.8 mm long and 2.2 mm wide while in *O. eichingeri*, panicles intermediately open; chlorophyllous veins across the length of the immature spikelets; mature spikelets 4.5–6.2 mm long and 1.6–2.8 mm wide. The anthers of *O. rhizomatis* are 2.3–4 mm long while *O. eichingeri*, 1.5–3.3 mm long. The chromosome number of both species *O. rhizomatis* and *O. eichingeri* are equal (24).

Figures 23.2 and 23.3 show certain characters of *O. rhizomaits* grown in a Plant House at the Plant Genetic Resource Centre, Sri Lanka.

The comparative molecular (AFLP analysis) studies performed between *O. eichingeri, O. officinalis,* and *O. rhizomatis* revealed that accessions collected from different localities indicated closest similarities between *O. rhizomatis* to a population of *O. eichingeri* (Bautista et al. 2006). *Oryza officinalis* accessions from Assam, India, and China were genetically less diverged from *O. eichingeri* and *O. rhizomatis* than two accessions of *O. officinalis* from Kerala state, India (Bautista et al. 2006). Further, molecular marker analysis has revealed that populations of *O. rhizomatis* from Northern and Southeastern Sri Lanka are genetically differentiated. One accession of *O. rhizomatis* was aligned with *O.eichingeri*.

The results of Subudhi et al. (2013) were in accordance with the observation made by Zhang and Song (2007) in which *O.eichingeri* was genetically more similar to *O. rhizomatis* than to *O. officinalis*. Retrotransposon-based probe for fingerprinting and evolutionary studies in rice,

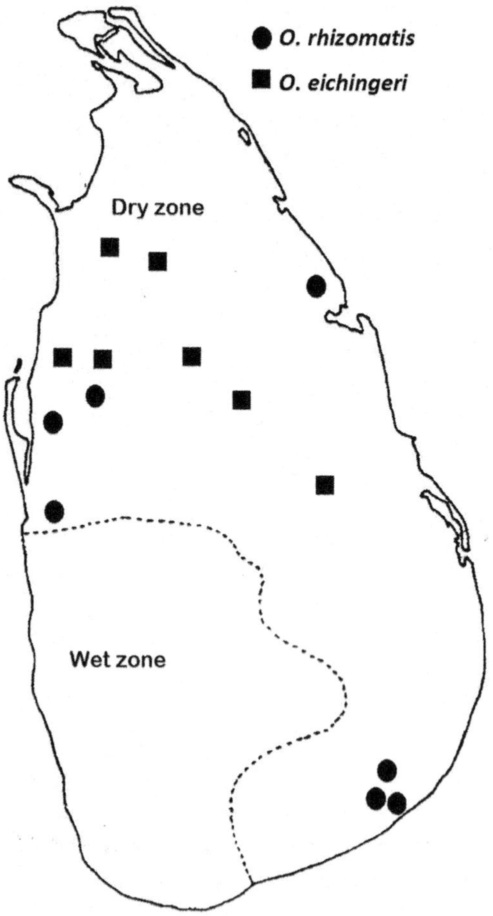

Fig. 23.1 Distribution map of *O. rhizomatis* and *O. eichingeri* in Sri Lanka

by Subudhi et al. (2013) reported that banding patterns of Sri Lankan accessions were identical but different from Ugandan accessions supporting the distinct species status of the Sri Lankan accession and African *O.eichingeri* as proposed by Zang et al. (2011). These findings will resolve the disjunctive distribution of *O. eichingeri* in Sri Lanka and Africa.

23.5 Uses and Conservation Status

23.5.1 Uses

The extensive thick root system and rhizome character is important to keep plant in live even during heavy drought period. After the rain comes, plant can produce new sprout and grow again. This character can be used in rice breeding programs to produce perennial and drought tolerant rice (Liyanage 2010). Moreover, the diploid CC genome species have useful genes such as resistance to brown plant hopper, white backed plant hopper, bacterial blight, and yellow dwarf virus disease (Brar and Khush 1997; Heinrichs et al. 1985).

23.5.2 Conservation Status

Oryza *rhizomatis* is endemic to Sri Lanka; therefore, conservation is important. This species is at present in protected areas as well as in many other unprotected areas. Ex situ conservation is carried

Table 23.1 Prominent morphological characters of *O. eichingeri* and *O. rhizomatis*

Character	*O. eichingeri*	*O. rhizomatis*
Habit	Stream sides, forest floor; semi-shade	Seasonally dry; open
Ligule	Purple or green in color	Purple in color
Panicle branches	Panicles open to intermediately open	Open panicles without whorled basal panicle branches
Awn	Awn flexuous with fine bristles	Often awnless
Distribution	East Africa and Sri Lanka	Endemic to Sri Lanka

Fig. 23.2 Adult *O. rhizomaits* plants with rhizomes

out in seed gene bank of the Plant Genetic Resources Centre (PGRC) and Rice Research Development Institute (RRDI), Bathalagoda, Sri Lanka.

23.6 Future Directions and Research Gap

The use of morphological and molecular data in taxonomic studies greatly relies on the sampling strategy and sampling effort. Taxonomy is exclusively employed on comparative approach which requires the investigation of larger number of specimens/samples in order to capture entire range of natural variation. Therefore, the quality of taxonomic studies partially depends on a thorough sampling of specimens/samples to be surveyed, and a biased sampling may cause erroneous conclusions. The inadequately sampling might lead to premature decisions. Therefore, *O. rhizomatis* and other related wild rice species in Sri Lanka require thorough molecular/morphological studies with a large number of samples from diverse ecological regions of the country. Further, it is worthwhile to conduct studies on genomic sequencing of *O. rhizomatis* to explore the novel genomic resources (markers, useful genes). This information is essential in plant breeding programs such as marker-assisted breeding for disease-resistant, drought-resistant, and salinity-resistant rice varieties.

Fig. 23.3 Close-up picture of ligule of *O. rhizomatis* indicated by arrow

Acknowledgements Authors wish to thank the Director, Mr. A.S.U. Liyanage and Mr. Anuradha Amarasinghe, Plant Genetic Resources Centre (PGRC), Gannoruwa, Sri Lanka for providing *O. rhizomatis* plant specimens to obtain photographs.

References

Brar DS, Khush GS (2003) Utilization of wild species of genus *Oryza* in rice improvement. In: Nanda JS. Sharma SD (eds.), Monograph on gnus *Oryza*, pp 283–309

Bautista NS, Vaughan DA, Jayasuriya AHM, Liyanage ASU, Kaga A, Tomooka N (2006) Genetic diversity in AA and CC genome *Oryza species* in southern South Asia. Genet Resour Crop Evol 53:631–640

Biswal J, Sharma DS (1987) Taxonomy and phylogeny of *Oryza collina*. Oryza 24:24–29

Bor NL (1960) The grasses of Burma, Ceylon, India and Pakistan. Pergamon Press, Oxford

Brar DS, Khush GS (1997) Alien introgression in rice. Plant Mol Biol 35(1–2):35–47

Dally AM, Second G (1990) Chloroplast DNA diversity in wild and cultivated species of rice (genus *Oryza* section *Oryza*). Cladistic-mutation and genetic distance analysis. Theor Appl Genet 80:209–222

Harriman NA (1994) *Oryza*. In: Dassanayake MD, Fosberg FR, Clayton WR (eds) Revised handbook to the flora of Ceylon, vol. 8. Poaceae. Amerind Publishing Co. Pvt. Ltd., New Delhi, India

Heinrichs EA, Medrano FG, Rapusas HR (1985) Genetic evaluation for insect resistance. IRRI. 356 p

Jean KK, Khush GS (1985) Genome analysis of *Oryza officinalis* complex. Rice Genetics Newsl 2:44–45

Jun CJ, Mastunaka S (1990) The propanil hydrolyzing enzyme aryl acylamidase in the wild rices of genus *Oryza*. Pestic Biochem Physiol 38(1):26–33

Liyanage ASU (2010) Eco-geographic survey of crop wild relatives. Plant Genetic Resources Centre, Gannoruwa, Peradeniya, Sri Lanka

Morishima H (1984) Wild plants and domestication. Chapter 1. In: Tsunoda S, Takahashi N (eds) Biology of rice, vol 7. Elsevier, Amsterdam, pp 3–30

Provan J, Corbett G, McNicol JW, Powell W (1997) Chloroplast DNA variability in wild and cultivated rice (*Oryza* spp.) revealed by polymorphic chloroplast simple sequence repeats. Genome 40(104–1):10

Sharma SD, Shastry SVS (1965) Taxonomic studies in genus *Oryza L*. IV. The Ceylonese *Oryza spp*. Affin. *O. officinalis* Wall. Ex Watt. Ind J Genetics Plant Breed 25:168–172

Subudhi PK, Magpantay GB, Karan R (2013) A retrotransposon-based probe for finger printing and evolutionary studies in rice (*Oryza sativa*). Genet Resour Crop Evol 60:1263–1273

Tateoka T (1962a) Taxonomic studies of *Oryza*. I. O. latifolia complex. Bot Mag 75:418–427

Tateoka T (1962b) Taxonomic studies of *Oryza*. II. Several species complexes. Bot Mag 75:455–461

Tateoka T (1963) Taxonomic studies of *Oryza* III. Key to the species and their enumeration. Bot Mag Tokyo 76:165–173

Tateoka T (1965) Taxonomy and chromosome numbers of African representatives of the *Oryza officinalis* complex. Bot Mag Tokyo 78:198–201

Trimen H (1889) Additions to the flora of Ceylon. J Bot 27:161–172

Vaughan DA (1989) Two species of the *Oryza officinalis* complex are present in Sri Lanka. Int Rice Res Newsl 14(4):5

Vaughan DA (1990) A new rhizomatous *Oryza* species (Poaceae) from Sri Lanka. Bot J Linn Soc 103:159–163

Vaughan DA, Morishma H (2002) Biosystematics of the genus *Oryza*. In: Wayne Smith C (ed) Rice: origin, history, technology and production. Wiley, New York

Vaughan DA, Morishimay H, Kadowaki K (2003) Diversity in the *Oryza* genus. Curr Opin Plant Biol 6:139–146

Wang ZY, Second G, Tanksley SD (1992) Polymorphism and phylogenetic relationships among species in the genus *Oryza* as determined by analysis of nuclear KFLPs. Theor Appl Genet 83:565–581

Zhang L, Song G (2007) Multilocus analysis of nucleotide variation and speciation in *Oryza officinalis* and its close relatives. Mol Biol Evol 24(3):769–783

Zang LL, Zou XH, Zhang FM, Yang Z, Song GE (2011) Phylogeny and species delimitation of the C-genome diploid species in *Oryza*. J Syst Evol 49(5):386–395

Oryza ridleyi Hook. F.

24

Mostafa Mamdouh Elshenawy
and Walid Hassan Elgamal

Abstract
Although rice is life of human from the civilization and conventional breeding has contributed immensely to improve the rice cultivar. Yet due to the climate change and high demand of paddy for food security, improved cultivar of rice is required. While extensive research has been conducted with the exciting gene pool of rice, wild species of rice is now of paramount importance. Thus, in this chapter we tried to describe one such little-known species *Oryza ridleyi* and its importance and economic benefits, which gives us the opportunity to search more accurately within the genes and benefit from the pursuit of environmental development.

24.1 Economic and Academic Importance

Being a wild species, although *Oryza ridleyi* is considered to have low economic importance due to the lack of higher productivity, it has several economic importance due to the presence of several tolerant genes for various biotic and abiotic stress. For an example, Padhi and Sen (2002) reported that *O. ridleyi* is tolerant to yellow stem borer. *Oryza ridleyi* has various agronomically important genes such as adaptation to aerobic soil and salinity tolerance, and resistance to bacterial blight, blast and stem borer (Khush 1997; Brar and Khush 1997; Ge et al. 1999).

M. M. Elshenawy (✉)
Rice Research and Training Center, Field Crops Research Institute, Agricultural Research Center, Giza, Egypt
e-mail: mostafarrtc@gmail.com

W. H. Elgamal
Institute of Food Crops, Yunnan Academy of Agricultural Sciences, Kunming Shi, China

24.2 Origin, Distribution and Taxonomy

The species is found in Southeast Asia such as Cambodia, Indonesia, Laos, Malaysia, Myanmar, and Thailand. It is also found in some areas such as the Sepik River and around the Chambri Lakes of Papua New Guinea (Fig. 24.1). It is commonly found in moist shady forests, or semi-shade of primary or well-developed secondary forest. It is also found in swamps or close to large rivers that seasonally inundated with water to considerable depth. It is an allotetraploid with HHJJ genome with chromosome number of $2n = 4x = 48$. The genome size is estimated to be 1283 Mb, highest among the 24 *Oryza* species (Zuccolo et al. 2007).

Because of strong crossability barriers between Meyeriana and Ridleyi complexes, hybrids are difficult to produce and assigning genomes based on meiotic pairing could not be

T. K. Mondal and R. J. Henry (eds.), *The Wild Oryza Genomes*,
Compendium of Plant Genomes, https://doi.org/10.1007/978-3-319-71997-9_24

Fig. 24.1 Geographical distribution of *O. ridleyi* plant (Blue dot indicates the place of availability)

Fig. 24.2 Close-up view of a young seedling. *Photograph courtesy* Dr. Gireesh C, DRR, India

carried out. Under such situations, an alternate approach based on total genomic DNA hybridization and molecular divergence analysis has been used; the genome GG has been proposed for the diploids of the *Oryza meyeriana* complex and HHJJ for the allotetraploids of the *O. ridleyi* complex (Aggarwal et al. 1997). Results of random fragment length polymorphism (RFLP) analysis (Wang et al. 1992), amplified fragment length polymorphism (AFLP) analysis (Aggarwal et al. 1999), sequence analysis of genes (Ge et al. 1999), and seed protein analysis (Sarkar and Raina 1992) support the genomic classification based on morphological and cytological data.

24.3 Botanical Descriptions

Oryza ridleyi complex has two tetraploid species, i.e., *Oryza longiglumis* Janson and *O. ridleyi* Hook. Available information about *O. longiglumis* (Fig. 24.2) is very limited. A detail description of Morphological characteristics is tabulated in Table 24.1.

24.4 Ability of Crossing with *Oryza sativa*

Of the most important and vital goals in rice breeding is to produce improved varieties and lines through breeding with the favorable allele of wild species, and there have been many attempts through direct hybridization between *Oryza sativa* and *O. ridleyi,* though until now, there is no successful hybrid between *O. sativa* and *O. ridleyi*. This may be due to the genetic barriers between the two species. The results show that the differences in the hybridization profiles were genome-specific but not species-specific. Thus, different species carrying similar genomes were indistinguishable from each other on genomic probing. A similar situation was observed for species/accessions belonging to the *O. ridleyi* and *O. meyeriana*

Table 24.1 Morphological characteristics of *O. ridleyi*

Plant height	Leaf length	Leaf width	Panicle shape	Description of spikelet	Pest resistance	Origin
1–2 m	Soft leaves 15--30 cm	1.5–2.5 cm	Panicle open with spikelets 7–13 mm long and 2–3 mm wide	One fertile floret and two basal sterile florets	Bacterial blight, rice blast, whorl maggot, and stem borers	Asia and New Guinea

complexes, indicating that the genomes of species in each of the two complexes are identical. Southern hybridization revealed that species belonging to the *O. ridleyi* and *O. meyeriana* complexes are highly diverged from all other *Oryza* species (Aggarwal et al. 1997).

24.5 In Vitro Regeneration

Anther culture is one of the best in vitro techniques with numerous advantages. They are (i) shortening breeding cycle by immediate fixation of homozygosity, (ii) increasing in selection efficiency, (iii) widening of genetic variability through the production of gametoclonal variants, and (iv) allowing early expression of recessive genes (Zapata 1992). Anthers of this species were cultured on three basal media (NK, N6, and MS medium) with four different combinations of auxin and cytokinin for their effects on callus induction. The best medium for callus induction from anthers of *O. ridleyi* was found to be MS fortified with 2, 4-D (2 mg/1) along with kinetin (2 mg/1). Around 0.59% calli could be produced from anthers which regenerated to *albino* plantlets (Tang et al. 1998). In that study, though anthers were derived from several species such as *O. punctata* (BB), *O. officinalis* (CC), *Oryza alta* (CCDD) and *O. ridleyi* (HHJJ), and anther-derived plants were regenerated from *O. ridleyi* only. It is suggested from this study that wild *Oryza* species with genomes B, C, E, BC, and CD exhibit a lower regeneration ability than wild Oryza species with the A genome.

24.6 Molecular Mapping of Genes and QTLs

Molecular markers are developed to track the desired alien trait(s) for marker-assisted selection (MAS). The strategy used to transfer genes from wild species into rice depending on the nature of the target trait(s), relatedness of the wild species, and incompatibility barriers. The transfer of genetic information from one species to another is mainly done by hybridization between them and repeated backcrossing. The tetraploid *O. ridleyi* complex comprises two species: *O. ridleyi* and *O. longiglumis*. *Oryza ridleyi* shows strong resistance to all the 10 Philippine races of BB. Hybrids between rice cv. IR56 and *O. ridleyi* (accession 100821) have been produced; however, the cross showed a strong necrosis phenotype. Thus, only few introgression lines (BC_3F_3) from this cross have been produced, but no introgression could be detected.

24.7 Mitochondrial and Chloroplast Genomes

Organelle genomes such as mitochondrial and chloroplast are usually maternally inherited, so results provide evidence of the donors of the maternal genome to allotetraploid species. A small fragment located 35 bp upstream of a repeat that consists of eight nucleotides (5-CATTCTAT-3′) in *RCt3* was found in all accessions of HHJJ. Species of the *O. ridleyi* complex all form distinct clades, as would be expected from species in different sections of the genus Oryza. The relationships among accessions of *O. ridleyi* and *O. longiglumis* of the *O. ridleyi* complex suggest that these two closely related species might have diverged quite recently. Based on the complete dataset of DNA sequences from the chloroplast and mitochondrial, SSR were mined analyzed, *O. ridleyi* mitochondrial DNA, containing SSR region (RMt10) (Tomotaro et al. 2005).

24.8 Genome Comparison

Comparative genomics has emerged as a powerful tool to decipher gene and genome evolution. Multiple species comparisons reveal novel insights into genome evolution (Bennetzen 2007), genome duplication (Ilic and Bennetzen, 2003) and can also identify previously unknown or poorly characterized genome components, such as novel transposable elements (Lai et al. 2005) and novel functional elements (Stark et al. 2007). Comparative genomics has given novel insights into Oryza genome evolution (Zhang et al. 2007; Ma et al. 2007) and genome size variation (Piegu et al. 2006; Ammiraju et al. 2007). Moreover, the allotetraploidy in some *Oryza* species allows for study of the evolutionary dynamics of duplicated genes in polyploids. Allotetraploid Oryza genomes, pseudogenization of duplicated genes, resulted from nonsense mutations, frameshift mutations, and sequence deletions. In the *O. ridleyi* JJ subgenome, 3 kb was deleted between gene 25 and gene 26, which eliminated exons in both genes and led to a double pseudogenization. In allotetraploids, each gene has two copies, one from each subgenome. One gene copy remains functional through purifying selection, and the other copy usually accumulates deleterious mutations due to relaxation of selection constraints. An alternative gene duplication model is the degeneration divergence complementation (DDC) model, which emphasizes that degenerative mutations facilitate preservation of duplicate genes (Force et al. 1999). Some Oryza allotetraploidizations support the DDC model, and others support the classical Ohno model (Ohno, 1970). In *O. minuta* and *O. alta*, which appear to have evolved as polyploids less than 2 Mya, only $\approx$ 5% (2 of 38), duplicated genes were identified as pseudogenes. In contrast, one-third *Oryza coarctata* and *O. ridleyi*-duplicated genes were observed to be pseudogenized (Fei et al. 2008). Comparative sequence analysis revealed extensive gene collinearity across the genus *Oryza*. This gene collinearity unambiguously revealed orthologous genes. However, repetitive sequences, such as retrotransposons and DNA transposons, have also shaped the *Oryza* genome landscape dramatically. Some orthologous genes reside in highly repetitive DNA blocks in some species, such as gene 23 in *O. ridleyi.*

24.9 Conclusion and Future Direction

Oryza ridleyi is understudied species. Except organelle genome, not much information is available. Although it has been reported to have few important traits, extensive screening is required to know its ability to withstand of tolerance to various other traits. Because, it may be used to transfer some good traits to the cultivated rice for improvement of certain traits, i.e., resistance to abiotic stress such as drought, salinity, heat or/and biotic stress such as pathogens and insect pests like yellow stem borer. Besides, there is an urgent need to do the whole genome sequencing of this species, which not only generate several genomic resources but also will reveal the evolutionary relationship with other species of *Oryza*. Decoding of genome will also enhance the marker assistant breeding to introgress the QTLs of this species to cultivated one. Although there are few attempts on in vitro culture, there is no report for a generic regeneration system along with transgenic protocol. This will be beneficial to manipulate the genetic background of this species.

References

Aggarwal RK, Brar DS, Khush GS (1997) two new genomes in the Oryza complex identified on the basis of molecular divergence analysis using total genomic DNA hybridization. Mol Gen Genet 254:1–12

Aggarwal RK, Brar DS, Nandi S, Huang N, Khush GS (1999) Phylogenetic relationships among *Oryza* species revealed by AFLP markers. Theor Appl Genet 98: 1320–1328

Aggarwal RK, Majumdar KC, Lang JW, Singh L (1994) Genetic affinities among crocodilians as revealed by DNA fingerprinting with a Bkm-derived probe. Proc Natl Acad Sci USA 91:10601–10605

Ammiraju JS et al (2006) The *Oryza* bacterial artificial chromosome library resource: construction and analysis of 12 deep-coverage large-insert BAC libraries that represent the 10-genome types of the genus *Oryza*. Genome Res 16:140–147

Ammiraju JS et al (2007) Evolutionary dynamics of an ancient retrotransposon family provides insights into evolution of genome size in the genus *Oryza*. Plant J. 52:342–351

Ammiraju JSS. Fan C, Yu Y, Song XS, Cranston KS, Pontaroli AC, Lu F, Sanyal A, Jiang J, Rambo T, Currie J, Collura K, Talag J, Bennetzen J, Chen M, Jackson S, Wing RA (2010) Spatio-temporal patterns of genome evolution in allotetraploid species of the genus *Oryza*. Plant J 63:430

Bennetzen JL (2007) Patterns in grass genome evolution. Curr Opin Plant Biol 10:176–181

Cheng ZK, Gu MH (1994) Karyotype analysis for pachytene chromosome of indica, japonica rice and their hybrid. Chin J Genet 21:182–187

Chung MC, Wu HK (1987) Karyotype analysis of ‘IR36’ and two trisomic lines of rice. Bot Bull Acad Sin 28:289–304

Cronn RC, Small RL, Wendel JF (1999) Duplicated genes evolve independently after polyploid formation in cotton. Proc Natl Acad Sci USA 96:14406–14411

Davenport RJ (2001) Syngenta finishes, consortium goes on. Science 291:807

Fei L, Jetty SSA, Abhijit S, Shengli Z, Rentao S, Jinfeng C, Guisheng L, Yi S, Xiang S, Zhukuan C, Antonio CO, Jeffrey LB, Scott AJ, Rod AW, Mingsheng C (2008) Comparative sequence analysis of MONOCULM1-orthologous regions in 14 *Oryza* genomes. PNAS 106:2071–2076

Force A et al (1999) Preservation of duplicate genes by complementary, degenerative mutations. Genetics 151:1531–1545

Ge SST, Lu BR, Hong DY (1999) Phylogeny of rice genomes with emphasis on origins of allotetraploid species. Proc Natl Acad Sci USA 96:14400–14405

Ge S, Sang T, Lu BR, Hong DY (2001) Phylogeny of the genus Oryza revealed by molecular approaches. In: Khush GS, Brar DS, Hardy B (eds) Rice genetics IV Proceedings of 4th International Rice Genetics Symposium, 22–27 Oct 2000. International Rice Research Institute, Los Baos, the Philippines, pp 89–105

Harland JR, De Wet MJ (1971) Towards rational classification of cultivated plants. Taxon 20:509–517

Ilic K, San MPJ, Bennetzen JL (2003) A complex history of rearrangement in an orthologous region of the maize, sorghum, and rice genomes. Proc Natl Acad Sci USA 100:12265–12270

Khush GS (1997) Origin, dispersal, cultivation and variation of rice. Plant Mol Biol 35:25–34

Khush GS, Kinoshita T (1991) Rice karyotype, marker genes, and linkage groups. In: Khush GS, Toenniessen GH (eds) Rice biotechnology. CAB International, Wallingford, UK, pp 83–108

Kurata N, Omura T, Iwata N (1981) Studies on centromere, chromomere and nucleolus in pachytene nuclei of rice, *Oryza sativa*, microsporocytes. Cytologia 46:791–800

Lai J, Li Y, Messing J, Dooner HK (2005) Gene movement by Helitron transposons contributes to the haplotype variability of maize. Proc Natl Acad Sci USA 102:9068–9073

Ma J, Wing RA, Bennetzen JL, Jackson SA (2007) Evolutionary history and positional shift of a rice centromere. Genetics 177:1217–1220

Morishima H, Oka HI (1970) A survey of genetic variations in the populations of wild *Oryza* species and their cultivated relatives. Jpn J Genet 45:371–385

Nayar NM (2014) Origins and phylogeny of rices. Elsevier, Amsterdam. https://doi.org/10.1016/b978-0-12-417177-0.12001-7

Ohno S (1970) Evolution by gene duplication. Springer, London

Padhi G, Sen P (2002) Evaluation of wild rice species against Yellow stem borer (*Scripophaga incertulas*, Walk). J Appl Zool Res 11(2 & 3):103–104

Piegu B, Guyot R, Picault N, Roulin A, Sanyal A, Kim H, Collura K, Brar DS, Jackson S, Wing RA, Panaud O (2006) Doubling genome size without polyploidization: dynamics of retrotransposition-driven genomic expansions in *Oryza australiensis*, a wild relative of rice. Genome Res 16:1262–1269

Sarkar R, Raina SN (1992) Assessment of genome relationships in the genus *Oryza* L. based on seed-protein pro®le analysis. Theor Appl Genet 85:127–132

Shastry SVS, Rangao Rao DR, Mistra RN (1960) Pachytene analysis in *Oryza*. I. Chromosome morphology in *Oryza sativa* L. Ind J Genet Breed 20:15–21

Stark A, Lin MF, Kheradpour P, Pedersen JS, Parts L, Carlson JW, Crosby MA, Rasmussen MD, Roy S, Deoras AN, Ruby JG (2007) Discovery of functional elements in 12 Drosophila genomes using evolutionary signatures. Nature 450:219–232

Tang K, Sun X, He Y, Zhang Z (1998) Anther culture response of wild *Oryza* species. Plant Breeding 117:443–446
Tomotaro N, Duncan A, Vaughan KK (2005) Phylogenetic analysis of *Oryza* species, based on simple sequence repeats and their flanking nucleotide sequences from the mitochondrial and chloroplast genomes. Theor Appl Genet 110:696–705
Wang ZY, Second G, Tanksley SD (1992) Polymorphism and phylogenetic relationships among species in the genus *Oryza* as determined by analysis of nuclear RFLPs. Theor Appl Genet 83:565–581
Wing RA, Ammiraju JS, Luo M, Kim H, Yu Y, Kudrna D, Goicoechea JL, Wang W, Nelson W, Rao K, Brar D, Mackill DJ, Han B, Soderlund C, Stein L, San MP, Jackson S (2005) The *Oryza* map alignment project: the golden path to unlocking the genetic potential of wild rice species. Plant Mol Biol 59(1):53–62
Zapata FJ (1992) IRRI anther culture: Procedure, progress, problem and prospects. In: Workshop on anther culture for rice breeders 12–24 Oct 1992. China National Rice Research Institute
Zhang S, Gu YQ, Singh J, Coleman DD (2007) new insights into *Oryza* genome evolution: High gene colinearity and differential retrotransposon amplification. Plant Mol Biol 64:589–600
Zuccolo A, Sebastian A, Talag J, Yu Y, Kim H, Collura K, Kudrna D, Wing RA (2007) Transposable element distribution, abundance and role in genome size variation in the genus *Oryza*. BMC Evol Biol 7:152–167

Oryza rufipogon Griff.

25

Kumari Neelam, Palvi Malik, Karminderbir Kaur,
Kishor Kumar, Sahil Jain, Neha and Kuldeep Singh

Abstract

Oryza rufipogon, the progenitor of present-day cultivated rice, *O. sativa,* is one of the most studied wild species of rice. It is a perennial plant commonly found in a marsh or aquatic habitats of eastern and southern Asia. It has partial outcrossing behavior and is photoperiod sensitive. The flowering time usually ranges between September and November. It has been and is being exploited as a source of valuable genes and QTLs for yield components as well as resistance against biotic and abiotic stresses. A number of populations like chromosome segment substitution lines, backcross inbred lines, near-isogenic lines, and recombinant inbred lines have been developed from crosses between *O. rufipogon* and *O. sativa* as a prebreeding resource. These are being employed for broadening the genetic base of cultivated rice and diversify the breeder's pool. With the advent of sequencing technologies, a number of phylogenetic studies have been conducted to reveal the evolutionary relationship of *O. rufipogon* with cultivated rice *O. sativa.* Further, transcriptomic studies characterizing the effect of various abiotic stresses have been conducted on this wild species. Role of miRNA under stress reaction has also been studied. Though the genetic, genomic, and transcriptomic resources are abundant, the proteomic resources for *O. rufipogon* are limited.

25.1 Introduction

Oryza rufipogon, a wild species of the genus *Oryza*, also known as *O. perennis*, or *O. balunga*, is a diploid, aquatic plant. It has adaptable habit, erect and lax panicles that stand narrow and oblique, and beaked spikelets with awns (Sarkar et al. 2017). The perennial *O. rufipogon* grows in swamps, channels, marshes, and by the boundaries of ponds and lakes. *O. rufipogon* has a running habit, creeping on the ground and rooting at internodes. Morphologically, *O. rufipogon* is an erect, perennial tufted grass, 150–400 cm tall, with spongy culms which have glabrous and hollow nodes (Pang and Wang 1996). It is a close relative of cultivated rice, *O. sativa*, and shares many of the characteristics of the crop. The inflorescence is terminal and paniculate with many bisexual awned spikelets, each on a pedicel up to 2 mm

K. Neelam (✉) · P. Malik · K. Kaur · K. Kumar · S. Jain · Neha
Punjab Agricultural University, Ludhiana, Punjab 141004, India
e-mail: kneelam@pau.edu

K. Singh
National Bureau of Plant Genetic Resources, New Delhi 110012, India

T. K. Mondal and R. J. Henry (eds.), *The Wild Oryza Genomes*,
Compendium of Plant Genomes, https://doi.org/10.1007/978-3-319-71997-9_25

long. Lemma and palea are yellowish green, often dark red at apex. The floret has six stamens and two pistils. The caryopsis is narrow, reddish brown and is enclosed by a stiff lemma and palea (http://www.cabi.org/isc/datasheet/37960). The seeds have a tendency to shatter as soon as they mature. *O. rufipogon* is photosensitive in nature and flowers during short days during the months of November and December, in the Northern Hemisphere, in its natural habitat (Sarkar et al. 2017). It is tolerant of flooding and acidic soils (Mandal and Gupta 1997).

25.2 Distribution

O. rufipogon is widely distributed throughout southern China, South and Southeast Asia, Papua New Guinea, and northern Australia (Morishima et al. 1984; Sang and Ge 2007; Vaughan and Tamooka 2008). It's natural habitat includes different rice cropping systems in South and Southeast Asia, including dry seeded and deepwater rice in Bangladesh; dry seeded, wet seeded, seedling nurseries, transplanted, and upland rice in India; transplanted rice in Malaysia and the Philippines; wet seeded rice in Sri Lanka; deepwater, transplanted, and wet seeded rice in Thailand; and dry seeded and transplanted rice in Vietnam (Moody 1989). Evidence shows that climatic fluctuations accompanied by temperature and sea level changes during glacial ages in the quaternary are the most important processes shaping the population genetic structure of this species (Miller et al. 2005; Berg et al. 2010). Huang and Schaal (2012) conducted comprehensive analyses on *O. rufipogon* populations using phylogeographic and SDM approaches. These studies detected two genetic groups of *O. rufipogon* populations and claimed that three potential factors contributed to the current genetic subdivision of this species, including palaeoclimatic conditions, introgression between species and migration-drift balance over the range of *O. rufipogon*. Thus, both intrinsically biological and extrinsically environmental factors shaped the pattern of geographic subdivision in *O. rufipogon*.

25.3 *O. rufipogon* as Progenitor of Cultivated Rice

O. rufipogon is widely accepted as the progenitor of Asian cultivated rice, *O. sativa* (Oka and Chang 1959). Various strategies ranging from numerological taxonomy, comparison of organellar genomes, and sequencing of genome-wide loci have been used to understand the mechanism of origin of Asian cultivated rice from *O. rufipogon*. Two alternative hypotheses have been proposed to account for the existence of divergent groups such as *japonica* and *indica*. The polyphyletic/diphyletic hypothesis suggests that *japonica* and *indica* were domesticated independently from different ecotypes of *O. rufipogon* as different events. Various lines of evidence show that *japonica* and *indica* are more closely related to different *O. rufipogon* accessions than they are to each other (Second 1982; Ishii et al. 1988; Bautista et al. 2001; Cheng et al. 2003; Kovach et al. 2007; Rakshit et al. 2007; He et al. 2011; Yang et al. 2012). The *japonica* group is more closely related to the perennial *O. rufipogon*, whereas *indica* is more closely related to the annual *O. rufipogon*. Also, the estimated divergence time of *indica* and *japonica* predates the date of domestication (Kovach et al. 2007).

In contrast, the monophyletic hypothesis proposes that Asian cultivated rice arose from a single wild rice lineage and then diverged into several groups, including *japonica* and *indica* (Morishima and Sano 1992; Sang and Ge 2007). Several studies supporting the monophyletic hypothesis indicate that *O. sativa* had a single origin from within a region of the Yangtze River valley and that *japonica* may have arisen subsequently from *indica* (Lu et al. 2002; Gao and Innan 2008; Vaughan et al. 2008; Molina et al. 2011). However, some recent large-scale sequencing studies have yielded conflicting results (He et al. 2011; Molina et al. 2011).

Rakshit et al. (2007) studied genetic relationships between *O. sativa* and *O. rufipogon* by constructing a tree using a 20 kb sequence consisting of 22 loci. Their findings support the hypothesis that *O. sativa* cultivars were independently domesticated from wild *O. rufipogon* in multiple occasions. It was also shown that *indica* contains more than an order of magnitude larger genetic diversity as compared to *japonica*. Yang et al. (2012) examined particular groups of *O. rufipogon* and investigated the phylogenetic relationships between wild and cultivated rice. Results indicate that the wild ancestors of *japonica* and *indica* may have arisen from divergent subpopulations of *O. rufipogon*. During the independent domestications of *japonica*-like (perennial) and *indica*-like (annual) *O. rufipogon*, domestication-related genes appear to have been artificially selected in both *japonica* and *indica*. Later, in an early stage of domestication, some useful *japonica* genes were introduced to *indica*, possibly by humans. This series of artificial selection and hybridization created the present Asian rice cultivars.

Londo et al. (2006) used a phylogeographic approach to study domestication of cultivated rice by analyzing DNA sequence variation in three genic regions. Results point to two independent domestication events originating from different *O. rufipogon* populations, leading to the evolution of the cultivated rice subspecies, *O. sativa indica* and *O. sativa japonica*. The analysis has revealed a southern Himalayan mountain region and southern China to be the area of domestication of *O. sativa indica* and *O. sativa japonica*, respectively.

25.4 *O. rufipogon* as Source of Favorable Alleles

All the wild species in the *Oryza* genus serve as a valuable gene pool that can be used to broaden the genetic background of cultivated rice in breeding programs.

Backcrossed and advanced backcrossed populations (AB-QTL) and introgression lines made from *O. rufipogon* can be used for QTL mapping. Using the AB-QTL approach, Moncada et al. (2001) identified eight yield-related traits (days to heading, plant height, panicle length, number of panicles per plant, grains per plant, 1000 grain weight, days to heading, and plant yield) in BC_2F_2 population of Caiapo and *O. rufipogon*. Li et al. (2004) fine mapped a grain weight QTL, *gw3.1*, using a set of near-isogenic lines (NILs) developed from *O. sativa*, and a cv. Jefferson X *O. rufipogon* (IRGC105491) population based on five generations of backcrossing and seven generations of selfing. Tian et al. (2006) mapped a QTL for grain number (*gpa7*) on chromosome 7 by using F_3 population derived from SIL040 (an introgression line derived from *O. rufipogon* in *O. sativa* background) with Guichao 2. However, the contributing alleles from *O. rufipogon* were not favorable for grain number. McCouch et al. (2007) used AB-QTL technique to detect and transfer yield-related QTLs from *O. rufipogon* using different parallel populations and under diverse environments. The favorable alleles from *O. rufipogon* were found to improve recurrent parent performance by 5–20% for most of the characters studied. Fu et al. (2010) detected a total 26 QTLs in BC_4F_2 and BC_4F_4 populations derived from *O. rufipogon* and 93–11 by employing single-point analysis and interval mapping in both generations. Of the 26 QTLs, the alleles of 10 (38.5%) QTLs originating from *O. rufipogon* showed a beneficial effect for yield-related traits in the 93–11 genetic background. Yun et al. (2016) mapped rice grain quality and starch viscosity QTLs from a population of 96 *O. rufipogon* ILs. *O. rufipogon* is also the source of various fertility restoring QTLs for various CMS systems (Hu et al. 2016). Various other yield-related QTLs have also been mapped in *O. rufipogon* (Table 25.1).

O. rufipogon is also a source of genes and QTLs for tolerance and resistance against biotic and abiotic stresses. *O. rufipogon* contains QTLs that can be integrated into cultivated rice to improve cold tolerance (Koseki et al. 2010). A mapping population of F_2 plants derived from a cold-tolerant wild rice, W1943 (*O. rufipogon*), and a sensitive *indica* cultivar, Guanglu- ai 4, was used to identify QTLs associated with cold tolerance at the seedling stage. Three QTLs were detected on chromosomes 3 (*qCtss 3*), 10 (*qCtss 10*), and 11 (*qCtss 11*) with *qCtss11* explaining about 40% of the phenotypic variation. Six genes for BPH resistance, viz *bph18(t)*, *bph19(t)*,

Table 25.1 Yield-related QTLs from *Oryza rufipogon*

Trait	QTLs	Chromosome	Population used	References
Grain Number	*gpl*1.1, *gpl*2.1, *gpl*4.1, *gpl*5.1, *gpl*8.1, *gpl*8.2	1, 2, 4, 5, 8	BC_2	Xiao et al. (1998)
	*gpl*1.1, *gpl*2.1, *gpl*11.1	1, 2, 11	BC_2	Moncada et al. (2001)
	*gpl*1.1	1	BC_2F_2	Septiningsih et al. (2003)
	*gnp*2.1, *gnp*2.2, *gnp*5.1	2, 5	BC_2F_1	Marri et al. (2005)
	*gn*9.1	9	BC_3F_4	Xie et al. (2008)
	*gpp*8	8	F_2:3	Jin et al. (2009)
	*gpp*1.1, *gpp*3.1, *gpp*7.1, *gpp*12.1	1, 3, 7, 12	BC_2F_4	Fu et al. (2010)
Grain size	*gw*1.1, *gw*3.1, *gw*3.2	1, 3	BC_2F_2	Septiningsih et al. (2003)
	*gw*8.1	8	BC_3F_3	Xie et al. (2006a, b)
Spikelet number per panicle	*spp*2.1, *spp*3.1, *spp*9.1	2, 3, 9	BC_2F_2	Septiningsih et al. (2003)
	*snp*2.1, *snp*5.1, *snp*5.2	2, 5	BC_2F_1	Marri et al. (2005)
	*qspp*1, *qspp*11	1, 11	ILs	Liu et al. (2009)
	qSPP5	5	BC_5F_4	Luo et al. (2013)
Thousand grain weight	*gy*2.1, *gy*2.2, *gy*2.3, *gy*9.2	2, 9	BC_2F_1	Marri et al. (2005)
	*kgw*1.1, *kgw*2.1, *kgw*3.1, *kgw* 4.1, *kgw*7.1, *kgw*11.1	1, 2, 3, 4, 7, 11	BC_2F_4	Fu et al. (2010)
	*qTGW*5	5	BC_5F_4	Luo et al. (2013)
Yield	*yld*1.1, *yld*1.2, *yld*2.1	1, 2	BC_2F_2	Septiningsih et al. (2003)
	*yldp*2.1, *yldp*2.2, *yldp*9.1	2, 9	BC_2F_1	Marri et al. (2005)
	*yld*1.1, *yld*2.1, *yld*8.1, *yld*12.1	1, 2, 8	BC_2F_4	Fu et al. (2010)

bph22(t), *bph23(t)*, *bph24(t)*, and *Bph27,* have been mapped in *O. rufipogon* (Li et al. 2006, Hou et al. 2011, Deen et al. 2010, Huang et al. 2013). Tian et al. (2006) developed a set of 159 introgression lines derived from the cross between an *indica* cultivar Guichao 2 and an accession of common wild rice collected from Dongxiang County, Jiangxi Province, China. *O. rufipogon* serves as a source of bacterial blight resistance, tungro resistance, and resistance to various other diseases (Table 25.2).

25.5 Resequencing of Whole Genomes

Traditional method of plant breeding, involving careful selection followed by crossing, has been successful to a great extent in developing improved varieties. The green revolution is an explicit example of successful plant breeding, whereby a careful selection of dwarf varieties of rice and wheat followed by hybridization

Table 25.2 Resistance and tolerance genes and QTLs from *O. rufipogon*

Trait	Gene (s)	References
Bacterial blight resistance	*Xa23(t), Xa30(t)*	Zhang et al. (1998), Jin et al. (2007)
	Pirf2-1(t)	Utami et al. (2008)
Brown planthopper resistance	*bph18(t), bph19(t), bph22(t), bph23 (t), bph24(t), Bph27*	Li et al. (2006), Hou et al. (2011), Deen et al. (2010), Huang et al. (2013)
White backed planthopper resistance	QTLs—*qWbph2, qWbph5 and qWbph9*	Chen et al. (2010)
	Unknown	Hoan et al. (1997)
Tungro tolerance	Unknown	Kobayashi et al. (1993)
Tolerance to iron toxicity	Unknown	Brar and Khush (2006)
Cold tolerance	QTLs—*qCtss 3, qCtss 10,qCtss 11, qCTS4-1, qCTS12a, qCTS12b, qCTS*-6, *qCTS*-11, *qCTS*-12 *qCTS*-12-2 12 *qCTS*-12-3,*qCTS*-12-4,	Koseki et al. (2010), Andaya and Mackill (2003), Luo et al. (2016)
Tolerance to acidic conditions	Unknown	Brar and Khush (2006)
Tolerance to P-deficiency	Unknown	Brar and Khush (2006)

experiments with cultivars prevalent at that time saved a billion people from a hunger crisis. The available germplasm has been thoroughly exploited by plant breeders for crop improvement, but domestication and artificial selection have led to the concentration of a few alleles in the breeder's pool, which has led to a yield ceiling. In order to overcome the yield ceiling and solve the nine billion people question (9BPQ), it is hoped that the large repertoire of variation present in wild relatives that has been underutilized and underexploited should be brought into the breeder's pool. Over time, better techniques and strategies like advanced backcross-QTL and genome-wide association studies utilizing wild relatives have been developed. In order to fully utilize the potential of a crop and its wild relatives, it is necessary to look at their genome composition.

Sequencing has greatly facilitated analysis of the genome at the structural and functional levels. Further, the emergence of high-throughput sequencing technologies and bioinformatics tools has hastened the process of crop improvisation. With the cost of sequencing going down and amount of sequencing data per run going up, reduced representation sequencing and resequencing of whole genomes have seen an upward trend.

From understanding the basic physiological aspects to dissecting the biology of expression of traits, resequencing of whole genomes, region specific sequencing, reduced representation sequencing have played a pivotal role. Availability of a gold standard reference sequence of the rice (IRGSP 2005) has immensely supplemented the efforts of breeders. Resequencing of genomes has played a key role in genome-wide discovery of markers, construction of high-density genetic maps, genome-wide association studies, comparative analysis, evolutionary studies, etc. One of the key examples is the 3000 genome project where a core collection of 3000 rice accessions from 89 countries was sequenced using Illumina HiSeq 2000 platform (3K RGP 2014). When aligned to the gold standard reference genome of Nipponbare, approximately 18.9 million single nucleotide polymorphisms (SNPs) were discovered.

Being a member of the primary gene pool and progenitor of cultivated rice, *O. rufipogon* can be easily crossed with *O. sativa* and thus resequencing of *O. rufipogon* accessions is of prime interest to rice molecular biologists. In another study by Xu et al. (2012), 40 cultivated and 10 wild accessions (*O. rufipogon* and *O. nivara*) were resequenced using an Illumina GA2 sequencing platform. Genome-wide variation patterns were investigated, and 6.5 million high-quality SNPs were obtained after excluding sites with missing data.

The knowledge of extent of linkage disequilibrium (LD) is essential to dissect complex traits at the molecular levels. To determine LD values, Mather et al. (2007) assessed variation in six genomic regions of ~500 Kb in 82 rice accessions comprising of *O. sativa* and *O. rufipogon*. LD was found to be highest in temperate *japonica* (probably >500 kb), trailed by tropical *japonica* (~150 kb), *indica* (~75 kb), and *O. rufipogon* (≪40 kb). Since LD prevails over longer distance in cultivated varieties, modest number of SNPs are expected to be sufficient for carrying out genome-wide association studies. Conversely, a large number of markers are required to exploit *O. rufipogon* accessions.

Brozynska et al. (2016) have reported a draft genome for two Australian wild A genome taxa: *O. rufipogon*-like population, referred to as Taxon A, and *O. meridionalis*-like population, referred to as Taxon B. Illumina and PacBio platforms were used for sequencing, and assembly was done by integration of short- and long-read next-generation sequencing (NGS) data to create a genomic platform for a wider rice gene pool. The genome sizes were estimated in silico to be about 390 and 370 Mb for Taxon A and Taxon B, respectively. The nuclear genome of *O. rufipogon*-like population (Taxon A) has been reported to have a sequence that is much closer to that of domesticated rice (*O. sativa*) than to the other Australian wild populations, but chloroplast genomes are quite distinct. Admixture studies have revealed possible introgression into the *O. rufipogon*-like population from the Asian wild *indica*/*O. nivara* clade. The research group has laid the possibility that Northern Australia may be the center of diversity of the A genome *Oryza* and thus the center of origin of this group and represent an ultimate resource for cultivated rice improvement. This is in congruence with results of Waters et al. (2012).

Reduced representation sequencing method was used to assess nuclear variation, and chloroplast variation was studied using Sanger sequencing in a study conducted by Kim et al. (2016). A set of 286 diverse *O. rufipogon* accessions were classified into six wild subpopulations, and 25% were classified as admixed. Out of these six subgroups, three were found to be associated with subpopulations of *O. sativa, i.e., indica, aus* and *japonica*, and *O. sativa* introgressions. The remaining groups were found to be diverse at the genetic level, and the potential of these diverse groups is still to be harnessed.

Various studies have focused on domestication of present-day cultivated rice from *O. rufipogon,* but a study conducted by Zhang et al. (2016) focused on revealing significant changes at the physiological and morphological levels between *O. sativa L.* and *O. rufipogon* Griff. Unassembled Illumina sequencing reads of Dongxiang wild rice (DXWR) were compared with the genome sequence of Nipponbare. An attempt was made to identify those genes deleted from DXWR that have been acquired by Nipponbare during the course of domestication, based on structural variations in two genomes. This study was further supplemented by comparing transcriptome sequences of DXWR and Nipponbare. Nipponbare acquired a total of 1591 genes during domestication events. Based on the results of the comparative genomic analysis, structural variations (SVs) between DXWR and Nipponbare were determined to locate deleted genes that could have been acquired by Nipponbare during the course of domestication. Apart from this, the role of transposable elements in genome evolution from wild to cultivated rice has been confirmed. An adaptation of the photophosphorylation and oxidative phosphorylation system in cultivated rice has also been implied.

25.6 Repetitive Sequences

Repetitive DNA forms a major part of plant genomes and has been predicted to play a key role in evolution. Repetitive DNA content tends to increase with increasing complexity of genomes. It is estimated that the rice genome contains ~40% repetitive DNA (Gill et al. 2010). On the basis of genomic organization and chromosomal location, repetitive DNA sequences in higher plants have been categorized into two groups: the first one consisting of satellite DNAs, telomeric repeats, and rDNA located at pericentromeric, subtelomeric, telomeric, or intercalary regions of the chromosomes in tandem repeat units, and the transposable elements (TEs) constitute the second group. TEs are further divided into two classes (Flavell 1994), namely Class I which includes long terminal repeat (LTR) retrotransposons that constitute 20% of the rice genome (i.e., ~80 Mb) and is distributed genome wide with a high copy number (Mao et al. 2009). Class I TEs follow "copy-and-paste" mechanism of RNA intermediates (Vitte et al. 2004). The class II transposable elements transpose via DNA intermediates, usually resulting in relatively low copy numbers (usually 100 copies per genome) (Kunze et al. 1997). It has been reported that LTR retrotransposons are inserted into the rice genome less than 15 MYA and completed the transposition events within the last three million years (SanMiguel et al. 1996; Ma et al. 2004).

Insertions and eliminations of TEs have generated numerous transposon insertion polymorphisms (TIPs). Xiao et al. (2011) attempted to find subspecies-specific (SS) markers that can distinguish *O. sativa* ssp. *indica* and ssp. *japonica*. These markers were derived from LTR sequences of AA genome-specific RIRE retrotransposon of *O. rufipogon*. A total of twenty-two SS markers were developed, out of which nine are *indica*-specific types and thirteen are *japonica*-specific types. The average marker accuracy in differentiating two subspecies was reported to be over 85%.

Gill et al. (2010) performed comparative analysis of repetitive sequences across ten genome types of rice including *O. rufipogon*. BAC-end sequences that represented 8–17% (Kim et al. 2008) of each of the ten *Oryza* genome types were analyzed for their repetitive DNA content using both homology-based (RepeatMasker, Blast) and de novo (Tallymer, RECON) method. It was observed that LTR-RTs were the most abundant type of repetitive DNA Sequences. The size variation among different species of *Oryza* genome has been attributed to LTR-RTs. BES pairs were assembled so as to obtain one sequence/BAC clone. The clones were categorized into low repetitive (0–40% repetitive), mid repetitive (40–70% repetitive), and high repetitive (70–100% repetitive) clones on the basis of frequencies of overlapping 20 mers for each BAC clone. The majority of clones in all the diploid species, i.e., >60%, are low repetitive except *O. officinalis* [CC], *O. australiensis* [EE], and *O. granulata* [GG]. These three species have the highest percentage of mid repetitive clones (47.2–58.8%) among all the diploids. For other diploids like *O. rufipogon*, only 0.5% clones have 70–100% repetitive content. The total repetitive content of entire *O. rufipogon* genome has been estimated to be 51.39%. Study of repetitive elements can lead to better understanding of their DNA sequences and their role in shaping *Oryza* genomes. Apart from this, events like domestication, speciation, polyploidy, size variation can also be thoroughly investigated.

25.7 Organellar Genome

Along with nuclear genome, various studies on *O. rufipogon* have focused on chloroplast and mitochondrial genomes. These have also been used to study evolutionary lineages (Ishii et al. 1988; Bautista et al. 2001; Sun et al. 2002; Kawakami et al. 2007).

RFLP analysis of the chloroplast (cp) genome has revealed that the *japonica* ecotype contains 1 type of cp genome, whereas 3 types of cp genome are present in the *indica* ecotype (Ishii et al. 1988). These results also suggest that rice domestication was polyphyletic. Phylogenetic studies of the cp genome also support a di- or polyphyletic hypothesis for rice domestication (Chen et al. 1993; Nakamura et al. 1998; Bautista et al. 2001; Park et al. 2003; Vitte et al. 2004; Garris et al. 2005; Kawakami et al. 2007). Waters et al. (2012) studied cp genome sequences of Australian and Asian *O. rufipogon* accessions. They found that Australian *O. rufipogon* cp sequences are closer to cp sequence of *O. meridionalis,* and then, they are to Asian *O. rufipogon* accessions. Similar results were reported by Brozynska et al. (2016).

Genetic analyses of the rice mitochondrial genome from *O. rufipogon* accessions along with *indica* and *japonica* cultivars revealed that genetic diversity is higher in the *indica* ecotype than the *japonica* ecotype (Kadowaki et al. 1988). While studying the mitochondrial genome of *O. rufipogon* strains in comparison with cultivated rice, Sun et al. (2002) detected 99 polymorphic fragments. Their findings indicate that mitochondrial DNA of *O. rufipogon* is more variable than cultivated rice. Igarashi et al. (2013) obtained an RT98A CMS line and an RT98C fertility restorer line by successive backcrossing between *O. rufipogon* W1109 and *O. sativa* cultivar Taichung 65. The findings of this study pointed specifically to *orf113* as the gene responsible for cytoplasmic male sterility.

25.8 Genome Initiatives

The International *Oryza* Map Alignment Project (Jacquemin et al. 2013) came with an objective of solving 9BPQ. It aims to generate reference genomes and transcriptomes for all the 23 species of *Oryza* genus. Under International *Oryza* Map Alignment Project (I-OMAP), National Center for Gene Research (NCGR), Chinese Academy of Sciences (CAS), independently finished the deep sequencing of *O. rufipogon* genome and produced the world's first highly heterozygous draft genome in August 2010. The *O. rufipogon* W1943 was sequenced at the whole-genome level using the Illumina platform, and assembly was performed. Large insertion mate-pair sequences were obtained to build sequence scaffolds and pseudomolecules of *O. rufipogon* W1943. Results demonstrated the genome size of *O. rufipogon* to be about 370 million base pairs with a total of 40,000 genes (http://www.ebi.ac.uk/ena/data/view/GCA_000817225.1). The drafts have been generated but not published yet. In addition to this, the Donxiang genotype of *O. rufipogon* is sequenced (NCBI SRA database with ID SRP070627) in 2016 by Zhang et al. *Oryza* base is an excellent genotype–phenotype association resource for analyzing rice functional and structural evolution. *Oryza* genome is a united repository that has integrated genotypic and phenotypic information. It contains imputed genotypic information of 446 *O. rufipogon* accessions and seventeen imputation-free accessions (Ohyanagi et al. 2015; Table 25.3).

25.9 Comparative Genomics of *O. rufipogon* and *O. sativa*

O. rufipogon, being the progenitor of cultivated rice, shares various genomic features with it. Studies show that *O. rufipogon* shares various genetic features and clusters closely with *O. nivara* in addition to cultivated rice (Lu et al. 2002; Ren et al. 2003). Ammiraju et al. (2008) have shown that a single shared intact element (J-ILTR1) between *indica* and *japonica* was also shared with *O. rufipogon* and *O. nivara.* Hurwitz et al. (2010) carried out comparative study of *O. sativa* and *O. rufipogon.* O. sativa was shown to have 206 contractions, 171 expansions, and 40 inverted regions compared to *O. rufipogon.* Of these, 83% of contractions and 99% of expansions were confirmed by validation. Song et al. (2011) studied synteny between *O. rufipogon* and *O. sativa* using 17 OTS/STS-PCR markers in eight accessions, which revealed 95–100% identity.

25.10 Transcriptomic Resources

The total of all transcripts produced at a specific developmental stage of a plant at a given time is known as the transcriptome, and study of the transcriptome is known as transcriptomics. Being

Table 25.3 Assembly statistics of *O. rufipogon* (W1943) genome sequence

Features[a]	Statistics (bp)
Total length	339,177,042
Ungapped length	332,710,542
Spanned gaps	64,665
Scaffolds	3818
Scaffold N50	27,785,585
Contigs	68,481
Contig N50	34,232

[a]*Source* Submitted by National Center for Gene Research (NCGR), Chinese Academy of Sciences (CAS), to European Nucleotide Archive. (http://www.ebi.ac.uk/ena/data/view/GCA_000817225.1)

a model crop for monocots, the first transcriptome approach was to collect and characterize the expressed sequence tags (ESTs) and full-length complementary DNA (FL-cDNA) for gene discovery and genome study. ESTs are single-pass sequences of randomly chosen cDNA. Approximately 1.2 million ESTs and >50,000 full-length cDNA sequences of cultivated rice are currently available in public databases (Yang et al. 2010). These ESTs have also been successfully mapped along the rice genome (Wu et al. 2002; Zhou et al. 2003; Zhang et al. 2005). A total of 5211 leaf ESTs were generated from *O. minuta* (Cho et al. 2004). However, ESTs are random sequences of cDNA and do not represent the full open reading frame (ORF) of a transcribed gene. Therefore, full-length cDNA (FL-cDNA) is regarded as the most critical component of transcriptome resources. Hence, a full-length cDNA consortium of Japan was launched in 2000, and around 28,000 FL-cDNA clones from japonica rice were annotated and characterized (Kikuchi et al. 2003). Eventually, this collection FL-cDNA was further expanded to 578,000 clones (available online at http://cdna01.dna.affrc.go.jp/cDNA/). There is also a collection of 10 096 FL-cDNAs of *indica* rice. Collection, characterization, and annotation of FL-cDNA have also been performed in *O. rufipogon* and O. *longistaminata* (Lu et al. 2008; Yang et al. 2010).

Being a perennial wild progenitor of modern cultivated rice, *O. rufipogon* is well known for many agronomic traits including drought, salt, and high temperature (Jenks and Hasegawa, 2005). Drought-responsible candidate genes have been identified using de novo transcriptome assembly of *O. rufipogon* using polyethylene glycol (PEG)-treated plants at different developmental stage (Tian et al. 2015). A transcriptome resource developed in *O. rufipogon* has been also utilized to identify salt tolerance genes from leaf and root samples at seedling stage (Zhou et al. 2016). Zhou et al. (2016) reported 57 upregulated genes in shoots, out of which two genes, late embryogenesis abundant gene (*OsLEA3-2*, LOC_Os01g21250) (Duan and Cai 2012) and *JcLEA* in *Arabidopsis* (Liang et al. 2013), have previously been proved in salt tolerance.

25.11 miRNA

Micro-RNAs (miRNAs) are a class of noncoding, endogenous small RNAs that regulate plant development, phase transition, and environmental stress (Sunkar and Zhu 2004; Jones-Rhoades et al. 2006; Khraiwesh et al. 2012; Sunkar et al. 2012). Initially, miRNAs were identified in *Arabidopsis* based on sequence conservation (Jones-Rhoades and Bartel 2004). More recently, conserved miRNAs have been identified in rice using direct cloning and traditional sequencing approaches (Liu et al. 2005; Sunkar et al. 2005; Luo et al. 2006).

The role of micro-RNA (miRNA) in domestication has been elucidated in *O. rufipogon* by high-throughput sequencing using the Illumina platform (Wang et al. 2012). Wang et al. (2012)

reported a total of 387 miRNA reads, out of which 259 miRNAs were absent and 48 miRNAs were novel in the cultivated rice, suggesting that they were potential targets of domestication selection. Flowering-related miRNAs have also been identified and validated from *O. rufipogon* suggesting that specific set of candidate miRNAs were involved in a complicated network that regulates the expression of flowering genes that control the induction of flowering. Variation in these miRNAs laid strong foundation of domestication in rice (Chen et al. 2013). The conserved miRNA and their expression patterns under drought stress have been identified and characterized from *O. rufipogon* using high-throughput sequencing (Zhang et al. 2015). Differentially expressed miRNAs have been identified and characterized under drought stress from *O. rufipogon* (Zhang et al. 2017). Zhang et al. (2017) predicted a total of 200 differentially expressed miRNAs, out of which 26 novel candidate miRNAs from the shoot and 43 from the root were differentially expressed during the drought stress (Table 25.4).

25.12 Proteomic and Metabolomic Studies

After the high-quality genome sequencing of rice, it has been a major challenge to identify the functions of all protein-coding genes. Proteome analysis in rice under different contexts was reviewed by many researchers (Komatsu and Tanaka 2004; Agrawal et al. 2006; Deng et al. 2013). Proteins have also been studied when a suspension culture of *O. meyeriana* was treated with *Xanthomonas oryzae* (Chen et al. 2016). Proteome analysis was done for plasma membrane-bound proteins which play important role in elicitor-mediated plant defense in rice by Chen et al. (2007). Differential proteome analysis has been done to study pedigree of elite rice hybrids and to study the hybrid vigor by Xie et al. (2006a, b). Xie et al. (2006a, b) reported proteome analysis could be a very powerful tool when a trait or phenotype cannot be correlated with its molecular information. Very little progress has been made in the proteomic and metabolomic analysis of *O. rufipogon*. Seed storage proteins such as rice glutelin and globulins which are major determinants of nutritional quality have been analyzed in three wild species of rice including *O. rufipogon* and compared with two cultivated rices (Jiang et al. 2014).

25.13 Impact on Plant Breeding and Prebreeding

Substantial "prebreeding" research can provide a platform to identify and transfer favorable alleles from wild and unadapted sources into elite rice cultivars (McCouch et al. 2007). Several types of mapping populations have been generated from interspecific crosses between *O. rufipogon* and *O. sativa* to transfer favorable alleles into elite cultivars. Advanced mapping populations like backcross inbred lines (BILs), near-isogenic lines (NILs), chromosome segment substitution lines (CSSLs), MAGIC populations, and association

Table 25.4 Transcriptomic resources generated in *O. rufipogon*

Transcriptome resources	Trait	References
FL-cDNA	–	Lu et al. (2008)
Microarray	Yield	Thalapati et al. (2012)
De novo transcriptome analysis	Drought	Tian et al. (2015)
	Salinity	Zhou et al. (2016)
miRNA	Domestication	Wang et al. (2012)
	Flowering	Chen et al. (2013)
	Drought	Zhang et al. (2015, 2017)

mapping panels are powerful tools for identifying the naturally occurring favorable alleles in unadapted germplasm (Ali et al. 2010; Jacquemin et al. 2013). Various workers have worked toward developing CSSLs and BILs with donor segments of *O. rufipogon* into background of *O. sativa.* There are a number of reports of using BILs to introgress QTLs from the wild rice species into cultivated rice, some of which are alleles from *O. rufipogon* associated with the increased number of grains (Tian et al. 2006) and improved drought tolerance (Zhang et al. 2006) in the background of the *indica* cultivar Guichao 2; alleles associated with increased yield and its components in the background of the *indica* cultivar IR64 (Cheema et al. 2008); alleles associated with early flowering (Maas et al. 2010) in the background of the tropical *japonica* cultivar Jefferson.

Under the OMAP, a set of 105 CSSLs with donor segments from *O. rufipogon* (IRGC105491) in the background of Zhenshan 97B has been developed (Wing et al. 2007). A companion mapping population (BC_2F_5) derived from the same cross-combination was also developed. More than 50% of the alleles derived from the *O. rufipogon* donor contributed to the increase in yield-related traits, such as spikelet number per panicle, grain weight, and panicle length.

Qiao et al. (2016) developed a set of 198 CSSLs from a cross between recurrent parent *indica* var. 9311 and an *O. rufipogon* donor parent. These were then genotyped using SSR markers evenly across the 12 rice chromosomes. The segments collectively covered 84.9% of the wild rice genome. In this study, 25 QTLs controlling 10 agronomic traits were identified. Seven CSSLs were subjected to a whole-genome single nucleotide polymorphism chip assay, and two QTLs, *qSH4-1* and *qDTH10-1*, were detected. In addition, a QTL associated with the heading date was detected on chromosome 1. Both these studies demonstrated the advantage of CSSLs as prebreeding resource as well as a source of identifying new QTLs. Ogawa et al. (2016) identified 13 QTLs for root system architecture, NDT, and morpho-physiological traits using chromosome segment substitution lines of *O. rufipogon.*

McCouch et al. (2007) used advanced backcross lines or introgression lines to simultaneously identify, map, and introgress novel alleles from the wild progenitor of cultivated rice, *O. rufipogon,* into the genetic background of several adapted cultivars. Haritha et al. (2017) analyzed *O. rufipogon* ILs and found significant variations in net photosynthetic rate (*Pn*), transpiration rate (*E*), transpiration efficiency (*Pn*/*E*), and carboxylation efficiency (*Pn*/*C*i). The ILs with enhanced *Pn* can serve as a potential source for developing rice varieties and hybrids with higher biomass and yield. *O. rufipogon* serves as a valuable genetic resource for developing a prebreeding gene pool, which in turn is being employed for broadening the genetic base of cultivated rice and to diversify the breeder's pool.

25.14 Future Prospectives

A remarkable progress has been made toward the introgression of important genes and QTLs from *O. rufipogon* into cultivated rice against biotic and abiotic stresses using classical breeding approaches. Though a large number of *O. rufipogon* accessions have been sequenced since the inception of the *Oryza* Map Alignment, *Oryza* Genome Evolution project (OMAP), and other resources, still their utilization is hindered due to lack of availability of the draft genome sequences to the public at large. Further, the BAC libraries of 12 *Oryza* species representing 10 genomes of *Oryza* constructed under OMAP project should be further utilized for physical mapping, positional cloning, integration of genetic maps, and comparative studies. Also, there is a need to intensify the research toward the identification of new gene (s)/QTLs for tapping the variability for biotic stress like sheath blight, neck blast, and stem borer for which there is no perfect resistance source is available. The research areas like proteomics, epigenomics, and metabolomics should be further strengthened for maximizing the usefulness of *O. rufipogon* species for the improvement of elite cultivars.

25.15 Conclusions

The wild species of rice, *O. rufipogon,* are comprehensively utilized in rice prebreeding and crop improvement programs. Recent resequencing of *O. rufipogon* accessions has led to a better understanding of this species as the progenitor of cultivated rice. The polyphyletic origin of rice as suggested by these studies explains the differences in present-day rice subspecies. The development of genomic resources has expedited molecular breeding approaches for the improvement of elite rice cultivars. Simultaneous, transcriptomic studies have helped in better understanding of the genes involved in response to abiotic and biotic stresses. There is a need to develop proteomic and metabolomic resources for developing an in-depth knowledge base of *O. rufipogon.* A well-coordinated database for all the omics approaches should go hand in hand with the conventional techniques to fully mine *O. rufipogon* accessions as a powerful breeding resource (Figs. 25.1, 25.2 and 25.3).

Fig. 25.1 Vegetative and reproductive stages of *O. rufipogon* plant along with the diversity present in the grains for color, awn, and size

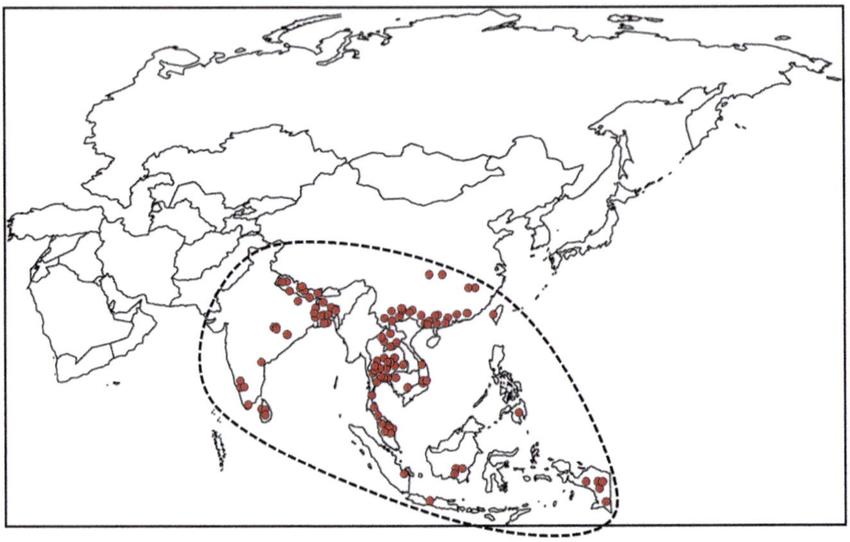

Fig. 25.2 Geographical distribution of Asian wild rice *O. rufipogon* (adopted from Huang et al. 2012)

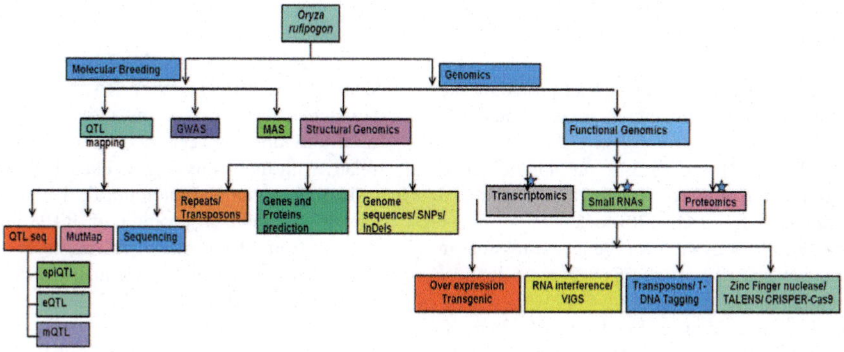

Fig. 25.3 Schematic representation of the overall omics approaches being used currently for the utilization of *O. rufipogon.* The stars represent research area that still needs to be emphasized

References

Agrawal GK, Jwa NS, Iwahash Y, Yonekura M, Iwahashi H, Rakwal R (2006) Rejuvenating rice proteomics: facts, challenges and visions. Proteomics 6:5549–5576

Ali ML, Sanchez PL, Yu S, Lorieux Eizenga GC (2010) Chromosome segment substitution lines: a powerful tool for the introgression of valuable genes from *Oryza* wild species into cultivated rice (*O. sativa* L.). Rice 3:218–234

Ammiraju JS, Lu F, Sanyal A, Yu Y, Song X, Jiang N, Pontaroli AC, Rambo T, Currie J, Collura K, Talag J (2008) Dynamic evolution of *Oryza* genomes is revealed by comparative genomic analysis of a genus-wide vertical data set. Plant Cell 20:3191–3209

Andaya VC, Mackill DJ (2003) Mapping of QTLs associated with cold tolerance during the vegetative stage in rice. J Exp Bot 54:2579–2585

Bautista NS, Solis R, Kamijima O, Ishii T (2001) RAPD, RFLP and SSLP analyses of phylogenetic relationships between cultivated and wild species of rice. Genes Genet Syst 76:71–79

Berg MP, Kiers E, Driessen G, Der Heijden Van, Kooi BW, Kuenen F, Liefting M, Verhoef HA, Ellers J (2010) Adapt or disperse: understanding species persistence in a changing world. Global Change Biol 16:587–598

Brar DS, Khush GS (2006) Cytogenetic manipulation and germplasm enhancement of rice (*Oryza sativa* L.). In: Singh RJ, Jauhar PP (eds) Genetic resources, chromosome engineering and crop improvement. CRC, Boca Raton, FL, pp 115–158

Brozynska M, Copetti D, Furtado A, Wing RA, Crayn D, Fox G, Ishikawa R, Henry RJ (2016) Sequencing of Australian wild rice genomes reveals ancestral relationships with domesticated rice. Plant Biotechnol. https://doi.org/10.1111/pbi.12674

Cheema KK, Grewal NK, Vikal Y, Das A, Sharma R, Lore JS, Bhatia D, Mahajan R, Gupta V, Singh K (2008) A novel bacterial blight resistance gene from *Oryza nivara* mapped to 38 Kbp region on chromosome 4L and transferred to *O. sativa* L. Genet Res 90:397–407

Chen WB, Nakamura I, Sato YI, Nakai H (1993) Distribution of deletion type in cpDNA of cultivated and wild rice. Jap J Genet 68:597–603

Chen F, Li Q, He Z (2007) Proteomic analysis of rice plasma membrane associated protein in response to chitooligosaccharide elicitors. J Integr Plant Biol 49:863–870

Chen J, Huang DR, Wang L, Liu GJ, Zhuang JY (2010) Identification of quantitative trait loci for resistance to whitebacked planthopper, *Sogatella furcifera*, from an interspecific cross *O. sativa* × *O. rufipogon*. Breed Sci 60:153–159

Chen Z, Li F, Yang S, Dong Y, Yuan Q et al (2013) Identification and functional analysis of flowering related microRNAs in common wild rice (*Oryza rufipogon* Griff.). PLoS One 8:e82844. https://doi.org/10.1371/journal.pone.0082844

Chen X, Dong Y, Yu C, Fang X, Deng Z, Yan C, Chen J (2016) Analysis of the proteins secreted from the *Oryza meyeriana* suspension-cultured cells induced by *Xanthomonas oryzae pv. oryzae*. PLoS One 11:e0154793. https://doi.org/10.1371/journal.one.0154793

Cheng C, Motohashi R, Tsuchimoto S, Fukuta Y, Ohtsubo H, Ohtsubo E (2003) Polyphyletic origin of cultivated rice: based on the interspersion pattern of SINEs. Mol Biol Evol 20:67–75

Cho SK, Ok SH, Jeung JU, Shim KS, Jung KW, You MK, Kang KH, Chung YS, Choi HC, Moon HP, Shin JS (2004) Comparative analysis of 5211 leaf ESTs of wild rice (*Oryza minuta*). Plant Cell Rep 22:839–847

Deen R, Ramesh K, Gautam SK, Rao YK, Lakshmi VJ, Viraktamath BC, Brar DS, Ram T (2010)

Identification of new gene for BPH resistance introgressed from *O. rufipogon*. Rice Genet Newsl 25:70–71
Deng ZY, Gong CY, Wang T (2013) Use of proteomics to understand seed development in rice. Proteomics 13:1784–1800
Duan J, Cai W (2012) OsLEA3-2, an abiotic stress induced gene of rice plays a key role in salt and drought tolerance. PLoS One 7(9):e45117
Flavell RB (1994) Inactivation of gene expression in plants as a consequence of specific sequence duplication. Proc Nat Acad Sci 91:3490–3496
Fu Q, Zhang P, Tan L, Zhu Z, Ma D, Fu Y, Zhan X, Cai H, Sun C (2010) Analysis of QTL for yield-related traits in Yuanjiang common wild rice (*Oryza rufipogon* Griff.). J Genet Genomics 37(2):147–157
Gao LZ, Innan H (2008) Nonindependent domestication of the two rice subspecies, *Oryza sativa* ssp. *indica* and ssp. *japonica*, demonstrated by multilocus microsatellites. Genetics 179:965–976
Garris AJ, Tai TH, Coburn J, Kresovich S, McCouch S (2005) Genetic structure and diversity in *Oryza sativa L.* Genetics 169:1631–1638
Gill N, Phillip S, Dhillon BDS, Abernathy B, Kim H, Stein L, Ware D, Wing R, Jackson SA (2010) Dynamic *Oryza* genomes: repetitive DNA sequences as genome modeling agents. Rice 3:251–269
Haritha G, Vishnukiran T, Yugandhar P, Sarla N, Subrahmanyam D (2017) Introgressions from *Oryza rufipogon* increase photosynthetic efficiency of KMR3 rice lines. Rice Sci 24:85–96
He Z, Zhai W, Wen H, Tang T, Wang Y, Lu X, Greenberg AJ, Hudson RR, Wu CI, Shi S (2011) Two evolutionary histories in the genome of rice: the roles of domestication genes. PLoS Genet 7:e1002100
Hoan NT, Sarma NP, Siddiq EA (1997) Identification and characterization of new sources of cytoplasmic male sterility in rice. Plant Breed 116:547–551
Hou LY, Ping YU, Qun XU, Yuan XP, Yu HY, Wang YP, Wang CH, Wan G, Tang SX, Peng ST, Wei XH (2011) Genetic analysis and preliminary mapping of two recessive resistance genes to brown planthopper, Nilaparvata lugens Stål in Rice. Rice Sci 18:238–242
Hu BL, Xie JK, Wan Y, Zhang JW, Zhang FT, Li X (2016) Mapping QTLs for fertility restoration of different cytoplasmic male sterility types in rice using two *Oryza sativa* × *O. rufipogon* backcross inbred line populations. Biomed Res. https://doi.org/10.1155/2016/9236573
Huang P, Schaal BA (2012) Association between the geographic distribution during the last glacial maximum of asian wild rice, *Oryza rufipogon* (Poaceae), and its current genetic variation. Am J Bot 99:1866–1874
Huang D, Qiu Y, Zhang Y, Huang F, Meng J, Wei S, Li R, Chen B (2013) Fine mapping and characterization of BPH27, a brown planthopper resistance gene from wild rice (*Oryza rufipogon* Griff.). Theor Appl Genet 126:219–229
Hurwitz BL, Kudrna D, Yu Y, Sebastian A, Zuccolo A, Jackson SA, Ware D, Wing RA, Stein L (2010) Rice structural variation: a comparative analysis of structural variation between rice and three of its closest relatives in the genus *Oryza*. Plant J 63:990–1003
Igarashi K, Kazama T, Motomura K, Toriyama K (2013) Whole genomic sequencing of RT98 mitochondria derived from *Oryza rufipogon* and northern blot analysis to uncover a cytoplasmic male sterility-associated gene. Plant Cell Physiol 54:237–243
IRGSP (2005) International rice genome sequencing project: the map based sequence of the rice genome. Nature 436:793–800
Ishii T, Terachi T, Tsunewaki K (1988) Restriction endonuclease analysis of chloroplast DNA from A-genome diploid species of rice. Jap J Genet 63:523–536
Jacquemin J, Bhatia D, Singh K, Wing RA (2013) The international *Oryza* map alignment project: development of a genus-wide comparative genomics platform to help solve the 9 billion-people question. Curr Opin Plant Biol 16:147–156
Jenks MA, Hasegawa PM (2005) Plant abiotic stress. Blackwell, UK
Jiang C, Cheng Z, Zhang C, Yu T, Zhong Q, Shen JQ, Huang X (2014) Proteomic analysis of seed storage proteins in wild rice species of the *Oryza* genus. Proteome Sci 12:51. https://doi.org/10.1186/s12953-014-0051-4
Jin XW, Wang CL, Yang Q, Jiang QX, Fan YL, Liu GC, Zhao KJ (2007) Breeding of near-isogenic line CBB30 and molecular mapping of Xa30(t), a new resistance gene to bacterial blight in rice. Sci Agri Sin 40:1094–1100 (Chinese with English abstract)
Jin FX, Kim DM, Ju HG, Ahn SN (2009) Mapping quantitative trait loci for awnness and yield component traits in isogenic lines derived from an *Oryza sativa/O. rufipogon* cross. J Crop Sci Biotech 12:9–15
Jones-Rhoades MW, Bartel DP (2004) Computational identification of plant MicroRNAs and their targets, including a stress-induced miRNA. Mol Cell 14:787–799
Jones-Rhoades MW, Bartel DP, Bartel B (2006) MicroRNAs and their regulatory roles in plants. Annu Rev Plant Biol 57:19–53
Kadowaki K, Yazaki K, Osumi T, Harada K, Katsuta M, Nakagahra M (1988) Distribution of mitochondrial plasmid-like DNA in cultivated rice (*Oryza sativa L.*) and its relationship with varietal groups. Theor Appl Genet 76:809–814
Kawakami SI, Ebana K, Nishikawa T, Sato YI, Vaughan DA, Kadowaki KI (2007) Genetic variation in the chloroplast genome suggests multiple domestication of cultivated Asian rice (*Oryza sativa L.*). Genome 50:180–187

Khraiwesh B, Zhu JK, Zhu J (2012) Role of miRNAs and siRNAs in biotic and abiotic stress responses of plants. Biochem Biophys Acta 1819:137–48

Kikuchi S, Satoh K, Nagata T (2003) Collection, mapping, and annotation of over 28,000 cDNA clones from japonica rice. Science 300:1566–1569

Kim H, Hurwitz B, Yu Y, Collura K, Gill N, SanMiguel P, Mullikin JC, Maher C, Nelson W, Wissotski M, Braidotti M, Kudrna D, Goicoechea JL, Stein L, Ware D, Jackson SA, Soderlund C, Wing RA (2008) Construction, alignment and analysis of twelve framework physical maps that represent the ten genome types of the genus *Oryza*. Genome Biol. https://doi.org/10.1186/gb-2008-9-2-r45

Kim HJ, Jung J, Singh N, Greenberg A, Doyle JJ, Tyagi W, Chung JW, Kimball J, Hamilton RS, McCouch SR (2016) Population dynamics among six major groups of the *Oryza rufipogon* species complex, wild relative of cultivated Asian rice. Rice 9:56

Kobayashi N, Ikeda R, Domingo IT, Vaughan DA (1993) Resistance to infection of rice tungro viruses and vector resistance in wild species of rice (*Oryza* spp.) Jpn J Breed 43:377–387

Komatsu S, Tanaka N (2004) Rice proteome analysis: A step toward functional analysis of the rice genome. Proteomics 4:938–949

Koseki M, Kitazawa N, Yonebayashi S, Maehara Y, Wang ZX, Minobe Y (2010) Identification and fine mapping of a major quantitative trait locus originating from wild rice, controlling cold tolerance at the seedling stage. Mol Genet Genomics 284:45–54

Kovach MJ, Sweeney MT, McCouch SR (2007) New insights into the history of rice domestication. Trends Genet 23:578–587

3K RGP (2014) The 3000 rice genomes project. GigaScience 3:7

Kunze R, Saedler H, Lönnig WE (1997) Plant transposable elements. In: Callow JA (ed) Advances in botanical research, vol 27. Academic Press, San Diego/London, pp 331–470

Li J, Thomson M, McCouch SR (2004) Fine mapping of a grain-weight quantitative trait locus in the pericentromeric region of rice chromosome 3. Genetics 168 (4):2187–2195

Li R, Li L, Wei S, Wei Y, Chen Y, Bai D, Yang L, Huang F, Lu W, Zhang X, Li X, Yang X, Wei Y (2006) The evaluation and utilization of new genes for brown planthopper resistance in common wild rice (*Oryza rufipogon* Griff.). Mol Plant Breed 4:365–371

Liang J, Zhou M, Zhou X, Jin Y, Xu M, Lin J (2013) JcLEA, a novel LEA-like protein from Jatropha curcas, confers a high level of tolerance to dehydration and salinity in Arabidopsis thaliana. PLoS One 8(12): e83056. https://doi.org/10.1371/journal.pone.0083056

Liu B, Li P, Li X, Liu C, Cao S, Chu C, Cao X (2005) Loss of function of *OsDCL1* affects microRNA accumulation and causes developmental defects in rice. Plant Physiol 139:296–305

Liu TM, Mao DH, Zhang SP, Xing YZ (2009) Fine mapping *SPP1*, a QTL controlling the number of spikelets per panicle, to a BAC clone in rice (*Oryza* sativa). Theor Appl Genet 118:1509–1517

Londo JP, Chiang YC, Hung KH, Chiang TY, Schaal BA (2006) Phylogeography of Asian wild rice, *Oryza rufipogon,* reveals multiple independent domestications of cultivated rice, *Oryza sativa*. Proc Nat Acad Sci 103:9578–83

Lu H, Liu Z, Wu N, Berné S, Saito Y, Liu B, Wang L (2002) Rice domestication and climatic change: phytolith evidence from East China. Boreas 31:378–385

Lu T, Yu S, Fan D, Mu J, Shangguan Y, Wang Z, Minobe Y, Lin Z, Han B (2008) Collection and comparative analysis of 1888 full-length cDNAs from wild rice *Oryza rufipogon* Griff. W1943. DNA Res 15:285–295

Luo YC, Hui Z, Yan L, Jun-Yu C, Jian-Hua Y, Yue-Qin C, Liang-Hu Q (2006) Rice embryogenic calli express a unique set of microRNAs, suggesting regulatory roles of microRNAs in plant post-embryogenic development. FEBS Lett 580:5111–5116

Luo X, Ji SD, Yuan PR, Lee HS, Kim DM, Balkunde S, Kang JW, Ahn SN (2013) QTL mapping reveals a tight linkage between QTLs for grain weight and panicle spikelet number in rice. Rice 6:33

Luo XD, Jun Z, Dai LF, Zhang FT, Yi Z, Yong W, Xie JK (2016) Linkage map construction and QTL mapping for cold tolerance in *Oryza rufipogon* Griff. at early seedling stage. J Intg Agri 15:2703–2711

Ma J, Devos KM, Bennetzen JL (2004) Analyses of LTR-retrotransposon structures reveal recent and rapid genomic DNA loss in rice. Genome Res 14:860–869

Maas LF, Mcclung A, McCouch S (2010) Dissection of a QTL reveals an adaptive, interacting gene complex associated with transgressive variation for flowering time in rice. Theor Appl Genet 120:895–908

Mandal N, Gupta S (1997) Anther culture of an interspecific rice hybrid and selection of fine grain type with submergence tolerance. Plant Cell, Tissue Organ Cult 51:79–82

Mao Y, Ge X, Frank GL, Madsion JM, Koehl- ner AN, Doud MK, Tassa C, Berry EM, Soda T, Singh KK (2009) Disrupted in schizophrenia 1 regulates neuronal progenitor proliferation via modulation of GSK3β/β-Catenin signaling. Cell 136:1017–1031

Marri PR, Sarla N, Reddy LV, Siddiq EA (2005) Identification and mapping of yield and yield related QTL from an Indian accession of *Oryza rufipogon*. BMC Genet 6:33

Mather KA, Caicedo AL, Polato NR, Kenneth M. Olsen KM, McCouch S, Purugganan MD (2007) The Extent of Linkage Disequilibrium in Rice *(Oryza sativa L.)*. Genetics 177:2223–2232

McCouch SR, Sweeney M, Li J, Jiang H, Thomson M, Septiningsih E, Edwards J, Moncada P, Xiao J, Garris A, Tai T, Martinez C, Tohme J, Sugiono M, McClung A, Yuan L, Ahn SN (2007) Through the genetic bottleneck: *O. rufipogon* as a source of trait-enhancing alleles for *O. sativa*. Euphytica 154:317–339

Miller KG, Kominz MA, Browning JV, Wright JD, Mountain GS, Katz ME, Sugarman PJ, Cramer BS, Christie-Blick N, Pekar SF (2005) The Phanerozoic record of global sea-level change. Science 310:1293–1298

Molina J, Sikora M, Garud N, Flowers JM, Rubinstein S, Reynolds A, Huang P, Jackson S, Schaal BA, Bustamante CD, Boyko AR (2011) Molecular evidence for a single evolutionary origin of domesticated rice. Proc Nat Acad Sci 10:8351–8356

Moncada P, Martinez CP, Borrero J, Chatel M, Gauch H, Guimaraes E, Tohme J, McCouch SR (2001) Quantitative trait loci for yield and yield components in an *Oryza* sativa x *Oryza rufipogon* BC2F2 population evaluated in an upland environment. Theor Appl Genet 102(1):41–52

Moody K (1989) Weeds reported in Rice in South and Southeast Asia. International Rice Research Institute, Manila, Philippines

Morishima H, Sano Y (1992) Evolutionary studies in cultivated rice. Oxford Surveys Evol Biol 8:135

Morishima H, Sano Y, Oka HI (1984) Differentiation of perennial and annual types due to habitat conditions in the wild rice *Oryza perennis*. Plant Syst Evol 144:119–135

Nakamura I, Urairong H, Kameya N, Fukuta Y, Chitrakon S, Sato YI (1998) Six different plastid subtypes were found in *O. sativa-O. rufipogon* complex. Rice Genet Newslett 15:80–82

Ogawa S, Valencia MO, Lorieux M, Arbelaez JD, McCouch S, Ishitani M, Selvaraj MG (2016) Identification of QTLs associated with agronomic performance under nitrogen-deficient conditions using chromosome segment substitution lines of a wild rice relative, *Oryza rufipogon*. Acta Physiol Plant 38:1–10

Ohyanagi H, Ebata T, Huang X, Gong H, Fujita M, Mochizuki T, Toyoda A, Fujiyama A, Kaminuma E, Nakamura Y, Feng Q, Wang ZX, Han B, Kurata N (2015) *Oryza* Genome: genome diversity database of Wild *Oryza* Species. Plant Cell Physiol 57:1–7

Oka HI and Chang WT (1959) The impact of cultivation on populations of wild rice, *Oryza sativa f. spontanea*. Phyton 13:105–17

Pang HH, Wang XK (1996) A study on the annual *O. rufipogon* Griff. in China. Crop Genetic Resources 3:8–11

Park KC, Kim NH, Cho YS, Kang KH, Lee JK, Kim NS (2003) Genetic variations of AA genome *Oryza* species measured by MITE-AFLP. Theor Appl Genet 107:203–209

Qiao W, Qi L, Cheng Z, Su L, Li J, Sun Y, Ren J, Zheng X, Yang Q (2016) Development and characterization of chromosome segment substitution lines derived from *Oryza rufipogon* in the genetic background of *O. sativa* spp. *indica* cultivar 9311. BMC Genom 17:580

Rakshit S, Rakshit A, Matsumura H, Takahashi Y, Hasegawa Y, Ito A, Ishii T, Miyashita NT, Terauchi R (2007) Large-scale DNA polymorphism study of *Oryza sativa* and *O. rufipogon* reveals the origin and divergence of Asian rice. Theor Appl Genet 114:731–743

Ren F, Lu BR, Li S, Huang J, Zhu Y (2003) A comparative study of genetic relationships among the AA-genome *Oryza* species using RAPD and SSR markers. Theor Appl Genet 108:113–120

Sang T, Ge S (2007) The puzzle of rice domestication. J Integr Plant Biol 49:760–768

SanMiguel P, Tikhonov A, Jin Y-K, Motchoulskaia N, Zakharov D, Melake-Berhan A (1996) Nested retrotransposons in the intergenic regions of the maize genome. Science 274:765–768

Sarkar S, Bhattacharyya S, Gantait S (2017) Cytological analysis for meiotic patterns in wild rice (*Oryza rufipogon* Griff.). Biotechnol Reports 13:26–29

Second G (1982) Origin of the genic diversity of cultivated rice (*Oryza* spp.): study of the polymorphism scored at 40 isozyme loci. Jap J Genet 57:25–57

Septiningsih EM, Prasetiyono J, Lubis E, Tai TH, Tjubaryat T, Moeljopawiro S, McCouch SR (2003) Identification of quantitative trait loci for yield and yield components in an advanced backcross population derived from the *Oryza sativa* variety IR64 and the wild relative O. rufipogon. Theor Appl Genet 107(8):1419–1432

Song BK, Waugh R, Marshall D, Nadarajah K, Ratnam W (2011) Comparative physical mapping using overgo-tagged site reveals strong conservation of synteny between cultivated and common wild rice in the 1.2 Mb yld1. 1 region. Asia-Pacific J Mol Biol Biotechnol 19:157–168

Sun C, Wang X, Yoshimura A, Doi K (2002) Genetic differentiation for nuclear, mitochondrial and chloroplast genomes in common wild rice (*Oryza rufipogon Griff.*) and cultivated rice (*Oryza sativa L.*). Theor Appl Genet 104: 1335–1345

Sunkar R, Zhu JK (2004) Novel and stress-regulated microRNAs and other small RNAs from Arabidopsis. Plant Cell. 16:2001–2019

Sunkar R, Girke T, Jain PK, Zhu JK (2005) Cloning and characterization of microRNAs from rice. Plant Cell 17:1397–1411

Sunkar R, Li YF, Jagadeeswaran G (2012) Functions of microRNAs in plant stress responses. Trends Plant Sci 17:196–203

Thalapati S, Batchu AK, Neelamraju S, Ramanan R (2012) Os11Gsk gene from a wild rice, *Oryza rufipogon* improves yield in rice. Funct Integr Genomics 12:277–289

Tian F, Li DJ, Fu Q, Zhu ZF, Fu YC, Wang XK, Sun CQ (2006) Construction of introgression lines carrying wild rice (*Oryza rufipogon* Griff.) segments in cultivated rice (*Oryza sativa* L.) background and characterization of introgressed segments associated with yield-related traits. Theor Appl Genet 112:570–580

Tian X, Long Y, Wang J, Zhang J-w, Wang Y, Li W, Peng Yu, Yuan Q, Pei X (2015) De novo transcriptome assembly of common wild rice (*Oryza rufipogon Griff.*) and discovery of drought-response genes in root tissue based on transcriptomic data. PLoS One. 10: e0131455. https://doi.org/10.1371/journal.pone.0131455

Utami DW, Moeljopawiro S, Aswidinnoor H, Setiawan A, Hanarida I (2008) Blast resistance genes in wild rice *Oryza rufipogon* and rice cultivar IR64. Indones J Agr 1:71–76

Vaughan DA, Lu BR, Tomooka N (2008) The evolving story of rice evolution. Plant Sci 174:394–408

Vitte C, Ishii T, Lamy F, Brar D, Panaud O (2004) Genomic paleontology provides evidence for two distinct origins of Asian rice (*Oryza sativa L.*). Mol Genet Genomics 272:504–511

Wang Y, Bai X, Yan C, Gui Y, Wei X, Zhu QH, Guo L, Fan L (2012) Genomic dissection of small RNAs in wild rice (*Oryza rufipogon*): lessons for rice domestication. New Phytol 196:914–925

Waters DL, Nock CJ, Ishikawa R, Rice N, Henry RJ (2012) Chloroplast genome sequence confirms distinctness of Australian and Asian wild rice. Ecology and evolution 2:211–217

Wing R, Kim H, Goicoechea J, Yu Y, Kudrna D, Zuccolo A, Ammiraju J, Luo M, Nelson W, Ma J, SanMiguel P (2007) The *Oryza* map alignment project (OMAP): a new resource for comparative genome studies within *Oryza*. Rice Funct Genomic 395–409

Wu J, Maehara T, Shimokawa T, Yamamoto S, Harada C (2002) A comprehensive rice transcript map containing 6591 expressed sequence tag sites. Plant Cell. 14:525–535

Xiao JH, Li JM, Grandillo S, Ahn SN, Yuan LP, Tanksley SD, McCouch SR (1998) Identification of trait-improving quantitative trait loci alleles from a wild rice relative. Oryza rufipogon. Genetics 150 (2):899–909

Xiao N, Sun G, Hong Y, Xia R, Zhang C, Su Y, Chen J (2011) Cloning of genome-specific repetitive DNA sequences in wild rice (*O. rufipogon Griff.*), and the development of Ty3-gypsy retrotransposon-based SSAP marker for distinguishing rice (*O. sativa L.*) *indica* and *japonica* subspecies. Genet Resour Crop Evol 58:1177–1186

Xie X, Song MH, Jin F, Ahn SN, Suh JP, Hwang HG, McCouch SR (2006a) Fine mapping of a grain weight quantitative trait locus on rice chromosome 8 using near-isogenic lines derived from a cross between *Oryza sativa* and *Oryza rufipogon.* Theor Appl Genet 113(5):885–894

Xie Z, Wang J, Cao M, Zhao C, Zhao K, Shao J, Lei T, Xu N, Liu S (2006b) Pedigree analysis of an elite rice hybrid using proteomic approach. Proteomics 6:474–486

Xie X, Jin F, Song MH, Suh JP, Hwang HG, Kim YG, McCouch SR, Ahn SN (2008) Fine mapping of a yield-enhancing QTL cluster associated with transgressive variation in an *Oryza sativa* x *O. rufipogon* cross. Theor Appl Genet 116(5):613–622

Xu X, Liu X, Ge S, Jensen JD, Hu F, Li X, Dong Y, Gutenkunst RN, Fang L, Huang L, Li J, He W, Zhang G, Zheng X, Zhang F, Li Y, Yu C, Kristiansen K, Zhang X, Wang J, Wright M, McCouch S, Nielsen R, Wang J, Wang W (2012) Resequencing 50 accessions of cultivated and wild rice yields markers for identifying agronomically important genes. Nat Biotechnol 30:105–111

Yang H, Hu L, Hurek T, Hurek BR (2010) Global characterization of the root transcriptome of a wild species of rice, *Oryza longistaminata*, by deep sequencing. BMC Genom 11:705. https://doi.org/10.1186/1471-2164-11-705

Yang CC, Kawahara Y, Mizuno H, Wu J, Matsumoto T, Itoh T (2012) Independent domestication of Asian rice followed by gene flow from *japonica* to *indica*. Mol Biol Evol 29:1471–1479

Yun YT, Chung CT, Lee YJ, Na HJ, Lee JC, Lee SG, Lee KW, Yoon YH, Kang JW, Lee HS, Lee JY (2016) QTL mapping of grain quality traits using introgression lines carrying *Oryza rufipogon* chromosome segments in *japonica* rice. Rice 9:62

Zhang Q, Lin SC, Zhao BY, Wang CL, Yang WC, Zhou YL, Li DY, Chen CB, Zhu LH (1998) Identification and tagging a new gene for resistance to bacterial blight (*Xanthomonas oryzae* pv. *oryzae*) from *O. rufipogon*. Rice Genet Newsl 15: 138–142

Zhang J, Feng Q, Jin C, Qiu D, Zhang L (2005) Features of the expressed sequences revealed by a large-scale analysis of ESTs from a normalized cDNA library of the elite indica rice cultivar Minghui 63. Plant J 42:772–780

Zhang X, Zhou S, Fu Y, Su Z, Wang X, Sun C (2006) Identification of a drought tolerant introgression line derived from Dongxiang common wild rice (*O. rufipogon* Griff.). Plant Mol Biol 62:247–259

Zhang F, Luo X, Zhou Y, Xie J (2015) Genome-wide identification of conserved microRNA and their response to drought stress in Dongxiang wild rice (*Oryza rufipogon* Griff.). Biotechnol. https://doi.org/10.1007/s10529-015-2012-0

Zhang F, Xu T, Mao L, Yan S, Chen X, Wu Z, Chen R, Luo X, Xie J, Gao S (2016) Genome-wide analysis of Dongxiang wild rice (*Oryza rufipogon* Griff.) to investigate lost/acquired genes during rice domestication. Plant Biol 16:103

Zhang J, Long Y, Xue M, Xiao X, Pei X (2017) Identification of microRNAs in response to drought in common wild rice (*Oryza rufipogon* Griff.) shoots and roots. PLoS One 12:e0170330. https://doi.org/10.1371/journal.pone.0170330

Zhou Y, Tang J, Walker MG, Zhang X, Wang J (2003) Gene identification and expression analysis of 86,136 Expressed Sequence Tags (EST) from the rice genome. Genomics Proteomics Bioinform 1:26–42

Zhou Y, Yang P, Cui F, Zhang F, Luo X, Xie J (2016) Transcriptome analysis of salt stress responsiveness in the seedlings of Dongxiang wild rice (*Oryza rufipogon* Griff.). PLoS One 11(1):0146242. https://doi.org/10.1371/journal.pone.0146242

An Account of Unclassified Species (*Oryza schlechteri*), Subspecies (*Oryza indandamanica* Ellis and *Oryza sativa* f. *spontanea* Baker), and Ortho-group Species (*Leersia perrieri*) of *Oryza*

26

Apurva Khanna, Ranjith Kumar Ellur, S. Gopala Krishnan, Tapan K. Mondal and Ashok Kumar Singh

Abstract

The *Oryza* genus comprises of 24 well-recognized, morphologically distinct species that are classified into four different major complexes, consisting of ten different types of genomes. Apart from that, there are some species which are considered to be subspecies due to some similarities with the existing species and lack of characters that clearly distinguish them from well-defined species. Wild relatives of rice are untapped reservoirs of valuable genes which can combat various stresses and improve the productivity of cultivated rice. *Oryza* L. and *Leersia* Sw. are the two largest genera of the Oryzae tribe. *Oryza schlechteri* Pilger is an allotetraploid ($2n = 48$) with HHKK genome of 1568 Mb and is distributed throughout the Finisterre Mountains of Papua New Guinea. *Leersia perrieri* is a diploid species from the nearest out-group of *Oryza* with an estimated genome size of 323 Mb which is found in wet and marshlands of Africa and Madagascar. The genome sequence of *L. perrieri* has been assembled. Both these species are perennials and are stoloniferous in plant habit. On the other hand, *O. indandamanica* is morphologically quite similar to *O. meyeriana,* while *O. sativa* f. *spontanea* is a subspecies of *O. sativa.* However, the systematic study needs to be carried out to identify the unique characteristics of these species for their effective utilization.

26.1 Introduction

Oryzeae (Poaceae) encompassing approximately 12 genera and less than 70 species is the largest tribe in the subfamily Ehrhartoideae (Clayton and Renvoize 1986; Vaughan 1994). Of the 12 genera in Oryzeae tribe, *Oryza* L. with 23 species and *Leersia* Sw. with 18 species form the largest genera with pantropical distributions (Watson and Dallwitz 1999; Terrell et al. 2001). *Oryza schlechteri* Pilger is an allotetraploid ($2n = 48$) (Naredo and Vaughan 1992) member of the tertiary gene pool with HHKK genome (Sanchez et al. 2014). The genome of *O. schlechteri* is highly diverged from the rest of the genomes of *Oryza* (Aggarwal et al. 1996).

L. perrieri is a diploid species from the nearest out-group of Oryza. The genome size of *L. perrieri* is estimated to be 323 Mb. Based on mitochondrial gene (*matK* and *GPA1*) sequence

A. Khanna · R. K. Ellur · S. Gopala Krishnan
A. K. Singh (✉)
Division of Genetics, ICAR-Indian Agricultural Research Institute, Pusa, New Delhi 110 012, India
e-mail: aks_gene@yahoo.com

T. K. Mondal
ICAR-National Research Centre on Plant Biotechnology, Pusa, New Delhi 110 012, India

T. K. Mondal and R. J. Henry (eds.), *The Wild Oryza Genomes*,
Compendium of Plant Genomes, https://doi.org/10.1007/978-3-319-71997-9_26

analysis, the *Oryza–Leersia* clade diverged from the remaining genera at approximately 20.5 to 22.1 MYA while *Oryza* and *Leersia* separation happened around approximately 14.2 MYA (Gua and Ge 2005).

O. indandamanica is argued as a separate species (Sharma et al. 1988) while others are considered to be subspecies due to high morphological similarity to *O. meyeriana* (Zoll. & Mor.) Baill. var. *meyeriana* (Backer s.n.) and *O. meyeriana* var. *granulate* (Watt). It is a diploid species with 24 chromosomes. *O. sativa* f. *spontanea* or weedy rice is also a diploid with 24 chromosomes and is classified taxonomically in the same species as cultivated rice (*O. sativa*), but shows significant variation with respect to high seed shattering and dormancy. Spontaneous crossing with rice during rice cultivation might have evolved this weedy rice. In areas, where it coexists with *O. rufipogon*, it is genetically similar to wild ancestors due to spontaneous crossing (Oka 1988; Vaughan et al. 2003), whereas in other areas, it is genetically more aligned to cultivated rice (Cao et al. 2006).

26.2 Distribution

The first collection of *O. schlechteri* was made by R. Schlechteri in 1907, at the Jamu Gorge in the Finisterre Mountains of Papua New Guinea, and later on, it was collected from Irian Jaya, Indonesia, in 1912, by A.A. Pulle. It was also collected by K.J. White in the Kikori District of the Gulf Province, Papua New Guinea, while Aro again reported a second largest population of *O. schlechteri* at higher elevation in the Finisterre Mountains in 1991. It has a unique distribution pattern in hotwet sites (Atwell et al. 2014), undisturbed forest areas, and shaded or semi-shaded regions (Vaughan 1994; Fig. 26.1). *Leersia perrieri (*previously known as *A. Camus)* Launertisa, is a diploid species of the genus *Leersia,* which usually occurs in the wet and marshlands of Africa and Madagascar (Raimondo et al. 2009).

O. indandamanica was originally discovered in Rutland (11.50°N, 92.25°E), Andaman Islands, India (Sharma et al. 1988) although thereafter, it is reported in several other parts of the world. On the other hand, *O. sativa* f. *spontanea* has been reported to be distributed in the rice-growing areas across the world (Vaughan et al. 2001; Gealy et al. 2002). Using population structure analyses of South Asian and US weedy rice, Huang et al. (2017) showed that weeds in South Asia have highly heterogeneous genetic backgrounds, with ancestry contributions from both cultivated varieties (*aus* and *indica*) and wild rice. Moreover, the two main groups of weedy rice in the USA, which are also related to *aus* and *indica* cultivars, constitute a separate

(a) **(b)**

Fig. 26.1 Geographical distribution of **a** *O. schlechteri* and **b** *L. perrieri*

origin from those of Asian weeds. Weedy rice populations in South Asia largely converge in some traits with the presence of red pericarps and awns and ease of shattering (Huang et al. 2017).

26.3 Botanical Description

O. schlechteri is vegetatively similar to *Leersia* sp. with a perennial (Menguer et al. 2017), stoloniferous, and profusely scrambling plant habit (Naredo et al. 1993). It has a short plant stature (30–50 cm) with pubescent nodes from which roots are readily formed. It is a shade-loving plant with short, narrow leaves (leaf length 30.4 cm and leaf width of 0.7 cm), high vein density (6.4 mm^{-1}), closer vein spacing (153.43 μm) with relatively small-sized mesophyll cells (90.12 μm). The mesophyll cells of *O. schlechteri* are without cell wall lobing and possess wider mesophyll cells (9.7 μm) in longitudinal axis with six mesophylls between two adjacent minor veins (Chatterjee et al. 2016). It has a glabrous cupule with unpaired sterile lemma extending across the anterior part of the palea with no siliceous triads and epidermal striations on the spikelets (Naredo et al. 1993). This is a distinguishing feature due to which O. schlechteri has been included in Oryza rather than Leersia.

The plants of *L. perrieri* are stoloniferous in nature with a profusely scrambling habit (Sanchez et al. 2014; Naredo et al. 1993). It has a short culm length (39 cm), moderate tillering habit (10 tillers per plant), tiny panicles (3.5 cm) with a very small flag leaf (length 2.5 cm and width 0.4 cm). It possesses a very small awned spikelet with a glume length of 3.5 mm and glume width of 1.5 mm but lacks sterile lemma (Katayama 1995). It has uniform lemmatal and paleal abaxial epidermal surfaces with two-celled micro-hairs, prickle hairs, siliceous tubercles, and orbicular bodies (Naredo et al. 1993).

O. indandamanica is a perennial, herbaceous species which grows on rocky slopes over light loam soil, on stream banks. It bears ~10 tillers/plant with an average shoot weight of 3.8 g (fresh) and root weight of 0.8 g (fresh). The mean basal internode diameter is 1.2 cm, while the average leaf area per plant is 0.24 m^2. Panicles are unbranched producing an average of eight awnless spikelets/panicles. It possesses four leaves per culm, and the flag leaf is 4.53 cm long and 0.50 cm broad (Sharma et al. 1988).

Morphologically, *O. sativa* f. *spontanea* is very similar to cultivated rice except for the seed shattering and seed dormancy. Prathepha (2009a) reported wide variation for seed morphological traits, including amylose content in this weedy rice, validating the fact that they are genetically diverse.

26.4 Nuclear and Chloroplast Genome Sequence

O. schlechteri is an allotetraploid with $2n = 48$ and an estimated genome size 1568 Mb. However, genome sequencing has not been carried out. *L. perrieri* is the nearest out-group of the genus Oryza with 12 chromosomes with an estimated genome size of 323 Mb. A total of 266.7 Mb of *L. perrieri* genome sequence has been assembled (Copetti et al. 2015) and has been predicted to possess 40,521 genes (Kersay et al. 2016). Genome-wide analysis of transposable elements showed that *L. perrieri* comprises of 810 long terminal repeat retro-transposons, 6 terminal repeat retro-transposons in miniature, 110 SINEs, 115 MULE DNA transposons, and 196 Helitrons (Copetti et al. 2015). The chloroplast genome of *L. perrieri* is 136196 bp with a pair of inverted repeat regions of 21321 bp each. The overall GC content of the circular genome is 38.9% with 70 genes located on the negative strand and 60 genes located on the positive strand. Among the 130 annotated genes, 85 genes are protein coding, 37 genes code for tRNA, and 8 genes code for rRNA. Interestingly, certain intron containing genes, namely *atpF, petD, petB, rpl2, ndhB, ndhA, rpl16, rps12, rps16,* and *ycf3* were identified (Hutang and Gao 2017).

BADH2 homologous to betaine aldehyde dehydrogenase is responsible for aroma in fragrant rice varieties. A deletion leading to recessive mutation of BADH2 allele encoding betaine aldehyde dehydrogenase induces the synthesis of

2-acetyl-1- pyrroline (2AP), a potent flavor component in rice fragrance. Prathepha (2009b) reported the defunct allele of BADH2 gene in the fragrant weedy rice from a population of 215 plants in the rice-growing area of northeastern Thailand.

The sizes of the genomes of *O. indandamanica* and *O. sativa* f. *spontanea* are not known, and their decoding is yet to be done. But being a subspecies of *O. meyeriana* and *O. sativa,* it is suggested that their genome size may be approximately around 727 and 420 Mb, respectively.

26.5 Importance in Plant Breeding

O. schlechteri is prolific in its natural habitat, but it does not set seeds in other than its natural habitat. No attempts have been made to cross *O. schlechteri* with any of the *Oryza* species. *L. perrieri* was earlier included in the genus *Oryza* (Tateoka 1963); however, it was reclassified and shifted to the genus *Leersia* (Launert 1965). The F_1 seeds from the crosses between *O. punctata* and *L. perrieri* as well as *O. latifolia* and *L. perrieri* were reported to be trihaploid with 36 chromosomes and showed no genomic relationship with any of the *Oryza* species (Katayama 1995). *L. perrieri* is a shade-tolerant genotype, and further systematic study is needed to identify the unique characteristics of these species for their utilization in introgression programs.

O. sativa f. *spontanea* is known to flower simultaneously across many rice growing regions, and spontaneous hybridization between them is more frequent (Langevin et al. 1990; Chen et al. 2004; Burgos et al. 2008; Shivrain et al. 2008; Krishnan et al. 2014). Genetic diversity of weedy rice populations was assessed by several workers with or without reporting their spatial distribution pattern (Vaughan et al. 2001; Yu et al. 2005; Cao et al. 2006; Londo and Schaal 2007; Xi et al. 2011; Wongtamee et al. 2015). Gealy et al. (2009) reported that the genetic diversity and structure of weedy rice are associated with geographic differentiation of USA. Genetic diversity of 99 weedy rice accessions from four populations in Thailand and Laos showed that while there was genetic variability among weedy rice individuals was more than variability among populations in all the populations (Prathepha et al. 2011). The genotypic and phenotypic variation indicated that high levels of genetic diversity as well as a wide range of variation in morphology of weedy rice may be due to frequent spontaneous crossing between common wild rice and cultivated rice. Continuous coexistence of the weedy forms with cultivated rice aids a situation similar to backcrossing leading to reduced genetic diversity and genetic variation in weedy rice, resulting in the genetic convergence and formation of region-specific structured weedy rice populations based on their co-cultivated rice varieties.

In conclusion, the work on these species is just at the initial stage. Although none of them are domesticated species, they are important. While being tetraploid *O. schlechteri* and the nearest out-group species, *L. perrieri,* these two species can be valuable biological material for evolutionary studies of *Oryza* species. The other two species, being members of primary gene pool of rice, may be useful for transferring useful traits into cultivated species. However, their detailed characterization to identify potential traits is needed together with "omics"-based study to decode their genome.

References

Aggarwal RK, Brar DS, Khush GS, Jackson MT (1996) *Oryzaschlechteri* Pilger has a distinct genome based on molecular analysis. Rice Genet Newslett 13:58–59

Atwell BJ, Wang H, Scafaro AP (2014) Could abiotic stresstolerance in wild relatives of rice be used to improve *Oryza sativa*? Plant Sci 215–216:48–58

Burgos NR, Norsworthy JK, Scott RC, Smith KL (2008) Red rice (*Oryzasativa*) status after 5 years of imidazolinone-resistant rice technology in Arkansas. Weed Technol 22:200–208

Cao QJ, Lu BR, Xia H, Rong J, Sala F et al (2006) Genetic diversity and origin of weedy rice (*Oryzasativa* f. spontanea) populations found in North-eastern China revealed by simple sequence repeat (SSR) markers. Ann Bot-London 98:1241–1252

Chatterjee J, Dionora J, Elmido-Mabilangan A, Wanchana S, Thakur V, Bandyopadhyay A,

Brar DS, Quick WP (2016) The evolutionary basis of naturally diverse rice leaves anatomy. PLoS ONE 11 (10):e0164532

Chen LJ, Lee DS, Song ZP, Suh HS, Lu BR (2004) Gene flow from cultivated rice (*Oryzasativa*) to its weedy and wild relatives. Ann Bot-London 93:67–73

Clayton WD, Renvoize SA (1986) Genera Graminum. Kew Bull Addit Ser XIII:1–389

Copetti D, Zhang J, Baidouri ME, Gao D, Wang J, Barghini E, Cossu RM, Angelova A, Maldonado LCE, Roffler S, Ohyanagi H, Wicker T, Fan C, Zuccolo A, Chen M, Oliveira, Bin Han AC, Henry R, Hsing Y, Kurata N, Wang W, Jackson SA, Panaud O, Wing RA (2015) RiTE database: a resource database for genus-wide rice genomics and evolutionary biology. BMC Genom 16:538

Gealy DR, Agrema HA, Eizenga GC (2009) Exploring genetic and spatial structure of U.S. weedy red rice (*Oryza sativa*) in relation to rice relatives worldwide. Weed Sci 57:627–643

Gealy DR, Tai TH, Sneller CH (2002) Identification of red rice, rice, and hybrid populations using microsatellite markers. Weed Sci 50:333–339

Guo Y, Ge S (2005) Molecular phylogeny of Oryzeae (poaceae) based on DNA sequences from chloroplast, mitochondrial, and nuclear genomes. Amer J Bot 92 (9):1548–1558

Huang Z, Young ND, Reagon M, Hyma KE, Olsen KM, Jia Y, Caicedo AL (2017) All roads lead to weediness: patterns of genomic divergence reveal extensive recurrent weedy rice origins from South Asian Oryza. Mol Eco. https://doi.org/10.1111/mec.14120

Hutang G, Gao L (2017) The complete chloroplast genome sequence of *Leersia perrieri* of the rice tribe Oryzeae, Poaceae. Conserv Genet Resour https://doi.org/10.1007/s12686-017-0729-x

Katayama (1995) Cytological studies on the genus oryza. XIV. Intergeneric hybridizations between tetraploid Oryza species and diploid *Leersia* species. Jpn J Genet 70:47–55

Kersey PJ, Allen JE, Armean I, Boddu S, Bolt BJ, Carvalho-Silva D, Christensen M, Davis P, Falin LJ, Grabmueller C, Humphrey J, Kerhornou A, Khobova J, Aranganathan NK, Langridge N, Lowy E, McDowall MD, Maheswari U, Nuhn M, Ong CK, Overduin B, Paulini M, Pedro H, Perry E, Spudich G, Tapanari E, Walts B, Williams G, Tello-Ruiz M, Stein J, Wei S, Ware D, Bolser DM, Howe KL, Kulesha E, Lawson D, Maslen G, Staines DM (2015) Ensembl Genomes 2016: moregenomes, morecomplexity. Nucleic Acids Res 44(D1):D574–D580

Krishnan SG, Waters DLE, Henry RJ (2014) Australian wild rice reveals pre-domestication origin of polymorphism deserts in rice genome. PLoS ONE 9(6): e98843. https://doi.org/10.1371/journal.pone.0098843

Langevin SA, Clay K, Grace JB (1990) The incidence and effects of hybridization between cultivated rice and its related weed red rice (*Oryzasativa* L). Evolution 44:1000–1008

Launert E (1965) A survey of the genus *Leersia* in Africa. Senckenb Biol 46(2):29–153

Londo JP, Schaal BA (2007) Origins and population genetics of weedy red rice in the USA. MolEcol 16:4523–4535

Menguer PK, Sperotto RA, Ricachenevsky FK (2017) A walk on the wild side: Oryza species as source for rice abiotic stress tolerance. Genet MolBiol 40(1):238–252 (Suppl)

Naredo E, Vaughan DA (1992) The chromosome number of *Oryza schlechteri* Pilger. Int Rice Res Newsl 17:5

Naredo MEB, Vaughan DA, Cruz FS (1993) Comparative spikelet morphology of *OryzaschlechteriPilger* and related species of *Leersia* and *Oryza* (Poaceae). J Plant Res 106:109–112

Oka HI (1988) Origin of cultivated rice. Elsevier, Amsterdam

Prathepha P (2009a) Seed morphological traits and genotypic diversity of weedy rice (*Oryza sativa f. spontanea*) populations found in the Thai Hom Mali rice fields of north-eastern Thailand. Weed Biol Manag 9:1–9

Prathepha P (2009b) The *badh2* allele of the fragrance (*fgr*/BADH2) gene is present in the gene population of weedy rice (*Oryza sativa* f. *spontanea*) from Thailand. Am-Eurasian J Agricand Environ Sci 5:603–608

Prathepha P (2011) Microsatellite analysis of weedy rice (*Oryza sativa* f. *spontanea*) from Thailand and Lao PDR. Australian. J Crop Sci 5(1):49–54

Raimondo D, Staden L, Foden W, Victor J, Helme N, Turner R, Manyama P (2009) Red list of South African plants. Strelitzia 25:19–40

Sanchez PL, Wing RA, Brar DS (2014) The wild relative of rice: genomes and genomics. In: Zhang Q, Wing RA (eds) Genetics and genomics of rice, plant genetics and genomics: crops and models. Springer Science Business Media, New York, pp 9–25. https://doi.org/10.1007/978-1-4614-7903-1_2

Sharma TVRS, Majumder ND, Ram T, Mandal AB (1988) Characteristics of *Oryza indandamanica* Ellis, a newly discovered wild species. 4 IRRN 13:5

Shivrain VK, Burgos NR, Gealy DR, Moldenhauer KAK, Baquireza CJ (2008) Maximum outcrossing rate and genetic compatibility between red rice (*Oryzasativa*) biotypes and clearfield (TM) rice. Weed Sci 6:807–813

Tateoka T (1963) Taxonomic studies of *Oryza*. III. Key to the species and their enumeration. Bot Mag Tokyo 76:165–173

Terrell EE, Peterson PM, Wergin WP (2001) Epidermal features and spikelet micromorphology in *Oryza* and related genera (Poaceae: Oryzeae). Smithson Contrib Botany 91:1–50

Vaughan DA (1994) The wild relative of rice: a genetic resources handbook. International Rice Research Institute, Manila

Vaughan DA, Morishima H, Kadowaki K (2003) Diversity in the *Oryza* genus. Curr Opin Plant Biol 6:139–146

Vaughan LK, Ottis BV, Prazak-Havey AM, Bormans CA, Sneller C et al (2001) Is all red rice found in commercial rice really Oryzasativa? Weed Sci 49:468–476

Watson L, Dallwitz MJ (1999) Grass genera of the world: descriptions, illustrations, identification, and information retrieval; including synonyms, morphology, anatomy, physiology, phytochemistry, cytology, classification, pathogens, world and local distribution, and references. Version: 18 Aug 1999, website http://biodiversity.uno.edu/delta/

Wongtamee A, Maneechote C, Pusadee T, Rerkasem B, Jamjod S (2015) The dynamics of spatial and temporal population genetic structure of weedy rice (*Oryzasativa* f. spontanea Baker). Genet Resour Crop Evol https://doi.org/10.1007/s10722-015-0330-7

Xia HB, Wang W, Xia H, Zhao W, Lu BR (2011) Conspecific crop-weed introgression influences evolution of weedy rice (*Oryzasativa* f. spontanea) across a geographical range. PLoS ONE 6(1):e16189. https://doi.org/10.1371/journal.pone.0016189

Yu GQ, Bao Y, Shi CH et al (2005) Genetic diversity and population differentiation of Liaoning weedy rice detected by RAPD and SSR markers. Biochem Genet 43:261–270

Printed by Printforce, the Netherlands